TODAY'S TECHNICIAN ™

SHOP MANUAL
FOR MANUAL TRANSMISSIONS & TRANSAXLES

SIXTH EDITION

TODAY'S TECHNICIAN ™

SHOP MANUAL
FOR MANUAL TRANSMISSIONS & TRANSAXLES

JACK ERJAVEC

MICHAEL RONAN

JACK ERJAVEC,
SERIES EDITOR

SIXTH EDITION

CENGAGE
Learning™

Australia · Canada · Mexico · Singapore · Spain · United Kingdom · United States

Today's Technician™: Shop Manual for Manual Transmissions and Transaxles, Sixth Edition

Jack Erjavec
Michael Ronan

SVP, GM Skills & Global Product Management: Dawn Gerrain

Product Team Manager: Erin Brennan

Senior Director, Development: Marah Bellegarde

Senior Product Development Manager: Larry Main

Content Developer: Mary Clyne

Product Assistant: Maria Garguilo

Vice President, Marketing Services: Jennifer Ann Baker

Marketing Manager: Linda Kuper

Senior Production Director: Wendy Troeger

Production Director: Andrew Crouth

Senior Content Project Manager: Cheri Plasse

Senior Art Director: Bethany Casey

Cover image(s): © cla78/Shutterstock

Library of Congress Control Number: 2014940367

Book-only ISBN: 978-1-3052-6177-8
Package ISBN: 978-1-3052-6178-5

Cengage Learning
20 Channel Center Street
Boston MA 02210
USA

Cengage Learning is a leading provider of customized learning solutions with office locations around the globe, including Singapore, the United Kingdom, Australia, Mexico, Brazil, and Japan. Locate your local office at **www.cengage.com/global**

Cengage Learning products are represented in Canada by Nelson Education, Ltd.

To learn more about Cengage Learning, visit **www.cengage.com**

Purchase any of our products at your local college store or at our preferred online store **www.cengagebrain.com**

Notice to the Reader

Publisher does not warrant or guarantee any of the products described herein or perform any independent analysis in connection with any of the product information contained herein. Publisher does not assume, and expressly disclaims, any obligation to obtain and include information other than that provided to it by the manufacturer. The reader is expressly warned to consider and adopt all safety precautions that might be indicated by the activities described herein and to avoid all potential hazards. By following the instructions contained herein, the reader willingly assumes all risks in connection with such instructions. The publisher makes no representations or warranties of any kind, including but not limited to, the warranties of fitness for particular purpose or merchantability, nor are any such representations implied with respect to the material set forth herein, and the publisher takes no responsibility with respect to such material. The publisher shall not be liable for any special, consequential, or exemplary damages resulting, in whole or part, from the readers' use of, or reliance upon, this material.

Printed in the United States of America
Print Number: 02   Print Year: 2017

CONTENTS

CONTENTS

CONTENTS

Photo Sequences

PHOTO SEQUENCES

JOB SHEETS

JOB SHEETS

PREFACE

Thanks to the support the *Today's Technician* series has received from those who teach automotive technology, Cengage Learning, the leader in automotive-related textbooks, is able to live up to its promise to provide new editions of the series every few years. We have listened and responded to our critics and our fans and present this updated and revised sixth edition. By revising this series on a regular basis, we can respond to changes in the industry, changes in technology, changes in the certification process, and to the ever-changing needs of those who teach automotive technology.

The *Today's Technician* series by Cengage features textbooks that cover all mechanical and electrical systems of automobiles and light trucks (while the heavy-duty trucks portion of the series does the same for heavy-duty vehicles). Principally, the individual titles correspond to the main areas of ASE (National Institute for Automotive Service Excellence) certification.

This new edition, like the last, was designed to give students a chance to develop the same skills and gain the same knowledge that today's successful technician has. This edition also reflects the changes in the guidelines established by the National Automotive Technicians Education Foundation (NATEF) in 2013.

The purpose of NATEF is to evaluate technician training programs against standards developed by the automotive industry and recommend qualifying programs for certification (accreditation) by ASE (National Institute for Automotive Service Excellence). Programs can earn ASE certification upon the recommendation of NATEF. NATEF's national standards reflect the skills that students must master. ASE certification through NATEF evaluation ensures that certified training programs meet or exceed industry-recognized, uniform standards of excellence.

Additional titles include remedial skills and theories common to all of the certification areas and advanced or specific subject areas that reflect the latest technological trends. Each text is divided into two volumes: a Classroom Manual and a Shop Manual.

The technician of today and for the future must know the underlying theory of all automotive systems and be able to service and maintain those systems. Dividing the material into two volumes provides the reader with the information needed to begin a successful career as an automotive technician without interrupting the learning process by mixing cognitive and performance learning objectives into one volume.

The design of Cengage's *Today's Technician* series was based on features that are known to promote improved student learning. The design was further enhanced by a careful study of survey results, in which the respondents were asked to value particular features. Some of these features can be found in other textbooks, while others are unique to this series.

Each Classroom Manual contains the principles of operation for each system and subsystem. The Classroom Manual also contains discussions on design variations of key components used by the different vehicle manufacturers. It also looks into emerging technologies that will be standard or optional features in the near future. This volume is organized to build upon basic facts and theories. The primary objective of this volume is to allow the reader to gain an understanding of how each system and subsystem operates. This understanding is necessary to diagnose the complex automobiles of today and tomorrow. Although the basics contained in the Classroom Manual provide the knowledge needed for diagnostics, diagnostic procedures appear only in the Shop Manual. An understanding of the underlying theories is also a requirement for competence in the skill areas covered in the Shop Manual.

A spiral-bound Shop Manual covers the "how-to's." This volume includes step-by-step instructions for diagnostic and repair procedures. Photo Sequences are used to illustrate some of the common service procedures. Other common procedures are listed and are accompanied with fine line drawings and photos that allow the reader to visualize and conceptualize the finest details of the procedure. This volume also contains the reasons for performing the procedures, as well as when that particular service is appropriate.

The two volumes are designed to be used together and are arranged in corresponding chapters. Not only are the chapters in the volumes linked together, the contents of the chapters are also linked. This linking of content is evidenced by marginal callouts that refer the reader to the chapter and page that the same topic is addressed in the other volume. This feature is valuable to instructors. Without this feature, users of other two-volume textbooks must search the index or table of contents to locate supporting information in the other volume. This is not only cumbersome but also creates additional work for an instructor when planning the presentation of material and when making reading assignments. It is also valuable to the students; with the page references they also know exactly where to look for supportive information.

Both volumes contain clear and thoughtfully selected illustrations, many of which are original drawings or photos specially prepared for inclusion in this series. This means that the art is a vital part of each textbook and not merely inserted to increase the numbers of illustrations.

The page layout, used in the series, is designed to include information that would otherwise break up the flow of information presented to the reader. The main body of the text includes all of the "need-to-know" information and illustrations. The wide side margins of each page contain many of the special features of the series. These features present truly "nice-to-know" information such as: simple examples of concepts just introduced in the text, explanations or definitions of terms that are not defined in the text, examples of common trade jargon used to describe a part or operation, and exceptions to the norm explained in the text. This type of information is placed in the margin, out of the normal flow of information. Many textbooks attempt to include this type of information and insert it in the main body of text; this tends to interrupt the thought process and cannot be pedagogically justified. By placing this information off to the side of the main text, the reader can select when to refer to it.

Jack Erjavec

Series Advisor

Highlights of this Edition-Classroom Manual

The text was updated throughout to include the latest developments. Some of these new topics include dual-clutch systems, various limited-slip differential designs, six-speed transmissions, constantly variable transmissions, and self-shifting manual transmissions. New coverage of high performance topics is integrated throughout the classroom manual to reflect recent advances in the fast-growing automotive performance industry. Chapter 11, Electronically Controlled and Automated Transmissions, covers one of the latest trends in manual transmissions. These transmissions are manual transmissions without a driver-operated clutch. They rely on electronics and hydraulics and provide increased performance and a reduction in fuel consumption and emissions while providing the driver with the conveniences of an automatic transmission.

Chapter 1 introduces the purpose of the main system and how it links to the rest of the vehicle. The chapter also describes the purpose and location of the subsystems, as well as the major components of the system and subsystems. The goal of this chapter is to establish a basic understanding that students can base their learning on. All systems and subsystems that will be discussed in detail later in the text are introduced and their primary purpose described. The second chapter covers the underlying basic theories of operation for the topic of the text. This is valuable to the student and the instructor because it covers the theories that other textbooks assume the reader knows. All related basic physical, chemical, and thermodynamic theories are covered in this chapter.

The order of the topics reflects the most common reviewer suggestions. Many of this edition's updates include current electronic applications. Current model transmissions are used as examples throughout the text. Some are discussed in detail. This includes six- and seven-speed and constantly variable transmissions. This new edition also has more information on nearly all manual transmission-related topics. Finally, the art has been updated throughout the text to enhance comprehension and improve visual interest.

HIGHLIGHTS OF THIS EDITION-SHOP MANUAL

Along with the Classroom Manual, the Shop Manual was updated to match current trends. Service information related to the new topics covered in the Classroom Manual is included in this manual. New coverage of high performance topics is integrated throughout the shop manual to reflect recent advances in the fast-growing automotive performance industry. Twenty-seven detailed photo sequences show students what to expect when they perform the same procedure. They can also provide a student with familiarity of a system or type of equipment they may not be able to perform at their school. Although the main purpose of the textbook is not to prepare someone to successfully pass an ASE exam, all of the information required to do so is included in the textbook.

To stress the importance of safe work habits, Chapter 1 is dedicated to safety. Included in this chapter are common shop hazards, safe shop practices, safety equipment, and legislation concerning the safe handling of hazardous materials and wastes. This chapter also includes precautions that must be adhered to when servicing a hybrid vehicle. Chapter 2 covers the basic skills a transmission technician uses to earn a living, including basic diagnostics. Also included in this chapter are those tools and procedures that are commonly used to diagnose and service manual transmissions and drivelines. In summary, this chapter describes what it takes to be a successful technician, typical pay plans for technicians, service information sources, preparing repair orders, ASE certification, and the laws and regulations a technician should be aware of.

The rest of the chapters have been thoroughly updated. Much of the updating focuses on the diagnosis and service to new systems, as well as those systems instructors have said they need more help in. Currently accepted service procedures are used as examples throughout the text. These procedures also served as the basis for new job sheets that are included in the text. Finally, the art has been updated throughout the text to enhance comprehension and improve visual interest.

CLASSROOM MANUAL

Features of this manual include:

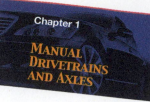

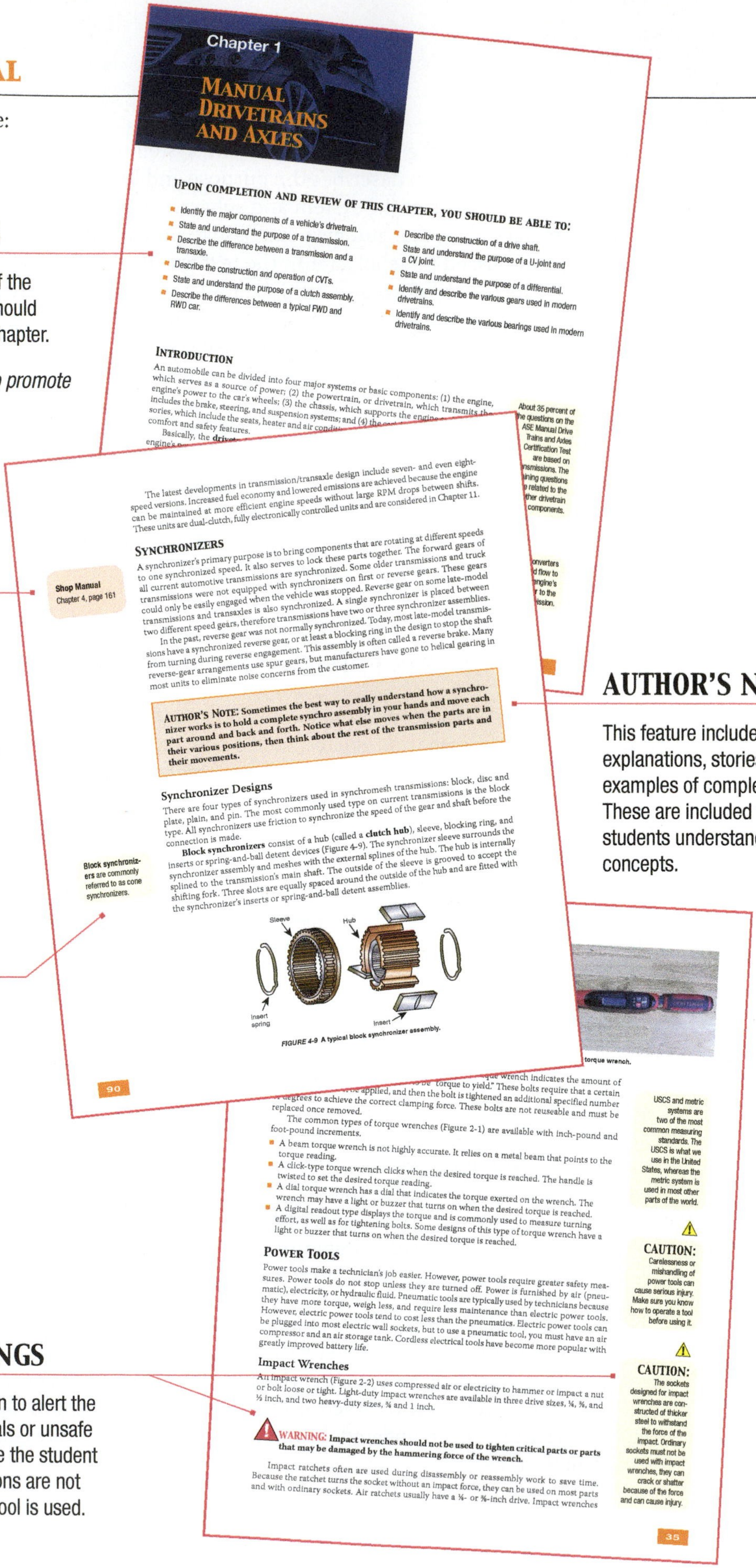

Chapter 1
MANUAL DRIVETRAINS AND AXLES

UPON COMPLETION AND REVIEW OF THIS CHAPTER, YOU SHOULD BE ABLE TO:

- Identify the major components of a vehicle's drivetrain.
- State and understand the purpose of a transmission.
- Describe the difference between a transmission and a transaxle.
- Describe the construction and operation of CVTs.
- State and understand the purpose of a clutch assembly.
- Describe the differences between a typical FWD and RWD car.

- Describe the construction of a drive shaft.
- State and understand the purpose of a U-joint and a CV joint.
- State and understand the purpose of a differential.
- Identify and describe the various gears used in modern drivetrains.
- Identify and describe the various bearings used in modern drivetrains.

INTRODUCTION

An automobile can be divided into four major systems or basic components, which serves as a source of power; (2) the powertrain, or drivetrain, which engine's power to the car's wheels; (3) the chassis, which transmits the includes the brake, steering, and suspension systems; and (4) the sories, which include the seats, heater and air conditi comfort and safety features.

Basically, the **drive** engine's p

About 35 percent of the questions on the ASE Manual Drive Trains and Axles Certification Test are based on transmissions. The remaining questions are related to other drivetrain components.

COGNITIVE OBJECTIVES

These objectives define the contents of the chapter and define what the student should have learned upon completion of the chapter.

Each topic is divided into small units to promote easier understanding and learning.

The latest developments in transmission/transaxle design include seven- and even eight-speed versions. Increased fuel economy and lowered emissions are achieved because the engine speed can be maintained at more efficient engine speeds without large RPM drops between shifts. These units are dual-clutch, fully electronically controlled units and are considered in Chapter 11.

SYNCHRONIZERS

A synchronizer's primary purpose is to bring components that are rotating at different speeds to one synchronized speed. It also serves to lock these parts together. The forward gears of all current automotive transmissions are synchronized. Some older transmissions and truck transmissions were not equipped with synchronizers on first or reverse gears. These gears could only be easily engaged when the vehicle was stopped. Reverse gear on some late-model transmissions and transaxles is also synchronized. A single synchronizer is placed between two different speed gears, therefore transmissions have two or three synchronizer assemblies.

In the past, reverse gear was not normally synchronized. Today, most late-model transmissions have a synchronized reverse gear, or at least a blocking ring in the design to stop the shaft from turning during reverse engagement. This assembly is often called a reverse brake. Many reverse-gear arrangements use spur gears, but manufacturers have gone to helical gearing in most units to eliminate noise concerns from the customer.

Shop Manual
Chapter 4, page 161

CROSS-REFERENCES TO THE SHOP MANUAL

Reference to the appropriate page in the Shop Manual is given whenever necessary. Although the chapters of the two manuals are synchronized, material covered in other chapters of the Shop Manual may be fundamental to the topic discussed in the Classroom Manual.

AUTHOR'S NOTE: Sometimes the best way to really understand how a synchronizer works is to hold a complete synchro assembly in your hands and move each part around and back and forth. Notice what else moves when the parts are in their various positions, then think about the rest of the transmission parts and their movements.

AUTHOR'S NOTES

This feature includes simple explanations, stories, or examples of complex topics. These are included to help students understand difficult concepts.

Synchronizer Designs

There are four types of synchronizers used in synchromesh transmissions: block, disc and plate, plain, and pin. The most commonly used type on current transmissions is the block type. All synchronizers use friction to synchronize the speed of the gear and shaft before the connection is made.

Block synchronizers consist of a hub (called a **clutch hub**), sleeve, blocking ring, and inserts or spring-and-ball detent devices (Figure 4-9). The synchronizer sleeve surrounds the synchronizer assembly and meshes with the external splines of the hub. The hub is internally splined to the transmission's main shaft. The outside of the sleeve is grooved to accept the shifting fork. Three slots are equally spaced around the outside of the hub and are fitted with the synchronizer's inserts or spring-and-ball detent assemblies.

Block synchronizers are commonly referred to as cone synchronizers.

converters of flow to engine's to the ssion.

Sleeve   Hub

Insert spring   Insert

FIGURE 4-9 A typical block synchronizer assembly.

MARGINAL NOTES

These notes add "nice-to-know" information to the discussion. They may include examples or exceptions, or may give the common trade jargon for a component.

90

que wrench indicates the amount of be applied, and then the bolt is tightened an additional specified number degrees to achieve the correct clamping force. These bolts are not reuseable and must be replaced once removed.

The common types of torque wrenches (Figure 2-1) are available with inch-pound and foot-pound increments.

- A beam torque wrench is not highly accurate. It relies on a metal beam that points to the torque reading.
- A click-type torque wrench clicks when the desired torque is reached. The handle is twisted to set the desired torque reading.
- A dial torque wrench has a dial that indicates the torque exerted on the wrench. The wrench may have a light or buzzer that turns on when the desired torque is reached.
- A digital readout type displays the torque and is commonly used to measure turning effort, as well as for tightening bolts. Some designs of this type of torque wrench have a light or buzzer that turns on when the desired torque is reached.

torque wrench.

USCS and metric systems are two of the most common measuring standards. The USCS is what we use in the United States, whereas the metric system is used in most other parts of the world.

Power Tools

Power tools make a technician's job easier. However, power tools require greater safety measures. Power tools do not stop unless they are turned off. Power is furnished by air (pneumatic), electricity, or hydraulic fluid. Pneumatic tools are typically used by technicians because they have more torque, weigh less, and require less maintenance than electric power tools. However, electric power tools tend to cost less than the pneumatics. Electric power tools can be plugged into most electric wall sockets, but to use a pneumatic tool, you must have an air compressor and an air storage tank. Cordless electrical tools have become more popular with greatly improved battery life.

Impact Wrenches

An impact wrench (Figure 2-2) uses compressed air or electricity to hammer or impact a nut or bolt loose or tight. Light-duty impact wrenches are available in three drive sizes, ¼, ⅜, and ½ inch, and two heavy-duty sizes, ¾ and 1 inch.

⚠ CAUTION:
Carelessness or mishandling of power tools can cause serious injury. Make sure you know how to operate a tool before using it.

⚠ CAUTION:
The sockets designed for impact wrenches are constructed of thicker steel to withstand the force of the impact. Ordinary sockets must not be used with impact wrenches, they can crack or shatter because of the force and can cause injury.

CAUTIONS AND WARNINGS

Throughout the text, warnings are given to alert the reader to potentially hazardous materials or unsafe conditions. Cautions are given to advise the student of things that can go wrong if instructions are not followed or if a nonacceptable part or tool is used.

⚠ **WARNING:** Impact wrenches should not be used to tighten critical parts or parts that may be damaged by the hammering force of the wrench.

Impact ratchets often are used during disassembly or reassembly work to save time. Because the ratchet turns the socket without an impact force, they can be used on most parts and with ordinary sockets. Air ratchets usually have a ¼- or ⅜-inch drive. Impact wrenches

35

A BIT OF HISTORY

This feature gives the student a sense of the evolution of the automobile. This feature not only contains nice-to-know information, but also should spark some interest in the subject matter.

TERMS TO KNOW LIST

A list of new terms appears next to the Summary.

REVIEW QUESTIONS

Short answer essay, fill-in-the-blank, and multiple-choice questions are found at the end of each chapter. These questions are designed to accurately assess the student's competence in the stated objectives at the beginning of the chapter.

SUMMARIES

Each chapter concludes with a summary of key points from the chapter. These are designed to help the reader review the chapter contents.

Energy Conversion

Energy conversion occurs when one form of energy is changed to another form. Because energy is not always in the desired form, it must be converted to a form we can use. Some of the most common automotive energy conversions are discussed here.

Chemical to Thermal Energy. Chemical energy in gasoline or diesel fuel is converted to thermal energy when the fuel burns in the engine cylinders.

Chemical to Electrical Energy. The chemical energy in a battery (Figure 2-4) is converted to electrical energy to power many of the accessories on an automobile. In some hybrid and all electric vehicles, the battery is used to power the drive wheels.

Electrical to Mechanical Energy. In an automobile, the battery supplies electrical energy to the starting motor, and this motor converts the electrical energy to mechanical energy to crank the engine.

Thermal to Mechanical Energy. The thermal energy that results from the burning of the fuel in the engine is converted to mechanical energy. The firing impulses on the pistons create rotational motion of the crankshaft. That motion is transferred to the drive train and is used to move the vehicle.

Mechanical to Electrical Energy. The generator is driven by mechanical energy from the engine. The generator converts this energy to electrical energy, which powers the electrical accessories on the vehicle and recharges the battery.

Electrical to Radiant Energy. Radiant energy is light energy. In the automobile, electrical energy is converted to thermal energy, which heats up the inside of light bulbs so that they illuminate and release radiant energy.

Solar to Electrical Energy. Solar energy can be converted to electrical energy through the use of photovoltaic panels. Automotive applications include roof panels that power cooling fans while the vehicle is parked.

Kinetic to Thermal Energy. To stop a vehicle, the brake system must change the kinetic energy of the moving vehicle to kinetic and static thermal or heat energy. This is the result of friction, which is discussed later in this chapter.

Kinetic to Mechanical to Electrical Energy. Hybrid vehicles have a systive braking, that uses the energy of the moving vehicle(s) mechanical energy used to operate the the batteries (Figure 2-4)

A BIT OF HISTORY

A Kinetic Energy Recovery System (KERS) has been used by some Formula 1 racing teams beginning in 2008. The system captures energy from braking forces and stores it in a high-RPM flywheel assembly (mechanical system) or a supercapacitor (electrical system). The energy is reused on acceleration. Various automakers are researching KERS systems for production use.

TERMS TO KNOW

(continued)

Helical gear
Horsepower
Hypoid gear
Idler gear
Journal
Lug nut
Mild hybrid
Overdrive
Parallel hybrid
Planetary carrier
Planetary gear
Planetary pinions
Radial load
Rear-wheel drive (RWD)
Ring gear
Roller bearing
Series hybrid
Spur gear
Stud
Sun gear
Thrust bearing
Torque
Transaxle
Transfer case
Underdrive
Universal joint (U-joint)
Worm gear

SUMMARY

- The rotating or turning effort of the engine's crankshaft is called engine torque.
- Gears are used to apply torque to other rotating parts of the drive train and to multiply torque.
- Transmissions offer various gear ratios through the meshing of various-sized gears.
- Reverse gear is accomplished by adding a third gear to a two-gear set. This gear, the reverse idler gear, causes the driven gear to rotate in reverse.
- The operation of a CVT is based on a steel belt linking two variable pulleys.
- Self-shifting manual transmissions are available. They are manual transmissions that use electronic or hydraulic actuators to shift the gears and work the clutch.
- Connected to the rear of the crankshaft is the flywheel, which serves many functions, including acting as the driving member of the clutch assembly.
- The clutch assembly is comprised of another driving disc, the pressure plate, and a driven disc, the clutch disc.
- The clutch disc is mounted to the input shaft of the transmission and carries the engine's torque to the transmission when the clutch assembly is engaged.
- In FWD cars, the transmission and drive axle is located in a single assembly called a transaxle. In RWD cars, the drive axle is connected to the transmission through a drive shaft.
- The drive shaft and its joints are called the drive line of the car.
- Universal joints allow the drive shaft to change angles in response to movements of the car's suspension and rear axle assembly.
- The rear axle housing encloses the entire rear-wheel driving axle assembly.
- The primary purpose of the differential is to allow a difference in driving wheel speed when the vehicle is rounding a corner or curve. The ring and pinion in the drive axle also multiples the torque it receives from the transmission.
- On FWD cars, the differential is part of the transaxle assembly.
- The drive axles on FWD cars extend from the sides of the transaxle to the drive wheels. The drive axles on FWD cars are fitted to the axles to allow the axles to move with the car's suspension. CV joints are fitted to the axles to allow the axles to move with the car's suspension.
- 4WD vehicles typically use a transfer case, which relays engine torque to both a front and rear driving axle.
- Bearings are used to reduce the friction caused by something rotating within something else.

REVIEW QUESTIONS

Short Answer Essays

1. What are the primary purposes of a vehicle's drive train?
2. What is the basic advantage of a limited-slip differential?
3. What is the purpose of an idler gear?
4. Why are transmissions equipped with many different forward gear ratios?
5. What is the primary difference between a transaxle and a transmission?
6. Why are U-joints and CV joints used in the drive line?
7. What does a differential (final drive) unit do to the torque it receives?
8. What is the purpose of the clutch assembly? How does it work?
9. What kind of gears are commonly used in today's automotive drive trains?
10. When are ball- or roller-type bearings used?

24

To stress the importance of safe work habits, the Shop Manual dedicates one full chapter to safety. Other important features of this manual include:

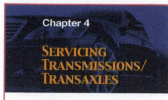

BASIC TOOLS LISTS

Each chapter begins with a list of the basic tools needed to perform the tasks included in the chapter.

PERFORMANCE-BASED OBJECTIVES

These objectives define the contents of the chapter and define what the student should have learned upon completion of the chapter. These objectives also correspond with the list of required tasks for NATEF certification. *Each NATEF task is addressed.*

Although this textbook is not designed to simply prepare someone for the certification exams, it is organized around the NATEF task list. These tasks are defined generically when the procedure is commonly followed and specifically when the procedure is unique for specific vehicle models. Imported and domestic model automobiles and light trucks are included in the procedures.

CAUTIONS AND WARNINGS

Throughout the text, warnings are given to alert the reader to potentially hazardous materials or unsafe conditions. Cautions are given to advise the student of things that can go wrong if instructions are not followed or if a non-acceptable part or tool is used.

CUSTOMER CARE

This feature highlights those little things a technician can do or say to enhance customer relations.

CROSS-REFERENCES TO THE CLASSROOM MANUAL

Reference to the appropriate page in the Classroom Manual is given whenever necessary. Although the chapters of the two manuals are synchronized, material covered in other chapters of the Classroom Manual may be fundamental to the topic discussed in the Shop Manual.

SPECIAL TOOLS LISTS

Whenever a special tool is required to complete a task, it is listed in the margin next to the procedure.

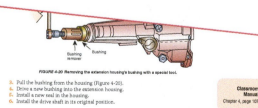

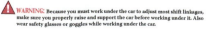

PHOTO SEQUENCES

Many procedures are illustrated in detailed Photo Sequences. These detailed photographs show the students what to expect when they perform particular procedures. They also can provide the student a familiarity with a system or type of equipment, which the school may not have.

SERVICE TIPS

Whenever a short-cut or special procedure is appropriate, it is described in the text. These tips are generally those things commonly done by experienced technicians.

MARGINAL NOTES

These notes add "nice-to-know" information to the discussion. They may include examples or exceptions, or may give the common trade jargon for a component.

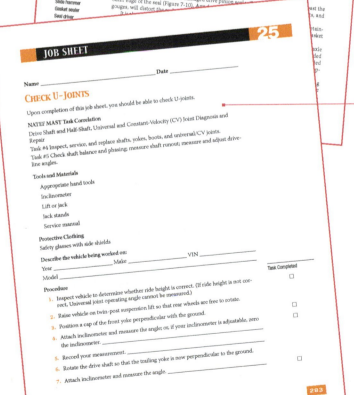

JOB SHEETS

Located at the end of each chapter, the Job Sheets provide a format for students to perform procedures covered in the chapter. A reference to the NATEF Task addressed by the procedure is referenced on the Job Sheet.

TERMS TO KNOW LIST

Terms in this list can be found in the Glossary at the end of the manual.

ASE-STYLE REVIEW QUESTIONS

Each chapter contains ASE-style review questions that reflect the performance-based objectives listed at the beginning of the chapter. These questions can be used to review the chapter as well as to prepare for the ASE certification exam.

CASE STUDIES

Case Studies concentrate on the ability to properly diagnose the systems. Beginning with Chapter 3, each chapter ends with a case study in which a vehicle has a problem, and the logic used by a technician to solve the problem is explained.

ASE CHALLENGE QUESTIONS

Each technical chapter ends with five ASE challenge questions. These are not more review questions, rather they test the students' ability to apply general knowledge to the contents of the chapter.

ASE PRACTICE EXAMINATION

A 50-question ASE practice exam, located in the appendix, is included to test students on the contents of the Shop Manual.

TERMS TO KNOW

(continued)

Lapping
Locking
Lugging
Neutral rollover rattle
Peening
Pitting
Rolling
Scoring
Spalling
Tail shaft

CASE STUDY

A customer brought his 2008 Toyota into the shop with a complaint of the transaxle jumping out of fifth gear. The technician verified the complaint, then proceeded to check the adjustment of the shift linkage. No problem was found after she checked the adjustment and inspected the linkage. She proceeded to check the alignment of the transaxle to the engine, and again the cause of the problem was not found.

The transaxle was removed and disassembled; again no cause for the problem was identified. The technician carefully reassembled the transaxle and installed it back in the car. During the road test, she again experienced the problem. Because she could not find a cause for the problem, she returned it to the customer and told him the problem could not be fixed. Taking this approach is a good way to lose customers and your job. True, she did check all of the right things. However, she probably did not check the internal parts of the transaxle carefully enough. Wear on the back taper of the dog clutch teeth of the speed gear and/or slider will cause this problem. Only a careful inspection will reveal this.

ASE-STYLE REVIEW QUESTIONS

1. When discussing shift problems,
   Technician A says that broken or worn engine and transaxle mounts can cause a transaxle to have shifting problems.
   Technician B says that poor shift boot alignment can cause a transaxle to jump out of gear.
   Who is correct?
   A. A only
   B. B only
   C. Both A and B
   D. Neither A nor B

2. When inspecting a transaxle's gears,
   ... that a wear pattern on the center of ...

4. When discussing transmission disassembly,
   Technician A says that the alignment of the parts of each synchronizer must be marked prior to disassembling the synchronizer assembly.
   Technician B says that most synchronizer hubs are splined to the shaft and are pressed on and off the shaft.
   Who is correct?
   A. A only
   B. B only
   C. Both A and B
   D. Neither A nor B

5. When disassembling a transaxle, a severely worn second gear synchronizer blocker ring is found.
   Technician A says that this should have caused the ... out of gear.

ASE CHALLENGE QUESTIONS

1. A vehicle makes a loud clicking sound during a left turn; during a right turn the clicking noise is still apparent but is less pronounced.
   Technician A says that the left inner CV joint may be the source of the noise.
   Technician B says that the left outer CV joint may be causing the noise.
   Who is correct?
   A. A only
   B. B only
   C. Both A and B
   D. Neither A nor B

2. A vehicle vibrates when under acceleration at all speeds above 10 mph; it does not vibrate during coast down.
   Technician A says that a worn inner CV joint could cause this problem.
   Technician B says that an out-of-balance tire is the likely problem.
   Who is correct?
   A. A only
   B. B only
   C. Both A and B
   D. Neither A nor B

3. A discussion concerning CV joint testing is being held.
   Technician A says that a recommended method to check CV joints is to raise the vehicle so that the lower control arms are hanging freely; at this point the engine can be accelerated while noise checks can be performed.
   Technician B says that the vehicle should be road tested so that the CV joints ...

4. A half shaft is being removed from a vehicle. The outer CV joint appears to be pressed into the hub.
   Technician A says that a brass punch or hammer can be used to remove the CV joint from the hub.
   Technician B says that a puller should be used to remove the CV joint from the hub.
   Who is correct?
   A. A only
   B. B only
   C. Both A and B
   D. Neither A nor B

5. The CV joint boots on the left side of a vehicle appear to be squeezed together; the boots on the right side of the vehicle appear to be stretched apart.
   Technician A says that one or more of the engine/transaxle's mounts may be broken.
   Technician B says that one or more of the engine/transaxle's subframe mounts may be broken.
   Who is correct?
   A. A only
   B. B only
   C. Both A and B
   D. Neither A nor B

APPENDIX A

ASE PRACTICE EXAMINATION

41. When discussing viscous clutch service and inspection,
    Technician A says that viscous clutches can be tested on a bench.
    Technician B says that a viscous clutch can be serviced in the shop by adding oil to its reservoir.
    Who is correct?
    A. Technician A
    B. Technician B
    C. Both A and B
    D. Neither A nor B

42. When discussing the unsprung weight of a vehicle,
    Technician A says that installing lightweight aluminum wheels on a vehicle will reduce its sprung weight.
    Technician B says that replacing a steel hood with an aluminum hood will reduce a vehicle's unsprung weight.
    Who is correct?
    A. Technician A
    B. Technician B
    C. Both A and B
    D. Neither A nor B

43. Transfer case problems are being discussed,
    Technician A says that incorrect drive shaft angles can result in harsh engagement when shifting into 4WD.
    Technician B says that an overheated viscous coupling can cause binding when making a sharp turn on dry pavement.
    Who is correct?
    A. Technician A
    B. Technician B
    C. Both A and B
    D. Neither A nor B

44. The resistance of an electromagnetic clutch coil is 200 ohms; it should be 4 ohms.
    Technician A says that this will result in a blown fuse.
    Technician B says that the magnetic field developed by the clutch coil will be excessive.
    Who is correct?
    A. Technician A
    B. Technician B
    C. Both A and B
    D. Neither A nor B

45. A poorly operating fan motor circuit is being tested. A voltmeter that is placed across the power and ground terminals of connector of the motor indicates 0.0 volts when the fan switch is turned on.
    Technician A says that this test reading indicates that the connector is probably okay.
    Technician B says that this test is referred to as an available voltage test.
    Who is correct?
    A. Technician A
    B. Technician B
    C. Both A and B
    D. Neither A nor B

46. The amount of current being drawn by ... is about to be measured.
    Technician A says that the ammeter can be connected in the positive side of the circuit.
    Technician B says that the amperage can be measured by connecting the ammeter to the negative side of the circuit.
    Who is correct?
    A. Technician A
    B. Technician B
    C. Both A and B
    D. Neither A nor B

47. A digital ammeter being used to measure current drain is indicating 101 mA.
    Which of the following represents this reading?
    A. .101 amps
    B. .00101 amps
    C. 1.01 amps
    D. 10.1 amps

48. A vehicle towed into the shop because of a no-start condition is found to have an inoperative starter. A voltmeter connected to both of the terminals of the clutch safety switch indicates 8 volts when the clutch pedal is depressed and the ignition switch is placed in the Start position.
    Technician A says that the clutch safety switch is faulty.
    Technician B says that the resistance of the circuit is higher than normal.
    Who is correct?
    A. Technician A
    B. Technician B
    C. Both A and B
    D. Neither A nor B

SUPPLEMENTS

INSTRUCTOR RESOURCES

A robust set of Instructor Resources is available on line through Cengage's Instructors Companion Website, and on DVD. This powerful suite of classroom preparation tools includes PowerPoint slides with images and video clips that coincide with each chapter's content coverage, an image gallery with pictures from the text, theory-based worksheets in Word that provide homework or in-class assignments, the Job Sheets from the Shop Manual in Word, a NATEF correlation chart, and an Instructor's Guide in electronic format. Cengage testing powered by Cognero is available through the Instructor Companion Website; in addition, the complete set of test bank questions is available on the Instructor Resources DVD in Word format.

MindTap Automotive for Today's Technician™: Manual Transmissions and Transaxles, 6th edition

MindTap Automotive for Manual Transmissions and Transaxles provides a customized learning experience with relevant assignments that will help students learn and apply concepts while it allows instructors to measure skills and outcomes with ease.

MindTap Automotive meets the needs of today's automotive classroom, shop and student. Within the MindTap, faculty and students will find editable and submittable job sheets based on NATEF tasks. MindTap also offers students the opportunity to practice diagnostic techniques in a safe environment while strengthening their critical thinking and troubleshooting skills with the inclusion of diagnostic scenarios from Delmar Automotive Training Online (DATO). Additional engaging activities include videos, animations, matching exercises and gradable assessments.

REVIEWERS

The authors and publisher would like to extend a special thanks to the following instructors for their contributions to this text:

Ronald Alexander
Professor, Automotive Technology
Morrisville State College, Morrisville, New York

Jeffrey P. Nidiffer
University of Northwestern Ohio
Lima, OH

Christopher Parrott
Automotive Program Director
Vatterott College Inc., Wichita Campus

William Schaefer
Pennsylvania College of Technology
Williamsport, PA

Lonnie Schulz
University of Northwestern Ohio
Lima, OH

William White
University of Northwestern Ohio
Lima, OH

Chapter 1

SAFETY

UPON COMPLETION AND REVIEW OF THIS CHAPTER, YOU SHOULD ABLE TO:

- Explain how safety practices are part of professional behavior.
- Explain the basic principles of personal safety, including protective eyewear, clothing, gloves, shoes, and hearing protection.
- Inspect equipment and tools for unsafe conditions.
- Work around batteries and high-voltage hybrid systems properly.
- Understand the importance of safety and accident prevention in an automotive shop.
- Explain the procedures and precautions for safely using tools and equipment.

- Explain the precautions that need to be followed to safely raise a vehicle on a lift.
- Lift heavy objects properly.
- Explain the procedures for responding to an accident.
- Recognize fire hazards and extinguish the common varieties of fires.
- Identify substances that could be regarded as hazardous materials.
- Dispose of hazardous waste materials following state and federal regulations.
- Describe the purpose of the laws concerning hazardous wastes and materials, including the Right-To-Know Laws.

INTRODUCTION

Safety is everyone's job. You should work safely to protect yourself and the people around you. Perhaps the best single safety rule is "Think before you act." Too often, people working in shops gamble by working in an unsafe way. Often they win and no one gets hurt, but all it takes is one accident and all of their past winnings are lost. By gambling and, perhaps, saving 5 minutes, you can lose an eye or a hand. By acting first and thinking last, you can ruin your back and lose your career. Accidents in a shop can be prevented by others and by YOU. Safe work habits also prevent damage to the vehicles and equipment in the shop.

Working on automobiles can be dangerous. It also can be fun and very rewarding. To keep the fun and rewards rolling in, you need to try to prevent accidents by working safely. In an automotive repair shop, there is great potential for serious accidents, simply because of the nature of the business and the equipment used. Through carelessness, the automotive repair industry can be one of the most dangerous occupations.

However, the chances of you being injured when working on a car are close to nil if you learn to work safely and use common sense. Shop safety is the responsibility of everyone in the shop: you, your fellow students or employees, and your employer or instructor. Everyone must work together to protect the health and welfare of all who work in the shop.

It has been said that 50 percent of all shop accidents could have been prevented by a single individual, the technician.

An accident is something that happens unintentionally.

PERSONAL SAFETY

Personal safety simply involves those precautions you take to protect yourself from injury. This includes wearing protective gear (Figure 1-1), dressing for safety, working professionally, and handling tools and equipment correctly.

When you have neat work habits, you display a professional attitude. Actually, neat habits also are safe habits. Cleaning up spills and keeping equipment and tools out of the path of others prevents accidents. True professionals take time to clean their tools and work area. With a professional attitude, you will not clown around in the shop, will not throw items in the shop, and will not create an unsafe condition for the sake of saving time. Rather than ignore basic safety rules to save time, a professional saves time by effective diagnostics and repair procedures.

Professionals also take pride in their work, treat customers and their vehicles with respect, and try to stay current with all technical, safety, and environmental concerns.

Eye Protection

Your eyes can become infected or permanently damaged by many things in a shop. Some procedures, such as grinding, result in tiny particles of metal and dust that are thrown off at very high speeds. These metal and dirt particles can easily get into your eyes, causing scratches or cuts on your eyeball. Pressurized gases and liquids escaping a hose or hose fitting can spray over a great distance. If these chemicals get into your eyes, they can cause blindness. Dirt and sharp bits of corroded metal can easily fall into your eyes when you are working under a vehicle.

Eye protection should be worn whenever you are exposed to these risks. To be safe, you should wear eye cover whenever you are working in the shop. There are many types of eye protection (Figure 1-2). Safety glasses have lenses and some sort of side protection. Regular prescription glasses should not be worn as a substitute for safety glasses. When prescription glasses are worn in a shop, they should be fitted with side shields.

It is a good habit to wear safety glasses at all times. Choose glasses that fit well and feel comfortable. Some shops require that everyone wear eye protection whenever they are in the work area (Figure 1-3).

Some procedures may require that you wear additional eye protection. When you are cleaning parts with a pressurized spray, you should wear a face shield (Figure 1-4). The face shield not only gives added protection to your eyes but also protects the rest of your face.

If chemicals such as battery acid, fuel, or solvents get into your eyes, flush them continuously with clean water. Have someone call a doctor and get medical help immediately.

FIGURE 1-1 Proper dress prevents injuries.

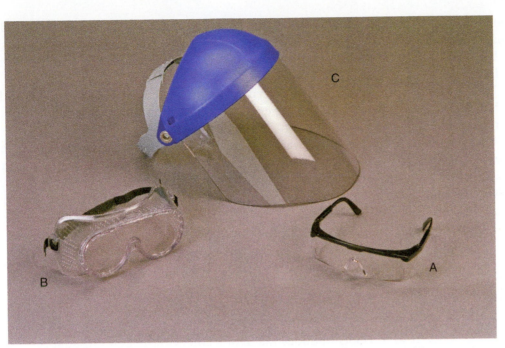

FIGURE 1-2 Different types of eye protection worn by automotive technicians: (A) safety glasses; (B) goggles; and (C) face shield.

CAUTION

AUTHORIZED PERSONNEL ONLY

Service Bays Are A Safety Area.
Eye Protection Is Required At All Times.

Insurance regulations prohibit customers
in the service bay area during work hours.
We suggest you check out our everyday low prices until
your technician has completed the work on your car.
Thank you for your cooperation.

FIGURE 1-3 A typical sign reminding everyone they should wear eye protection when they are in the shop.

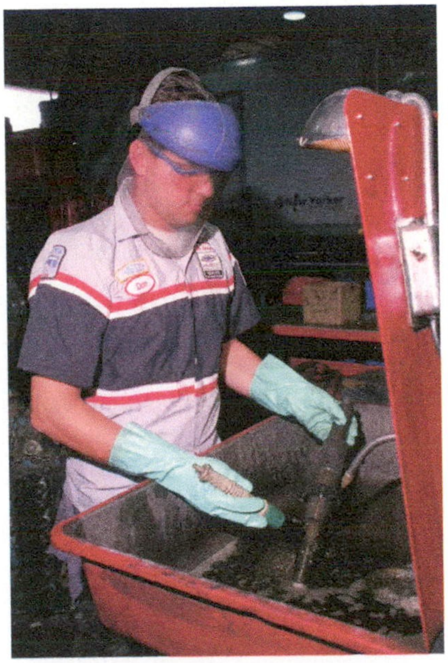

FIGURE 1-4 A face shield should be used in addition to safety glasses when doing some operations.

Many shops have eyewash stations or safety showers that should be used whenever you or someone else has been sprayed or splashed with a chemical.

Welding and brazing operations require special eye protection. Always wear the proper dark eyewear or welding helmet when performing these tasks.

Clothing

Clothing that hangs out freely, such as shirttails, can create a safety hazard and cause serious injury. Nothing you wear should be allowed to dangle in the engine compartment or around equipment. Shirts should be tucked in and buttoned and long sleeves buttoned or carefully rolled up. Your clothing should be well fitted and comfortable but made of strong material. Loose, baggy clothing can be caught easily in moving parts and machinery. Neckties should not be worn. Some technicians prefer to wear coveralls or shop coats to protect their personal clothing.

Long hair and loose, hanging jewelry can create the same type of hazard as loose-fitting clothing. They can get caught in moving engine parts and machinery. If you have long hair, tie it back or tuck it under a cap.

Rings, necklaces, bracelets, and watches should not be worn when working. A ring can rip your finger off, a watch or bracelet can cut your wrist, and a necklace can choke you. This is especially true when working with or around electrical wires. The metal used to make jewelry conducts electricity very well and can easily cause a short, through you, if it touches a bare wire.

Keep your clothing clean. If you spill gasoline or oil on yourself, change that item of clothing immediately. Oil against your skin for a prolonged period of time can produce rashes or other allergic reactions. Gasoline can irritate cuts and sores.

Foot Protection

You also should protect your feet. Tennis and jogging shoes provide little protection if something falls on your foot. Boots or shoes made of leather or a material that approaches the strength of leather offer much better protection from falling objects. There are many designs of safety shoes and boots that also have steel plates built into the toe and **shank**. Many also have soles that are designed to resist slipping on wet surfaces. Foot injuries are not only quite painful but also can put you out of work for some time.

Hand Protection

Good hand protection is often overlooked. A scrape, cut, or burn can seriously impair your ability to work for many days. A well-fitted pair of heavy work gloves should be worn when grinding, welding, or when handling chemicals or high-temperature components. Polyurethane or vinyl gloves should be worn when handling strong and dangerous **caustic** chemicals. These chemicals can easily burn your skin.

Many technicians wear thin, surgical-type latex gloves whenever they are working on vehicles (Figure 1-5). These offer little protection against cuts but do offer protection against disease and grease buildup under and around your fingernails. These gloves are comfortable and are quite inexpensive. Nitrile gloves can also be worn as protection against disease to keep

An electrical short is basically an alternative path for the flow of electricity.

The **shank** of a shoe is that portion of the shoe that protects the ball of your foot. It is the narrow part of a shoe between the heel and sole.

A **caustic** material has the ability to destroy or eat through something. Caustic materials are considered extremely corrosive.

FIGURE 1-5 Many technicians wear thin, surgical-type latex gloves while working on vehicles.

4

FIGURE 1-6 Heavy-duty professional technician's gloves are often worn to protect hands.

grease from building up under and around your fingernails. Latex gloves are more comfortable to wear but weaken when they are exposed to gas, oil, and solvents. Nitrile gloves are not as comfortable as latex gloves, but they are not affected by gas, oil, and solvents. Other types of protective hand gear are available that reduce the chance of injury or discomfort. Many technicians use professional technician's gloves while performing repairs (Figure 1-6). Your choice of hand protection should be based on what you are doing.

Disease Prevention. When you are ill with something that may be contagious, see a doctor and do not go to work or school until the doctor says there is little chance of someone else contracting the illness from you.

You also should be concerned with and protect yourself and others from blood-borne pathogens. **Blood-borne pathogens** are pathogenic microorganisms that are present in human blood and can cause disease. These pathogens include, but are not limited to, hepatitis B virus (HBV) and human immunodeficiency virus (HIV). For everyone's protection, any injury that causes bleeding should be dealt with as a threat to others. You should avoid contact with the blood of another person. If you need to administer some form of first aid, make sure that you wear hand protection before you do so. You also should wear gloves and other protection when handling the item that caused the cut. This item should be sterilized immediately. Most important, as with all injuries, report the accident to your instructor or supervisor.

You also need to protect your hands and other body parts when working with pressurized fluids. There is a risk of blood poisoning (sepsis) due to the inadvertent introduction of toxins (solvents, diesel fuel, etc.) into the bloodstream from a pressurized source. Common-rail diesel fuel systems pose a particular risk to the technician, as pressures in these systems can reach over 30,000 psi. Always follow factory procedures.

Ear Protection

Exposure to very loud noise levels for extended periods of time can lead to a loss of hearing. Air wrenches, air hammers, engines running under a load, and vehicles running in enclosed areas can all generate annoying and harmful levels of noise. Simple earplugs or earphone-type protectors should be worn in environments that are constantly noisy.

Respiratory Protection

It is not uncommon for a technician to work with chemicals that have toxic fumes. Air or respiratory masks should be worn whenever you will be exposed to toxic fumes. Cleaning parts with solvents and painting are the most common times when respiratory masks should be worn. Masks also should be worn when handling hazardous materials.

⚠️ **WARNING: The use of asbestos in automotive components has been reduced dramatically in recent years. Industry sources warn that some components manufactured outside of the United States still use asbestos, however. Never use air pressure to blow off brake or clutch components, and use a respiratory mask when the possibility of exposure is present.**

Work Area Safety

Your entire work area should be kept clean and safe. Any oil, coolant, or grease on the floor can make it slippery. To clean up oil, use oil-absorbent pads or commercial oil absorbent (Figure 1-7). Keep all water off the floor. Water is slippery on smooth floors, and electricity flows well through water. Aisles and walkways should be kept clean and wide enough to move through easily. Make sure the work areas around machines are large enough to operate machines safely.

Disconnecting the vehicle's battery before working on the electrical system or before welding can prevent fires caused by a vehicle's electrical system. To disconnect the battery, remove the negative or ground cable from the battery and position it away from the battery.

FIRE HAZARDS AND PREVENTION

The presence of gasoline is so common that its dangers often are forgotten. A slight spark or an increase in heat can cause a fire or explosion. Gasoline fumes are heavier than air. Therefore, when an open container of gasoline is sitting about, the fumes spill out over the sides of the container. These fumes are more flammable than liquid gasoline and can easily explode.

FIGURE 1-7 Oil-absorbent mats are an efficient and safe way to clean up spills.

FIGURE 1-8 Flammable liquids should be stored in safety-approved containers.

FIGURE 1-9 Dirty rags, such as soiled with gasoline, should be put into a safety container.

Never smoke around gasoline or in a shop filled with gasoline fumes. If the vehicle has a gasoline leak or if you have caused a leak by disconnecting a fuel line, wipe it up immediately and stop the leak. Make sure that any grinding or welding that may be taking place in the area is stopped until the spill is totally cleaned up and the floor has been flushed with water. The rags used to wipe up the gasoline should be taken outside to dry, then stored in an approved dirty rag container. If vapors are present in the shop, have the doors open and turn on the ventilating system. It takes only a small amount of fuel mixed with air to cause combustion.

Gasoline should always be stored in approved containers (Figure 1-8) and never in a glass bottle or jar. If the glass jar was knocked over or dropped, a terrible explosion could take place.

Diesel Fuel. Diesel fuel is not as **volatile** as gasoline, but it should be stored and handled in the same way. It also is not as refined as gasoline and tends to be a very dirty fuel. It normally contains many impurities, including active microscopic organisms that can be highly infectious. If diesel fuel happens to get on an open cut or sore, thoroughly wash it immediately.

Solvents. Cleaning solvents also are not as volatile as gasoline, but they are still **flammable**. Handle all solvents (and any liquids) with care to avoid spillage. Keep all solvent containers closed, except when pouring. Proper ventilation is very important in areas where volatile solvents and chemicals are used. Solvent and other combustible materials must be stored in approved and designated storage cabinets or rooms with adequate ventilation. Never light matches or smoke near flammable solvents and chemicals, including battery acids.

Rags. Oily or greasy rags also can be a source for fires. These rags should be stored in an approved container (Figure 1-9) and never thrown out with normal trash. Like gasoline, oil is a **hydrocarbon** and can ignite with or without a spark or flame. This ignition without an external source of heat or fire is called **spontaneous combustion**.

Fire Extinguishers

In case of a fire, you should know the location of the fire extinguishers and fire alarms in the shop and also should know how to use them. You also should be aware of the different types of fires and the fire extinguishers (Figure 1-10) used to put out these types of fires. Never put water on a gasoline fire. The water will just spread the fire; the proper fire extinguisher smothers the flames. During a fire, never open doors or windows unless it is absolutely necessary; the extra draft will only make the fire worse. Make sure the fire department is contacted before or during your attempt to extinguish a fire.

Basically, there are four types of fires: **Class A fires** are those in which wood, paper, and other ordinary materials are burning. **Class B fires** are those involving flammable liquids, such as gasoline, diesel fuel, paint, grease, oil, and other similar liquids. **Class C fires** are electrical fires. **Class D fires** are a unique type of fire, for the material burning is a metal. An example of this is

⚠️

CAUTION:
Never siphon gasoline or diesel fuel with your mouth. These liquids are poisonous and can make you sick or fatally ill.

The **volatility** of a substance is a statement of how easily the substance vaporizes or explodes.

The **flammability** of a substance is a statement of how well the substance supports combustion.

A **hydrocarbon** is a substance composed of hydrogen and carbon molecules.

FIGURE 1-10 Different types of fire extinguishers.

a burning "mag" wheel, bell housing, or transfer case housing; the magnesium used in the construction of the wheel is a flammable metal and will burn brightly when subjected to high heat.

Using the wrong type of extinguisher may cause the fire to grow instead of putting it out. All extinguishers are marked with a symbol or letter to signify which class of fire they were intended for (Table 1-1). You should know where each fire extinguisher is in the shop and what their ratings are before you need one.

Using a Fire Extinguisher

Remember, during a fire, never open doors or windows unless it is absolutely necessary; the extra draft will only make the fire worse. Make sure the fire department is contacted before or during your attempt to extinguish a fire. To extinguish a fire, stand 6–10 feet from the fire. Hold the extinguisher firmly in an upright position. Aim the nozzle at the base and use a side-to-side motion, sweeping the entire width of the fire. Stay low to avoid inhaling the smoke. If it gets too hot or too smoky, get out. Remember, never go back into a burning building for anything. To help remember how to use an extinguisher, remember the word "PASS" (Figure 1-11):

Pull the pin from the handle of the extinguisher.

Aim the extinguisher's nozzle at the base of the fire.

Squeeze the handle.

Sweep the entire width of the fire with the contents of the extinguisher.

If there is not a fire extinguisher handy, a blanket or fender cover may be used to smother the flames. Be careful when doing this, as the heat of the fire may burn you and the blanket. If the fire is too great to smother, move everyone away from the fire and call the local fire

TABLE 1-1 GUIDE TO EXTINGUISHER SELECTION

	Class of Fire	Typical Fuel Involved	Type of Extinguisher
Class **A** Fires (green)	**For Ordinary Combustibles** Put out a class A fire by lowering its temperature or by coating the burning combustibles.	**Wood** Paper Cloth Rubber Plastics Rubbish Upholstery	**Water***[1] Foam* Multipurpose dry chemical[4]
Class **B** Fires (red)	**For Flammable Liquids** Put out a class B fire by smothering it. Use an extinguisher that gives a blanketing, flame-interrupting effect; cover whole flaming liquid surface.	**Gasoline** Oil Grease Paint Lighter fluid	**Foam*** Carbon dioxide[5] Halogenated agent[6] Standard dry chemical[2] Purple K dry chemical[3] Multipurpose dry chemical[4]
Class **C** Fires (blue)	**For Electrical Equipment** Put out a class C fire by shutting off power as quickly as possible and by always using a nonconducting extinguishing agent to prevent electric shock.	**Motors** Appliances Wiring Fuse boxes Switchboards	**Carbon dioxide[5]** Halogenated agent[6] Standard dry chemical[2] Purple K dry chemical[3] Multipurpose dry chemical[4]
Class **D** Fires yellow	**For Combustible Metals** Put out a class D fire of metal chips, turnings, or shavings by smothering or coating with a specially designed extinguishing agent.	**Aluminum** Magnesium Potassium Sodium Titanium Zirconium	**Dry power extinguishers**

*Cartridge-operated water, foam, and soda-acid types of extinguishers are no longer manufactured. These extinguishers should be removed from service when they become due for their next hydrostatic pressure test.

Notes:

(1) Freezes in low temperatures unless treated with antifreeze solution, usually weighs over 20 pounds (9 kilograms) and is heavier than any other extinguisher mentioned.

(2) Also called ordinary or regular dry chemical (sodium bicarbonate).

(3) Has the greatest initial fire-stopping power of the extinguishers mentioned for class B fires. Be sure to clean residue immediately after using the extinguisher so sprayed surfaces will not be damaged (potassium bicarbonate).

(4) The only extinguishers that fight A, B, and C classes of fires. However, they should not be used on fires in liquefied fat or oil of appreciable depth. Be sure to clean residue immediately after using the extinguishers so sprayed surfaces will not be damaged (ammonium phosphates).

(5) Use with caution in unventilated, confined spaces.

(6) May cause injury to the operator if the extinguishing agent (a gas) or the gas produced when the agent is applied to a fire is inhaled.

FIGURE 1-11 When using an extinguisher, aim the nozzle at the base of the fire.

department. A simple under-the-hood fire can cause the total destruction of the car and the building and can even take some lives. You must be able to respond quickly and precisely to avoid a disaster.

ROTATING PULLEYS AND BELTS

Be very careful around belts, pulleys, wheels, chains, or any other rotating mechanism. When working around an engine's drive belts and pulleys, make sure your hands, shop towels, or loose clothing do not come in contact with the moving parts. Hands and fingers can be quickly pulled into a revolving belt or pulley even at engine idle speeds. Other rotating components to stay away from include drive shafts and half shafts.

⚠️ **WARNING:** Be careful when working around electric engine cooling fans. These fans are controlled by a thermostat and can come on without warning, even when the engine is not running. Whenever you must work around these fans, disconnect the electrical connector to the fan motor before reaching into the area around the fan.

The thermostatic switch for the electric cooling fans also may be disconnected to prevent the fan from coming on.

TOOL AND EQUIPMENT SAFETY

When you work with any equipment, make sure you use it properly and that it is set up according to the manufacturer's instructions. All equipment should be properly maintained and periodically inspected for unsafe conditions. Frayed electrical cords or loose mountings can cause serious injuries. All electrical outlets should be equipped to allow for the use of three-pronged electrical cords. The third prong allows for a safety ground connection (Figure 1-12). All equipment with rotating parts should be equipped with safety guards that reduce the possibility of the parts coming loose and injuring someone (Figure 1-13). Do not depend on someone else to inspect and maintain equipment. Check it out before you use it! If you find the equipment unsafe, disconnect it from the power source and put a sign on it to warn others and notify the person in charge.

Hand Tool Safety

Hand tools should always be kept clean and be used only for the purpose for which they were designed. Oily hand tools can slip out of your hand and cause broken fingers or at least cut or skinned knuckles. Your tools also should be inspected for cracks, broken parts, or other

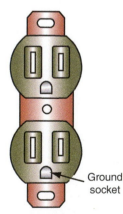
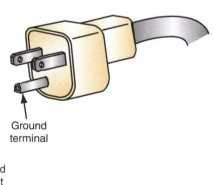

FIGURE 1-12 Make sure all three prongs on an electrical connector are in good condition before plugging in the cord.

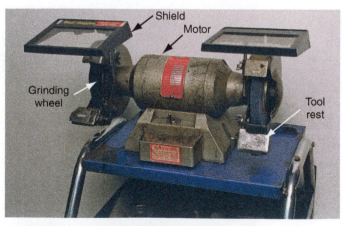

FIGURE 1-13 Equipment such as grinding wheels should be equipped with safety shields and tool guards.

dangerous conditions before you use them. Punches and chisels are maintained by "dressing" them by filing or grinding mushroomed ends, and by maintaining the surface of the business end of the tool.

Knives, chisels, and scrapers must be used in a motion that will keep the point or blade moving away from your body. Always hand a pointed or sharp tool to someone else with the handle directed toward the person.

Power Tool Safety

Power tools are operated by an outside source of power, such as electricity, compressed air, or hydraulic pressure. Safety around power tools is very important. Serious injury can result from carelessness. Always wear safety glasses when using power tools.

If the tool is electrically powered, make sure it is properly grounded. Also, when using electrical power tools, never stand on a wet or damp floor. Disconnect the power source before doing any work on the machine or tool. Before using the tool, check the power cord for bare wires or cracks in the insulation. If the cord is damaged, do not plug it into the wall outlet. Repair the cord before using the tool or use another tool. Before plugging in any electric tool, make sure its switch is in the OFF position. When you are done using the tool, turn it off and unplug it. Never leave a running power tool unattended.

When using power equipment on a small part, never hold the part in your hand. Always mount the part in a bench vise or use vise grip pliers. Never try to use a machine or tool beyond its stated capacity or for operations requiring more than the rated power of the tool.

When working with larger power tools, such as a bench or floor grinding wheel, check the machine and the grinding wheels for signs of damage before using them. If the wheels are damaged, it should be replaced and not used. Check the speed rating of the wheel and make sure it matches the speed of the machine. Never spin a grinding wheel at a speed higher than it is rated for. Be sure to place all safety guards in position. A safety guard is a protective cover over a moving part. Although the safety guards are designed to prevent injury, you should still wear safety glasses or a face shield when using the machine. Make sure there are no people or parts around the machine before starting it. Keep your hands and clothing away from the moving parts. Maintain a balanced stance when using the machine.

Compressed Air Equipment Safety

Compressed air is used to inflate tires, apply paint, and drive tools. Compressed air can be dangerous when it is not used properly. When using compressed air, safety glasses or a face shield should be worn. Particles of dirt and pieces of metal, blown by the high-pressure air, can penetrate your skin or get into your eyes.

Before using a compressed air tool, check all hose connections. Pneumatic tools must always be operated at the pressure recommended by the manufacturer.

Always hold an air nozzle or air control device securely when starting or shutting off the compressed air. A loose nozzle can whip suddenly and cause serious injury. Never point an air nozzle at anyone. Never use compressed air to blow dirt from your clothes or hair. Never use compressed air to clean the floor or workbench. Also, never spin bearings with compressed air. If the bearing is damaged, one of the steel balls or rollers might fly out and cause serious injury.

Lift Safety

Always be careful when raising a vehicle on a lift or a hoist. Adapters and hoist plates must be positioned correctly on twin post and rail-type lifts to prevent damage to the underbody of the vehicle. There are specific lift points that allow the weight of the vehicle to be evenly supported by the adapters or hoist plates. The correct lift points can be found in the vehicle's service manual. Figure 1-14 shows typical locations for unibody and frame cars. Always follow the manufacturer's instructions. Before operating any lift or hoist, carefully read the operating manual and follow the operating instructions.

Once the lift supports are properly positioned under the vehicle, raise the lift until the supports contact the vehicle. Then check the supports to make sure they are in full contact with the vehicle. Shake the vehicle to make sure it is securely balanced, then raise it to the desired working height. Before working under a vehicle, make sure the lift's locking devices are fully engaged.

The Automotive Lift Institute (ALI) is an association concerned with the design, construction, installation, operation, maintenance, and repair of automotive lifts. Their primary concern is safety. Every lift approved by ALI has the label shown in Figure 1-15. It is a good idea to read through the safety tips included on this label before using a lift. Many shops require their technicians to go through an ALI training course to become certified in lift safety and inspection.

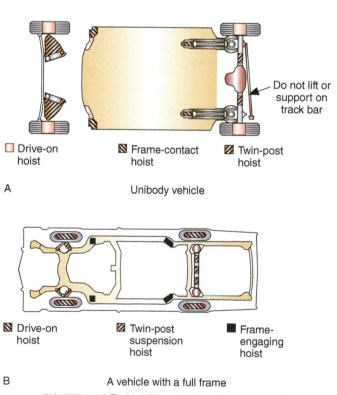

FIGURE 1-14 Typical lift points for (A) unibody and (B) frame/body vehicles.

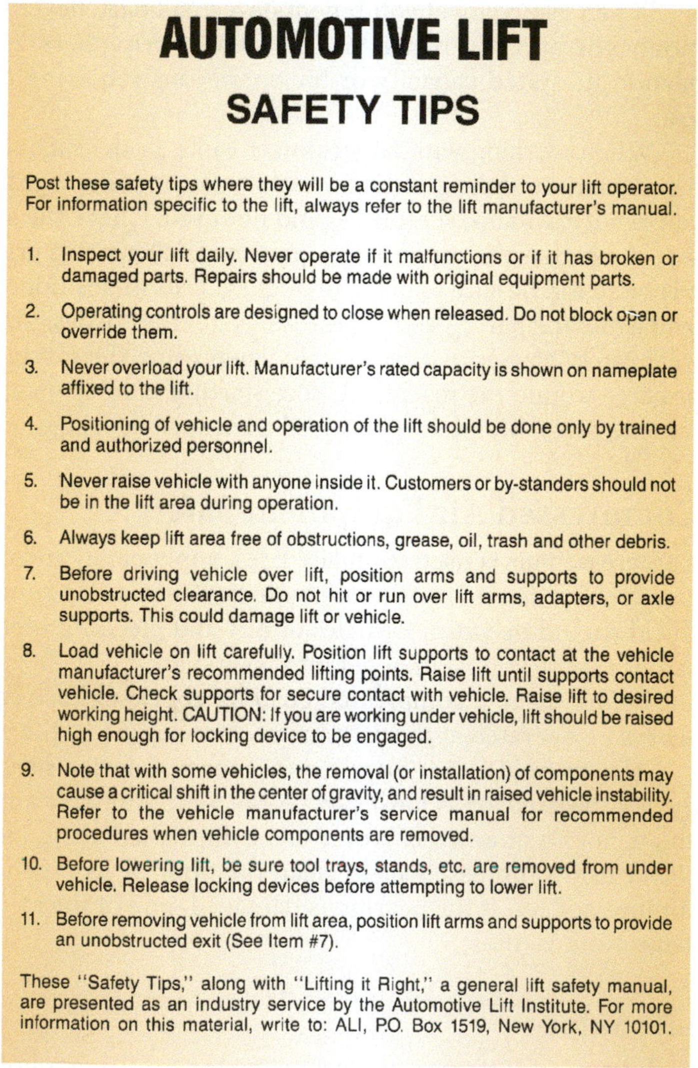

AUTOMOTIVE LIFT
SAFETY TIPS

Post these safety tips where they will be a constant reminder to your lift operator. For information specific to the lift, always refer to the lift manufacturer's manual.

1. Inspect your lift daily. Never operate if it malfunctions or if it has broken or damaged parts. Repairs should be made with original equipment parts.

2. Operating controls are designed to close when released. Do not block open or override them.

3. Never overload your lift. Manufacturer's rated capacity is shown on nameplate affixed to the lift.

4. Positioning of vehicle and operation of the lift should be done only by trained and authorized personnel.

5. Never raise vehicle with anyone inside it. Customers or by-standers should not be in the lift area during operation.

6. Always keep lift area free of obstructions, grease, oil, trash and other debris.

7. Before driving vehicle over lift, position arms and supports to provide unobstructed clearance. Do not hit or run over lift arms, adapters, or axle supports. This could damage lift or vehicle.

8. Load vehicle on lift carefully. Position lift supports to contact at the vehicle manufacturer's recommended lifting points. Raise lift until supports contact vehicle. Check supports for secure contact with vehicle. Raise lift to desired working height. CAUTION: If you are working under vehicle, lift should be raised high enough for locking device to be engaged.

9. Note that with some vehicles, the removal (or installation) of components may cause a critical shift in the center of gravity, and result in raised vehicle instability. Refer to the vehicle manufacturer's service manual for recommended procedures when vehicle components are removed.

10. Before lowering lift, be sure tool trays, stands, etc. are removed from under vehicle. Release locking devices before attempting to lower lift.

11. Before removing vehicle from lift area, position lift arms and supports to provide an unobstructed exit (See Item #7).

These "Safety Tips," along with "Lifting it Right," a general lift safety manual, are presented as an industry service by the Automotive Lift Institute. For more information on this material, write to: ALI, P.O. Box 1519, New York, NY 10101.

FIGURE 1-15 Automotive lift safety tips.

Jack and Jack Stand Safety

A vehicle can be raised by a hydraulic jack. A handle on the jack is moved up and down to raise part of a vehicle and a valve is turned to release the hydraulic pressure in the jack to lower the part. At the end of the jack is a lifting pad. The pad must be positioned under an area of the vehicle's frame or at one of the manufacturer's recommended lift points. Never place the pad under the floor pan or under steering and suspension components. Always position the jack so the wheels of the vehicle can roll as the vehicle is being raised.

⚠ **WARNING: Never use a lift or jack to move something heavier than it is designed for. Always check the rating before using a lift or jack. If a jack is rated for 2 tons, do not attempt to use it for a job requiring 5 tons. It is dangerous for you and the vehicle.**

Safety (jack) stands are placed under a sturdy chassis member, such as the frame or axle housing, to support the vehicle. Once the safety stands (Figure 1-16) are in position, the hydraulic pressure in the jack should be released slowly until the weight of the vehicle is on the stands. Jack stands have a capacity rating. Always use the correct rating of jack stand.

Never move under a vehicle when it is only supported by a jack; rest the vehicle on safety stands before moving under the vehicle. The jack should be removed after the jack stands are set in place. This eliminates a hazard, such as a jack handle sticking out into a walkway.

Chain Hoist and Crane Safety

Heavy parts of the automobile, such as engines, are removed with chain hoists or cranes. Another term for a chain hoist is chain fall. To prevent serious injury, chain hoists and cranes must be properly attached to the parts being lifted. Always use bolts with enough strength to support the object being lifted. Place the chain hoist or crane directly over the assembly. Then attach the chain or cable to the hoist. The engine crane lift arm is often designed to locate in different positions to adjust length. Each adjustment will be labeled with the rated capacity for that position.

Cleaning Equipment Safety

Cleaning automotive parts can be divided into four basic categories: chemical cleaning, thermal cleaning, abrasive cleaning, and steam cleaning. Regardless of what method you use, you should always follow the precautions given by the manufacturer and you should always wear the recommended protection gear.

FIGURE 1-16 Jack or safety stands are placed under the vehicle after it has been raised by a hydraulic jack and before work is done under the vehicle.

BATTERIES

When possible, you should disconnect the vehicle's battery before disconnecting any electrical wire or component. Most transmission and engine removal procedures begin with disconnecting the battery. This prevents the possibility of a fire or electrical shock. It also eliminates the possibility of an accidental short, which can ruin the car's electrical system. Disconnect the negative or ground cable first (Figure 1-17), then disconnect the positive cable. Because electrical circuits require a ground to be complete, by removing the ground cable you eliminate the possibility of a circuit accidentally becoming completed. When reconnecting the battery, connect the positive cable first, then the negative.

The hydrogen gases that form in the top of a battery when it is being charged are very explosive. Never smoke or introduce any form of heat around a charging battery. An explosion will not only destroy the battery but also may spray sulfuric acid all over you, the car, and the shop. When connecting a battery charger to a battery, leave the charger off until all its leads are connected. This will prevent electrical sparks and prevent a possible explosion.

The most dangerous battery is one that has been overcharged. It is hot and has been, or still may be, producing large amounts of hydrogen. Allow the battery to cool before working with or around it. Also, never use or charge a battery that has frozen electrolyte. Extreme amounts of hydrogen may be released from the electrolyte as it thaws.

Working Safely on High-Voltage Systems

Electric drive vehicles (battery operated, hybrid, and fuel cell electric vehicles) have high-voltage electrical systems (from 42 volts to 650 volts). These high voltages can kill you! Most high-voltage circuits are identifiable by size and color. The cables have thicker insulation and are colored orange for systems operating at 144–600 volts or higher (Figure 1-18). The connectors are also colored orange. 42 volt systems use yellow- or blue-colored wires. These do not present a shock hazard, but an arc will be maintained if a circuit is opened. On some vehicles, the high-voltage cables are enclosed in an orange shielding or casing. In addition,

FIGURE 1-17 Always disconnect a battery by removing the negative cable first.

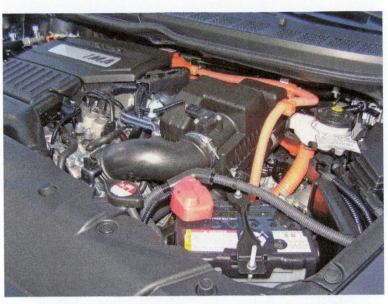

FIGURE 1-18 Most high-voltage circuits are identifiable by size and are typically colored orange.

DANGER ⚠ HIGH VOLTAGE: 300 VOLTS
HAUTE TENSION : 300 VOLTS

SHIELD EYES. EXPLOSIVE GASES CAN CAUSE BLINDNESS OR INJURY.
SE PROTÉGER LES YEUX. LES GAZ EXPLOSIFS PEUVENT CAUSER DES BLESSURES OU LA CÉCITÉ.

DO NOT REMOVE BATTERY COVERS OR CABLES.
NE PAS ENLEVER LES COUVERCLES OU LES CÂBLES DE LA BATTERIE.

ALKALINE ELECTROLYTE CAN CAUSE BLINDNESS OR SEVERE BURNS.
L'ÉLECTROLYTE BASIQUE PEUT CAUSER DES BRÛLURES GRAVES OU LA CÉCITÉ.

KEEP OUT OF THE REACH OF CHILDREN.
TENIR HORS DE LA PORTÉE DES ENFANTS.

NO SPARKS OR FLAMES. DO NOT INCINERATE.
ÉVITER LES ÉTINCELLES ET LES FLAMMES. NE PAS METTRE AU FEU.

FOR SERVICE, RETURN TO FORD DEALER. DISPOSE OF PROPERLY. DO NOT DISCARD.
POUR LA RÉPARATION, RETOURNER LA BATTERIE AU CONCESSIONNAIRE FORD.
METTRE AU REBUT DE FAÇON APPROPRIÉE. NE PAS JETER.

TO PROTECT THE BATTERY, DO NOT FLOOD WITH WATER.
POUR PROTÉGER LA BATTERIE, NE PAS L'INONDER D'EAU.

⚠ WARNING: TO REDUCE THE RISK OF PERSONAL INJURY SEE OWNER GUIDE FOR JUMP START INSTRUCTIONS.
⚠ AVERTISSEMENT : POUR RÉDUIRE LES RISQUES DE BLESSURES, VOIR LE GUIDE DU PROPRIÉTAIRE QUI DONNE LA MÉTHODE DE DÉMARRAGE-SECOURS À L'AIDE DE CÂBLES VOLANTS.

Ford   Dearborn, Michigan   4USA-10661-EC

FIGURE 1-19 High-voltage components may have caution notices to alert technicians of the possible dangers.

the high-voltage battery pack and most high-voltage components have "High Voltage" caution labels (Figure 1-19). Be careful not to touch these wires and parts.

Wear insulating gloves, commonly called "lineman's gloves," when working on or around the high-voltage system (Figure 1-20). These gloves must be class "0" rubber insulating gloves, rated at 1000 volts as rated by the American National Standards Institute (ANSI) and the American Society for Testing and Materials (ASTM). Also, to protect the integrity of the insulating gloves, wear leather gloves over the insulating gloves while doing any service or repair to the high-voltage system (Figure 1-21). Occupational Safety and Health Administration (OSHA) regulations require that HV gloves get inspected every 6 months by a qualified glove inspection laboratory.

Make sure they have no tears, holes, or cracks and are dry. Electrons can enter through the smallest of holes in your gloves. The integrity of the gloves must be checked every time before using them. To check the gloves, blow enough air into each one so they balloon out. Then fold the open end over to seal the air in. Continue to slowly fold that end of the glove toward the fingers. This will compress the air. If the glove continues to balloon as the air is compressed, it has no leaks. If any air leaks out, the glove should be discarded. All gloves, new and old, should be checked before they are used.

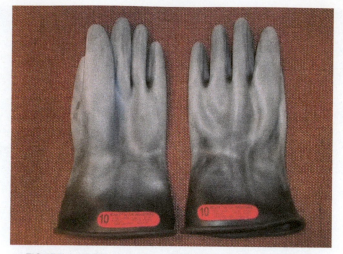

FIGURE 1-20 Class 0 rubber linesman's gloves are rated to protect to up to 1000 volts.

FIGURE 1-21 Leather gloves over the linesman's gloves help prevent damage to the rubber gloves.

Other safety precautions that should always be adhered to when working on an electric drive vehicle:

- Always adhere to the safety guidelines given by the vehicle's manufacturer.
- Obtain the necessary training before working on these vehicles.
- Perform each operation following the procedures defined by the manufacturer.
- Disable or disconnect the high-voltage system before servicing those systems.
- Anytime the engine is running, the generator is producing high voltage, and care must be taken to prevent being shocked.
- Systems may have a large capacitor that must be discharged after the high-voltage system has been isolated. Wait the prescribed amount of time (normally about 10 minutes) before working on or around the high-voltage system.
- After removing a high-voltage cable, cover the terminal with electrical tape.
- Always use insulated tools.
- Alert other technicians that you are working on the high-voltage system with a warning sign such as "High-Voltage Work: Do Not Touch."
- Always install the correct circuit protection device into a high-voltage circuit.
- Many electric motors have a strong permanent magnet in them; individuals with a pacemaker should not handle these parts.

- When an electric drive vehicle needs to be towed into the shop for repairs, make sure it is not towed on its drive wheels. Doing this will drive the generator(s), which can overcharge the batteries and cause them to explode. Always tow these vehicles with the drive wheels off the ground or move them on a flat bed.
- Never assume the vehicle is shut off just because the engine is off. Many hybrids have a READY indicator on the dash display. The vehicle is off when the READY indicator is off.
- Many hybrids use an automatic engine restart to charge the battery any time the key is in the ignition or the smart key is within range. Keep the key at least 20 feet from the vehicle to ensure the engine does not start while under-hood repairs or inspection is taking place.

If the electric motor is sandwiched between the engine and transmission, make sure you follow all procedures. The permanent magnet used in the motor is very strong and requires special tools to remove and install it.

VEHICLE OPERATION

When the customer brings a vehicle in for service, certain driving rules should be followed to ensure your safety and the safety of those working around you. For example, before moving a car into the shop, buckle your safety belt. Make sure no one is near, the way is clear, and there are no tools or parts under the car before you start the engine. Check the brakes before putting the vehicle in gear. Then drive slowly and carefully in and around the shop.

When road-testing the car, obey all traffic laws. Drive only as far as is necessary to check the automobile and verify the customer's complaint. Never make excessively quick starts, turn corners too quickly, or drive faster than conditions allow.

If the engine must be running when working on the car, block the wheels to prevent the car from moving. Place the transmission into park for automatic transmissions or in neutral for manual transmissions. Set the parking (emergency) brake. Never stand directly in front of or behind a running vehicle.

Run the engine only in a well-ventilated area to avoid the danger of poisonous **carbon monoxide (CO)** in the engine exhaust. CO is an odorless but deadly gas. Most shops have an exhaust ventilation system (Figure 1-22); always use it. Connect the hose from the vehicle's tailpipe to the intake for the vent system. Make sure the vent system is turned on before running the engine.

Exhaust contains an odorless, colorless, and deadly gas: carbon monoxide. This poisonous gas gives very little warning to the victim and can kill in just a few minutes.

FIGURE 1-22 When running an engine in a shop, always connect the exhaust to the ventilation system.

LIFTING AND CARRYING

When lifting a heavy object such as a transmission, use a hoist or have someone else help you. If you must work alone, *always* lift heavy objects with your legs, not your back. Bend down with your legs, not your back, and securely hold the object you are lifting; then stand up, keeping the object close to you. Photo Sequence 1 shows this procedure. Trying to "muscle" something with your arms or back can result in severe damage to your back and may end your career and limit what you do for the rest of your life!

ACCIDENTS

Your work area should be kept clean and safe. The floor and bench tops should be kept clean, dry, and orderly. Any oil, coolant, or grease on the floor can make it slippery. Slips can result in serious injuries. To clean up oil, use commercial oil absorbent or absorbent mats (Figure 1-23). Most shops use absorbent mats in work areas to limit the effects of oil and fluid spills. Keep all water off the floor. Water is slippery on smooth floors, and electricity flows well through dirty water. Aisles and walkways should be kept clean and wide enough to easily move through. Make sure the work areas around machines are large enough to safely operate the machine.

Make sure all drain covers are snugly in place. Open drains or covers that are not flush to the floor can cause toe, ankle, and leg injuries.

Keep an up-to-date list of emergency telephone numbers clearly posted next to the telephone. These numbers should include a doctor, a hospital, and fire and police departments.

Handle all solvents (or any liquids) with care to avoid spillage. Keep all solvent containers closed, except when pouring. Proper ventilation is very important in areas where volatile solvents and chemicals are used. Solvent and other combustible materials must be stored in approved and ventilated storage cabinets or rooms.

Be extra careful when transferring flammable materials from bulk storage. Static electricity can build up enough to create a spark that could cause an explosion. Discard or clean all empty

FIGURE 1-23 If any oil, coolant, or grease is on the floor, use an absorbent mat to clean it up.

LIFTING HEAVY OBJECTS

P1-1 Heavy items such as transmissions are often lifted and carried by technicians. Lifting heavy items can cause injury if you are not careful.

P1-2 Before lifting, position yourself at the correct angle for lifting the transmission and place your feet close to it.

P1-3 Bend your knees, keep your back straight, and lower yourself enough to firmly grab the transmission.

P1-4 Begin to lift the transmission by straightening your knees. Keep your elbows as straight as possible and hold the transmission close to your body.

P1-5 Continue lifting by straightening your legs while keeping your back straight.

P1-6 To place the transmission on the bench, turn your entire body. Do not twist at the hips. Lower the transmission and place one end of it on the bench.

P1-7 While keeping your back straight, slide the rest of the transmission on the bench.

P1-8 It will be easier to work on the transmission now that it is on the workbench.

FIGURE 1-24 A typical first-aid kit and its contents.

solvent containers. Never light matches or smoke near flammable solvents and chemicals, including battery acids.

Also, accidents can be prevented simply by the way you act. The following are some additional guidelines for working safely in a shop. This list does not include everything you should or shouldn't do; it merely gives some things to think about.

- Never smoke while working on a vehicle or working with any machine in the shop.
- Playing around is not fun when it sends someone to the hospital.
- To prevent serious burns, keep your skin away from hot metal parts such as the radiator, exhaust manifold, tailpipe, catalytic converter, and muffler.
- Always disconnect electric engine cooling fans when working around the radiator. These can turn on without warning and can easily chop off a finger or hand. Make sure you reconnect the fan after you have completed your repairs.
- When working with a hydraulic press, make sure the pressure is applied in a safe manner. It is generally wise to stand to the side when operating the press.
- Properly store all parts and tools by putting them away in a place where people will not trip over them. This practice not only cuts down on injuries, it also reduces time wasted looking for a misplaced part or tool.

If there is an accident, the quicker you respond to it, the less damage there will be. Your supervisor should be immediately informed of all accidents that occur in the shop. The work area should have a first-aid kit (Figure 1-24) for treating minor injuries. Facilities for flushing eyes also should be near or in the shop area. Know where they are.

First Aid

It also is a good idea for you to know first aid. The knowledge of how to treat certain injuries can save someone's life. The American Red Cross offers many low-cost but thorough courses on first aid. You will realize the importance of these classes the first time you have to give first aid to someone or when someone must give it to you.

You should find out if there is a resident nurse in the shop or at the school, and know where the nurse's office is. If there are specific first-aid rules in your school or shop, make sure you are aware of them and follow them.

If someone is overcome by carbon monoxide, get him or her fresh air immediately. Burns should be cooled immediately by rinsing them with water. Whenever there is severe bleeding from a wound, try to stop the bleeding by applying pressure with clean gauze on or around the wound, and get medical help. Never move someone who may have broken bones unless the person's life is otherwise endangered. Moving that person may cause additional injury. Call for medical assistance.

HAZARDOUS MATERIALS

A typical shop contains many potential health hazards for those working in it. These hazards can be classified as:

- Chemical hazards are caused by high concentrations of vapors, gases, or solids in the form of dust.
- Hazardous wastes are those substances that are the result of a service.
- Physical hazards include excessive noise, vibration, pressures, and temperatures.
- Ergonomic hazards are conditions that impede normal or proper body position and motion.

There are many government agencies charged with ensuring safe work environments for all workers. These include the **Occupational Safety and Health Administration (OSHA)**, the Mine Safety and Health Administration (MSHA), and the National Institute for Occupational Safety and Health (NIOSH). These, in addition to state and local governments, have instituted regulations that must be understood and followed. Everyone in a shop has the responsibility for adhering to these regulations.

OSHA

OSHA (Occupational Safety and Health Administration) is a branch of the U.S. government's Department of Labor. This branch was formed to write and enforce operational guidelines for all businesses to ensure that employees work in safe and healthy conditions. The guidelines include standards for cleanliness, air ventilation, fire prevention, emergency measures, equipment condition, and personal protective equipment.

It is the employers' responsibility to provide a place of employment that is free from all recognized safety and health hazards. OSHA controls all safety and health issues of the automotive industry. Businesses may be inspected periodically by OSHA personnel. The owner of the business may be cited for any violations of the standards. The owner will be given a period of time to bring the work area into compliance. If the owner does not correct the situation or has been a repeat violator of the standards, OSHA will fine the owner.

Right-To-Know Law

Every employee in a shop is protected by "Right-To-Know Laws." The general intent of these laws is for employers to provide a safe working place as it relates to hazardous materials. All employees must be trained about their rights under the legislation, the nature of the hazardous chemicals in their workplace, the labeling of chemicals, and information about each chemical listed and described on **Material Safety Data Sheets (MSDSs)**. These sheets (Figure 1-25) are available, in writing or on their Web sites, from the manufacturers and suppliers of the chemicals. They detail the chemical composition and precautionary information for all products that can present health or safety hazards. The Canadian equivalents of the MSDS are called Workplace Hazardous Materials Information Systems (WHMIS).

Employees must be familiar with the intended purposes of the substance, the recommended protective equipment, accident and spill procedures, and any other information regarding the safe handling of hazardous materials. This training must be given annually to employees and provided to new employees as part of their job orientation.

CAUTION:
When handling any hazardous material, always wear the appropriate safety protection. Always follow the correct procedures when using the material, and be familiar with the information given on the MSDS for that material.

HEXANE

==

MSDS Safety Information

==

Ingredients

==

Name: HEXANE (N_HEXANE)
% Wt: >97
OSHA PEL: 500 PPM
ACGIH TLV: 50 PPM
EPA Rpt Qty: 1 LB
DOT Rpt Qty: 1 LB

==

Health Hazards Data

==

LD50 LC50 Mixture: LD50:(ORAL,RAT) 28.7 KG/MG
Route Of Entry Inds _ Inhalation: YES
Skin: YES
Ingestion: YES
Carcinogenicity Inds _ NTP: NO
IARC: NO
OSHA: NO
Effects of Exposure: ACUTE:INHALATION AND INGESTION ARE HARMFUL AND MAY BE FATAL. INHALATION AND INGESTION MAY CAUSE HEADACHE, NAUSEA, VOMITING, DIZZINESS, IRRITATION OF RESPIRATORY TRACT, GASTROINTESTINAL IRRITATION AND UNCONSCIOUSNESS. CONTACT W/SKIN AND EYES MAY CAUSE IRRITATION. PROLONGED SKIN MAY RESULT IN DERMATITIS (EFTS OF OVEREXP)
Signs And Symptions Of Overexposure: HLTH HAZ:CHRONIC:MAY INCLUDE CENTRAL NERVOUS SYSTEM DEPRESSION.
Medical Cond Aggravated By Exposure: NONE IDENTIFIED.
First Aid: CALL A PHYSICIAN. INGEST:DO NOT INDUCE VOMITING. INHAL:REMOVE TO FRESH AIR. IF NOT BREATHING, GIVE ARTIFICIAL RESPIRATION. IF BREATHING IS DIFFICULT, GIVE OXYGEN. EYES:IMMED FLUSH W/PLENTY OF WATER FOR AT LEAST 15 MINS. SKIN:IMMED FLUSH W/PLENTY OF WATER FOR AT LEAST 15 MINS WHILE REMOVING CONTAMD CLTHG & SHOES. WASH CLOTHING BEFORE REUSE.

==

Handling and Disposal

==

Spill Release Procedures: WEAR NIOSH/MSHA SCBA & FULL PROT CLTHG. SHUT OFF IGNIT SOURCES:NO FLAMES, SMKNG/FLAMES IN AREA. STOP LEAK IF YOU CAN DO SO W/OUT HARM. USE WATER SPRAY TO REDUCE VAPS. TAKE UP W/SAND OR OTHER NON-COMBUST MATL & PLACE INTO CNTNR FOR LATER (SU PDAT)
Neutralizing Agent: NONE SPECIFIED BY MANUFACTURER.
Waste Disposal Methods: DISPOSE IN ACCORDANCE WITH ALL APPLICABLE FEDERAL, STATE AND LOCAL ENVIRONMENTAL REGULATIONS. EPA HAZARDOUS WASTE NUMBER:D001 (IGNITABLE WASTE).
Handling And Storage Precautions: BOND AND GROUND CONTAINERS WHEN TRANSFERRING LIQUID. KEEP CONTAINER TIGHTLY CLOSED.
Other Precautions: USE GENERAL OR LOCAL EXHAUST VENTILATION TO MEET TLV REQUIREMENTS. STORAGE COLOR CODE RED (FLAMMABLE).

==

Fire and Explosion Hazard Information

==

Flash Point Method: CC
Flash Point Text: _9F,_23C
Lower Limits: 1.2%
Upper Limits: 77.7%
Extinguishing Media: USE ALCOHOL FOAM, DRY CHEMICAL OR CARBON DIOXIDE. (WATER MAY BE INEFFECTIVE).

FIGURE 1-25 A sample of a Material Safety Data Sheet.

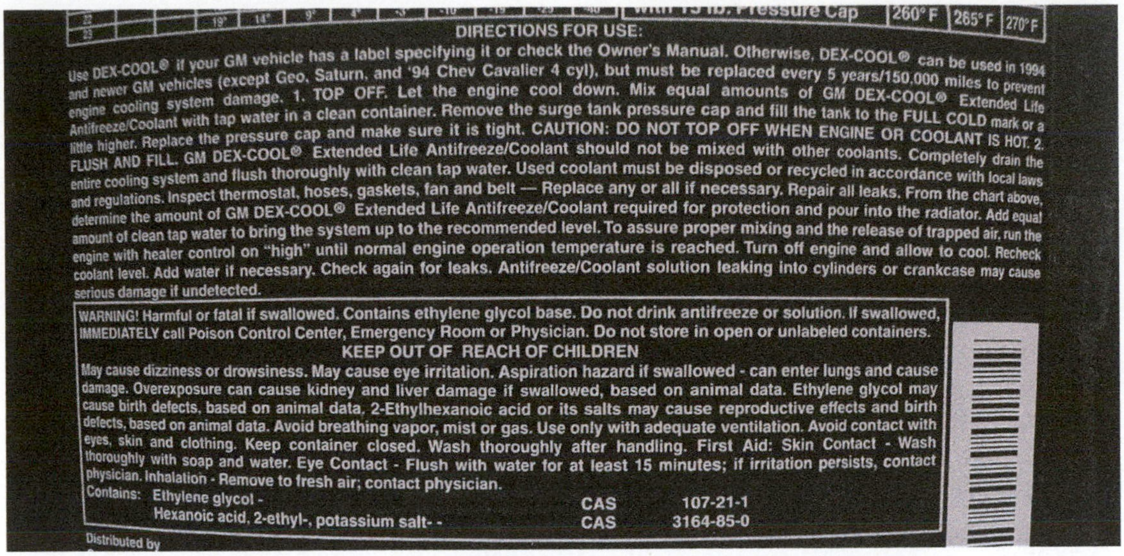

| | | | | | | | | | | | with 15 lb. Pressure Cap | 260° F | 265° F | 270° F |

DIRECTIONS FOR USE:

Use DEX-COOL® if your GM vehicle has a label specifying it or check the Owner's Manual. Otherwise, DEX-COOL® can be used in 1994 and newer GM vehicles (except Geo, Saturn, and '94 Chev Cavalier 4 cyl), but must be replaced every 5 years/150,000 miles to prevent engine cooling system damage. 1. TOP OFF. Let the engine cool down. Mix equal amounts of GM DEX-COOL® Extended Life Antifreeze/Coolant with tap water in a clean container. Remove the surge tank pressure cap and fill the tank to the FULL COLD mark or a little higher. Replace the pressure cap and make sure it is tight. CAUTION: DO NOT TOP OFF WHEN ENGINE OR COOLANT IS HOT. 2. FLUSH AND FILL. GM DEX-COOL® Extended Life Antifreeze/Coolant should not be mixed with other coolants. Completely drain the entire cooling system and flush thoroughly with clean tap water. Used coolant must be disposed or recycled in accordance with local laws and regulations. Inspect thermostat, hoses, gaskets, fan and belt — Replace any or all if necessary. Repair all leaks. From the chart above, determine the amount of GM DEX-COOL® Extended Life Antifreeze/Coolant required for protection and pour into the radiator. Add equal amount of clean tap water to bring the system up to the recommended level. To assure proper mixing and the release of trapped air, run the engine with heater control on "high" until normal engine operation temperature is reached. Turn off engine and allow to cool. Recheck coolant level. Add water if necessary. Check again for leaks. Antifreeze/Coolant solution leaking into cylinders or crankcase may cause serious damage if undetected.

WARNING! Harmful or fatal if swallowed. Contains ethylene glycol base. Do not drink antifreeze or solution. If swallowed, IMMEDIATELY call Poison Control Center, Emergency Room or Physician. Do not store in open or unlabeled containers.
KEEP OUT OF REACH OF CHILDREN
May cause dizziness or drowsiness. May cause eye irritation. Aspiration hazard if swallowed - can enter lungs and cause damage. Overexposure can cause kidney and liver damage if swallowed, based on animal data. Ethylene glycol may cause birth defects, based on animal data. 2-Ethylhexanoic acid or its salts may cause reproductive effects and birth defects, based on animal data. Avoid breathing vapor, mist or gas. Use only with adequate ventilation. Avoid contact with eyes, skin and clothing. Keep container closed. Wash thoroughly after handling. First Aid: Skin Contact - Wash thoroughly with soap and water. Eye Contact - Flush with water for at least 15 minutes; if irritation persists, contact physician. Inhalation - Remove to fresh air; contact physician.

Contains: Ethylene glycol -   CAS   107-21-1
      Hexanoic acid, 2-ethyl-, potassium salt- -   CAS   3164-85-0

Distributed by

FIGURE 1-26 Carefully read the labels on all chemicals before using them.

A MSDS must include the following information about the product:

- The trade and chemical name of the product.
- The manufacturer of the product.
- All of the ingredients of the product.
- Health hazards such as headaches, skin rashes, nausea, and dizziness.
- The product's physical description; this information may include the product's color, odor, permissible exposure limit (PEL), threshold limit value (TLV), specific gravity, boiling point, freezing point, evaporation data, and volatility rating.
- The product's explosion and fire data, such as flash point.
- The reactivity and stability data.
- The product's weight compared to air.
- Protection data including first aid and proper handling.

Shops must maintain documentation on the hazardous chemicals in the workplace, proof of training programs, records of accidents or spill incidents, satisfaction of employee requests for specific chemical information via the MSDS, and a general right-to-know compliance procedure manual utilized within the shop.

All hazardous material must be properly labeled, indicating what health, fire, or reactivity hazard it poses (Figure 1-26) and what protective equipment is necessary when handling each chemical. The manufacturer of the hazardous waste materials must provide all warnings and precautionary information, which must be read and understood by the user before application. Attention to all label precautions is essential for the proper use of the chemical and for prevention of hazardous conditions.

Hazardous Wastes

Many repair and service procedures generate what are known as hazardous wastes. Dirty solvents and cleaners are good examples of hazardous wastes. Something is classified as a hazardous waste by the Environmental Protection Agency (EPA) if it is on the EPA list of known harmful materials or has one or more of the following characteristics:

Ignitability. If it is a liquid with a flash point below 140° F (60° C) or a solid that can spontaneously ignite.

Corrosivity. If it dissolves metals and other materials or burns the skin.

Reactivity. Any material that reacts violently with water or other materials or releases cyanide gas, hydrogen sulfide gas, or similar gases when exposed to low pH acid

Reactivity is a statement of how easily a substance can cause or be part of a chemical reaction.

solutions. This also includes material that generates toxic mists, fumes, vapors, and flammable gases.

EP toxicity. Materials that leach one or more of eight heavy metals in concentrations greater than 100 times primary drinking water standard concentrations.

A complete EPA list of hazardous wastes can be found in the Code of Federal Regulations. It should be noted that no material is considered hazardous waste until the shop is finished using it and ready to dispose of it.

In the United Sates, OSHA regulates the use of many of these materials. The EPA regulates the disposal of **hazardous waste**. A summary of these regulations follows:

- All businesses that generate hazardous waste must develop a hazardous waste policy.
- Each hazardous waste generator must have an EPA identification number.
- When waste is transported for disposal, a licensed waste hauler must be used to transport and dispose of the waste. A copy of a written manifest (an EPA form) must be kept by the shop.

Regulations on hazardous waste handling and generation have led to the development of equipment that is now commonly found in shops. Examples of these are thermal cleaning units, closed-loop steam cleaners, waste oil furnaces, oil filter crushers, refrigerant recycling machines, engine coolant recycling machines, and highly absorbent cloths.

> ⚠️ **WARNING:** **The shop is ultimately responsible for the safe disposal of hazardous wastes, even after the waste leaves the shop. Only licensed waste removal companies should be used to dispose of the waste. Make sure you know what the company is planning to do with the waste. Make sure you have a written contract stating what is supposed to happen with the waste. Leave nothing to chance. In the event of an emergency hazardous waste spill, contact the National Response Center (1-800-424-8802) immediately. The Center also has an online reporting mechanism. Failure to do so can result in a $10,000 fine, a year in jail, or both.**

OSHA and the EPA have other strict rules and regulations that help to promote safety in the auto shop. These are described throughout this text whenever they are applicable. Maintaining a vehicle involves handling and managing a wide variety of materials and wastes. Some of these wastes can be toxic to fish, wildlife, and humans when improperly managed. No matter the amount of waste produced, it is to the shop's legal and financial advantage to manage the wastes properly and, even more important, to prevent pollution.

Hazardous Waste Disposal

Hazardous wastes must be properly stored and disposed of. It is the shop's responsibility to hire a reputable, financially stable, and state-approved hauler who will dispose of the shop wastes legally. Select a licensed hazardous waste hauler after seeking recommendations and reviewing the firm's permits and authorizations. If hazardous waste is dumped illegally, your shop may be held responsible.

Always keep hazardous waste separate, properly labeled, and sealed in the recommended containers. The storage area should be covered and may need to be fenced and locked if vandalism could be a problem.

Handling Shop Wastes

The following is a summary of the proper method of preparing and disposing of common hazardous wastes. These are general guidelines; always follow the specific state regulations for disposing of these items.

Oil. Recycle oil. Set up equipment, such as a drip table or screen table with a used oil collection bucket, to collect oils dripping off parts. Place drip pans under vehicles that are leaking fluids. Recycle the oil according to local regulations. Do not mix other wastes with used oil, except as allowed by your recycler.

Oil Filters. Drain for at least 24 hours, crush, and recycle used oil filters.

Batteries. Recycle batteries by sending them to a reclaimer or back to the distributor. Store batteries in a watertight, acid-resistant container. Inspect batteries for cracks and leaks when they come in. Treat a dropped battery as if it were cracked. Acid residue is hazardous because it is corrosive and may contain lead and other toxics. Neutralize spilled acid by using baking soda or lime, and dispose of as hazardous material.

Nickel-metal hydride (NiMH) hybrid batteries are recycled at special centers. There is a toll-free number located under the hood or on the battery pack that can be called for recycling information.

Metal Residue from Machining. Collect metal filings when machining metal parts. Keep separate and recycle if possible. Prevent metal filings from falling into a storm sewer drain.

Other Solids. Store materials such as scrap metal, old machine parts, and worn tires under a roof or tarpaulin to protect them from the elements and to prevent the potential to create contaminated runoff.

Refrigerants. Recover and/or recycle refrigerants during the service and disposal of motor vehicle air conditioners and refrigeration equipment. It is not allowable to knowingly vent refrigerants to the atmosphere. Recovery and/or recycling during service must be performed by an EPA-certified technician using certified equipment and following specified procedures.

Solvents. Replace hazardous chemicals with less toxic alternatives that have equal performance. For example, substitute water-based cleaning solvents for petroleum-based solvent degreasers (Figure 1-27). To reduce the amount of solvent used when cleaning parts, use a two-stage process: dirty solvent followed by fresh solvent. Hire a hazardous waste

FIGURE 1-27 An aqueous (water-based) solution parts cleaner.

management service to clean and recycle solvents. Store solvents in closed containers to prevent evaporation. Evaporation of solvents contributes to ozone depletion and smog formation. Properly label spent solvents and store on drip pans or in diked areas and only with compatible materials.

Liquid Recycling. Collect and recycle coolants from radiators. Store transmission fluids, brake fluids, and solvents containing chlorinated hydrocarbons separately, and recycle or dispose of them properly.

Containers. Cap, label, cover, and properly store above ground and outdoors any liquid containers and small tanks within a diked area and on a paved impermeable surface to prevent spills from running into surface or ground water.

Shop Towels/Rags. Keep waste towels in a closed container marked "Contaminated shop towels only." To reduce costs and liabilities associated with disposal of used towels, which can be classified as hazardous waste, investigate using a laundry service that is able to treat the wastewater generated from cleaning the towels.

Waste Storage. Always keep hazardous waste separate, properly labeled, and sealed in the recommended containers. The storage area should be covered and may need to be fenced and locked if vandalism could be a problem. Select a licensed hazardous waste hauler after seeking recommendations and reviewing the firm's permits and authorizations.

SUMMARY

- You have an important role in creating an accident-free work environment. Your appearance and work habits will go a long way toward preventing accidents. Common sense and concern for others will result in fewer accidents.
- It is everyone's job to check periodically for safety hazards in a shop.
- Dressing safely for work is very important. This includes snug-fitting clothing, eye and ear protection, protective gloves, strong shoes, and caps to cover long hair.
- Safety glasses should be worn whenever you are working under a vehicle or when you are using machining equipment, grinding wheels, chemicals, compressed air, or fuels.
- There are two areas of housekeeping for which you are responsible: your work area and the rest of the shop.
- Dirty and oily rags should be stored in approved containers.
- All equipment should be inspected for safety hazards before being used.
- Gasoline and diesel fuel are highly flammable and should be kept in approved safety cans.
- It is important to know when to use each of the various types of fire extinguishers. When fighting a fire, aim the nozzle at the base and use a side-to-side sweeping motion.
- Fire, heat, and sparks should be kept away from a battery. These could cause the battery to explode.
- Always strictly follow the service recommendations and precautions given by the manufacturer when working on a hybrid vehicle.
- Heavy objects should be lifted with your legs, not your back.
- Special care should be taken whenever using power tools, such as impact wrenches, gear pullers, and jacks.
- Always observe all relevant safety rules when operating a vehicle lift or hoist. Jacks, jack stands, chain hoists, and cranes also can cause injury if not operated safely.
- Use care whenever it is necessary to remove a vehicle in the shop. Carelessness and horseplay can lead to a damaged vehicle and serious injury.

TERMS TO KNOW

Blood-borne pathogens

Carbon monoxide (CO)

Caustic

Class A fire

Class B fire

Class C fire

Class D fire

Corrosivity

EP toxicity

Flammability

Hazardous waste

Hydrocarbon

Ignitability

Material Safety Data Sheet (MSDS)

Occupational Safety and Health Administration (OSHA)

Reactivity

- Carbon monoxide gas is a poisonous gas present in engine exhaust fumes. It must be vented properly from the shop using tailpipe hoses or other reliable methods.
- Accidents can be prevented by not having anything dangle near rotating equipment and parts.
- Accidents can happen and, when they do, you should respond immediately to prevent further injury or damage.
- All employees have the right to know what hazardous materials they are using to perform their job. Most of the needed information is contained on an MSDS.
- Hazardous wastes must be properly disposed of. Typically, shops hire full-service waste management firms to remove and dispose of these wastes.

TERMS TO KNOW

(continued)

Shank

Spontaneous combustion

Volatility

ASE-STYLE REVIEW QUESTIONS

1. When discussing ways to prevent fires,
   Technician A says that all dirty and oily rags should be stored in approved containers.
   Technician B says that all dirty and oily rags should be kept in a pile until the end of the day, and then they should be moved to a suitable container.
   Who is correct?
   A. A only
   B. B only
   C. Both A and B
   D. Neither A nor B

2. *Technician A* says that accidents can be prevented by not having anything dangle near rotating equipment and parts.
   Technician B says that accidents can be prevented by having common sense.
   Who is correct?
   A. A only
   B. B only
   C. Both A and B
   D. Neither A nor B

3. When discussing what to do first when an accident does occur,
   Technician A says that technicians should immediately determine the cause of the accident.
   Technician B says that technicians should respond immediately to prevent further injury or damage.
   Who is correct?
   A. A only
   B. B only
   C. Both A and B
   D. Neither A nor B

4. When discussing safety concerns involving gasoline,
   Technician A says that gasoline spills should be immediately cleaned up.
   Technician B says that gasoline should only be stored in approved containers.
   Who is correct?
   A. A only
   B. B only
   C. Both A and B
   D. Neither A nor B

5. When discussing shop safety procedures,
   Technician A says that unsafe equipment should have its power disconnected and marked with a sign to warn others.
   Technician B says that all equipment should be inspected for safety hazards before being used.
   Who is correct?
   A. A only
   B. B only
   C. Both A and B
   D. Neither A nor B

6. When discussing hazardous waste removal,
   Technician A says that all hazardous wastes must be properly disposed of.
   Technician B says that shops can hire full-service waste management firms to remove and dispose of hazardous wastes.
   Who is correct?
   A. A only
   B. B only
   C. Both A and B
   D. Neither A nor B

7. When discussing ways to create an accident-free work environment,
   Technician A says that everyone in the shop should take full responsibility for ensuring safe work areas.
   Technician B says that the appearance and work habits of technicians can help prevent accidents.
   Who is correct?
   A. A only
   B. B only
   C. Both A and B
   D. Neither A nor B

8. When discussing why exhaust fumes should be vented outdoors or drawn into a ventilation/filtration system,
   Technician A says that exhaust gases contain amounts of carbon monoxide.
   Technician B says that carbon dioxide is an odorless, colorless, and deadly gas.
   Who is correct?
   A. A only
   B. B only
   C. Both A and B
   D. Neither A nor B

9. When discussing simple first-aid procedures,

   Technician A says that if someone is overcome by carbon monoxide, get him or her fresh air immediately.

   Technician B says that burns should be cooled immediately by rinsing them in water.

   Who is correct?

   A. A only

   B. B only

   C. Both A and B

   D. Neither A nor B

10. When discussing the car's electrical system,

    Technician A says that you should always disconnect the positive battery cable first, then disconnect the negative cable.

    Technician B says that hybrid vehicle repair requires the use of latex gloves during service on the high-voltage system.

    Who is correct?

    A. A only

    B. B only

    C. Both A and B

    D. Neither A nor B

Name _____ Date _____

SHOP SAFETY SURVEY

As a professional technician, safety should be one of your chief concerns. This job sheet will increase your awareness of shop safety items. As you survey your shop and answer the following questions, you should learn how to evaluate the safety of any workplace.

Procedure

Evaluate your work environment and how you fit into it.

_____

Task Completed

1. Are you properly dressed for work?

   If yes, describe how you are dressed. _____

   If no, explain why you are not properly dressed. _____

2. Are your safety glasses OSHA approved?   ☐ Yes   ☐ No

   Do they have side shields?   ☐ Yes   ☐ No

3. Carefully inspect your shop; note any potential hazards.

NOTE: Although a hazard is not necessarily a safety violation, you must still be aware of it.

4. Are there safety areas marked around grinders and other machinery?   ☐ Yes   ☐ No

5. What is the air pressure in your shop? _____

   What is the condition of air hoses? _____

   Are the nozzles OSHA approved? _____

6. Where are the tools stored in your shop?

   Are they clean and neatly stored?   ☐ Yes   ☐ No

7. Explain how you could improve tool storage. _____

8. Describe the kind of hoists used in your shop. _____

9. Ask your instructor to demonstrate hoist usage. _____

10. Where is the first-aid kit in your shop? _____

11. Where are the fire extinguishers and eye washes?

    _____

    _____

    _____

12. Describe what you should do in the event of an accident. _____

_____

_____

_____

Instructor's Response: _____

_____

_____

Name _____ **Date** _____

FIRE HAZARD INSPECTION

Fire is always a danger in any automotive shop. The very nature of automotive work involves the use of many highly flammable chemicals. Because of this, a technician must be extremely vigilant. Notice and immediately correct all fire hazards.

Procedure

1. Are there any flammable liquids stored in your shop?   ☐ Yes   ☐ No

2. Are they properly stored?   ☐ Yes   ☐ No

   If not, explain why. _____

3. Where are the fire extinguishers located in your shop? _____
   _____

4. Against what class of fires can each extinguisher be used? _____
   _____
   _____

5. Explain how to use each fire extinguisher in your shop. _____
   _____
   _____

6. Does your shop have a fire blanket?   ☐ Yes   ☐ No

   If yes, where is it stored? _____
   _____
   _____

7. Where are the fire alarms in your shop located? _____
   _____
   _____

8. Where are the fire exits in your shop located? _____
   _____
   _____

9. Are the fire escape routes in your shop clearly posted?   ☐ Yes   ☐ No

Instructor's Response: _____
_____
_____

Name _____ Date _____

HAZARDOUS MATERIALS INSPECTION

Upon completion of this job sheet, you should be able to demonstrate your ability to identify hazardous wastes and explain how to handle them.

Procedure

Task Completed

_____

1. Inspect your shop. Identify and list all hazardous materials found.

   Solvents _____

   Gasoline _____

   Oils _____

   Diesel fuels _____

   Cleaners _____

   Other _____

2. Check the containers in which hazardous materials are stored. Are they clearly marked?   ☐ Yes   ☐ No

3. Check to see if your shop has a Materials Safety Data Sheet (MSDS) file. Is it located near the hazardous waste?   ☐ Yes   ☐ No

4. Make sure your shop has an MSDS list posted on a bulletin board where everyone can read it.

5. Read the MSDS bulletins on each of the materials you have found in the shop and explain to the instructor how you would handle a spill of each material.

Instructor's Response: _____

_____

_____

UPON COMPLETION AND REVIEW OF THIS CHAPTER, YOU SHOULD ABLE TO:

- Describe the use of common pneumatic, electrical, and hydraulic power tools found in an automotive service department.

- Describe some of the special tools used to service manual transmissions and the driveline.

- List the basic units of measure for length, volume, and mass in the two measuring systems.

- Identify the major measuring instruments and devices used by technicians.

- Explain what the common measuring instruments and devices measure and how to use them.

- Describe the proper procedure for measuring with a micrometer.

- Explain the procedure for using and measuring with a micrometer.

- Describe the measurements normally taken by a technician when working on a vehicle's drivetrain. Describe the different sources for service information that are available to technicians.

- Describe the different types of fasteners used in the automotive industry.

- Describe the requirements for ASE certification as an automotive technician and a master auto technician.

INTRODUCTION

Manual transmission and drivetrain technicians use a variety of tools. These include basic hand tools as well as special tools that a manual transmission and driveline technician uses every day. Common repair and diagnostic tools are discussed in this chapter.

TORQUE WRENCHES

Torque is the twisting force used to turn a fastener against the friction between the threads and between the head of the fastener and the surface of the component. The fact that practically every vehicle and engine manufacturer publishes a list of torque recommendations is ample proof of the importance of using proper amounts of torque when tightening nuts or bolts. The amount of torque applied to a fastener is measured with a torque indicating or torque wrench.

A **torque wrench** measures how tight a nut or bolt is. Many fasteners should be tightened to a certain amount and have a torque specification that is expressed in foot-pounds (USCS) or Newton-meters (metric). A foot-pound is the work or pressure accomplished by a force of 1 pound through a distance of 1 foot. A Newton-meter is the work or pressure accomplished by a force of 1 kilogram through a distance of 1 meter.

A torque wrench is basically a ratchet or breaker bar with some means of displaying the amount of torque exerted on a bolt when pressure is applied to the handle. Torque wrenches are available with the various drive sizes. Sockets are inserted onto the drive and then placed

The USCS system is also known as the English or Imperial system.

The SI system is normally called the metric system.

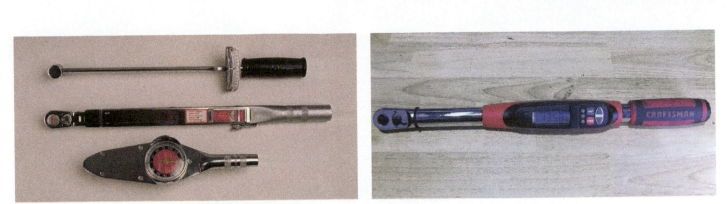

FIGURE 2-1 The (A) Basic types of torque indicating wrenches. (B) Digital torque wrench.

over the bolt. As pressure is exerted on the bolt, the torque wrench indicates the amount of torque. Many fasteners are designed to be "torque to yield." These bolts require that a certain torque specification be applied, and then the bolt is tightened an additional specified number of degrees to achieve the correct clamping force. These bolts are not reuseable and must be replaced once removed.

The common types of torque wrenches (Figure 2-1) are available with inch-pound and foot-pound increments.

- A beam torque wrench is not highly accurate. It relies on a metal beam that points to the torque reading.
- A click-type torque wrench clicks when the desired torque is reached. The handle is twisted to set the desired torque reading.
- A dial torque wrench has a dial that indicates the torque exerted on the wrench. The wrench may have a light or buzzer that turns on when the desired torque is reached.
- A digital readout type displays the torque and is commonly used to measure turning effort, as well as for tightening bolts. Some designs of this type of torque wrench have a light or buzzer that turns on when the desired torque is reached.

POWER TOOLS

Power tools make a technician's job easier. However, power tools require greater safety measures. Power tools do not stop unless they are turned off. Power is furnished by air (pneumatic), electricity, or hydraulic fluid. Pneumatic tools are typically used by technicians because they have more torque, weigh less, and require less maintenance than electric power tools. However, electric power tools tend to cost less than the pneumatics. Electric power tools can be plugged into most electric wall sockets, but to use a pneumatic tool, you must have an air compressor and an air storage tank. Cordless electrical tools have become more popular with greatly improved battery life.

Impact Wrenches

An impact wrench (Figure 2-2) uses compressed air or electricity to hammer or impact a nut or bolt loose or tight. Light-duty impact wrenches are available in three drive sizes, ¼, ⅜, and ½ inch, and two heavy-duty sizes, ¾ and 1 inch.

⚠ WARNING: Impact wrenches should not be used to tighten critical parts or parts that may be damaged by the hammering force of the wrench.

Impact ratchets often are used during disassembly or reassembly work to save time. Because the ratchet turns the socket without an impact force, they can be used on most parts and with ordinary sockets. Air ratchets usually have a ¼- or ⅜-inch drive. Impact wrenches

USCS and metric systems are two of the most common measuring standards. The USCS is what we use in the United States, whereas the metric system is used in most other parts of the world.

CAUTION:
Carelessness or mishandling of power tools can cause serious injury. Make sure you know how to operate a tool before using it.

CAUTION:
The sockets designed for impact wrenches are constructed of thicker steel to withstand the force of the impact. Ordinary sockets must not be used with impact wrenches, they can crack or shatter because of the force and can cause injury.

FIGURE 2-2 (A) An air impact wrench. (B) A battery-powered impact wrench.

and ratchets are not torque sensitive; therefore, a torque wrench should be used after installing a bolt with these tools.

Bench Grinder

This electric power tool is generally bolted to a workbench. The grinder should have safety shields and guards. Always wear face protection when using a grinder. Three types of wheels typically are available: a grinding wheel, a wire wheel brush, and a buffing wheel.

Shop Light

Adequate light is necessary when working under and around automobiles. A shop light can be battery powered or need to be plugged into a wall socket. Some shops have lights that pull down from a reel suspended from the ceiling. Shop lights use either incandescent bulbs, fluorescent tubes, or light-emitting diodes (LEDs). Because incandescent bulbs can pop and burn, it is highly recommended that you only use fluorescent bulbs or LEDs. Take extra care when using an electric shop light. Make sure the cord does not get caught in a rotating object. The bulb or tube is surrounded by a cage or enclosed in clear plastic to prevent accidental breaking and burning.

LIFTING TOOLS

Lifting tools are necessary tools for most drivetrain repair procedures. Typically, these tools are provided by the shop and are not the property of a technician. Correct operating and safety procedures should always be followed when using lifting tools.

Jacks

Jacks are available in two basic designs and in a variety of sizes. The most common jack is the hydraulic floor jack (Figure 2-3), which are classified by the weights they can lift: 1½, 2, and 2½ tons, and so on. The other design of portable floor jack uses compressed air. Pneumatic jacks are operated by controlling air pressure at the jack.

Safety Stands

When a vehicle is raised by a jack, it should be supported by **jack (safety) stands** (Figure 2-4). Never work under a car with only a jack supporting it; always use safety stands. Hydraulic seals in the jack can let go and allow the vehicle to drop.

Hydraulic Lift

The hydraulic floor lift (Figure 2-5) is the safest lifting tool and is able to raise the vehicle high enough to allow you to walk and work under it. Various safety features prevent a hydraulic lift from dropping if a seal does leak or if air pressure is lost. Before lifting a vehicle, make sure the lift is correctly positioned.

FIGURE 2-3 A hydraulic floor jack.

FIGURE 2-4 Safety stands are used to support a vehicle after it has been jacked up.

FIGURE 2-5 A typical hydraulic lift or hoist.

MEASURING TOOLS

Many of the procedures discussed in this manual require exact measurements of parts and clearances. Accurate measurements require the use of precision measuring devices that are designed to measure things in very small increments. Measuring tools are delicate instruments and should be handled with great care. Never strike, pry, drop, or force these tools. Also, make sure you clean them before and after every use.

SERVICE TIP:
It is a good idea to check all measuring tools against known good equipment or tool standards. This will ensure that they are operating properly and are capable of accurate measurement.

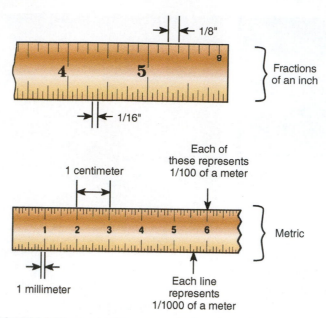

FIGURE 2-6 The graduations on an USCS and metric machinist rule.

Machinist's Rule

The **machinist's rule** looks very much like an ordinary ruler. Each edge of this basic measuring tool is divided into increments based on a different scale. A typical machinist's rule based on the Imperial system of measurement may have scales based on 1/8-, 1/16-, 1/32-, and 1/64-inch intervals. Of course, metric machinist rules also are available. Metric rules are usually divided into 0.5-mm and 1-mm increments (Figure 2-6).

Some machinist rules may be based on decimal intervals. These are typically divided into 1/10-, 1/50-, and 1/1,000-inch (0.1, 0.03, and 0.001) increments.

Micrometers

The **micrometer** is used to measure linear outside and inside dimensions. Both outside and inside micrometers are calibrated and read in the same way. Measurements on both are taken with the measuring points of the tool in contact with the surfaces being measured.

The major components and markings of a micrometer include the frame, anvil, spindle, locknut, sleeve, sleeve numbers, sleeve long line, thimble marks, thimble, and ratchet (Figure 2-7). Micrometers are calibrated in either inch or metric graduations and are available in a range of sizes.

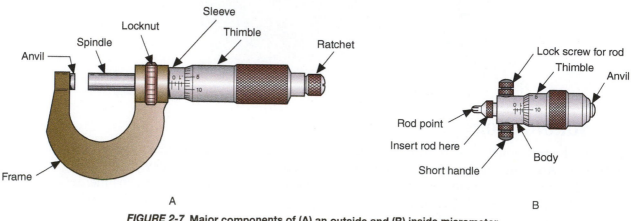

FIGURE 2-7 Major components of (A) an outside and (B) inside micrometer.

FIGURE 2-8 The correct way to hold a micrometer while measuring a small object.

To use and read a micrometer, choose the appropriate size for the object being measured. Typically, they measure an inch; therefore, the range covered by one size micrometer would be from 0 to 1 inch and another would measure 1 to 2 inches, and so on.

To measure small objects with an outside micrometer, open the jaws of the tool and slip the object between the spindle and the anvil. While holding the object against the anvil, turn the ratchet using your thumb and forefinger until the spindle contacts the object (Figure 2-8). The ratchet provides only enough pressure on the object to allow it to just fit between the tips of the anvil and spindle. The object should slip through with only a very slight resistance. When a satisfactory feel is reached, lock the micrometer.

Each graduation on the sleeve represents 0.025 inches (Figure 2-9). The graduations on the thimble assembly define the area between the lines on the sleeve; therefore, the number indicated on the thimble should be added to the measurement shown on the sleeve. The sum is the outside diameter of the object.

Micrometers are available to measure in 0.0001 (ten-thousandths) of an inch. Use this type of micrometer if the specifications call for this much accuracy.

To measure larger objects, hold the frame of the micrometer and slip it over the object. Turn the thimble when continuing to slip the micrometer over the object until you feel a very slight resistance. Rock the micrometer from side to side when doing this to make sure the spindle cannot be closed any further (Figure 2-10). Then, lock the micrometer and take a measurement reading.

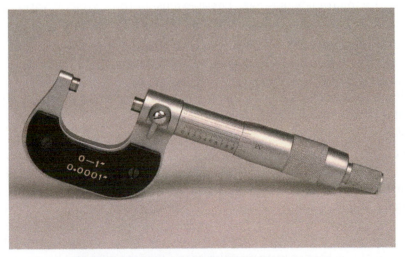

FIGURE 2-9 The graduations of an inch micrometer sleeve represents 0.025 inches.

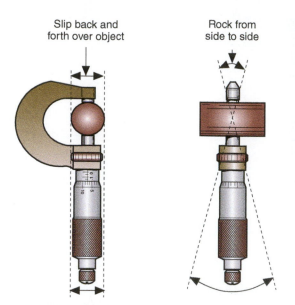

Slip back and forth over object

Rock from side to side

FIGURE 2-10 Slip and rock the micrometer to get a feel of the object in the micrometer. A correct measurement can only be had by a full contact of the object in the micrometer.

FIGURE 2-11 A micrometer with a digital readout.

Some technicians use a digital micrometer, which is easier to read. These tools do not have the various scales; rather, the measurement is read directly off the micrometer (Figure 2-11).

A metric micrometer is read in the same way except that the graduations are expressed in the metric system of measurement. Each number on the sleeve represents 5 millimeters (mm) or 0.005 meter (m). Each of the 10 equal spaces between each number, with index lines alternating above and below the horizontal line, represents 0.5 mm or five-tenths of an mm. Therefore, one revolution of the thimble changes the reading one space on the sleeve scale or 0.5 mm. The beveled edge of the thimble is divided into 50 equal divisions with every fifth line numbered: 0, 5, 10, . . . 45. Because one complete revolution of the thimble advances the spindle 0.5 mm, each graduation on the thimble is equal to one-hundredth of a millimeter. As with the inch-graduated micrometer, the separate readings are added together to obtain the total reading.

Inside Micrometers. Inside micrometers can be used to measure the inside diameter of a bore. To do this, place the tool inside the bore and extend the measuring surfaces until each end touches the bore's surface. If the bore is large, it might be necessary to use an extension rod to increase the micrometer's range. These extension rods come in various lengths. The inside micrometer is read in the same manner as an outside micrometer.

Depth Micrometers. A depth micrometer is used to measure the distance between two parallel surfaces. The sleeves, thimbles, and ratchet screws operate in the same way as other micrometers. Likewise, depth micrometers are read in the same way as other micrometers.

If a depth micrometer is used with a gauge bar, it is important to keep both the bar and the micrometer from rocking. Any movement of either part will result in an inaccurate measurement.

Telescoping Gauge. Telescoping gauges (Figure 2-12) are used for measuring bore diameters and other clearances. They may also be called snap gauges. Telescoping gauges are available in sizes ranging from fractions of an inch through 6 inches. Each gauge consists of two telescoping plungers, a handle, and a lock screw. Snap gauges are normally used with an outside micrometer.

To use a telescoping gauge, insert it into the bore and loosen the lock screw. This will allow the spring-loaded plungers to snap against the bore. Once the plungers have expanded, tighten the lock screw. Then remove the gauge and measure the expanse with a micrometer.

FIGURE 2-12 A set of telescoping gauges.

Small Hole Gauge. A small hole or ball gauge (Figure 2-13) works just like a telescoping gauge, but it is designed for small bores. After it is placed into the bore and expanded, it is removed and measured with a micrometer. Like the telescoping gauge, the small-hole gauge consists of a lock, a handle, and an expanding end. The end is made to expand or retract by turning the gauge handle.

Feeler Gauge. A **feeler gauge** is a thin strip of metal or plastic of known and closely controlled thickness. Several of these metal strips are often assembled together as a feeler gauge set that looks like a pocketknife (Figure 2-14). The desired thickness gauge can be pivoted

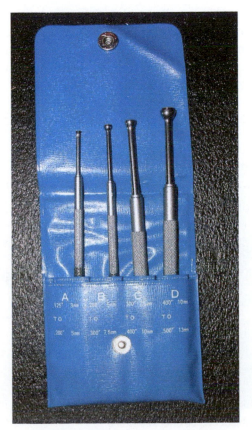

FIGURE 2-13 A small hole gauge set.

FIGURE 2-14 A typical feeler gauge pack.

away from others for convenient use. A feeler gauge set typically contains strips or leaves of 0.002- to 0.035-inch thickness (in steps of 0.001 inch).

A feeler gauge can be used by itself to measure gear and synchronizer clearances, end play, and other distances. It can also be used with a precision straightedge to check the flatness of a sealing surface and the alignment of parts.

Straightedge. A straightedge is no more than a metal bar machined to be totally flat and straight. Any surface that should be flat can be checked with a straightedge and feeler gauge set (Figure 2-15). The straightedge is placed across and at angles on the surface. At any low points on the surface, a feeler gauge can be placed between the straightedge and the surface. The size gauge that fills in the gap is the amount of warpage or distortion.

Dial Indicator

The **dial indicator** (Figure 2-16) is calibrated in 0.001-inch (one-thousandth inch) increments. Metric dial indicators also are available. Both types are used to measure movement. Common uses of the dial indicator include measuring backlash, end play (Figure 2-17), and flywheel and axle flange runout. Dial indicators are available with various face markings and measurement ranges to accommodate many measuring tasks.

To use a dial indicator (see Photo Sequence 2), position the indicator rod against the object to be measured. Then, push the indicator toward the work until the indicator needle

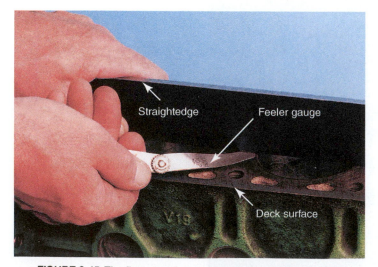

FIGURE 2-15 The flatness of an object can be checked with a straight edge and a feeler gauge set.

FIGURE 2-16 The face of a dial indicator.

FIGURE 2-17 This setup with a dial indicator and adaptive holding fixture setup can measure the end play of the axle.

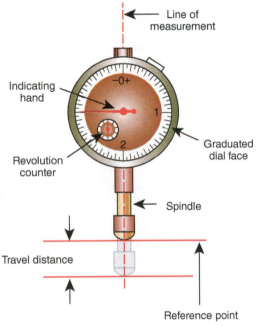

Line of measurement

Indicating hand

Graduated dial face

Revolution counter

Spindle

Travel distance

Reference point

FIGURE 2-18 The primary features of a dial indicator.

travels far enough around the gauge face to permit movement to be read in either direction (Figure 2-18). Move the object to one extreme of its travel, then zero the indicator needle on the gauge. Always be sure the range of the dial indicator is sufficient to allow the amount of movement being measured. For example, never attempt to use a 1-inch indicator on a component that will move 2 inches.

Calipers

Sliding **calipers** are versatile measuring tools that can take a direct reading of inside and outside diameters and depth. A vernier caliper is graduated similar to a micrometer. The inside jaws measure inner diameter, the outside jaws measure outside diameter or component thickness, and the depth probe measures depth (Figure 2-19). Dial calipers use a dial face that reads similar to that of a dial indicator. Digital calipers are easy to read and typically can be switched for SAE or metric measurement (Figure 2-20).

PHOTO SEQUENCE 2

SETTING UP A DIAL INDICATOR

What you will need:
A dial indicator
A magnetic base or vise grip-type mount for the indicator
Vee-blocks
A shaft (preferably from a manual transmission or an axle shaft) or pipe that is slightly distorted or bent

P2-1 To measure the runout of a shaft, place the ends into Vee-blocks.

P2-2 Place the dial indicator toward the center of the shaft so it can measure all deviations.

P2-3 Position the mounting base so the indicator can be properly placed at the center of the shaft.

P2-4 Place the indicator rod at the center of the shaft. Depress the indicator rod so the plunger is in contact with the shaft but has the freedom to move up or down.

P2-5 Once the mount and indicator are in position, secure both.

P2-6 Set the indicator to zero and tighten the tool's scale screw and all mounting brackets.

P2-7 As you rotate the shaft, any deviation or change from a pure horizontal plane will show up on the dial indicator.

P2-8 The total movement of the needle indicates the total out-of-roundness of the shaft. Ideally, all shafts will have zero runout.

P2-9 To measure end play of a shaft, position the dial indicator to the end of the shaft.

P2-10 Once in position, zero the indicator so it can measure all end-to-end movement.

P2-11 As you can see, the indicator displays all movement of the shaft. When measuring end play, compare the measured movement to the specifications to determine if there is a problem.

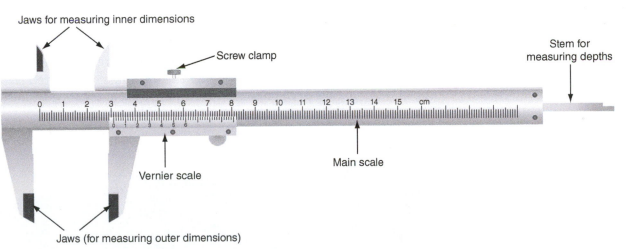

Jaws for measuring inner dimensions

Screw clamp

Stem for measuring depths

0 1 2 3 4 5 6 7 8 9 10 11 12 13 14 15 cm

0 1 2 3 4 5 6

Vernier scale

Main scale

Jaws (for measuring outer dimensions)

FIGURE 2-19 The components of a vernier caliper.

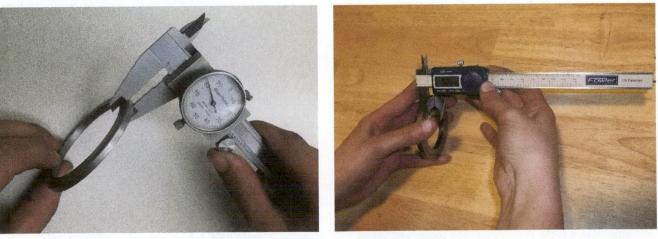

FIGURE 2-20 (A) Dial caliper. (B) Digital caliper.

MANUAL TRANSMISSION AND DRIVELINE TOOLS AND EQUIPMENT

Different tools are used to service transmissions and drivelines. The National Automotive Technicians Education Foundation (NATEF) has identified many of these and has specified that a manual transmission and driveline technician must know what they are and how and when to use them. The tools and equipment listed by NATEF are covered in the following discussion. Although you need to be familiar with and will be using common hand tools, they are not part of this discussion. You already should know what they are and how to use and care for them.

Portable Crane

Engine hoists often are referred to as "cherry pickers."

Often to remove and install a transmission, the engine will be moved out of the vehicle with the transmission. To remove or install an engine, a portable crane, frequently called a cherry picker, is used. A crane uses hydraulic pressure that is converted to a mechanical advantage and lifts the engine from the vehicle. To lift an engine, attach a pulling sling or chain to the engine. Some engines have eyeplates for use in lifting (Figure 2-21). If they are not available, the sling must be bolted to the engine. The sling attaching bolts must be large enough to support the engine and must thread into the block a minimum of 1½ times the bolt diameter. Connect the crane to the chain. Raise the engine slightly and make sure the sling attachments are secure. Carefully lift the engine out of its compartment. Lower the engine close to the floor so the transmission can be safely removed from the engine, if necessary.

Transmission Jacks

Transmission jacks are designed to help you when removing a transmission from under the vehicle. The weight of the transmission can make it difficult and unsafe to remove it without much assistance or a transmission jack. These jacks fit under the transmission (Figure 2-22) and are typically equipped with hold-down chains or straps that are used to secure the transmission to the jack. The transmission's weight rests on the jack's saddle.

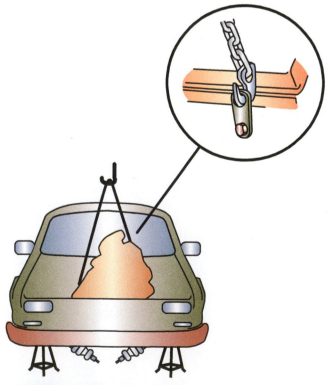

FIGURE 2-21 When using a chain hoist to pull an engine, the attachments to the engine should be secure and strong.

FIGURE 2-22 A typical transmission jack.

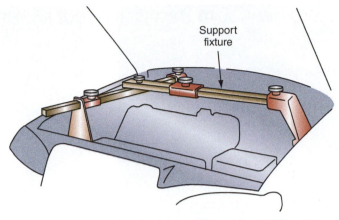

FIGURE 2-23 Typical engine support fixtures for a FWD vehicle.

Transmission jacks are available in two basic styles. One is used when the vehicle is sitting on jack stands. The other style is used when the vehicle is raised on a lift.

Transaxle Removal and Installation Equipment

The removal and replacement (R&R) of transversely mounted engines may require tools not required for removing a RWD engine. The engines of some FWD vehicles are removed by lifting them from the top. Others must be removed from the bottom, and the procedure requires different equipment. Make sure you follow the instructions given by the manufacturer and use the appropriate tools and equipment.

To remove the engine and transmission from under the vehicle, the vehicle must be raised. A crane or support fixture is used to hold the engine and transaxle assembly in place when the assembly is being readied for removal. The cradle is then used to lower the assembly. The cradle is similar to a hydraulic floor jack and is used to lower the assembly further so it can be rolled out from under the vehicle. The transaxle can be separated from the engine once it has been removed from the vehicle.

When the transaxle is removed by itself, the engine must be supported when it is in the vehicle before, during, and after transaxle removal. Special fixtures (Figure 2-23) mount to the vehicle's upper frame or suspension parts. These supports have a bracket that is attached to the engine. With the bracket in place, the engine's weight is now on the support fixture and the transmission can be removed.

Transmission/Transaxle Holding Fixtures

Special holding fixtures should be used to support the transmission or transaxle after it has been removed from the vehicle. These holding fixtures may be standalone units or may be bench mounted and allow the transmission to be easily repositioned during repair work. Removable carrier differentials can also be supported by holding fixtures (Figure 2-24).

Presses

Many transmission and driveline repairs require the use of a powerful force to assemble or disassemble parts that are **press-fit** together. Axle and final drive bearing removal and installation, universal joint replacement, and transmission assembly work are just a few of the examples. Presses can be hydraulic-, electric-, air-, or hand-driven. Capacities range up to 150 tons of pressing force, depending on the size and design of the press. Smaller arbor and C-frame presses can be bench- or pedestal-mounted, whereas high-capacity units are freestanding or floor-mounted (Figure 2-25).

CAUTION:
Always wear safety glasses when using a press.

FIGURE 2-24 A transmission/differential holding fixture.

FIGURE 2-25 A floor-mounted hydraulic press.

Blowgun

Blowguns are used for blowing off parts during cleaning. Never point a blowgun at yourself or someone else. A **blowgun** snaps into one end of an air hose and directs airflow when a button is pressed. Always use an OSHA-approved air blowgun (Figure 2-26). Before using a blowgun, be sure it has not been modified to eliminate air-bleed holes on the side.

Gear and Bearing Pullers

Many tools are designed for a specific purpose. An example of a special tool is a gear and bearing puller (Figure 2-27). Many gears and bearings have a slight interference fit when they are installed on a shaft or in a housing. Something that has a press-fit has an interference fit. For example, the inside diameter of a bore is 0.001-inch smaller than the outside diameter of

FIGURE 2-26 An OSHA-approved air blowgun.

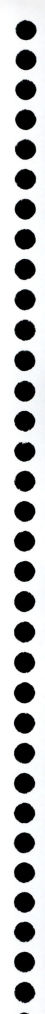

FIGURE 2-27 A final drive's side bearing being removed with a universal bearing puller.

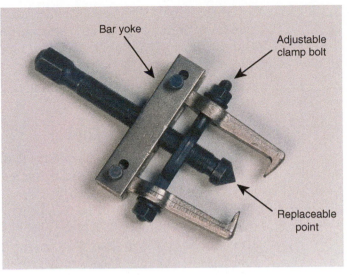

FIGURE 2-28 The main parts of a universal gear and bearing puller.

a shaft, when the shaft is fitted into the bore it must be pressed in to overcome the 0.001 inch interference fit. This press-fit prevents the parts from moving on each other. The removal of these gears and bearings must be done carefully to prevent damage to the gears, bearings, or shafts. Prying or hammering can break or bind the parts. A puller with the proper jaws and adapters should be used to remove gears and bearings (Figure 2-28). Using the proper puller, the force required to remove a gear or bearing can be applied with a slight and steady motion.

Bushing and Seal Pullers and Drivers

Another commonly used group of special tools is the various designs of bushing and seal drivers and pullers. Pullers are either threaded or slide hammer-type tools (Figure 2-29). Bushings and seals are easily damaged if the wrong tool or procedure is used. Car manufacturers and specialty tool companies work closely together to design and manufacture special tools required to repair cars. Most of these special tools are listed in the appropriate service manuals.

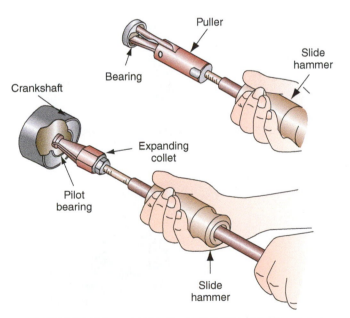

FIGURE 2-29 Using a seal puller and slide hammer to remove an extension housing seal.

A commonly used bearing puller is a clutch pilot bearing/bushing remover. Typically, the tool hooks into the back of the bearing. Once it is in place, the threaded portion is tightened and the bearing pulled out. The new pilot bearing or bushing is driven into the bore with a hammer and correctly sized driver.

Axle Pullers. Axle pullers are used to pull rear axles in RWD vehicles. Most rear axle pullers are slide hammer type.

Clutch Alignment Tool

To keep the clutch disc centered on the flywheel when assembling the clutch, a clutch alignment tool is used (Figure 2-30). The tool is inserted through the input shaft opening of the pressure plate and is passed through the clutch disc. The tool then is inserted into the pilot bushing or bearing. The outer diameter (OD) of the alignment tool that goes into the pilot must be only slightly smaller than the inner diameter (ID) of the pilot bushing. The OD of the tool that holds the disc in place also must be only slightly smaller than the ID of the disc's splined bore. The effectiveness of this tool depends on its diameters, which is why it is best to have various sizes of clutch alignment tools.

Universal Joint Tools

Although servicing universal joints can be done with hand tools and a vise, many technicians prefer the use of specifically designed tools. One such tool is a C-clamp modified to include a bore that allows the joint's caps to slide in when tightening the clamp over an assembled joint to remove it (Figure 2-31). Other tools are the various drivers used with a press to press the joint in and out of its yoke.

FIGURE 2-30 Examples of a clutch alignment tool.

FIGURE 2-31 A C-clamp-based universal joint.

FIGURE 2-32 An assortment of snap/retaining ring pliers.

Drive Shaft Angle Gauge

Critical to the durability of universal joints and vibration and free vehicle operation is the angle of the drive shaft. The angle of the drive shaft at the transmission should equal its angle at the drive axle. There are many ways to measure the angle; one way involves the use of an **inclinometer** or drive shaft angle gauge. Digital angle gauges are also available.

Retaining (Snap) Ring Pliers

Often, a transmission technician will run into many different styles and sizes of retaining or snaprings that hold subassemblies together or keep them in a fixed location. Using the correct tool to remove and install these rings is the only safe way to work with them (Figure 2-32). All transmission and driveline technicians should have an assortment of retaining ring pliers.

Special Tool Sets

Vehicle manufacturers and specialty tool companies work closely together to design and manufacture special tools required to repair transmissions. Most of these special tools are listed in the appropriate service manuals and are part of each manufacturer's "essential toolkit."

DIAGNOSTIC INFORMATION

The technician relies on various sources of information to diagnose and repair vehicles. Most technicians today rely on Internet-based sites that may be provided by a specific manufacturer or by private concerns such as All-Data or Mitchell. Printed service manuals and other electronic aids supplement the information available to the technician.

SERVICE MANUALS

Service manuals (Figure 2-33) are a necessary part of transmission and driveline service. There is no way a technician can remember all of the procedures and specifications needed to repair an automobile correctly. Thus, a good technician relies on service manuals and other sources for this information. Good information plus knowledge allows a technician to fix a problem with the least bit of frustration and at the lowest expense to the customer. Service manuals also provide drawings and photographs that show where and how to perform certain procedures on the particular vehicle you are working on. Precautions also are given to prevent injury or damage to parts. Perhaps the most important tools you will use are service manuals.

The primary source of repair and specification information for any car, van, or truck is the manufacturer. The manufacturer publishes service manuals each year, for every vehicle

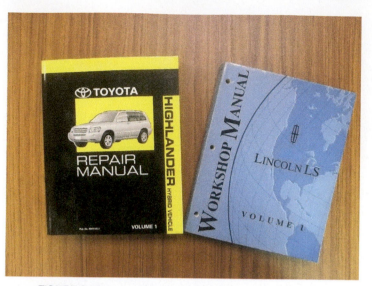

FIGURE 2-33 A couple of manufacturers' service manuals.

built. Because of the enormous amount of information, some manufacturers publish more than one manual per year per car model. They are typically divided into sections based on the major systems of the vehicle. In the case of transmissions, there is a section for each transmission that may be in the vehicle. Manufacturers' manuals cover all repairs, adjustments, specifications, detailed diagnostic procedures, and special tools required.

Computer-Based Information

The same information that is available in service manuals is now commonly found electronically on compact discs (CD-ROMs) (Figure 2-34), digital video discs (DVDs), and the Internet (Figure 2-35). A single compact disc can hold a quarter million pages of text, eliminating the need for a huge library to contain all of the printed manuals. Using electronics to find

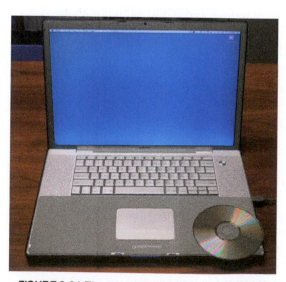

FIGURE 2-34 The use of CD-ROMs and computers make accessing information quick and easy.

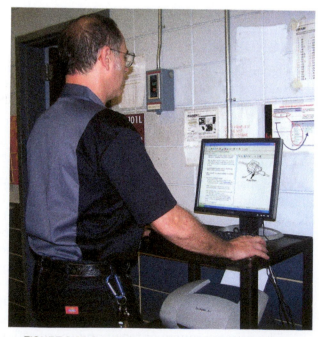

FIGURE 2-35 A technician using Internet-based service information.

information is also easier and quicker. The discs are normally updated quarterly and not only contain the most recent service bulletins but also engineering and field service fixes. DVDs can hold more information than CDs; therefore, fewer discs are needed with systems that use DVDs. The CDs and DVDs are inserted into a computer. All a technician needs to do is enter vehicle information and then move to the appropriate part or system. The appropriate information will then appear on the computer's screen.

Online data can be updated instantly and requires no space for physical storage. These systems are easy to use and the information is quickly accessed and displayed. The computer's keyboard, mouse, or light pen are used to make selections from the screen's menu. Once the information is retrieved, it can be read off the screen or printed out and taken to the service bay. Many technicians have a laptop computer at their fingertips in the shop. Most manufacturers and some private companies provide Internet connections to "flash," or update, vehicle control modules.

Technical Service Bulletins

Because many technical changes occur on specific vehicles each year, manufacturers' service information needs to be constantly updated. Updates are published as service bulletins (often referred to as **Technical Service Bulletins, (TSBs)**) that show the changes in specifications and repair procedures during the model year. These changes often do not appear in the service manual until the next year, but can instantly be available online. The car manufacturer provides these bulletins to dealers and repair facilities on a regular basis. TSBs are also published to help lead the technician to a likely cause of a vehicle problem. The experienced technician will often begin the diagnostic process by checking for any TSBs that might relate to the problem.

Service manuals are sometimes simply called shop manuals.

General and Specialty Repair Manuals

Service manuals also are published by independent companies rather than the manufacturers. However, they pay for and get most of their information from the carmakers. They contain component information, diagnostic steps, repair procedures, and specifications for several car makes in one book. Information is usually condensed and is more general in nature than the manufacturer's manuals. The condensed format allows for more coverage in less space and, therefore, is not always specific. They also may contain several years of models as well as several car makes in one book.

Many of the larger parts manufacturers have excellent guides on the various parts they manufacture or supply. They also provide updated service bulletins on their products. Other sources for up-to-date technical information are trade magazines and trade associations. Two popular drivetrain publications are *Transmission Digest* and *Gears* magazine.

Flat Rate Manuals

Flat rate manuals contain standards for the length of time a specific repair is supposed to require. Normally, they also contain a parts list with approximate or exact prices of parts. They are excellent for making cost estimates and are published by the manufacturers and independents. Most often, flat rate manuals are contained in the shop's management software, are accessed online, or are included in the service information available with the service manuals on a CD or DVD.

Using a Service Manual

Although the manuals from different publishers vary in presentation and arrangement of topics, all service manuals are easy to use after you become familiar with their organization. Most shop manuals are divided into a number of sections, each covering different aspects of the vehicle. The beginning sections commonly provide vehicle identification and basic maintenance information. The remaining sections deal with each different vehicle system in detail, including diagnostic, service, and overhaul procedures. Each section has an index indicating more specific areas of information.

To use a service manual:

1. Select the appropriate manual for the vehicle being serviced.
2. Use the table of contents to locate the section that applies to the work being done.
3. Use the index at the front of that section to locate the required information.
4. Carefully read the information and study the applicable illustrations and diagrams.
5. Follow all of the required steps and procedures given for that service operation.
6. Adhere to all of the given specifications and perform all measurement and adjustment procedures with accuracy and precision.

Using Electronic Service Data. Today most technicians rely on the Internet to find specific repair information. This is due to the fact that service information can be obtained very quickly and easily. Although much of the information on the Web is the same as that found on DVDs, the information on the Internet can be easily updated. With DVDs, the provider must send out updated versions to the service facility. This takes time and will only happen periodically, not when an update is necessary. Another advantage of the Internet is that all of the information is there, and there is no need to scan through DVDs to find what you need. A technician simply inputs what he or she needs, and a search engine finds the information. Also, to simplify things, the information is divided by specific systems.

The data available from most electronic service data providers include the following information for each system of a specific vehicle:

- Adjustments
- Component locators
- Description and operation
- Diagnostic flowcharts
- Diagnostic trouble codes
- Diagrams
- Fluid types and volumes
- Maintenance schedules
- OEM wiring diagrams
- Parts and labor information
- Recall notices
- Required tools and equipment
- Service and repair procedures
- Service precautions
- Shortcuts
- Specifications
- Technical Service Bulletins
- Testing and inspection procedures
- User tips

Vehicle Identification

Before performing any service, it is important for you to know exactly what type of vehicle you are working on. The best way to do this is to refer to the **vehicle identification number (VIN)**. The VIN (Figure 2-36) is given on a plate behind the lower corner of the driver's side of the windshield, as well as other locations on the vehicle. The VIN is made up of 17 characters and contains all pertinent information about the vehicle. The use of the number and letter code became mandatory beginning with 1981 vehicles and is used by all manufacturers of vehicles both domestic and foreign.

VIN SAMPLE (BRZ Limited)	J	F	1	Z	C	A	C	1	0	D	2	6	0	2	6	8	1
VIN position #	1	2	3	4	5	6	7	8	9	10	11	12–17					
Means	Manufacturer's code			Line	Body	Engine	Model/trim	Restraint systems	Check digit	Year	Factory, transmission	Sequential production number					

FIGURE 2-36 Basic makeup of a VIN. VIN makeup of a 2013 Subaru BRZ. Note that the engine ID is the sixth digit.

YEAR	CODE	YEAR	CODE	YEAR	CODE	YEAR	CODE
1980	A	1990	L	2000	Y	2010	A
1981	B	1991	M	2001	1	2011	B
1982	C	1992	N	2002	2	2012	C
1983	D	1993	P	2003	3	2013	D
1984	E	1994	R	2004	4	2014	E
1985	F	1995	S	2005	5	2015	F
1986	G	1996	T	2006	6	2016	G
1987	H	1997	V	2007	7	2017	H
1988	J	1998	W	2008	8	2018	J
1989	K	1999	X	2009	9	2019	K

FIGURE 2-37 Interpretation of VIN model year codes.

Each character of a VIN has a particular purpose. The first character identifies the country where the vehicle was manufactured, for example: 1 or 4 = United States, 2 = Canada, J = Japan, and W = Germany.

The second character identifies the manufacturer, for example: B = BMW, C = Chrysler, G = General Motors, H = Honda, and T = Toyota.

The third character identifies the vehicle type or manufacturing division (passenger car, truck, bus, and so on). The fourth through eighth characters identify the features of the vehicle, such as the body style, vehicle model, and engine type. Most American manufacturers use the eighth character to identify the engine in the vehicle.

The ninth character is used to identify the accuracy of the VIN and is a check digit. The 10th character identifies the model year. The 17-digit VIN was not required until 1981. However, 1980 was used as the first year in the VIN, and it is represented by the letter A. The following years are represented by the next letter in the alphabet (B = 1981), and that trend ended in 2000 with the letter Y. The years 2001 through 2009 are represented by the digits 1 through 9, and the subsequent years will begin in alphabetic order again (Figure 2-37).

The 11th character identifies the plant where the vehicle was assembled and the 12th to 17th characters identify the production sequence of the vehicle as it rolled off the manufacturer's assembly line.

The specifics needed for decoding the characters of the VIN can be found in the service information for the vehicle.

Hotline Services

Hotline services provide answers to service concerns by telephone. Manufacturers provide help by telephone for technicians in their dealerships. There are subscription services for independents to be able to get repair information by phone. Some manufacturers also have a phone modem system that can transmit computer information from the car to another location. The vehicle's diagnostic link is connected to the modem. The technician in the service bay runs a test sequence on the vehicle. The system downloads the latest repair information for the concern. If that does not repair the problem, a technical specialist at the manufacturer's location will review the data and propose a repair.

iATN

The International Automotive Technician's Network (iATN) is comprised of a group of thousands of professional automotive technicians from around the world. The technicians in this group exchange technical knowledge and information with other members. The Web address for this group is *www.iatn.net*. This network is one of many examples of Internet-based technical information–sharing networks.

FASTENERS

Many types and sizes of fasteners are used by the automotive industry. Each fastener is designed for a specific purpose and condition. Threaded fasteners are one of the most commonly used types of fasteners. Threaded fasteners include bolts, nuts, screws, and similar items that allow a technician to install or remove parts easily.

Threaded fasteners are available in many sizes, designs, and threads. The threads can be either cut or rolled into the fastener. Rolled threads are 30% stronger than cut threads. They also offer better fatigue resistance because there are no sharp notches to create stress points. These fasteners are available with coarse and fine threads.

Coarse threads are used for general-purpose work, especially when rapid assembly and disassembly is required. Fine threaded fasteners are used when greater holding force is necessary. They are also used when greater resistance to vibration is desired.

Bolts have a head on one end and threads on the other. Bolts are identified by defining the head size, shank diameter, thread pitch, length (Figure 2-38), and grade.

Studs are rods with threads on both ends. Most often, the threads on one end are coarse, whereas the other end is fine thread. One end of the stud is screwed into a threaded bore. A hole in the part to be secured is fitted over the stud and held in place with a nut. Studs are used when the clamping pressures of a fine thread are needed and a bolt will not work. If the material the stud is being screwed into is soft (such as aluminum) or granular (such as cast iron), fine threads will not withstand a great amount of pulling force. Therefore, a coarse thread is used to secure the stud in the work piece and a fine threaded nut is used to secure the other part to it. Doing this allows the clamping force of fine threads and the holding power of coarse threads.

Nuts are used with other threaded fasteners when the fastener is not threaded into a piece of work. Many different designs of nuts are found on today's cars. The most common one is the hex nut, which are used with studs and bolts and are tightened with a wrench.

Setscrews are used to prevent rotary motion between two parts, such as a pulley and shaft. Setscrews are either headless and require an Allen wrench or screwdriver to loosen and tighten them or have a square head.

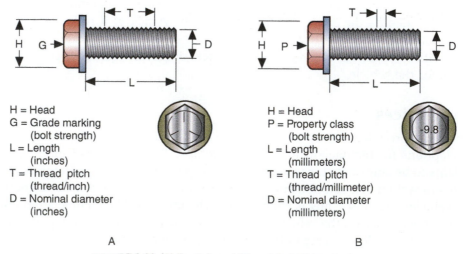

H = Head
G = Grade marking
  (bolt strength)
L = Length
  (inches)
T = Thread pitch
  (thread/inch)
D = Nominal diameter
  (inches)

H = Head
P = Property class
  (bolt strength)
L = Length
  (millimeters)
T = Thread pitch
  (thread/millimeter)
D = Nominal diameter
  (millimeters)

A   B

FIGURE 2-38 (A) English and (B) metric bolt terminology.

Bolt Identification

The **bolt head** is used to loosen and tighten the bolt. The size of the bolt head varies with the diameter of the bolt and is available in USCS and metric wrench sizes. Many confuse the size of the head with the size of the bolt. The size of a bolt is determined by the diameter of its shank. The size of the bolt head determines what size wrench is required to screw it.

Bolt diameter is the measurement across the major diameter of the threaded area or across the **bolt shank**. The length of a bolt is measured from the bottom surface of the head to the end of the threads.

The **thread pitch** of a bolt in the USCS system is determined by the number of threads that are within one inch of the threaded bolt length and is expressed in number of threads per inch. A bolt with a 1/10-inch diameter and 24 threads per inch would be a 3/8 × 24 bolt.

The distance, in millimeters, between two adjacent threads determines the thread pitch in the metric system. This distance will vary between 0.8 and 2.0 and depends on the diameter of the bolt. The lower the number, the closer the threads are placed and the finer the threads are.

The bolt's tensile strength, or grade, is the amount of stress or stretch it is able to withstand before it breaks. The type of material the bolt is made of and the diameter of the bolt determines its grade. In the Imperial system, the tensile strength of a bolt is identified by the number of radial lines (**grade marks**) on the bolt's head. More lines mean higher tensile strength (Table 2-1). Count the number of lines and add two to determine the grade of a bolt.

A property class number on the bolt head identifies the grade of metric bolts. This numerical identification is comprised of two numbers. The first number represents the tensile strength of the bolt. The higher the number, the greater the tensile strength. The second number represents the yield strength of the bolt. This number represents how much stress the bolt can take before it is not able to return to its original shape without damage. The second number represents a percentage rating. For example, a 10.9 bolt has a tensile strength of 1,000 MPa (145,000 psi) and a yield strength of 900 MPa (90% of 1,000).

Nuts are graded to match their respective bolts. For example, a grade 8 nut must be used with a grade 8 bolt. If a grade 5 nut were used, a grade 5 connection would result. Grade 8 and critical applications require the use of fully hardened flat washers. These will not dish out when torqued as soft washers will.

Bolt heads can pop off because of **fillet** damage. The fillet is the smooth curve where the shank flows into the bolt head (Figure 2-39). Scratches in this area introduce stress to the bolt head, causing failure. Removing any burrs around the edges of holes can protect the bolt head. Also, place flat washers with their rounded, punched side against the bolt head and their sharp side to the work surface.

TABLE 2-1   STANDARD BOLT STRENGTH MARKINGS

SAE Grade Markings					
DEFINITION	No Lines: unmarked indeterminate quality SAE grades 0-1-2	3 Lines: common commercial-quality automotive and bolts SAE grade 5	4 Lines: medium commercial-quality automotive and bolts SAE grade 6	5 Lines: rarely used SAE grade 7	6 Lines: best commercial-quality NAS and aircraft screws SAE grade 8
MATERIAL	Low-carbon steel	Medium carbon steel, tempered	Medium carbon steel, quenched and tempered	Medium carbon alloy steel	Medium carbon alloy steel, quenched and tempered
TENSILE STRENGTH	65,000 psi	120,000 psi	140,000 psi	140,000 psi	150,000 psi

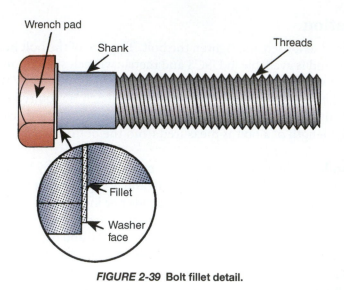

Wrench pad

Shank

Threads

Fillet

Washer face

FIGURE 2-39 Bolt fillet detail.

Fatigue breaks are the most common type of bolt failure. A bolt becomes fatigued from working back and forth when it is too loose. Undertightening the bolt causes this problem. Bolts also can be broken or damaged by overtightening, being forced into a nonmatching thread, or bottoming out, which happens when the bolt is too long.

Tightening Bolts

Any fastener is near worthless if it is not as tight as it should be. When a bolt is properly tightened, it will be "spring loaded" against the part it is holding. This spring effect is caused by the stretch of the bolt when it is tightened. Normally a properly tightened bolt is stretched to 70% of its elastic limit. The elastic limit of a bolt is that point of stretch where the bolt will not return to its original shape when it is loosened. Not only will an overtightened or stretched bolt not have sufficient clamping force, it also will have distorted threads. The stretched threads will make it more difficult to screw and unscrew the bolt or a nut on the bolt. Always check the service manual to see if there is a torque specification for a bolt before tightening it. If there is, use a torque wrench and tighten the bolt properly. Torque-to-yield (TTY) bolts are torqued beyond the elastic limit of the fastener and become permanently stretched. The tightening procedure for TTY fasteners requires tightening to a specific specification and turning the fastener a further specified number of degrees. TTY bolts are not intended to be reused.

BASIC GEAR ADJUSTMENTS

While a drivetrain is in operation, the gears, shafts, and bearings are subjected to loads and vibrations. Because of this, the parts of a drivetrain must be adjusted for proper fit. There are three basic adjustments that are made when reassembling a unit or when a problem suggests that readjustment is necessary: backlash, end play, and preload.

Backlash in Gears

Backlash is the clearance between two gears in mesh (Figure 2-40). Excessive backlash can be caused by worn gear teeth, improper meshing of teeth, or bearings that do not support the gears properly. Excessive backlash can result in severe impact on the gear teeth from sudden stops or direction changes of the gears, which can cause broken gear teeth and gears. Insufficient backlash causes excessive overload wear on the gear teeth and could cause premature gear failure.

Backlash is measured with a dial indicator mounted so that its stem is in line with the rotation of the gear and perpendicular to the angle of the teeth (Figure 2-41). The gear is

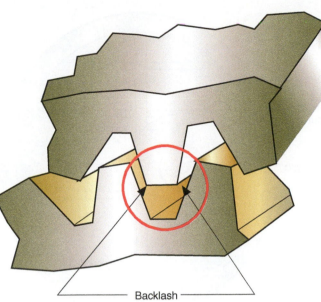

FIGURE 2-40 Backlash is the clearance between two gears in mesh.

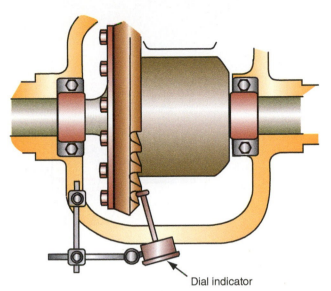

Dial indicator

FIGURE 2-41 A dial indicator is normally used to measure the backlash of a gear set. The plunger tip should be in line with gear tooth movement in order to get the most accurate reading.

moved in both directions, whereas the other gear it meshes with is held. The amount of movement on the dial indicator equals the amount of backlash present. Backlash can be adjusted by changing shim thickness or by using screw-type adjusters.

End Play in Gears and Shafts

End play refers to the measurable axial or end-to-end looseness of a bearing or shaft. End play is always measured in an unloaded condition. To check end play, a dial indicator is mounted against the side of a gear or the end of a shaft (Figure 2-42). The gear shaft is then pried in both directions and the readings noted. The difference between the two readings is the amount of end play. Shims, selective thrust washers, or adjusting nuts are widely used to adjust end play.

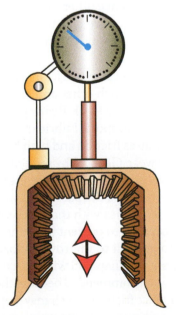

FIGURE 2-42 Measuring the end play of a final drive's side gears.

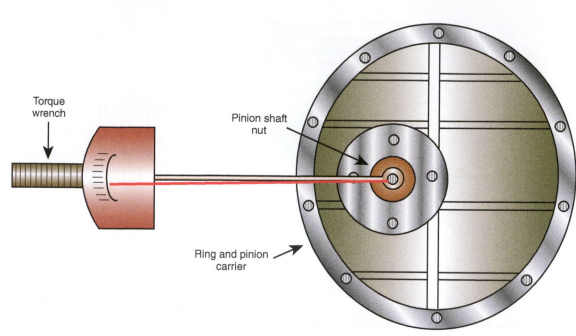

FIGURE 2-43 Checking preload on a pinion bearing.

Preloading of Geartrains

When normal operating loads are great, geartrains are often preloaded to reduce the deflection of parts. The amount of **preload** is specified in shop manuals and must be correct for the design of the bearings and the strength of the parts. If bearings are excessively preloaded, they will heat up and fail. When bearings are set too loose, the shaft or gear set will wear rapidly because of the great amounts of deflection it will experience. Geartrains are preloaded by shims, thrust washers, adjusting nuts, or by using double-race bearings. Preload adjustments are normally checked by measuring turning effort with a beam-torque wrench (Figure 2-43). A dial-type torque wrench is also acceptable, but click-type torque wrenches cannot be used to measure turning effort.

GENERAL MAINTENANCE

Most of the driveline needs little maintenance other than periodic oil changes. However, when you repair a transmission, examine the entire geartrain to locate worn or faulty parts and repair or replace them at that time. In this way, you may prevent a breakdown and the need to disassemble the transmission once again.

Transmissions and transaxles require clean lubricant to function properly and be durable. The moving metal parts must not touch each other and should be continuously separated by a thin film of lubricant to prevent excessive wear. Bearings rely on lubricant for smooth and low-friction rotation of the shafts they support. Lubricant is continuously being wiped away as meshed gears rotate. This action causes friction and high heat, and without a constant supply of lubricant, the gears would rapidly wear. Gear oil, by reducing the friction in the transmission, also limits the amount of power loss because of friction. Gear oil also protects the gears and related parts from rust and corrosion, keeps the parts cool, and keeps the internal parts clean.

The type and amount of lubricant varies with the different designs (Figure 2-44). Gear oil must have adequate load-carrying capacity to prevent the puncturing of the oil film on the gears. Chemical additives are mixed with gear oil to improve its load-carrying capacity. The typical lubricant for transmissions is a straight mineral oil with an extreme pressure (EP) additive.

In some cases, EP additives are not recommended because the synchronizers may not be effective with the additive. EP additives reduce friction, which may stop the synchros from working. Not all transmissions require heavy-gear oil; many use engine oil, automatic transmission fluid (ATF), or a manufacturer-specific oil containing friction modifiers to assist synchronizer operation.

EP additives increase the load-carrying capacity of oil.

CAUTION: Using incorrect oil will cause transmission failure.

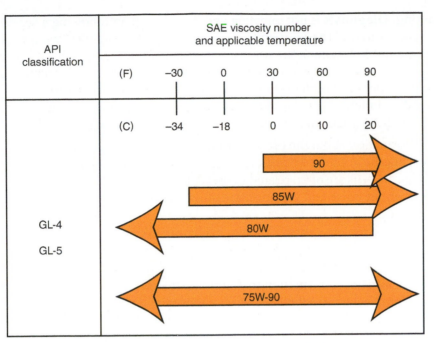

FIGURE 2-44 Typical transaxle/transmission gear oil classification and viscosity range data.

Thinner lubricants, such as ATF, are more commonly recommended for transmissions and transfer cases because these lubricants are better at lubricating the bearings inside the constantly meshing gears and they also absorb less engine power, which increases power output and increases fuel economy.

Clutches. Older external clutch linkages may require checking and adjustment at regular intervals. Vehicles with external clutch linkage require periodic lubrication. Most vehicles utilize hydraulic clutch release mechanisms, and there is another but smaller master cylinder close to the brake master cylinder. This is the clutch master cylinder. Its fluid level needs to be checked, which is done in the same way as brake fluid. In most cases, the clutch master cylinder uses the same type of fluid as the brake master cylinder. However, check this out before adding any fluid.

Drive Shafts. Most modern U-joints are of the extended life design, meaning they are sealed and require no periodic lubrication. If the U-joints are equipped with zerk or grease fittings, they should be periodically lubricated with the correct grease.

Final Drives and Drive Axles. Maintenance includes inspecting the level of and changing the gear lubricant. Proper lubrication is necessary for drive axle durability. Different applications require different gear lubes. The American Petroleum Institute (API) has established a rating system for the various gear lubes available. In general, rear axles use either SAE 80- or 90-weight gear oil for lubrication, meeting GL-5 specifications. Most newer vehicles use synthetic gear lube to reduce friction and improve component life. With limited-slip axles, it is very important that the proper gear lube be used. Most often, a special friction modifier fluid should be added to the fluid. If the wrong lubricant is used, damage to the clutch packs and grabbing or chattering on turns will result. If this condition exists, try draining the oil and refilling with the proper gear lube before servicing it.

LOGICAL DIAGNOSTICS

The true measure of a good technician is an ability to find and correct the cause of problems. Service manuals and other information sources will guide you through the diagnosis and repair of problems. However, those guidelines will not always lead you to the exact cause of the problem. To do this you must use your knowledge and take a logical approach when

The higher the numerical classification of an oil, the thicker the oil is.

Today's manual transmissions and transaxles require the use of single- and multiviscosity gear oils, engine oils, or automatic transmission fluids. Always refer to the appropriate service manual to determine the proper lubricant for the vehicle being serviced. Using the correct gear oil not only will prevent premature wear but also will allow the transmission/transaxle to shift smoothly and operate quietly.

troubleshooting. **Diagnosis** is not guessing and it is more than following a series of interrelated steps in order to find the solution to a specific problem. Diagnosis is a way of looking at systems that are not functioning the way they should and finding out why. It is knowing how the system should work and deciding if it is working correctly. Through an understanding of the purpose and operation of the system, you can accurately diagnose problems.

Most good technicians use the same basic diagnostic approach. Because this is a logical approach, it can quickly lead to the cause of a problem. Logical diagnosis follows these steps:

1. Gather information about the problem from the customer or service adviser. The technician must ask the important questions to gather the facts needed for systematic diagnosis. Drivetrain concerns are often related to speed, load, gear range, turning, and temperature. Questions about noise and vibration are often needed. Don't let the customer diagnose the car for you!
2. Verify that the problem exists. This step may mean taking a test drive with the customer to duplicate the conditions present when the problem occurs.
3. Thoroughly define what the problem is and when it occurs.
4. Research all available information to determine the possible causes of the problem. This step includes checking the vehicle's service history and any pertinent TSBs.
5. Isolate the problem by testing. The purpose here is to separate one system from another; for example, whether a vibration is caused by tires (one system) or a faulty U-joint (another system).
6. Continue testing to pinpoint the cause of the problem.
7. Get customer approval and repair the problem, then verify the repair.

WORKING AS AN AUTO TECHNICIAN

To be a successful automotive technician you need to have good training, possess a desire to succeed, and be committed. A good employee works well with others and strives to make the business successful. Good technicians need to have good reading, writing, and math skills. These skills will allow you to better understand and use the material found in service manuals and textbooks, as well as provide you with the basics for good communication with customers (Figure 2-45) and others.

Compensation

Technicians are typically paid according to their abilities. Most often, new or apprentice technicians are paid by the hour. When being paid, they are learning the trade and the business. Time usually is spent working with a master technician or doing low-skilled jobs. As an apprentice learns more, he or she can earn more and take on more complex jobs. Once technicians have demonstrated a satisfactory level of skills, they can go on flat rate.

FIGURE 2-45 Good customer relations is important for all who work in the automotive industry.

Flat rate is a pay system in which a technician is paid for the amount of work they do. Each job has a flat rate time established by the manufacturer. Pay is based on that time, regardless of how long it took to complete the job. To explain how this system works, let us look at a technician who is paid $15.00 per flat rate. If a job has a flat rate time of 3 hours, the technician will be paid $45.00 for the job, regardless of how long it took to complete it. If a technician turns in 60 hours of work in a 40-hour workweek, he or she actually earned $22.50 each hour worked. However, if he or she turned in only 30 hours in the 40-hour week, the hourly pay is $11.25.

The flat rate system favors good technicians that work in a shop that has a large volume of work. The use of flat rate times allows for more accurate repair estimates to the customers. It also rewards skilled and productive technicians.

Employer–Employee Relationships

Being a good employee requires more than learning the skills for the job. When you begin a job, you enter into a business agreement with your employer. When you become an employee, you sell your time, skills, and efforts. In return, your employer pays you for these resources.

As part of the employment agreement, your employer also has certain responsibilities. You should be told what is expected of you. A supervisor should observe your work, tell you if it is satisfactory, and offer ways to improve your performance. You should know how much you are to be paid, what your pay will be based on, and when you will be paid before accepting a job. When you are hired, you also should be told what benefits, such as paid vacations and employer contributions to health insurance and retirement plans, you can expect.

Employers also should give you a chance to succeed and possibly advance within the company. They also are responsible for providing safe working conditions including proper labeling and handling of hazardous waste. In addition, all employees should be treated equally, without prejudice or favoritism.

On the other side of this business transaction, employees have responsibilities to their employers. Your obligations as an employee include the following:

- **Regular attendance**—A good employee is reliable. Businesses cannot operate successfully unless their workers are on the job.
- **Following directions**—As an employee, you are part of a team, and doing things your way may not serve the best interests of the company.
- **Responsibility**—You must be willing to answer for your behavior and work. You also need to realize that you are legally responsible for the work you do.
- **Productivity**—Remember, you are paid for your time as well as your skills and effort. Keys to high productivity include quality workmanship and the practice of safe work habits.
- **Loyalty**—Loyalty is expected and by being loyal you will act in the best interests of your employer, both on and off the job.
- **Honesty**—Honesty and integrity are expected of each employee. Trust between the employee and the employer, as well as the customers, is critical to a good work atmosphere.

Another responsibility you have as an employee is good customer relations. Learn to listen and communicate clearly. Be polite and organized, particularly when dealing with customers.

Respect the vehicles on which you work. They are important to the lives of your customers. Always return the vehicle to the owner in a clean, undamaged condition. Remember, a car is the second largest expense a customer has. Treat it that way. It does not matter if you like the car. It belongs to the customer; treat it respectfully.

Explain the repair process to the customer in understandable terms. Whenever you are explaining something to a customer, make sure you do this in a simple way without making the customer feel stupid. Always show the customers respect and be courteous to them. Not only is this the right thing to do but also it leads to loyal customers.

ASE CERTIFICATION

The National Institute for **Automotive Service Excellence (ASE)** has established a voluntary certification program for automotive, heavy-duty truck, auto body repair, and engine machine shop technicians. In addition to these programs, ASE also offers individual testing in the areas of parts, alternate fuels, and advanced engine performance. This certification system combines voluntary testing with on-the-job experience to confirm that technicians have the skills needed to work on today's complex vehicles. ASE recognizes two distinct levels of service capability: the automotive technician and the master automotive technician. The master automotive technician is certified by ASE in all eight major automotive systems. An automotive technician may have certification in only certain areas.

To become ASE certified, a technician must pass one or more tests that stress diagnostic and repair problems. One of these areas is manual transmissions and drive axles. The A3 test (Figure 2-46) covers the diagnosis and repair of clutches, transmissions, transaxles, drive and half shaft universal and constant-velocity joints, rear axles, and 4WD/AWD components. You will find content on all of these topics in this Manual, as well as in the Classroom Manual.

After passing at least one exam and providing proof of 2 years of hands-on work experience, the technician becomes ASE certified. Retesting is necessary every 5 years to remain certified. A technician who passes one examination receives an automotive technician shoulder patch. The master automotive technician patch is awarded to technicians who pass all eight of the basic automotive certification exams (Figure 2-47).

You may receive credit for one of the 2 years of hands-on work experience by substituting relevant formal training in one, or a combination, of the following:

- High school training—Three years of training may be substituted for 1 year of experience.
- Post-high school training—Two years of post-high school training in a public or private trade school, technical institute, community or 4-year college or in an apprenticeship program may be counted as 1 year of work experience.
- Short courses—For shorter periods of post-high school training, you may substitute 2 months of training for 1 month of work experience.
- You may receive full credit for the experience requirement by satisfactorily completing a 3- or 4-year apprenticeship program.

Each certification test consists of 40–80 multiple-choice questions. The questions are written by a panel of technical service experts, including domestic and import vehicle manufacturers,

A3 MANUAL DRIVE TRAIN AND AXLES TEXT SPECIFICATIONS			
	Content Area	Questions in Text	Percentage of Test
A.	Clutch Diagnosis and Repair	6	16%
B.	Transmission Diagnosis and Repair	7	17%
C.	Transaxle Diagnosis and Repair	7	17%
D.	Drive Shaft/Half Shaft and Universal Joint/ Constant Velocity (CV) Joint Diagnosis and Repair (Front and Rear Wheel Drive)	5	13%
E.	Rear Axle Diagnosis and Repair	7	17%
	1. Ring and Pinion Gears (3) 2. Differential Case/Carrier Assembly (2) 3. Limited Slip/Locking Differential (1) 4. Axle Shafts and Housing (1)		
F.	Four-Wheel Drive/All Wheel Drive Component Diagnosis and Repair	8	20%
	Total	40	100%

FIGURE 2-46 The A3 manual drivetrain and axles test specifications.

FIGURE 2-47 ASE certification shoulder patches worn by (left) automotive technicians and (right) master automotive technicians.

repair and test equipment and parts manufacturers, working automotive technicians, and automotive instructors. All questions are pretested and quality-checked on a national sample of technicians before they are included in the actual test. Many test questions force the student to choose between two distinct repair or diagnostic methods. The knowledge and skills needed to pass the tests follows.

- Basic technical knowledge—What is it? How does it work? This requires knowing what is in a system and how the system works. It also calls for knowing the procedures and precautions to be followed in making repairs and adjustments.
- Repair knowledge and skill—What is a likely source of a problem? How do you fix it? This requires you to understand and to apply generally accepted procedures and precautions for inspecting, disassembling, rebuilding, replacing, or adjusting components within a particular system.
- Testing and diagnostic knowledge and skill—How do you find what is wrong? How do you know you corrected a problem? This requires that you be able to recognize that a problem does exist and to know what steps should be taken to identify the cause of the problem.

For further information on the ASE certification program, write to the National Institute for Auto Service Excellence (ASE), 101 Blue Seal Drive S.E., Suite 101, Leesburg, VA 20175 or go to *www.ase.com*.

SUMMARY

- Torque wrenches are used to tighten fasteners to a specified torque.
- Always use the correct tool, in the correct way, for the job.
- Special tools are for special purposes, and are not normally part of a basic tool set.
- Dial indicators are used to measure movement and are commonly used to measure the backlash and end play of a set of gears. Dial indicators are also used to measure component runout.
- A micrometer can be used to measure the diameter of objects.
- Calipers are versatile tools that can measure inside and outside diameter and depth.
- The primary source of repair and specification information for any vehicle is the manufacturer's service information.

TERMS TO KNOW

Backlash
Blowgun
Bolt head
Bolt shank
Calipers
Diagnosis
Dial indicator
End play
Feeler gauge
Fillet
Grade marks
Inclinometer
Jack (safety) stands

SUMMARY

TERMS TO KNOW
(continued)

Machinist's rule

Micrometer

Automotive Service Excellence (ASE)

Preload

Press-fit

Technical Service Bulletin (TSB)

Thread pitch

Torque wrench

Vehicle Identification Number (VIN)

- Flat rate manuals and software are ideal for making cost estimates. Published by manufacturers and independent companies, they contain figures showing how long specific repairs should take to complete, as well as a list of the necessary parts and their prices.
- Gear backlash is a statement of how tightly the teeth of two gears mesh.
- Whenever bolts are replaced, they should be replaced with exactly the same size, grade, and type as was used by the manufacturer.
- Diagnosis means finding the cause or causes of a problem. It requires a thorough understanding of the purpose and operation of the various automotive systems.
- Besides learning technical and mechanical skills, service technicians must learn to work as part of a team. As an employee, you will have certain responsibilities to both your employer and customers.
- Customer relations is an extremely important part of doing business. Professional, courteous treatment of customers and their vehicles is a must.
- The National Institute for Automotive Service Excellence (ASE) actively promotes professionalism within the industry. Its voluntary certification program for automotive technicians and master auto technicians helps guarantee a high level of quality service.

ASE-STYLE REVIEW QUESTIONS

1. When discussing the purpose of a torque wrench,
   Technician A says that a torque wrench is used to tighten fasteners to a specified torque.
   Technician B says that a click-type torque wrench should be used to measure turning effort.
   Who is correct?
   A. A only
   B. B only
   C. Both A and B
   D. Neither A nor B

2. When discussing gear removal,
   Technician A inserts a pry bar between two gears to remove one of the gears from the shaft.
   Technician B uses a gear puller to remove a gear from a shaft when the gear has an interference fit on the shaft.
   Who is correct?
   A. A only
   B. B only
   C. Both A and B
   D. Neither A nor B

3. When discussing automotive fasteners,
   Technician A says that bolt sizes are listed by their appropriate wrench size.
   Technician B says that whenever bolts are replaced, they should be replaced with exactly the same size and type as was installed by the manufacturer.
   Who is correct?
   A. A only
   B. B only
   C. Both A and B
   D. Neither A nor B

4. When discussing the purpose of micrometers,
   Technician A says that micrometers are used to measure the diameter of an object.
   Technician B says that micrometers are used to measure the runout of a shaft.
   Who is correct?
   A. A only
   B. B only
   C. Both A and B
   D. Neither A nor B

5. When discussing gear measurement,
   Technician A says that dial indicators are commonly used to measure the backlash of a set of gears.
   Technician B says that dial indicators are commonly used to measure the end play of a shaft.
   Who is correct?
   A. A only
   B. B only
   C. Both A and B
   D. Neither A nor B

6. When discussing gear backlash,
   Technician A says that excessive gear backlash causes overload wear on the gear teeth.
   Technician B says that gear backlash is a statement of how tightly the teeth of two gears mesh.
   Who is correct?
   A. A only
   B. B only
   C. Both A and B
   D. Neither A nor B

7. When discussing different types of service information,

Technician A says that the same information that is available in service manuals is now commonly found electronically on the Internet.

Technician B says that TSBs contain changes in specifications and repair procedures that occur during a vehicle's model year.

Who is correct?

A. A only

B. B only

C. Both A and B

D. Neither A nor B

8. When discussing VINs,

Technician A looks at the ninth digit of the VIN to determine the exact year of the vehicle.

Technician B says the information about the vehicle's engine can be found in the first three characters of the VIN.

Who is correct?

A. A only

B. B only

C. Both A and B

D. Neither A nor B

9. When discussing kinds of threads,

Technician A says that fine threaded fasteners are used where greater resistance to vibration is desired.

Technician B says that coarse threads are used for general-purpose work, especially when rapid assembly and disassembly is required.

Who is correct?

A. A only

B. B only

C. Both A and B

D. Neither A nor B

10. When discussing the ASE certification exam,

Technician A says that after an individual passes a particular ASE certification exam, he or she is automatically certified in that test area.

Technician B says that the questions on an ASE certification exam test are for theoretical knowledge of a system and not diagnosis of that system.

Who is correct?

A. A only

B. B only

C. Both A and B

D. Neither A nor B

Name _____ **Date** _____

FILLING OUT A WORK ORDER

Upon completion of this job sheet, you will be able to prepare a service work order based on customer input, vehicle information, and service history.

NATEF MAST Task Correlation

General: Drivetrain Diagnosis

Task #2 Research applicable vehicle and service information, fluid type, vehicle service history, service precautions, and technical service bulletins.

Tools and Materials

An assigned vehicle or the vehicle of your choice

Service work order or computer-based shop management package

Parts and labor guide

Describe the Vehicle Being Worked On

Year _____ Make _____ VIN _____

Model _____

Work Order Source: Describe the system used to complete the work order. If a paper repair order is being used, describe the source.

_____

_____

Procedure

_____

Task Completed

1. Prepare the shop management software for entering a new work order or obtain a blank paper work order. What will you be using?

   _____

   _____

2. Enter customer information, including name, address, and phone numbers onto the work order.

3. Locate and record the vehicle's VIN. What is the VIN?

   _____

4. Enter the necessary vehicle information, including year, make, model, engine type and size, transmission type, license number, and odometer reading. ☐

5. Does the VIN verify that the information about the vehicle is correct? _____

6. Normally, you would interview the customer to identify his or her concerns. However, to complete this job sheet, assume the only concern is that the clutch pedal is very spongy and the cause is a leaking clutch master cylinder, which must be replaced. This concern should be added to the work order. ☐

7. The history of service to the vehicle can often help diagnose problems as well as indicate possible premature part failure. Gathering this information from the customer can provide some of this information. For this job sheet, assume the vehicle has not had a similar problem and was not recently involved in a collision. Service history is further obtained by searching files for previous service. Often this search is done by customer name, VIN, and license number. Check the files for any related service work.

8. Search for technical service bulletins on this vehicle that may relate to the customer's concern. Were there any? If so, will they help with the repair?

_____

_____

9. Based on the customer's concern, service history, TSBs, and your knowledge, what is the likely cause of this concern?

_____

_____

10. Enter this information onto the work order.

11. Prepare to make a repair cost estimate for the customer. Identify all parts that may need to be replaced to correct the concern. List these here.

_____

_____

12. Describe the task(s) that will be necessary to replace the part.

_____

_____

13. Using the parts and labor guide, locate the cost of the parts that will be replaced and enter the cost of each item onto the work order at the appropriate place for creating an estimate. What is the total part cost?

_____

14. Now, locate the flat rate time for work required to correct the concern. List each task and its flat rate time.

_____

_____

15. Multiply the time for each task by the shop's hourly rate and enter the cost of each item onto the work order at the appropriate place for creating an estimate. What is the shop's labor rate?

_____

16. Many shops have a standard amount they charge each customer for shop supplies and waste disposal. For this job sheet, use an amount of $10 for shop supplies.

17. Add the total costs and insert the sum as the subtotal of the estimate. What is the subtotal due?

_____

_____

18. Taxes must be included in the estimate. What is the sales tax rate and does it apply to both parts and labor, or just one of these?

_____

_____

19. Enter the appropriate amount of taxes to the estimate; then add this to the subtotal. The end result is the estimate to give the customer. What is the final estimated cost?

_____

20. By law, how accurate must your estimate be?

_____

21. Generally speaking, the work order is complete and is ready for the customer's signature. However, some businesses require additional information; make sure you enter that information to the work order. On the work order there is a legal statement that defines what the customer is agreeing to. Briefly describe the contents of that statement.

_____

_____

_____

_____

Instructor's Response: _____

_____

_____

Name _____ Date _____

GATHERING VEHICLE INFORMATION

Upon completion of this job sheet, you should be able to gather service information about a vehicle and its drivetrain and manual transmission or transaxle.

NATEF MAST Task Correlation

General: Drivetrain Diagnosis

Task #2 Research applicable vehicle and service information, fluid type, vehicle service history, service precautions, and technical service bulletins.

Tools and Materials

Appropriate service manuals

Computer

Protective Clothing

Goggles or safety glasses with side shields

Describe the Vehicle Being Worked On

Year _____ Make _____ VIN _____

Model _____

Procedure

1. Visually inspect the vehicle and describe the major components and their location on the drivetrain.

   _____

   _____

   _____

   _____

2. How many forward speeds does the transmission/transaxle have?

   _____

   _____

3. Using the service manual or other information source, describe what each letter and number in the VIN for this vehicle represents.

   _____

   _____

4. Locate the Vehicle Emissions Control Information (VECI) label and describe where you found it.

_____

_____

5. Summarize what information you found on the VECI label.

_____

_____

6. When looking in the engine compartment or under the vehicle, locate the identification tag on the transmission or transaxle. Describe where you found it.

_____

_____

7. Summarize the information contained on this label.

_____

_____

_____

8. If the final drive housing is separate from the transmission or transaxle housing, locate the identification tag on the final drive unit. Describe where you found it.

_____

_____

9. Summarize the information found on the final drive identification tag.

_____

_____

10. Using a service manual or electronic database, locate the information about the vehicle's drivetrain. Describe the clutch linkage and any controls or sensors attached directly to parts of the drivetrain.

_____

_____

_____

_____

_____

11. Using a service manual or electronic database, locate and record all service precautions regarding the drivetrain noted by the manufacturer.

_____

_____

_____

_____

12. Using the information that is available, locate and record the vehicle's service history.

_____

_____

_____

_____

13. Using the information sources that are available, summarize all Technical Service Bulletins for this vehicle that relate to the transmission and the drivetrain.

_____

_____

_____

_____

Instructor's Response: _____

_____

_____

Name _____ **Date** _____

ASE CERTIFICATION

Upon completion of this job sheet, you should be able to describe the process for becoming ASE certified in Manual Transmissions and Drivelines.

Tools and Materials

ASE test booklet (available by mail or at their Web site).

Procedure

Answer the following questions using the ASE test booklet.

1. How many test areas are there for automotive repair?

   _____

   _____

   _____

   _____

2. How much documented experience must a technician have before he or she becomes certified by ASE?

   _____

   _____

   _____

   _____

3. Explain how your time in school applies to the requirement for experience.

   _____

   _____

   _____

   _____

4. Where do you take an ASE exam?

   _____

   _____

   _____

   _____

5. When do you take an ASE exam?

   _____

   _____

   _____

   _____

6. Fill out the sample ASE registration sheet in the ASE booklet and have your instructor look it over.

7. What are three areas that are emphasized on the exam?

_____

_____

_____

_____

8. How much does it cost to register for and take an ASE exam?

_____

_____

_____

_____

Instructor's Response: _____

_____

_____

Chapter 3

SERVICING CLUTCHES

BASIC TOOLS
Basic mechanic's tool set
Feeler gauge
Rule
Straightedge
Lithium grease
Transmission jack
Jack stands
Center punch

UPON COMPLETION AND REVIEW OF THIS CHAPTER, YOU SHOULD BE ABLE TO:

- Diagnose clutch-related problems.

- Inspect, adjust, and replace clutch pedal linkage, cables and automatic adjuster mechanisms, brackets, bushings, pivots, and springs.

- Inspect, adjust, repair, and replace clutch slave and master cylinders and lines; check for leaks.

- Bleed a hydraulic clutch system.

- Inspect, adjust, and replace release bearing, lever, and pivot.

- Inspect and replace clutch disc and pressure plate assembly.

- Inspect and replace pilot bearing.

- Inspect, repair, service, or replace flywheel and ring gear.

- Inspect engine block, clutch housing, and transmission case mating surfaces; determine needed repairs.

- Measure flywheel-to-block runout and crankshaft end play; determine needed repairs.

- Measure clutch housing bore-to-crankshaft runout and face squareness; determine needed repairs.

- Inspect, replace, and align powertrain mounts.

CLUTCH PROBLEM DIAGNOSIS

Diagnosing clutch problems is much like diagnosing problems in any automotive system. The more information you have about the problem, the easier it is to properly diagnose it. Diagnostics should begin with gathering as much information as possible from the owner by asking questions such as:

- When is the problem most noticeable?
- How long have you noticed the problem?
- What kind of driving do you normally do?
- Have you had a similar problem in the past? Was it corrected? What types of repairs were made?

AUTHOR'S NOTE: One of the best diagnostic tools you can have is your computer. A good technician will always look up the service history of a vehicle to see if there may be a relationship between past repairs or complaints to the current problem. While you are at the screen, check for any relevant technical service bulletins.

Always take careful notes of what the customer is saying. Written notes make it easier to remember details. It also gives the customer confidence that he's dealing with a professional because you are paying attention and are interested in solving the problem.

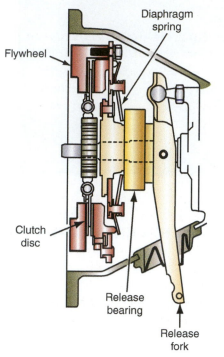

FIGURE 3-1 Typical components of a clutch assembly.

After you have gathered the information from the owner, take the vehicle on a road test and attempt to verify the customer's complaint. If possible, try to take a test drive with the customer. This lets you see and hear the problem as it happens. It also lets you watch the owner's driving habits, which can contribute to a clutch problem. For example, does the driver ride the clutch, downshift at high engine speeds, or miss shifts? The test drive gives you clues about the source of the problem and it gives you the opportunity later to suggest alternative driving techniques that will prolong the life of a clutch.

While you are driving the car, pay close attention to the action of the clutch during all phases of operation. Notice where the clutch engages. If a clutch with adjustable linkage grabs way up on top of the pedal travel, adjust the free-play back to specifications. If the clutch slips after this adjustment, it needs to be replaced. If the free-play is correct and the car can accelerate without slipping, but the clutch slips at higher engine speeds, suspect a weak pressure plate. Check the throw-out bearing for quiet, smooth operation. Listen for bearing and gear noises while downshifting to make sure that a clutch problem has not caused transmission problems, such as a damaged synchronizer. This type of information further defines the problem. With this information, determine what components are responsible for preventing this type of problem. These components are the most likely causes of the problem. Once these possible causes have been identified, inspect and/or test those parts. After the cause of the problem is located, it should be repaired and the repair verified by taking the vehicle on a test drive.

CUSTOMER CARE: One of the common driving techniques that can destroy a clutch is coasting down a hill with the clutch disengaged and the transmission in a low gear. This can cause the clutch disc to explode, potentially causing serious injury. This happens due to the high speeds that are generated by the multiplication of speed ratios that travel through the drive axle and transmission to the input shaft and clutch disc. By the time the speed ratio reaches the clutch, the speed could be as high as 10,000 rpm. This far exceeds the speed a clutch can withstand. These facts need to be thoughtfully explained to anyone guilty of driving this way.

WARNING: Many clutch testing procedures require that the vehicle be operated in the shop area. Always place wheel chocks against the drive wheels or the wheels that remain on the ground. Be sure the area is properly vented or a ventilating hose is attached to the vehicle's exhaust. Do *not* allow anyone to stand in front of the vehicle while it is running. Always test all-wheel-drive vehicles with all four wheels off the ground.

CUSTOMER CARE: Many clutch problems result from use or misuse. Many customers are accustomed to being able to use their vehicles for towing trailers but have not considered the changes in vehicle size and weight or the changes in engine and clutch sizes. Prior to buying a new vehicle, individuals should consider whether that vehicle will be used to pull a camper, boat trailer, or whatever. Most people just have a trailer hitch put on the car and begin pulling. They do not worry about weight and uses of the car or truck until the clutch fails. Customers should be advised as to the towing capabilities of a vehicle and should be given suggestions as to how the vehicle should be equipped for towing.

Often technicians use troubleshooting guides given in service manuals to aid them in diagnostics. To use them, the technicians must first describe the problem then refer to a chart or diagnostic tree to determine the most probable causes of the problem.

Clutch Slippage

Clutch slippage is evident when the driver has the clutch engaged and the engine's speed increases but the vehicle's road speed does not. Slipping is often most obvious during acceleration and shifting. A road test can determine if the clutch is slipping. Normal acceleration from a stop and several gear positions should provide the conditions necessary to witness slipping. Slippage may also be noticed when driving in a higher gear going up a hill.

Slippage can also be verified in the shop. Depress the clutch pedal, shift the transmission into high gear, and increase the engine's speed to approximately 2,000 rpm. Slowly release the clutch pedal until the clutch engages. The engine should stall immediately. If the engine does not stall within a few seconds, the clutch is slipping. If the clutch slips, depress the clutch pedal to end the test quickly.

If slippage is evident during either test, raise the vehicle and check the linkage release mechanism for anything that might cause the throw-out bearing to maintain pressure on the pressure plate fingers. Check for worn or binding parts. Also check for loose or worn engine mounts. Clutch slippage can also be caused by an overadjusted clutch. In this case, the clutch is always partially released and is never fully engaged. Overadjustment is possible on a cable system, so back off the adjustment and check for slippage. On hydraulic linkage systems, make sure the return port to the master cylinder is not blocked. This can prevent the slave cylinder from returning fully. If no problem is evident, the clutch should be disassembled and repaired. Clutch slippage is often caused by an oil-soaked or worn disc facing (Figure 3-2), warped flywheel or pressure plate, weak pressure plate springs, or the release bearing contacting the fingers of the pressure plate. The cause of the problem should be repaired. Other causes of clutch slippage are riding the clutch pedal with the vehicle in motion and holding the vehicle on an incline by using the clutch as a brake.

Oil and grease contamination on the disc's frictional material results in a loss in the coefficient of friction (Figure 3-3). This reduces the ability of the disc to remain tightly clamped between the flywheel and the pressure plate. When the clutch slips, it generates heat, which

CAUTION:
Severe or prolonged clutch slippage may cause grooving and/ or heat damage to the pressure plate or flywheel. Therefore, end all testing as soon as slippage is evident.

SERVICE TIP:
Clutch slippage on vehicles with a dual-mass flywheel can be very difficult to diagnose. The cause of the problem may be the clutch or a bad flywheel. If the customer's complaint is a slipping clutch, but the clutch assembly shows no signs of slippage, the cause of the problem is probably the dual-mass flywheel, which should be replaced.

FIGURE 3-2 A severely worn clutch disc.

FIGURE 3-3 Grease and oil on the hub of this disc is an indication that the disc may be contaminated with oil.

causes it to slip even more. Late-model cars are more prone to this problem because they use smaller clutch discs and discs with nonasbestos linings. Nonasbestos friction materials are prone to fail when subjected to the slightest amount of oil or grease. During a visual inspection, examine the clutch disc and the transmission's input shaft for oil residue. If oil is detected, look for leaks at the engine's rear main seal or at the transmission's front seal. Oil leaking from valve covers can also find its way into the clutch housing. Oil leaks must be corrected prior to installing a new clutch disc.

Any large particles that become trapped between the diaphragm spring and pressure plate cover can cause clutch slip by preventing the pressure plate from fully engaging the clutch. More than a few technicians have been mystified by a slipping condition until they find a mouse nest in the pressure plate assembly!

CUSTOMER CARE: An increasingly common cause of clutch slippage stems from engine modifications made to vehicles that can overwhelm the torque capabilities of an OEM clutch assembly. The technician should be ready to advise the customer of upgrades that will be able to withstand the higher output of a modified engine. Upgrades can include more aggressive clutch friction materials, dual-disc clutch setups, and pressure plates with stiffer springs or centrifugal assist. Remember that release mechanisms often have to be upgraded to go with performance clutch components.

Clutch Drag

Clutch dragging occurs when the clutch disc is not fully released when the clutch pedal is fully depressed. The pressure plate must have enough disengagement travel to be completely away from the disc, or gear clash will result. This can cause gear clashing, especially when shifting into reverse gear. The problem is often most noticeable shifting into reverse, which may not be synchronized and requires a reverse idler gear or reverse driven gear to be moved into position. The clutch disc, input shaft, and transmission gears should require no more than a few seconds

Classroom Manual

Chapter 3, page 77

to come to a stop after the clutch pedal is depressed. This time is called the "spindown time" and is normal. However, if the time exceeds 5 seconds, clutch drag is evident.

To check for clutch drag, start the engine and depress the clutch pedal. Shift the transmission into first gear but do not release the clutch. The transmission internal components should come to a stop. Now shift the transmission into neutral. Wait 5–10 seconds, then attempt to shift the transmission into reverse. If the shift causes gear clash, drag on the clutch disc or input shaft has caused the gears to spin, making reverse engagement difficult or impossible. Raise the vehicle and check the clutch linkage. If there are no problems with the linkage, the clutch must be disassembled and inspected. Clutch drag can be caused by air in a hydraulic system, a warped disc or pressure plate, a loose disc facing, incorrect clutch pedal adjustment, a defective release lever, or a warped pressure plate or disc. One of the most common causes of a bent disc occurs during transmission installation. If the technician lets the weight of the transmission hang on the input shaft after it has engaged the clutch hub but is not yet up to the block, the center of the disc will bend.

A worn or damaged pilot bearing or bushing could cause the input shaft to spin even though the clutch is fully disengaged, causing gear clash during engagement.

Pulsating Clutch Pedal

Pedal **pulsation** is a rapid up-and-down pumping movement of the clutch pedal as the clutch engages or disengages. The movement of the pedal is normally slight but can be felt through the pedal. To test for pulsations, start the engine and slowly depress the clutch pedal. Pay careful attention to the pedal as it is being depressed. Very slight pulsations are normal; however, if the pulsations are quite noticeable or severe, the clutch assembly needs to be disassembled and inspected. Pedal pulsation is normally caused by broken, bent, or warped release levers, a misaligned bell housing, or a warped pressure plate, flywheel, or clutch disc.

Pulsations are normally felt better than they are heard.

Binding Clutch

If the clutch pedal does not operate smoothly, check the clutch linkage or cables for binding. Also check the bearing retainer collar for wear (Figure 3-4) or grooves. Make sure the retainer is lightly lubricated with white lithium grease. If the clutch pedal moves but fails to disengage, check the linkage for wear or improper adjustment. Make sure the disc is not installed backward. Also check that the disc's hub splines are properly lubricated.

Examine the release fork to make sure it is on its pivot and that it is not warped, cracked, or excessively worn. Check the condition of the pilot bearing or bushing. A bad pilot bearing

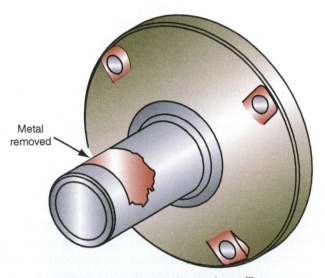

Metal removed

FIGURE 3-4 **A worn bearing retainer will cause noise, incomplete clutch release, and a hard clutch pedal.**

can cause the disc to bind by preventing smooth rotation of the input shaft or may allow the input shaft to wobble, causing the disc to be in continual contact with either the flywheel or the pressure plate. This problem can cause gear clashing, as well as poor disengagement.

Hard pedal effort may be caused by binding in the pressure plate assembly, particularly around the pivot ring and diaphragm spring. In some cases the effort to push the pedal down is so severe that clutch master or slave cylinder seals can be damaged.

If the clutch returns to its stop but the clutch does not engage, the clutch lining may have become separated from the cushion segments. This could be caused by a piece of the clutch lining wedging between the pressure plate's cover and the diaphragm spring, preventing the pressure plate from exerting pressure on the clutch disc. During missed shifts, the clutch disc spins free, which can cause the lining or cushion segments to peel off from the disc and be thrown about. This problem can also cause damage to the bell housing and to the driver.

NOISE AND VIBRATIONS

Vibrations are usually caused by something being out of balance or misaligned, and are felt throughout the entire vehicle.

Noise and vibration problems are best identified by road-testing the vehicle. Begin your observations while the transmission is cold. Pay attention to any change in noise as it warms up. During this check, write down all sounds and when they occur or change.

Start the engine with the transmission in neutral; pay attention to all noises. Then listen for a change in noise as the clutch pedal is depressed and released. Now repeat this cycle with the engine at a higher speed.

Drive the vehicle and shift through all of the gears, including reverse. Listen for any change in noise in a particular direction or gear. If the noise is most noticeable in a particular gear, drive the vehicle in that gear and then depress the clutch. Listen for any change in noise. If there is a change, the noise may be amplified by the engine's vibrations. Release the clutch and wait until the original noise returns. Then shift the transmission into neutral, release the clutch, and allow the vehicle to coast. If the noise remains, the cause is undoubtedly in the driveline.

General Diagnosis

Most clutch noises are caused by bearings and bushings. Release bearings make a whirring, grating, or grinding noise, which occurs when the clutch pedal is depressed and stops when the pedal is fully released. Pilot bushing noises sound like a squeal or howl and are most noticeable in cold weather.

A simple test can help you check the bearings. Set the parking brake, put the transmission in neutral, and allow the engine to idle. Leave the clutch engaged. Listen to all sounds from the engine and transmission. If you hear any bearing noise, the problem is the transmission's input shaft bearing, input pocket bearing, countershaft bearings, or speed gear needle bearings. All of these are in motion with the engine running, the transmission in neutral, and the clutch engaged.

Now with the transmission still in neutral and the engine running, push the clutch pedal only to the end of the free-play. This will lightly put the throw-out bearing in contact with the fingers of the pressure plate, but not enough to disengage the clutch. Bearing noise in this position indicates a problem with the throw-out bearing.

> **CUSTOMER CARE:** Regardless of what kind of noise seems to be coming from the transmission/clutch area, customers seem to always have the opinion that "it must be the throw-out bearing." This often turns out to be wishful thinking. The fact is that if the noise disappears when the clutch pedal is depressed, the problem is most certainly in the transmission. When internal components stop turning, the noise stops as well. A brief explanation of how this works can help the customer understand why the transmission probably needs to be disassembled.

Finally, with the transmission still in neutral and the engine running, push the clutch pedal fully to the floor to disengage the clutch, and then slowly allow the clutch to engage. If there was a brief squeal as the clutch was engaged, the pilot bearing or bushing is bad. During this time, the input shaft and the crankshaft are rotating at different speeds. This noise may be more noticeable as the clutch is being engaged in first or reverse gear. Damaged or worn transmission input shaft bearings and pilot bearings should be replaced as needed. Damaged release bearings should also be replaced, but only after the cause of failure is identified. Normally release bearings are damaged by misalignment, overheating, slippage, or component damage.

Clutch noise during engagement and disengagement may be caused by the disc's torsional springs contacting the flywheel bolts due to excessive machining of the flywheel. Loose torsional springs can cause a rattle sound coming from the bell housing area, often heard only at idle.

Flywheel Noises

A loose flywheel will cause a knocking sound similar to an engine bearing knock. Unlike an engine knock, the noise will not change when individual cylinders are shorted out. One test for a loose flywheel is to shut the ignition off, then immediately restart the vehicle. As the flywheel's momentum carries it forward, then the started engine throws it back, a "double-knock" will result.

Dual-mass flywheels may also make a clacking noise under certain conditions. Correct any rough engine idle condition before diagnosing these flywheels. Worn dual-mass flywheels that have excessive movement between the two sections can also make this sound.

Looking at the face of the release bearing and the center of the pressure plate can tell you if there are misalignment problems. A spiral or wide wear pattern on the bearing's face can be caused by misalignment between the flywheel and transmission input shaft, a warped pressure plate cover, and missing dowel pins between the engine and clutch housing. Trapped wires, cables, or mounting brackets lodged between the engine and clutch housing can also cause misalignment.

The transmission's input shaft must be in alignment with the crankshaft and at a right angle to the flywheel's friction surface. A worn input shaft bearing or pilot bushing or bearing will cause the input shaft to be out of alignment. A bent input shaft or clutch disc will also cause misalignment. These typically are caused by careless installation or removal of the transmission. If the weight of the transmission is not properly supported, the shaft or disc can be distorted.

Clutch noise during engagement and disengagement may be caused by the disc's torsional springs contacting the flywheel bolts due to excessive machining of the flywheel.

Warped pressure plates can cause bearing damage, disc slippage, and heat buildup. This is usually caused by disc slippage and overheating due to poor driving habits or incorrect adjustments.

Dual-mass flywheels can make a clacking noise if the engine has a rough idle. Before replacing the flywheel because of the excessive noise, check the idle of the engine. Correct that problem before diagnosing the flywheel for noise.

Clutch Chatter

Clutch **chatter** is a shaking or shuddering that is felt when the clutch is engaged. This normally occurs when the pressure plate (Figure 3-5) first makes contact with the clutch disc and will stop when the clutch is fully engaged.

To check for chatter, start the engine and completely depress the clutch pedal. Shift the transmission into first gear and increase the engine's speed to about 1,500 rpm. Slowly release the clutch pedal and listen as it begins to engage. If chattering is evident, depress the clutch pedal and reduce engine speed immediately to prevent damage to the clutch parts.

Clutch chatter is normally caused by broken engine mounts and glazed clutch and/or flywheel or pressure plate facings. Leaks onto the clutch can lead to glazing. Possible sources of leaks are the engine's rear main seal, transmission input shaft seal, and clutch slave cylinder. The engine and transmission mounts should also be checked for looseness, breakage, and

FIGURE 3-5 The marks on the surface of this pressure plate are caused by clutch chatter.

Image: Schaeffler

wear. Chatter can also result from uneven engagement, which can be caused by a worn front bearing retainer or release fork. Other possible causes for clutch chatter include:

- A bent clutch disc
- Burned or damaged disc facing
- Excessive heat (this can be caused by the driver riding the clutch)
- Damaged pressure plate
- Pressure plate has excessive runout
- Worn or damaged pilot bearing

Noise While Shifting

A grinding noise while shifting gears may be accompanied by difficult gear shifts. The most common causes are:

- Low fluid in the hydraulic system
- Air in the hydraulic system
- Clutch linkage out of adjustment
- Worn or damaged internal shift mechanisms
- Worn or damaged sliding gears and/or synchronizers
- Misaligned transmission housing
- Worn or damaged shafts.

Clutch Vibrations

Clutch vibrations are usually caused by something being out of balance or misaligned and can occur at any clutch pedal position. Normally it occurs at normal operating speeds and can be felt throughout the entire vehicle. To identify the cause of clutch vibrations, raise the vehicle and check the engine mounts by looking for signs that the engine or transmission is rubbing against the frame or body. Check the mountings for the engine's accessories. If all mountings are good, the vibration is probably caused by loose flywheel bolts, excessive flywheel runout, imbalanced flywheel and/or pressure plate assemblies, or a worn or damaged pilot bearing.

MISALIGNMENT

The transmission's input shaft must be in alignment with the crankshaft and at a right angle with the flywheel's friction surface. A worn input shaft bearing or pilot bushing or bearing will cause the input shaft to be out of alignment. Debris, wires, cables, or brackets trapped between the engine and clutch housing will also cause misalignment. Both dowels must be installed in the back of the block to maintain alignment. In some cases, usually found on a new vehicle, a clutch housing is not centered with the crankshaft. In any case of misalignment, taper wear on the ends of the clutch disc hub may provide the technician a clue to the problem (Figure 3-6).

VISUAL INSPECTION

Before proceeding with an inspection of the clutch assembly, do a careful inspection of the vehicle. Check all drivetrain parts for looseness and leaks. With the engine off, check the free-play and smoothness of the clutch. While working the clutch pedal, listen for noises. Carefully identify the source of any noise. If the noise is from the linkage, repair the linkage as is necessary and readjust the free-play to specifications.

To check the transmission and engine mounts, set the parking brake and start the engine. Place the transmission in first gear, then increase engine speed to about 1,500–2,000 rpm and gradually release the clutch pedal until engine torque causes tension at the drivetrain mounts. Watch the torque reaction of the engine. If the reaction appears to be excessive, broken or worn drivetrain mounts may be the cause.

The engine mounts on FWD cars are important to the operation of the clutch and transaxle (Figure 3-7). Any engine movement may change the effective length of the shift and clutch control cables and therefore may affect the engagement of the clutch and/or gears. A clutch may slip due to clutch linkage changes as the engine pivots on its mounts. A broken clutch cable may be caused by worn mounts and improper cable routing. Inspect all clutch and transaxle linkages and cables for kinks or stretching.

To check transaxle mounts, pull up and push down on the transaxle while watching the mounts. If a mount's rubber separates from the metal plate or if the case moves up but not down, replace the mount. If there is movement between the metal plate and its attaching point on the frame, tighten the attaching bolts to an appropriate torque.

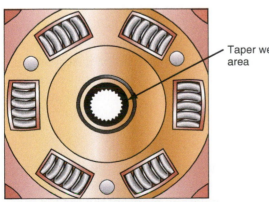

Taper wear area

FIGURE 3-6 Missalignment between the crankshaft and input shaft can cause the ends of the clutch hub splines to show taper wear.

FIGURE 3-7 The upper engine mount for a FWD vehicle with the engine mounted transversely.

Replacing Transaxle Mounts

If it is necessary to replace a transaxle mount, follow the recommended procedure for maintaining the alignment of the driveline. Failure to follow the correct installation procedure may result in poor gear shifting, poor clutch operation, and/or broken cables. Some manufacturers recommend that a holding fixture or special bolt be used to keep the unit in its proper location.

When removing the mount, begin by disconnecting the battery's negative cable. Disconnect any electrical connectors that may be located around the mount. Be sure to label any wires you remove to facilitate reassembly. Install the engine support fixture and attach it to an engine hoist. Lift the engine just enough to take the pressure off of the mounts. Remove the bolts attaching the mount to the frame. Remove the bolts attaching the mount to its bracket and then remove the mount. Some manufacturers require that these bolts be discarded and new ones installed during reassembly.

To install the new mount, position the mount in its correct location on the frame and tighten its attaching bolts to the proper torque. Install the bolts that attach the mount to the transaxle bracket. Prior to tightening these bolts, check the alignment of the mount. In the procedure for removing the mount, some manufacturers recommend the use of a special alignment bolt, which is installed in an engine mount. This bolt serves as an indicator of power train alignment. If excessive effort is required to remove the alignment bolt, the power train should be shifted to allow for proper alignment. Once you have confirmed that the alignment is correct, tighten all loosened bolts to their specified torque. Remove the engine hoist and fixture from the engine and reinstall all accessories and wires that may have been removed earlier.

 **WARNING:** Always wear eye protection when working under a vehicle.

MECHANICAL CLUTCH CONTROLS

Classroom Manual
Chapter 3, page 73

Before beginning to adjust any type of clutch linkage, refer to the manufacturer's procedures and specifications.

The following preliminary checks and adjustments should be made before assuming the clutch is in need of repair or replacement. Often clutch problems can be corrected by properly adjusting the linkage. Worn bushings, bent rods, broken springs, and damaged cotter pins can cause excessive pedal effort when operating the clutch. Clutch pedal free travel can often indicate the cause of a clutch problem.

Insufficient free-play can cause the release bearing to constantly ride against the pressure plate's fingers or thrust ring. This causes rapid release bearing and pressure plate wear and can cause clutch slippage. Excessive free-play may prevent the clutch from fully disengaging with the pedal fully depressed, causing gear clash and difficult shifting. This is usually most obvious in reverse gear, which is usually a non-synchronized or sliding gear.

The free-play normally decreases as the clutch disc wears. Over time the fingers or thrust ring of the pressure plate move closer to the release bearing, thereby decreasing the free-play of the pedal. Worn mechanical clutch linkages can make accurate free-play adjustments difficult due to deflections and worn bushings. This type of wear causes the free-play measured at the pedal to be greater than the actual free-play at the release bearing. The linkage should be inspected, repaired, and lubricated prior to making a free-play adjustment (Figure 3-8).

Free-Play Adjustments

Free-play is the clearance between the clutch release yoke fingers and the release bearing housing.

Free-play is the clearance between the clutch pressure plate release yoke fingers and the face of the release bearing housing. This clearance is felt at the top of clutch pedal travel. The amount of clutch pedal free-play varies according to the make and model of the car. On most cars and trucks with adjustable linkage, whenever free pedal travel falls below ½ inch (13 mm), it should be adjusted. Free-play is measured inside the vehicle at the clutch pedal (Figure 3-9).

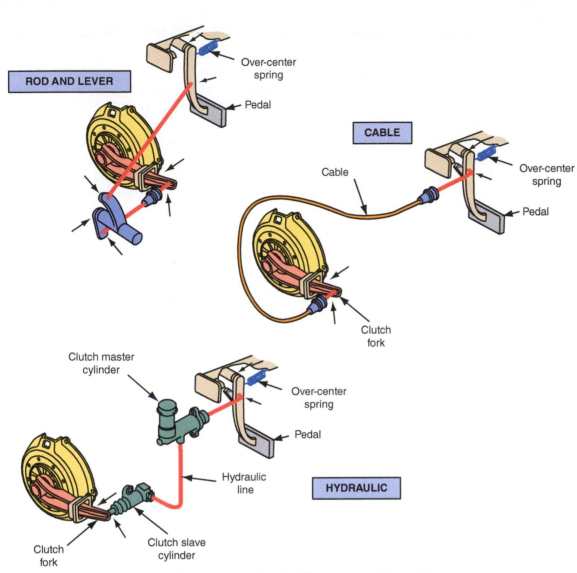

ROD AND LEVER

Over-center spring

Pedal

CABLE

Cable

Over-center spring

Pedal

Clutch fork

Clutch master cylinder

Over-center spring

Pedal

Hydraulic line

HYDRAULIC

Clutch fork

Clutch slave cylinder

FIGURE 3-8 Lubrication points for different types of clutch linkages.

This procedure is only performed on vehicles that have provisions for linkage free-play adjustments. If free pedal travel cannot be restored to about ½ to 1¾ inches (13–44 mm), the linkage may need to be repaired or replaced. While working the clutch pedal, observe the linkage for looseness that may be causing some lost motion due to worn clevis eyes and pins, or worn bushings and shafts at the clutch pedal or linkage arms.

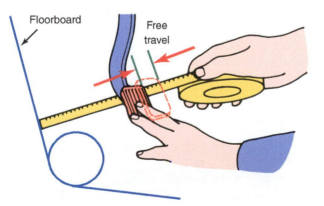

Floorboard

Free travel

FIGURE 3-9 Checking clutch pedal free-play.

A **high pedal** is one that has an excessive amount of pedal travel.

Total pedal travel is the total amount the pedal moves from no free-play to complete clutch disengagement.

A **pawl** is a lever that engages after its linkage has been moved a certain distance.

A **quadrant** is a section of a gear.

Gear clash is the noise that results when two gears are traveling at different speeds and are forced together.

A general procedure for making a free-play adjustment on a lever-type clutch linkage follows:

1. Disconnect the clutch return spring from the release lever.
2. Loosen the release lever rod locknut approximately three or four turns.
3. If there is no free travel, shorten the rod (by turning it at the square wrench area) until it is free of the clutch release lever.
4. Move the clutch release lever rearward until the release bearing lightly contacts the clutch pressure plate release fingers.
5. Adjust the rod length until the rod just contacts its seat in the release lever.
6. Adjust the locknut to obtain approximately 3/16-inch clearance between the nut and the rod sleeve end.
7. Turn the rod at the square wrench area until the nut just contacts the rod sleeve end.
8. Tighten the locknut against the sleeve while holding the rod with a wrench.
9. Install the clutch return spring.
10. Check the free travel at the pedal and compare it to the specification (normally 1 inch). Readjust the linkage if the free-play is not within the specified distance.
11. As a final check, measure the free travel with the engine idling. This dimension should not be less than ½ inch.
12. Measure the total pedal travel and compare this to the specifications. If the pedal travel is less than the specified amount, it can be increased by trimming the rubber pedal stop or by moving the pedal stop until the pedal travel is within the specified distance. A very **high pedal** may be caused by an incorrectly positioned, damaged, or missing pedal stop.

The clutch pedal free-play and **total pedal travel** should be checked whenever the clutch does not disengage or engage properly, or when new clutch parts are installed. Improper adjustment of the clutch pedal is one of the most frequent causes of clutch failure and can cause transmission failure.

If the vehicle is equipped with an overcenter assist spring, the spring will not normally require adjustment and should not be disturbed. However, if pedal effort is not correct or the pedal's return action is erratic, an adjustment may be necessary. Also, if the linkage is removed or replaced, free-play adjustments should be checked. Adjustable cable-type clutch controls are adjusted in a similar way as other mechanical controls. The end of the cable that fastens to the release fork at the bell housing is threaded and fitted with an adjusting nut and locknut (Figure 3-10). After pedal travel and free-play are adjusted, the locknut should be tightened to hold the adjustment.

Some vehicles have a self-adjusting, cable-type mechanical linkage (Figure 3-11). The cable is of a fixed length and often adjustments cannot be made. The self-adjusting mechanism consists of a toothed wheel, or quadrant, and a spring-loaded **pawl** (Figure 3-12). The quadrant is spring-loaded to keep just enough tension on the cable to position the release bearing lightly against the pressure plate. As the cable stretches, the **quadrant** rotates enough to allow the spring-loaded pawl to engage the next tooth when the clutch is applied. This maintains the specified amount of free-play. Regular service to the cable is not required. However, if the transmission is difficult to shift or if there is **gear clashing** during shifts, the quadrant and cable should be checked for binding (Figure 3-13). Also check for stripped teeth on the quadrant. Following the installation of a new clutch, lift the pedal to the top of its travel to reset the adjuster.

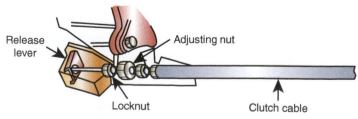

FIGURE 3-10 Pedal free-play adjustment on a cable-type clutch control.

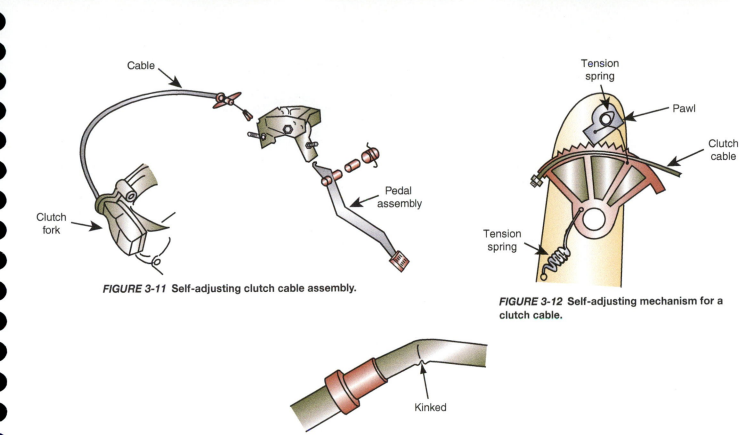

FIGURE 3-11 Self-adjusting clutch cable assembly.

FIGURE 3-12 Self-adjusting mechanism for a clutch cable.

FIGURE 3-13 Kinked or damaged clutch cables cause hard pedal effort.

Typically, to remove a clutch cable:

1. Support the clutch pedal upward against its bumper stop. This will allow the pawl to be released from its detent.
2. Disconnect the cable from the clutch release lever at the transmission or transaxle.
3. Disconnect the clutch cable from the detent end tangs.
4. Lift the locking pawl away from the detent, then slide the cable forward between the detent and locking pawl.
5. Pull the clutch cable out, disengaging it from the clutch pedal mounting bracket. Be careful not to lose any falling parts as the insulators, dampener, and washers may separate from the cable during removal (Figure 3-14).

SERVICE TIP:
A common cause of cable failure is a poor engine electrical ground. With a poor ground, current may flow through the clutch or shift cable to complete the circuit. The problem will weld up the cable in a short time. Some clutch cables may fail because the vehicle's fire wall flexes as tension is put on the cable.

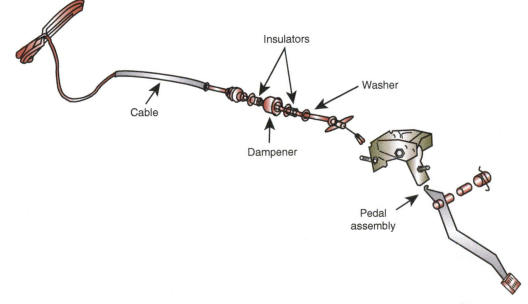

FIGURE 3-14 Location of washers and insulators on a typical clutch cable assembly.

Classroom Manual

Chapter 3, page 78

SPECIAL TOOLS

Line wrenches
Clean brake fluid
Container for old fluid

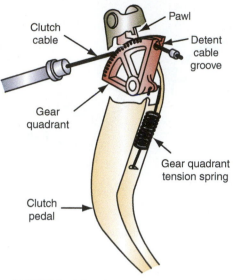

FIGURE 3-15 Routing of the clutch cable around the self-adjuster.

6. Disconnect the cable from the transaxle or transmission mounting bracket and remove the cable.

7. Inspect the clutch cable. Replace the cable if the inner cable is difficult to move within its outer covering, or if it is kinked, has frayed wires, or its ends are worn.

To install a clutch cable:

1. Install the cable into its insulators, dampener, and washers.

2. Lubricate the rear insulator with tire mounting lube or an equivalent to make it easier to install the assembly into the pedal mounting bracket.

3. From inside the car, attach one end of the cable to the detent, being sure to route the cable underneath the pawl and into the detent cable groove (Figure 3-15).

4. Support the clutch pedal upward against the bumper stop to release the pawl from the detent.

5. Install the other end of the cable to the clutch release lever and transmission mount bracket.

6. Check the operation of the clutch and adjust it by lifting the clutch pedal up to allow the self-adjusting mechanism to adjust the free-play.

7. Slowly depress the pedal several times to set the pawl into mesh with the detent teeth.

Maintenance

External clutch linkages should be lubricated on a regular basis, normally during chassis lubrication. The same grease is used to lubricate both. All sliding surfaces and pivot points in the linkage should be lubricated. After lubricating the parts, make sure the linkage moves freely. If it doesn't, check for damaged parts. Always refer to the manufacturer's recommendations before lubricating the linkage.

HYDRAULIC CLUTCH SYSTEMS

Hydraulic clutch linkage components (Figure 3-16) are serviced like hydraulic brake system components. Common service problems include fluid leaks, worn-out piston seals, air in the system, and corrosion buildup. Brake fluid is most often used in both brake and clutch hydraulic systems; however, some manufacturers recommend the use of hydraulic clutch fluid.

The proper level for the fluid in the reservoir is normally marked on the fluid reservoir. This reservoir is normally mounted to the top of the clutch master cylinder (Figure 3-17) or

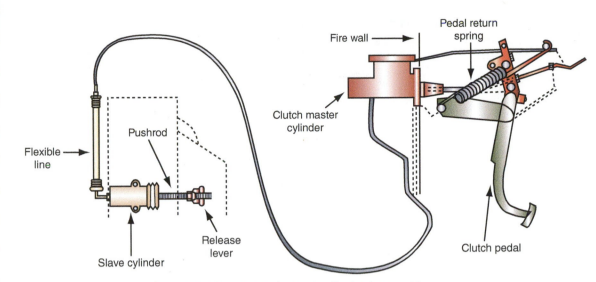

FIGURE 3-16 A typical hydraulic clutch assembly.

is part of the master cylinder assembly. Some vehicles have a remote reservoir with a line going to the clutch master cylinder. A few examples use a separate section of the brake master cylinder reservoir to store clutch fluid. The hydraulic system does not consume fluid, therefore if the reservoir is low, check for leaks at the master and slave cylinders and the connecting hydraulic lines. Fill the reservoir only to the fill line to allow the fluid to rise as the clutch disc wears. Overfilling the system will not allow the bearing to change position as the clutch wears, causing slipping and premature failure. Air can enter the system through the compensation and bleed ports if the fluid level in the reservoir is too low or if a seal leaks. The system must be bled to remove the trapped air.

System Diagnosis

Diagnostics of a hydraulic clutch system should begin with an inspection of the fluid. Check the fluid reservoir for dirt and contamination. Foreign matter in the fluid will destroy the seals and wear grooves in the master and slave cylinder's bores. Brake fluid is extremely **hygroscopic**. A good supply of fresh, new brake fluid is often recommended when servicing the hydraulic system.

Most of the diagnostic problems encountered with hydraulics are due to a misunderstanding of how the system works.

Hygroscopic means fluid has a tendency to absorb moisture from the atmosphere.

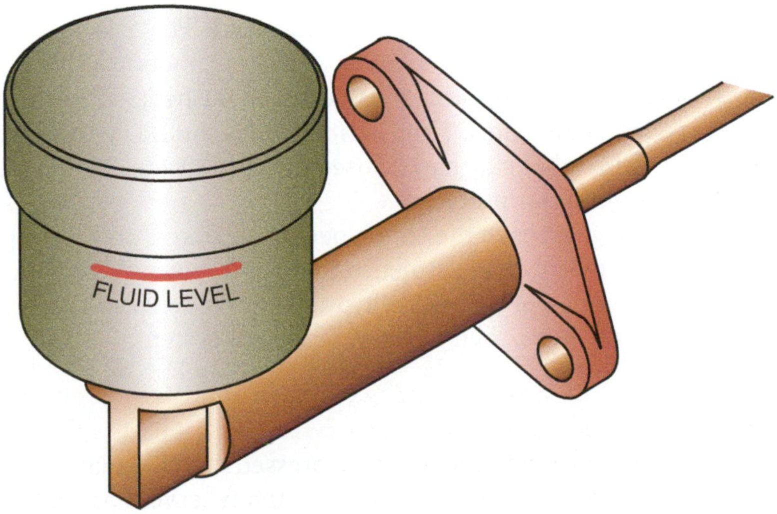

FIGURE 3-17 Fluid reservoir mounted to a clutch master cylinder.

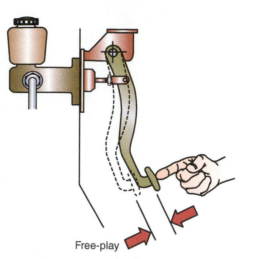

FIGURE 3-18 If the clutch fails to release when the clutch pedal is depressed, check the pedal adjustments.

A soft clutch pedal can be caused by low fluid in the reservoir. To correct this problem, refill the reservoir to the correct fluid level and then bleed the system. This problem can also be caused by a faulty or damaged primary or secondary seal in the master cylinder. A leaking secondary seal will be evident by external leaks and the technician should check the inside of the firewall for evidence of leaking. A primary seal leak will be internal. To correct either of these problems, replace or rebuild the master cylinder. Then refill and bleed the system. A leaking slave cylinder may also cause this problem. In this case, the slave cylinder should be replaced and the system refilled with fluid and bled. If there is excessive pedal travel, the fluid level in the reservoir could be too low. The system should be refilled and the system bled.

If the clutch fails to release when the pedal is depressed, suspect the same problems as those causing a soft pedal. Also check the pedal adjustments and adjust pushrod-to-piston clearance to manufacturer specifications (Figure 3-18).

If there is an extremely hard pedal, check the pedal mechanism and the release fork for binding. If there is evidence of binding, repair and lubricate the assembly to ensure free movement. A hard pedal can also be caused by a blocked compensation port in the master cylinder. The port may be blocked by improper pushrod adjustments or because the piston is binding in the master cylinder bore. If the piston is binding, the master cylinder should be replaced or rebuilt and the hydraulic system flushed, refilled, and bled. This problem may be caused by swollen cup seals or contamination in the master or slave cylinders. If this is the problem, the master or slave cylinder should be replaced and the system flushed, refilled, and bled. Restricted hydraulic lines also cause a hard pedal. The restricted lines should be replaced and the system flushed to remove the debris that may have caused the restriction. A worn clutch disc and/or pressure plate may also cause this problem.

If the clutch fails to engage when the clutch pedal is released, check the pedal and release assemblies for binding or improper adjustment and repair them as needed. A swollen primary cup, overfilled master cylinder, or restricted hydraulic lines can also cause a lack of engagement. Replace the defective parts, then flush and refill the system.

Clutch Reserve

Clutch reserve is the distance between a fully depressed clutch pedal and its position when the vehicle starts to move forward while in gear. Clutch reserve is adjusted by adjusting the free-play. Insufficient reserve can cause clutch drag and hard shifting.

To measure clutch reserve:

1. Apply the parking brake.
2. Make sure nothing is behind the clutch pedal that would prevent full movement.
3. Attach a cable tie to the lower part of the clutch pedal and connect one end of a measuring tape to it.
4. Fully depress the clutch and measure the distance between the pedal and a stationary reference point.
5. Start the engine.
6. Release the clutch pedal.
7. Slowly and gently move the shift lever toward the reverse or first gear position.
8. When gear clash is heard, hold the lever there. *Do not* move it further.
9. Slowly depress the clutch pedal.
10. When the clashing of gears stops, hold the clutch pedal in that position.
11. Measure the distance from the clutch pedal to the reference point.
12. The difference between the two measurements is the clutch reserve.
13. Compare that measurement to specifications.

Master Cylinders

The master cylinder (Figure 3-19) uses a single piston that reacts to the movement of the clutch pedal. Normally master cylinder problems are caused by external or internal fluid leaks that require that the unit be replaced. Rebuild kits are available for most designs of master cylinders. If a rebuild kit is available for a cast-iron master cylinder, the cylinder bore should be honed to remove any imperfections in the bore and new seals installed. The bores of aluminum or plastic master cylinders should never be honed.

Slave Cylinders

The slave cylinder (Figure 3-20) reacts to the pressure applied to it through the hydraulic line that connects it to the output of the master cylinder. Internal and external leaks are also typical problems for slave cylinders. They are usually replaced rather than rebuilt. If it appears that the piston of a slave cylinder is seized in its bore, check the movement of the release fork and lever at the clutch before replacing or repairing the slave cylinder. Leaks may also result from damaged or corroded hydraulic lines. If they are damaged, these lines should be replaced with the same type tube as originally installed.

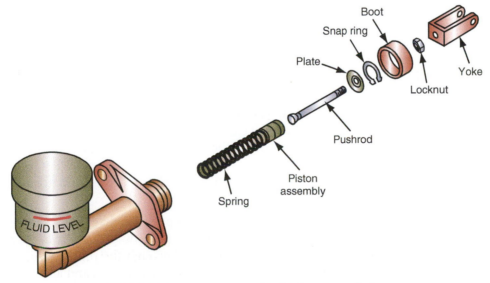

FIGURE 3-19 An exploded view of a clutch master cylinder.

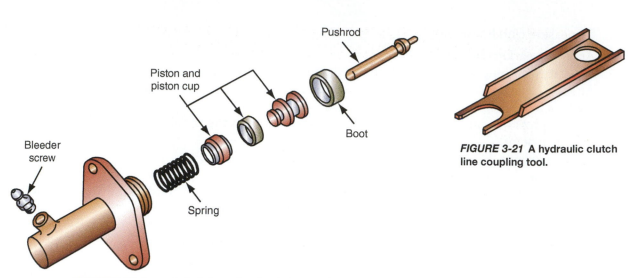

FIGURE 3-20 An exploded view of a clutch slave cylinder.

FIGURE 3-21 A hydraulic clutch line coupling tool.

If the slave cylinder leaks or does not build or maintain pressure, it should be replaced. Replacement is rather straightforward on most vehicles. Simply disconnect the hydraulic lines and unbolt the unit. On vehicles that use a concentric-type slave cylinder, the transmission must be removed to replace the cylinder. The hydraulic fluid line must be disconnected before the transmission can be removed. To do this, a special tool (Figure 3-21) must be used to unlock the quick disconnect.

Bleeding the Hydraulic System

Whenever the hydraulic system is opened for repair, the entire system should be bled. Bleeding is a process of moving fluid through the system and getting rid of any air that may be trapped in the fluid. **Bleeding** also may be necessary if the system has run low on fluid and air is trapped in the lines or cylinders. Bleeding can be accomplished through the use of a power bleeder (the same device used to bleed a brake system) or the use of a buddy. On most cars, it is impossible to pressurize the system and bleed the hydraulic lines at the same time; therefore, it is important that you have the proper equipment or someone to assist you. Photo Sequence 3 covers the basic procedure for bleeding a clutch hydraulic system, with an assistant.

Another procedure for bleeding the system follows:

1. Check the entire hydraulic circuit to make sure there are no leaks.
2. Check the clutch linkage for wear and repair any defects before continuing the procedure.
3. Make sure all mounting points for the master and slave cylinders are solid and do not flex under the pressure of depressing the pedal.
4. Press the clutch pedal to the floor.
5. Open the bleeder screw to purge the air from the system.
6. Close the bleeder screw and release the clutch pedal.
7. Repeat steps 4, 5, and 6 until all air is out of the system. Make sure to check the fluid level between bleeding.
8. Pump the clutch pedal several times; if clutch engagement is not satisfactory, repeat the bleeding procedure.

On some vehicles, it is very difficult to get all of the trapped air out of the hydraulic circuit. Sometimes a pressure bleeder kit will help. During pressure bleeding, the fluid in the reservoir is pressurized. The pressurized fluid is allowed to pass through the master cylinder to the bleeder valve. When the bleeder valve is opened, the pressurized fluid and air will be forced out the valve.

Bleeding is the process of removing air from a closed system.

CAUTION:

Brake fluid can destroy the finish on a car. Always be careful not to allow brake fluid to get on the car's painted surfaces. If it does, immediately flush the surface with water.

Most clutch release problems are caused by air in the hydraulic system.

BLEEDING A HYDRAULIC CLUTCH

P3-1 Check the entire hydraulic circuit to make sure there are no leaks.

P3-2 Check the clutch linkage for wear and repair any defects before continuing the procedure.

P3-3 Make sure all mounting points for the master and slave cylinders are solid and do not flex under the pressure of depressing the pedal.

P3-4 Fill the master cylinder with the approved fluid.

P3-5 Attach one end of a hose to the end of the bleeder screw and the other end into a catch can. Loosen the bleed screw at the slave cylinder approximately one-half turn.

P3-6 Fully depress the clutch pedal. Allow the fluid and air to exit the system.

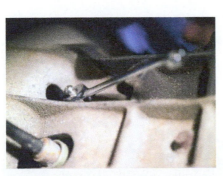

P3-7 Close the bleeder screw.

P3-8 Release the pedal rapidly.

P3-9 Once the air is removed from the system, refill the reservoir to the proper level.

Some technicians prefer to use a vacuum bleeder to bleed the system. During vacuum bleeding, negative pressure (vacuum) is exerted on the system from each of the bleeder valves. This negative pressure draws the fluid from the reservoir, through the system, and out the bleeder valve.

In difficult-to-bleed systems, the reverse bleed method may be the most effective approach. Reverse bleeding is the process of forcing fluid into the system at the lowest point. This forces fluid back and upward to the master cylinder fluid reservoir. When pressure-bleeding any hydraulic system, it is necessary to provide the best seal possible around the bleeder fitting threads so as not to inadvertently draw in air as the fluid is being pumped into the system. To achieve this, remove, clean, and carefully apply three to four wraps of Teflon tape to the thread area of the bleeder screw. Most recommend that the system be bled until the fluid is clean and void of air bubbles. Then tighten the bleeder screw, cycle the clutch pedal 50 times at varying rates and stroke depths, then perform one additional bleed before tightening the bleeder screw to specifications. On some model vehicles, it may be necessary to unbolt the slave cylinder and tip it so its bleeder valve is up. This allows the trapped air to find a path out. On vehicles with a concentric slave cylinder, bleeding the system when the slave cylinder is mounted in place on the vehicle is very difficult. This is caused by the fact that the master cylinder is mounted on a downward angle, which makes it very difficult for trapped air to escape into the reservoir. At times, it may be necessary to jack up the vehicle at different angles to get all of the air out when bleeding.

After bleeding the system, make sure the bleeder screw is tight and the bleeder screw cover is installed. Then, slowly pump the clutch pedal several times. Check the action of the pedal and then visually inspect the system to make sure there are no leaks. If there are no problems, make sure the fluid is at the correct level.

Systems with an Internal Slave Cylinder. A **concentric slave cylinder** is found on some cars and light trucks. These units serve as both the slave cylinder and the clutch release bearing. They are located around the transmission's input shaft (Figure 3-22). The cylinder is either bolted to the transmission's front bearing cover or is held by a pressed pin.

The procedure for bleeding these systems is similar to a conventional system, except the slave cylinder cannot be seen.

1. Remove the clutch fluid reservoir cap and diaphragm (Figure 3-23).
2. Check the fluid level. Fill the reservoir to or above the full mark.
3. Install the cap and diaphragm.
4. Put the transmission in neutral and raise the vehicle with an assistant in the vehicle.
5. Remove the bleeder screw cover and attach a vinyl hose to the bleeder hose.

SERVICE TIP:
Some technicians recommend pulling the clutch pedal up rapidly with your hand during the bleeding procedure to help pull fluid into the master cylinder and purge trapped air.

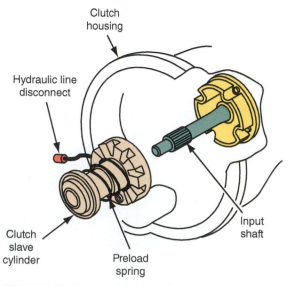

FIGURE 3-22 A concentric internal clutch slave cylinder.

Clutch housing
Hydraulic line disconnect
Clutch slave cylinder
Preload spring
Input shaft

FIGURE 3-23 To get an accurate view of the fluid's level, the diaphragm under the reservoir cap must be removed.

6. Place the other end of the vinyl hose into a clear container partially filled with fluid.
7. Loosen the bleeder screw.
8. Have the assistant press the pedal and hold it down.
9. Allow air to escape and then tighten the bleeder screw.
10. Repeat these steps until no air bubbles are seen in the container.

Bleeding the Master Cylinder

To bleed a typical clutch master cylinder:

1. Disconnect the plastic hose from the slave cylinder at the transmission.
2. Remove the master cylinder, remote reservoir (if so equipped), and the line to the slave cylinder.
3. Place the master cylinder in a vise. Be careful not to overtighten the vise. Doing this can damage the cylinder.
4. If the line to the slave cylinder has a check valve, the valve must be held open during bleeding. To do this, insert a small object to hold the end of the check valve open.
5. Insert the line into the fluid reservoir or a container of brake fluid. If there is a check valve, make sure it stays open while the end of the hose is in the reservoir.
6. Keep the brake fluid reservoir higher than the master cylinder (Figure 3-24) while slowly stroking the cylinder's pushrod.
7. Look for expelled air in the reservoir.
8. When there are no air bubbles, allow the check valve to close while the line is still submerged in the fluid.
9. Push on the master cylinder rod. If it is firm, the master cylinder can be installed in the vehicle.
10. After it is installed, bleed the slave cylinder.
11. Test-drive the vehicle and check the operation of the clutch to verify the repair.

Adjustments

Although basically self-adjusting, a hydraulic linkage sometimes needs adjustment after clutch service or repair. Some vehicles have an adjustment rod between the pedal assembly and the clutch master cylinder. There is usually a locknut against the adjustment nut or fork to help

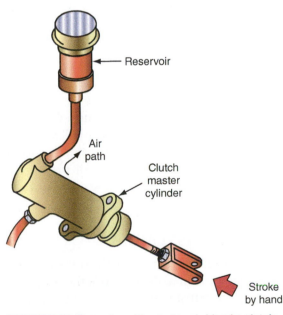

Reservoir

Air path

Clutch master cylinder

Stroke by hand

FIGURE 3-24 Correct position to bench-bleed a clutch master cylinder.

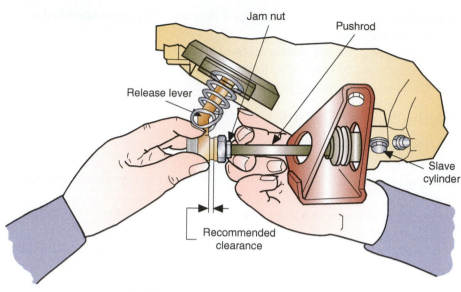

FIGURE 3-25 Checking free-play at a slave cylinder.

maintain the adjustment (Figure 3-25). The adjustment rod must not contact the master cylinder piston when the pedal is released. Always refer to the manufacturer's recommendations for linkage adjustments before proceeding to make an adjustment.

A general procedure for adjusting a hydraulic clutch linkage follows:

1. Measure and adjust the clutch pedal free travel according to the manufacturer's recommendations.
2. Visually inspect the system for leaking seals, low fluid level, broken or kinked lines, and leaking pistons in the cylinders.
3. Bleed the air from the system after repairing or replacing parts.
4. Check the fluid level in the reservoir and add fluid if necessary.
5. Adjust the master or slave cylinder rod as specified to set the clutch throw-out bearing in its proper position.

CLUTCH PEDAL AND RELEASE ASSEMBLIES

If a clutch problem is caused by a worn or damaged clutch pedal assembly, the assembly is normally removed to repair or replace it. To begin removal of the assembly, disconnect the clutch cable, linkage, or master cylinder from the pedal assembly. Disconnect all wires leading to switches on the assembly. Remove the clutch pedal pivot bolt and remove the clutch pedal assembly from its bracket. Inspect, clean, and replace all worn or damaged parts.

Stamped release forks may appear to be in normal condition but may be bent or out of alignment. Carefully inspect the fork and its pivot or ball stud for damage and wear. Release bearing retainer springs and clips; linkage rods and bushings also should be inspected for wear and damage. In general, approximately 0.050-inch of pressure plate travel is obtained by 4 inches of pedal movement and any wear or bending of the linkage can easily cause gear clash or a clutch not to release.

REMOVING THE CLUTCH ASSEMBLY

Servicing the clutch assembly requires removal of the transmission or engine from the vehicle. Begin by referring to a service manual to determine if the transmission can be removed without removing the engine. If the transmission and engine must be removed as a unit, follow the guidelines for removing the engine and separate the transmission from the engine after they are removed from the car. If the transmission can be removed, raise the vehicle on a hoist.

SERVICE TIP:
Whenever you disconnect more than one electrical connector, mark each connector in a way that will make it easy to identify where each should be reconnected.

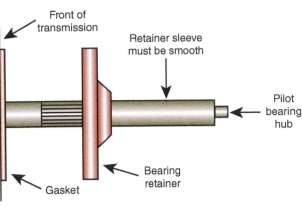

FIGURE 3-26 Inspect the front bearing retainer.

FIGURE 3-27 Inspect the splines and pilot bearing journal of the input shaft.

Before beginning to remove the engine or transmission, disconnect the negative battery cable and clean all dirt, grease, and debris from around the clutch housing and transmission.

After the area is clean, disconnect and remove the clutch linkage. Normally, this is done at the release fork and at the bell-housing. On rear-wheel-drive (RWD) vehicles, the drive shaft must be removed. Before doing this, it is advisable to drain the fluid from the transmission. Then unbolt the rear shaft flange from the rear axle unit and pull the shaft from the transmission.

When separating the transmission from the engine, *never* let the full weight of the transmission rest on the input shaft. Always use the proper equipment, hoists, and jacks when separating the assembly. Use a chain to secure the transmission to the jack.

On front-wheel-drive (FWD) vehicles, the drive axles are normally removed before the transaxle can be pulled from the engine. As parts are being disconnected to allow for the removal of the axles, suspend them with wire so that they do not freely hang by their own weight. This is especially true of brake parts.

After the transmission has been removed, check the condition of the transmission's front bearing retainer (Figure 3-26). Check for grooves or wear that would prevent the release bearing from moving freely along its surface. If the retainer is worn more on one side than the other, the release fork is undoubtedly not centered. If any defects or cracks are found on the retainer, it should be replaced. Check the input shaft for worn or damaged splines (Figure 3-27); these may cause the clutch disc to bind and not release properly. Rusty splines also can cause a clutch to fail to release. Also, check the pilot bearing journal on the shaft for wear and grooves. The shaft should be lightly lubricated. When removing a pressure plate, always lightly loosen all bolts holding the pressure plate to the flywheel before removing any of them. This helps to prevent the pressure plate from warping the clutch disc. The bolts that attach the pressure plate to the flywheel are manufactured to a specific hardness. If the pressure plate is not located to the flywheel by three dowels, the pressure plate bolts are made with precision shoulders to center the pressure plate precisely.

Make sure to thoroughly clean the pressure plate and flywheel mating surfaces as well as the transmission front bearing retainer to remove all oil, grease, and metal deposits. Also carefully check the flywheel for contamination and/or damage. Resurface or replace the flywheel as necessary.

PRESSURE PLATE

If the car and its clutch are subjected to normal use, the pressure plate can last the life of the car. When replacing a worn clutch disc, it is not uncommon to find the pressure plate to be reusable. However, because of the relatively low cost of a new pressure plate and the desire to provide long clutch life for the customer, most technicians replace the pressure plate whenever

SERVICE TIP:
Clutch linkage wear or breakage can result from the use of a heavy-duty clutch assembly with standard clutch linkages or hydraulic components. If a heavy-duty clutch assembly is installed, follow the manufacturer's recommendation for special requirements.

CAUTION:
When cleaning the underside of the vehicle and when working there, make sure the hoist is locked in that position and that you have safety glasses on.

Classroom Manual
Chapter 3, page 68

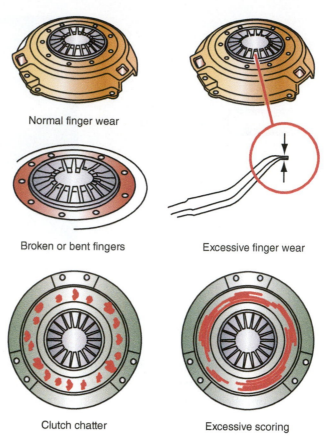

Normal finger wear

Broken or bent fingers

Excessive finger wear

Clutch chatter

Excessive scoring

FIGURE 3-28 **Pressure plates should be carefully inspected before they are reused.**

Hot spots are small areas on the friction surface that are a different color, normally blue, or are harder than the rest of the surface.

CAUTION:

Never substitute different bolts when installing the pressure plate to the flywheel. The bolts may shear at high engine rpm or at heavy loads if they are not the correct hardness. This could cause the clutch assembly to loosen and explode through the bell housing. Severe damage and personal injury to the car's occupants and nearby pedestrians may result. Care should also be taken to tighten the bolts according to the specified order and to the proper torque (Figure 3-29).

a disc is installed. A pressure plate can be quickly damaged if there is insufficient free-play, if it has been overheated due to clutch slippage, and if the clutch disc was worn to its rivets. Grooves will be ground into the fingers or thrust ring of the pressure plate if there is insufficient free-play. Clutch slippage can cause **hot spots** on the surface of the plate due to overheating (Figure 3-28). A severely worn disc will allow the rivets to score the pressure plate and/or flywheel.

WARNING: **The dust inside the bell housing and on the clutch assembly may contain asbestos fibers. Asbestos dust causes lung cancer. Remove the asbestos dust only with an approved vacuum collection or liquid cleaning system. *Never* use compressed air or a brush to clean off the dust. Always dispose of collected asbestos dust or liquid containing the dust in accordance to federal, state, and local laws.**

Typically, the flywheel and clutch assembly are balanced as a unit to prevent vibration and to allow for smooth operation. Some flywheels are fitted with pressure plate locating dowels. Mounting the pressure plate to the dowels allows the plate to be installed in one way only, precisely centered on the flywheel to maintain assembly balance (Figure 3-30). Others use alignment marks to indicate the correct mounting of the pressure plate to the flywheel. If there are no alignment marks on the pressure plate you are removing, use a center punch to mark the location of the pressure plate on the flywheel before removing it. The alignment of the pressure plate and the flywheel is necessary to maintain proper clutch assembly balance.

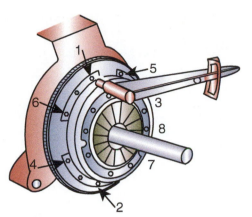

FIGURE 3-29 Pressure plate bolts must be tightened in a specific order and torque.

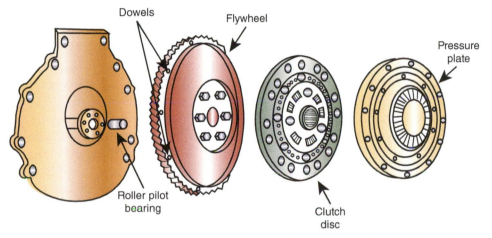

Dowels

Flywheel

Pressure plate

Roller pilot bearing

Clutch disc

FIGURE 3-30 Locating dowels in the flywheel allows the pressure plate to be installed in a position that maintains assembly balance.

⚠ WARNING: Pressure plates are assembled with very strong springs. Never remove any nuts or bolts used to hold the pressure plate assembly together.

The pressure plate can sometimes be inspected without removing it. If the bell housing has a removable flywheel inspection cover, it can be removed to allow visual inspection of the pressure plate, as well as the clutch disc and flywheel. If the pressure plate is found to be defective, replace it with a remanufactured or new one. Individual pressure plate parts are not available; never attempt to rebuild a pressure plate.

Distortion of the pressure plate will cause misalignment problems. Such distortion or warpage is caused by carelessness when removing or installing the pressure plate (Figure 3-31). The attaching bolts must be loosened or tightened evenly and in a staggered sequence, or the assembly may become distorted by the pressure of the springs.

Once the pressure plate has been removed, it should be carefully inspected for scoring, cracks, **blueing**, and hot spots, and the thrust ring or fingers should also be inspected for excessive wear. Excessive finger wear on all fingers indicates a release bearing failure. If the finger wear is uneven, the probable cause is improper tightening of the pressure plate to flywheel bolts. Also look for bent or uneven release levers, or check to see if the pressure ring is not parallel with the clutch cover. Replace the entire pressure plate assembly if any defects are found.

Blueing results from overheating.

FIGURE 3-31 Checking pressure plate warpage and distortion with a straightedge.

Classroom Manual
Chapter 3, page 66

Riding the clutch describes an improper driving technique in which the driver's foot is kept partially on the clutch pedal at all times.

CLUTCH DISC

The clutch disc transfers engine torque from the flywheel to the transmission input shaft when it is engaged. The disc is normally replaced when its friction facing wears thin. In most modern passenger cars, the clutch disc friction material is a woven or molded compound consisting of cotton, brass, copper, rubber, and phenolic resin. Racing and heavy-duty applications may use sintered metallic or ceramic friction material. The asbestos lining material and other lining materials are normally riveted or bonded to the disc. When the lining is riveted on, the disc should be replaced before the material is worn flush with the rivets. If the rivets contact the flywheel or pressure plate, rapid wear and grooves on the surface of either will occur (Figure 3-32).

The clutch disc will wear quickly whenever it is operated in a partially engaged position. This is usually caused by inadequate pressure plate spring force or incorrect clutching and declutching. When a driver "rides the clutch," the pressure plate is unable to apply full clamping pressure on the clutch disc, which causes the disc and release bearing to wear rapidly. Other conditions that cause rapid disc wear are insufficient free-play, binding clutch linkage, and high engine rpm starts. Overloading will also cause premature wear.

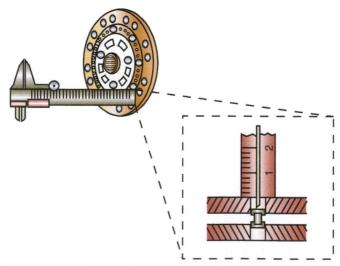

FIGURE 3-32 To measure the depth of the rivet heads in the lining, use a depth gauge or vernier caliper.

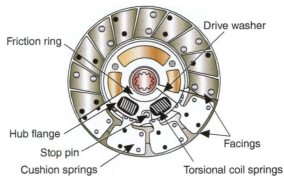

Friction ring
Drive washer
Hub flange
Stop pin
Cushion springs
Facings
Torsional coil springs

FIGURE 3-34 Carefully inspect the disc's torsional springs.

FIGURE 3-33 An old input shaft and some pilot bearing and clutch alignment tools.

> **CUSTOMER CARE:** If it appears that the cause of a clutch slippage problem is the driver, tactfully inform the customer about the driving habits that can damage the clutch. These habits include riding the clutch and holding the vehicle on an incline by using the clutch as a brake.

The clutch disc must remain dry and free of motor and transmission oil or other lubricants. A leaky front oil seal on a transmission or an engine rear main oil seal may oil-soak the clutch disc, causing the friction facing to glaze over and slip due to improper clamping.

Often a thorough inspection of the clutch disc can indicate the exact cause of failure. For example, if the hub is uniformly broken away from the disc, a defective or missing pilot bearing could be the cause. This would result from improper alignment of the transmission to the engine.

Although the disc must be removed for a complete inspection, you can quickly inspect it by removing the flywheel inspection cover. Look for signs of oil and metal or lining materials on the inspection cover or bell housing. Evidence of any of these indicates that the disc should be replaced. If the linings are oil-soaked, repair the oil leak before installing the new disc. Replace any disc that shows signs of overheating, indicated by a blueing of the steel disc backing or glazing of the linings.

If the disc passes these checks, inspect its torsion dampener springs (Figure 3-34) that dampen or cushion the input shaft and drivetrain from harsh engagement when the clutch is applied. These springs are located between the friction facing and the splines of the input shaft. Try to rotate the dampener springs with your finger. They should rotate, but not easily. If they rotate easily, replace the disc. Usually a clutch disc used for more than 50,000 miles that is removed for other vehicle work should be replaced, regardless of its condition. The time and labor saved by replacing the disc at this time will benefit the customer.

FLYWHEEL

Carefully inspect the flywheel. Make sure it is flat and relatively free from roughness and ridges. Also check the surface for hot spots, grooves, and scoring. It is recommended that the flywheel be resurfaced whenever a new disc is installed. Resurfacing is done through a grinding process

SPECIAL TOOLS
Torque wrench
An old input shaft, "dummy shaft," or a clutch alignment tool (Figure 3-33)
Flywheel locking tool

CAUTION:
A common source for oil on a clutch disc is a technician's hand. Never touch the frictional surfaces of a clutch assembly with greasy hands. Always clean your hands well before assembling the clutch and avoid touching the friction surfaces.

Classroom Manual
Chapter 3, page 64

FIGURE 3-35 An overheated flywheel before surface grinding.

at the machine shop (Figure 3-35). A scored flywheel normally can be saved by removing it and resurfacing the contact area. However, a scored pressure plate must be replaced.

Check the flywheel for hot spots. These could be the cause of clutch chatter. The flywheel should be machined at every clutch change. Resurfacing the flywheel will assure that it has the flatness and micro finish it needs. Normally, a flywheel is replaced only when it is so badly damaged that it cannot be resurfaced or cut so many times before that it is now too thin to be safe. A normal resurface cut is .010"–.040".

After the flywheel has been resurfaced and before it is reinstalled, it should be cleaned with soap and hot water or a solvent that will not leave a film or residue on the surface.

Ring Gear Service

Check the condition of the flywheel's ring gear. If the teeth are worn or if teeth are severely chipped or missing, the flywheel ring gear should be replaced. Ring gear replacement is a quick and simple operation, and can save the customer considerable expense compared to complete flywheel replacement. The ring gear is first drilled most of the way through, then split with a chisel (Figure 3-36). Because the gear has an interference fit, as soon as it splits it will pop off the flywheel. The new ring gear is heated to approximately 450°F (232.2°C) and placed into position using welding gloves. The technician should correct the starter problem that likely caused the ring gear failure before returning the vehicle to the owner.

FIGURE 3-36 A ring gear that has been drilled and split with a chisel.

Taper Wear

The clutch disc will often wear the flywheel more toward the center of the flywheel. This causes taper wear on the flywheel. It can be measured with a straightedge across the flywheel, with a feeler gauge being inserted toward the center (Figure 3-37). The flywheel will have to be surface ground if the taper wear is more than .004-inch to .006-inch, depending on manufacturer specifications.

Runout

The flywheel should be checked for excessive runout (Figure 3-38). Make sure the flywheel to crankshaft bolts are properly torqued before checking the runout of the flywheel. While checking runout, also check the end play of the crankshaft. To do this, attach the dial indicator

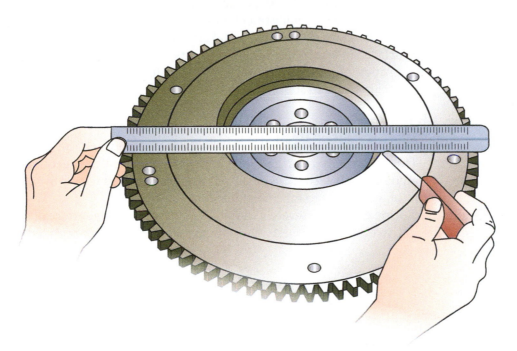

FIGURE 3-37 Measuring flywheel taper wear with a feeler gauge.

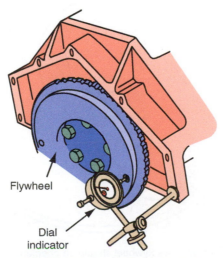

Flywheel

Dial indicator

FIGURE 3-38 The flywheel should be inspected and checked for excessive runout.

to the flywheel and push the flywheel toward the front of the engine. Set the dial indicator to zero, then pry the flywheel away from the block and observe the reading on the indicator. Runout can be checked with the same dial indicator setup. The flywheel should be rotated and the indicator observed. The amount of movement shown on the indicator is the amount of runout. Compare the reading with the specifications given in the service manual for crankshaft end play and flywheel runout.

Excessive machining of the flywheel can lead to a clutch release problem. This is because the hydraulic clutch system is a quick release system, which means the slave cylinder travels a short distance to disengage the clutch. Overmachining of the flywheel may also cause the disc's torsion springs to contact the flywheel bolts and cause noise and wear during engagement and disengagement. Several methods have been used to correct a release problem if too much has been machined from the flywheel. These methods involve replacing the flywheel, replacing the slave cylinder pushrod with a longer one, shaving off a section of the clutch pedal stop to provide additional distance for adjustment, or installing a precision shim between the flywheel and the crankshaft. Care should also be taken when tightening the pressure plate attaching bolts on resurfaced flywheels. Make sure the bolts do not bottom in the hole before the pressure plate is properly torqued. A loose pressure plate assembly can rip off of the flywheel, damaging the clutch and related components.

Dual-Mass Flywheels

Dual-mass flywheels should be checked for excessive looseness between the primary and secondary plates. This is done by rotating the primary plate against the secondary plate (Figure 3-39). The amount of allowed movement between plates varies widely by vehicle, so compare the amount of movement to specifications.

Dual-mass flywheels can become severely warped. This is because the surface of the main plate is steel. They are typically not resurfaced and when warped, they are replaced. Also, if

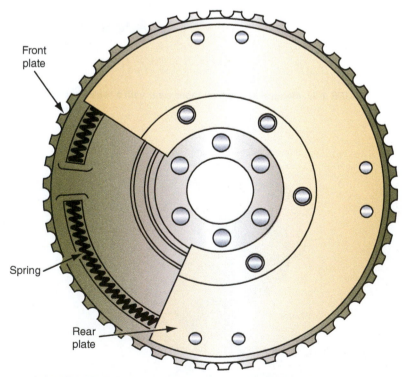

FIGURE 3-39 A dual-mass flywheel should be checked for looseness between the front and rear plates.

the flywheel has faced extreme loads, the friction ring will wear quickly. Again, the repair is replacement.

It is important to note that these flywheels require a special clutch disc. When they are replaced, the correct disc or an aftermarket damper yoke must be installed with it. At times, due to expense, dual-mass flywheels are replaced with conventional flywheels. Also, whenever a clutch disc and pressure plate is replaced, a new dual mass flywheel is also installed.

> **AUTHOR'S NOTE:** Modified engines, especially diesel engines in trucks that have been chipped and have had cold-air and exhaust modifications performed, end up with more torque than an OEM dual-mass flywheel can handle. The technician must have enough knowledge to recommend clutch components that can handle the extra torque. A good-quality dual-disc setup with a solid flywheel is a common solution to the customer's needs.

Removing the Flywheel

When removing the flywheel, scribe or punch marks into both the flywheel and the crankshaft to ensure that engine balance is maintained when the flywheel is installed (Figure 3-40).

Bell Housing

Check the bell housing for warpage and cracks. Also check the alignment of the clutch housing's face (Figure 3-41) and bore. To check both, use a dial indicator attached to a special tool that is installed in the pilot bearing or bushing. For most cars, the misalignment limit is less than 0.010-inch; however, always refer to the manufacturer's specifications. Misalignment beyond specifications normally requires bell housing replacement. To correct misalignment of the face of the clutch housing, shims must be installed between the housing

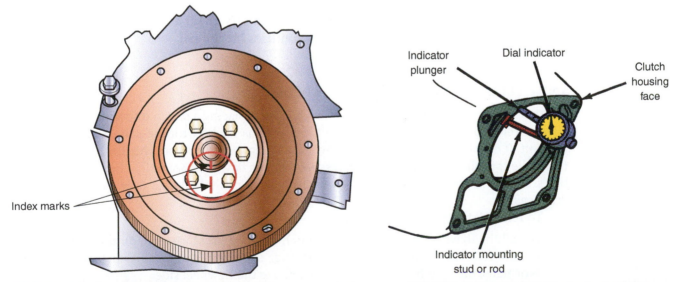

Index marks

Indicator plunger

Dial indicator

Clutch housing face

Indicator mounting stud or rod

FIGURE 3-40 The flywheel to crankshaft alignment should be marked before the flywheel is removed.

FIGURE 3-41 Using a dial indicator to check the alignment of a bell housing's face.

and the rear of the engine block. Some engines are equipped with offset dowel pins on the rear face, which can be adjusted to compensate for runout of the housing bore. Often when face alignment is corrected, bore alignment is also corrected. If the bore alignment remains out of specification, the bell housing must be replaced. The technician may find that the clutch disc hub demonstrates taper wear on the edge of the splines when misalignment is excessive (Figure 3-6).

> **CUSTOMER CARE:** Sometimes the type of care a technician can extend to a customer can amount to a matter of life or death. A knowledgeable technician is aware that very high rpm's can cause a clutch assembly to explode, potentially causing harm to the driver. In racing and extreme street conditions, the technician should be prepared to advise the customer that extra protection is required. If the bell housing is a separate unit that bolts to the transmission, it can be replaced with a steel unit, often called a scattershield. If the transmission or transaxle does not have a separate bell housing, the clutch area can be wrapped in an explosion-proof Kevlar transmission blanket.

Locating Dowels

All engines have two locating dowel pins that project into matching bores in the bell housing. Although bell housing bolts hold the housing to the block, the dowels are responsible for maintaining perfect alignment between the crankshaft and input shaft of the transmission. These pins help to maintain the mating of the engine with the transmission. These dowels should be carefully inspected for looseness and damage. The mating bores in the bell housing should also be inspected. There are also locating dowels on the flywheel for the mounting of the pressure plate. These should also be carefully inspected.

To remove damaged dowel pins, use a drift in punch to drive out the pin if it is installed in an open bore. If it is installed in a blind bore, use locking pliers to pull it out. Install the new dowel pins by driving them into their bores with a brass or plastic mallet.

The clutch may not release if the disc is installed backward.

CAUTION:
The use of a bolt that is too long may result in the bottoming-out of the bolts without the assembly being securely fastened. As a result, you may measure the correct amount of bolt torque without the assembly being tightly secured.

! **WARNING:** Be careful when loosening the flywheel attaching bolts. Flywheels are extremely heavy and can cause injuries if they fall.

INSTALLING THE CLUTCH ASSEMBLY

Photo Sequence 4 outlines the typical procedure for replacing a clutch disc and pressure plate. Always refer to the manufacturer's recommendations for bolt torque prior to reassembling the unit. It may be necessary to use a flywheel locking tool to prevent the flywheel from turning while torquing the bolts for the pressure plate or flywheel. Some flywheel-to-crankshaft bolts require a special sealer in cases where the crankshaft flange holes are open to the crankcase.

To reinstall the assembly:

1. Install the flywheel and position the clutch disc and pressure plate on the flywheel. The clutch disc is installed with the dampener springs offset toward the transmission. Normally, the clutch disc is marked to indicate the flywheel side.
2. Align the hub of the disc to the center of the pressure plate. This is best accomplished with a clutch alignment tool or an old transmission input shaft by inserting it through the clutch disc into the pilot bearing (Figure 3-42).

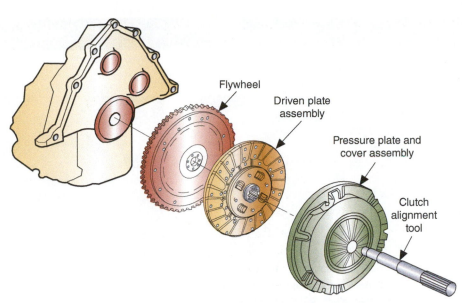

FIGURE 3-42 Aligning the clutch disc with a clutch alignment tool.

FIGURE 3-43 To remove or install the flywheel, it may be necessary to lock the flywheel.

3. Tighten the bolts attaching the pressure plate to the flywheel evenly in a crisscross pattern and to the correct torque, then remove the dummy shaft or alignment tool. Sometimes a flywheel holding tool must be used to tighten the pressure plate and flywheel bolts (Figure 3-43).

4. Lightly lubricate the splines of the input shaft, transmission front bearing retainer, pilot bearing surface, and clutch disc hub with white lithium grease.

5. Install a new release bearing into the release fork.

6. Install the transmission. Make sure the weight of the transmission does not rest on the splines of the clutch disc. This can cause the disc to warp.

7. Make sure the transmission is evenly seated against the engine, then tighten the attaching bolts.

8. Check the operation of the clutch and adjust the free-play using the recommended procedure.

SERVICE TIP:
A **modular clutch** includes the flywheel, pressure plate, and clutch disc in a riveted assembly. The clutch disc is prealigned by the manufacturer. The assembly is bolted to a plate similar to an automatic transmission flexplate (Figure 3-44).

SPECIAL TOOLS
The correct-sized bearing puller (Figure 3-45) and driver (Figure 3-46)
Ball-peen hammer
Brass hammer

PHOTO SEQUENCE 4

INSTALLING AND ALIGNING A CLUTCH DISC

P4-1 The removal and replacement of a clutch assembly can be completed while the engine is in or out of the car. The clutch assembly is mounted to the flywheel that is mounted to the rear of the crankshaft.

P4-2 Before disassembling the clutch, make sure alignment marks are present on the pressure plate and flywheel, if you are not replacing the pressure plate.

P4-3 The attaching bolts should be loosened before removing any of the bolts. With the bolts loosened, support the assembly with one hand while using the other to remove the bolts. The clutch disc will be free to fall as the pressure plate is separated from the flywheel. Keep it intact with the pressure plate.

P4-4 The surface of the pressure plate should be inspected for signs of burning, warpage, and cracks. Any faults normally indicate that the plate should be replaced.

P4-5 The surface of the flywheel should also be carefully inspected. Normally the flywheel surface can be resurfaced to remove any defects. The pilot bushing or bearing should also be inspected.

P4-6 The new clutch disc is placed into the pressure plate as the pressure plate is moved into its proper location. Make sure the disc is facing the correct direction. Most are marked to indicate which side should be seated against the flywheel surface.

P4-7 Install the pressure plate according to the alignment marks made during disassembly.

P4-8 Install the attaching bolts, but do not tighten. Then install the clutch alignment tool through the hub of the disc and the pilot bearing. This will center the disc on the flywheel.

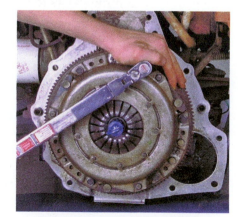

P4-9 With the disc aligned, tighten the attaching bolts according to the procedures outlined in the service manual. Hold the alignment tool in place so the disc cannot drop.

FIGURE 3-44 A riveted modular clutch assembly with flywheel, clutch disc, and pressure plate.

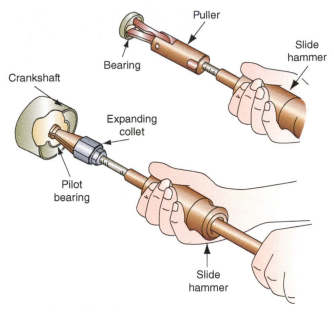

FIGURE 3-45 Typical clutch pilot bearing remover.

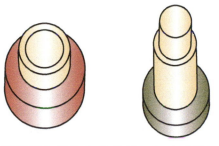

FIGURE 3-46 Typical pilot bearing installer.

INPUT SHAFT PILOT BEARING AND BUSHINGS

On most RWD vehicles, a pilot bushing or bearing (Figure 3-47) is used to support and locate the forward end of the transmission's input shaft. A pilot bushing is made of bronze and is vacuum-impregnated with a lubricant. A pilot bearing is either a ball- or roller-type bearing. The pilot bearing or bushing is fitted in a bore centered in the rear flange of the crankshaft or in the center of the flywheel. A pilot bearing should be replaced if it is noisy, worn, or damaged. There are two common ways to remove a pilot bearing. First, insert the hooked end of a slide hammer behind the bearing and pull it out, or use a puller specially designed for the removal of pilot bearings. This tool can also be used to remove a pilot bushing.

Classroom Manual
Chapter 3, page 66

FIGURE 3-47 Examples of new pilot bearings and bushings.

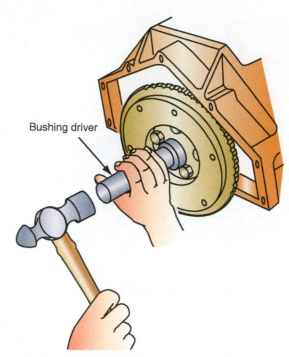

Bushing driver

FIGURE 3-48 Installing a clutch pilot bearing.

Normally, a pilot bushing can be removed by filling the bushing's bore with heavy grease, then driving a round, tight-fitting driver or dowel into the bushing. This should put a sufficient amount of pressure on the grease to force the bushing out. Some technicians will place a ball bearing in the crankshaft bore for the pilot bushing and then thread a large tap into the bushing. The tap bottoms against the ball bearing, and the bushing is pulled out by the tap.

To install a new bushing or bearing, clean the bore to make sure no material is present that would prevent the bushing from seating properly in the bore. The new bushing or bearing should be driven (Figure 3-48) or pressed into its bore with the correct-sized driver. Pilot bearings have an o-ring seal that should face away from the crankshaft. After installing the bushing or bearing, coat it and the transmission's input shaft lightly with engine oil.

Pilot bushings are impregnated with a special oil during manufacturing. Never coat a pilot bushing with grease or a heavy oil. This will block the pores in the bushing and actually increase the friction between the input shaft and the bushing. Use only low-viscosity (i.e., 30W) motor oil to lubricate the bushing. However, needle-bearing pilot bearings should be lubricated with white lithium grease.

CLUTCH RELEASE BEARING

Classroom Manual Chapter 3, page 72

The clutch release bearing (Figure 3-49) transmits the movement of the clutch linkage and release fork to the pressure plate. Its outer race depresses the pressure plate fingers or thrust ring and rotates with the pressure plate when the clutch pedal is depressed. This action causes the pressure plate to release its clamping force on the clutch disc. The inner race of the release bearing is pressed into a collar, which slides back and forth over the transmission's front bearing retainer. Most release bearings are constructed with either a sealed ball or roller bearing or have a low-friction carbon surface. Self-centering release bearings are often used to compensate for minor differences in the alignment of the pressure plate and release fork.

A common cause for a damaged release bearing is improper clutch adjustment. As the clutch disc's friction material gradually wears from use, the pressure plate moves closer to the flywheel and the clutch release fingers move outwardly. This forces the clutch release bearing backward and the clutch pedal with it. If the pedal is against its pedal stop, the bearing will contact the release fingers and turn at all times. This continuous pressure on the release bearing will tend to partially disengage the clutch, causing the friction facings to slip and wear rapidly.

FIGURE 3-49 Clutch release bearings should be checked by turning the bearing while applying force in the direction of rotation.

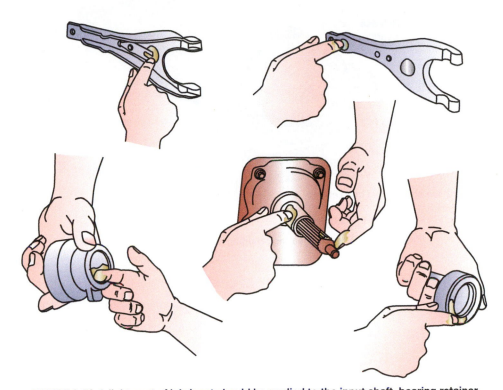

FIGURE 3-50 A light coat of lubricant should be applied to the input shaft, bearing retainer, release lever, lever pivot, and lever-to-release bearing surfaces when installing a throw-out bearing.

To remove the release bearing after the engine and transmission have been separated, simply disconnect it from the clutch release fork. While installing a new release bearing into the release fork, make sure the bearing is fully seated. Release bearings are lubricated and sealed during manufacturing; therefore, they require no additional lubrication when they are installed. However, to prevent the release bearing from binding on the bearing retainer or release fork, a very light coating of grease should be applied to the bearing's collar and on the release fork (Figure 3-50). Do not allow the grease to get on the friction surfaces of the disc, pressure plate, or flywheel.

A release bearing should never be washed in cleaning solvent because the solvent will remove the lubricant from the bearing and cause premature failure. The hub of the bearing and the transmission's front bearing retainer should be free of all defects and rough spots. If the bearing retainer has any defects, it should be replaced. When installing a release bearing, make sure its retaining clips are securely in position (Figure 3-51). Constant-running release bearings are installed with a preload that must be set during installation. Procedures for doing this can be found in the appropriate service manual.

Constant-running release bearings are used on all vehicles equipped with a self-adjusting clutch cable.

Concentric Slave Cylinders

If the vehicle is equipped with a concentric slave cylinder, the release bearing is part of that assembly (Figure 3-52). The entire assembly should be replaced whenever a new clutch and pressure plate is installed. However, on some models, the release bearing and clutch slave cylinder can be individually replaced. Normally the release bearing is held to the cylinder by a retaining ring.

Proper installation of these units is critical. Not only must they be positioned correctly, they also have special procedures due to the hydraulics involved. Some units do not have a bleeder screw, and therefore must be filled with fluid and bled before installation. In fact, it is recommended that all of these units be filled with fluid before installation. Always follow the manufacturer's instructions. Once the transmission is installed in the vehicle, the entire hydraulic system should be bled.

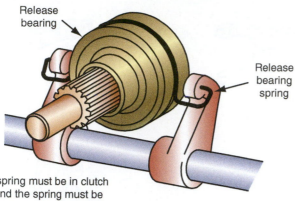

Release bearing

Release bearing spring

Ends of the spring must be in clutch fork holes and the spring must be seated in the groove of the bearing

FIGURE 3-51 Release bearing retaining springs.

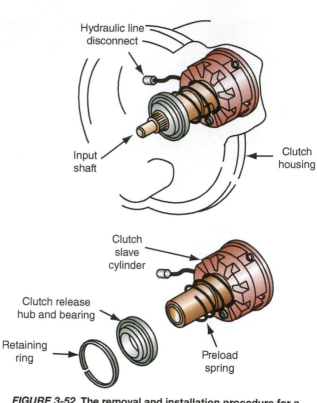

Hydraulic line disconnect

Input shaft

Clutch housing

Clutch slave cylinder

Clutch release hub and bearing

Retaining ring

Preload spring

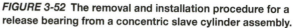

FIGURE 3-52 The removal and installation procedure for a release bearing from a concentric slave cylinder assembly.

SERVICING DUAL-CLUTCH UNITS

Service work on current **dual-clutch transmission** systems require special procedures that differ significantly from standard clutch repair practices. Dual-clutch units may be hydraulically activated wet clutches or solenoid-operated dry clutches. In either case, special tooling is necessary for removal, repair, and installation of these units (Figure 3-53). In some cases dual wet clutch assemblies are only serviceable as a unit. Because these clutch assemblies are

FIGURE 3-53 Special toolkits required to remove and install a Ford dual dry clutch assembly.

electronically controlled and rely on various inputs from the driver and vehicle, most manufacturers require a relearn procedure after repair to allow the adaptive logic of the controller to begin from a preset baseline.

GUIDELINES FOR SERVICING CLUTCH SYSTEMS

1. Take the vehicle on a test drive. Ride with the owner and, if necessary, educate the driver on the proper driving habits and how the clutch system functions.
2. When disconnecting the clutch linkage, check all rods and bushings for wear or damage.
3. When separating the transmission from the engine, *never* let the full weight of the transmission rest on the input shaft. Always use the proper equipment, hoists, and jacks while separating the assembly. Use a chain to secure the transmission to the jack.
4. Mark alignment points on the pressure plate assembly and flywheel before removing the pressure plate.
5. After removing the pressure plate assembly, check these four areas: pressure plate surface, clutch disc, flywheel surface, and the release bearing.
6. Never use an air hose to blow off dust and dirt during clutch removal. Always carefully wipe off all parts and use the proper personal protection to avoid inhaling any dust.
7. Make sure all surfaces are clean and any oil leaks from the transmission or engine are sealed.
8. Install good-quality new parts and make sure all parts are correct for the application.
9. Always use the procedures recommended by the manufacturer when removing or installing a clutch assembly.
10. Never disassemble a pressure plate assembly. Serious injury can result from the high-pressure springs.
11. Inspect the pressure plate for nicks, scores, and signs of overheating.
12. Inspect the fingers and thrust ring of the pressure plate for wear.
13. Check the spring tension of the pressure plate assembly to make sure the plate is contacting evenly.
14. Check for high and low spots of wear on the clutch facings. On riveted linings, use the level of material above the rivet head as an indicator of wear.
15. Replace the disc if the rivet heads are flush or just below the outer surface.
16. Replace the disc if it appears glazed or cracked.
17. Handle the pressure plate and disc assembly with clean hands.
18. Slide the new disc onto the input shaft of the transmission. It should slide easily without restriction; however, if the disc is able to rock back and forth or from side to side, replace it. Check the splines of the input shaft.
19. The engine flywheel surface should be flat and relatively free from roughness and ridges and should be resurfaced whenever a new disc is installed.
20. Inspect the flywheel for hot spots.
21. Resurface the flywheel—starting out with a flat flywheel will extend clutch life.
22. Install a new pilot bearing or bushing whenever doing clutch work.
23. Inspect the input shaft for score marks. Scoring may indicate that the pilot bearing is faulty.
24. Check the release bearing to be sure that it rotates freely.
25. Inspect the transmission bearing retainer for grooves and wear that may prevent the release bearing from moving freely along its surface.
26. Check the bell housing for warpage and cracks.
27. Check the pressure plate bolt holes on the flywheel. Make sure the bolts do not bottom-out in the bores. This is especially necessary if the flywheel has been resurfaced.
28. Install the disc properly with the side marked "flywheel side" toward the flywheel.

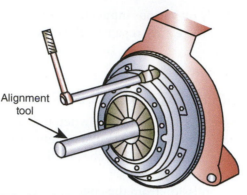

FIGURE 3-54 When removing the clutch assembly, insert a clutch alignment tool into the disc's hub. This prevents the disc from falling out when the pressure plate is loosened.

29. Use a clutch alignment tool when positioning the assembly on the flywheel (Figure 3-54).
30. Alternately and evenly using a star pattern, tighten the pressure plate to flywheel mounting bolts. Never use an air ratchet or impact wrench.
31. Install the transmission to the engine block and torque bell housing bolts to specifications. Never let the transmission hang on the clutch disc during installation.
32. Check the free play of the clutch before running the vehicle with a new clutch.
33. On vehicles with an adjustable clutch release mechanism, have the customer bring the vehicle back to the shop for a clutch adjustment after a short break-in period.

TERMS TO KNOW

Bleeding

Blueing

Chatter

Clutch drag

Clutch reserve

Clutch slippage

Concentric slave cylinder

Dual clutch transmission

Free-play

Gear clash

High pedal

Hot spots

Hygroscopic

Modular clutch

Pawl

Pulsation

Quadrant

Total pedal travel

Vibration

CASE STUDY

A customer brings in a car that has a series of driveability problems. It uses an excess amount of fuel and oil, it lacks acceleration, and seems to lose power while traveling on the highway.

The technician listens to the customer's complaints and asks the right questions. Then he prepares the vehicle for a test drive. While doing this, he checks the oil level and finds it to be low. This somewhat verifies that the engine is using excessive amounts of oil, so he removes the spark plugs from the engine to check for signs of oil burning. The spark plugs are clean and show no sign of burning oil. The technician then checks the engine for signs of leaks, hoping to find the cause of excessive oil loss. The only wet spot is to the rear of the oil pan and the front of the bell housing. A leaking rear main seal is the likely cause of the oil usage. However, it doesn't explain the other problems of excessive fuel consumption and the general lack of power. If the engine had been burning oil, all of the problems would have been related.

He then takes the vehicle on a test drive and finds that the clutch is slipping. When the car is in high gear, he steps hard on the accelerator and notices that the engine speed increases while the car's road speed stays the same. After some thought, he determines this is the cause of the poor fuel economy and performance complaints. He also determines that all of the customer's complaints are related to the same problem, a leaky rear main seal. Oil has been leaking out of the engine and onto the clutch disc, which allows it to slip. He notifies the customer and gives an estimate. The customer authorizes the repair, which includes a new rear main seal, release bearing, pilot bushing, clutch disc, and pressure plate, and resurfacing of the flywheel. A good technician realizes that problems in one area of the drivetrain can cause issues with other drivetrain components. These connections can be straightforward, as in the rear main seal leak, or can be more difficult to find. Electrical issues, for example, can often affect engine, transmission, and transfer case performance.

ASE-STYLE REVIEW QUESTIONS

1. When discussing common causes of problems with clutch systems with adjustable linkage,

   Technician A says that often problems can be corrected by properly adjusting the clutch linkage.

   Technician B says that excessive clutch pedal free travel can indicate an excessively worn clutch disc.

   Who is correct?
   - A. A only
   - B. B only
   - C. Both A and B
   - D. Neither A nor B

2. When discussing self-adjusting clutch cables,

   Technician A says that a toothed wheel and a pawl are used to adjust the cable for clutch disc wear.

   Technician B says that the self-adjusting unit maintains the desired amount of free-play.

   Who is correct?
   - A. A only
   - B. B only
   - C. Both A and B
   - D. Neither A nor B

3. When reviewing the procedure for adjusting a clutch pedal,

   Technician A says that 1 inch of pedal free travel is normal on most vehicles.

   Technician B says that the clutch master cylinder pushrod should be adjusted for constant light pressure on the piston.

   Who is correct?
   - A. A only
   - B. B only
   - C. Both A and B
   - D. Neither A nor B

4. When discussing the possible causes for damage to the surface of a pressure plate,

   Technician A says that a worn clutch disc can score the surface.

   Technician B says that insufficient pedal free travel or riding the clutch can cause it to overheat.

   Who is correct?
   - A. A only
   - B. B only
   - C. Both A and B
   - D. Neither A nor B

5. When installing a pressure plate,

   Technician A says that pressure plate bolts are tightened one at a time in a circular pattern.

   Technician B says that the dowel pins locate the pressure plate to ensure that proper balance of the clutch assembly is maintained.

   Who is correct?
   - A. A only
   - B. B only
   - C. Both A and B
   - D. Neither A nor B

6. While discussing the ways to eliminate the problems caused by removing too much material from a flywheel while it was resurfaced,

   Technician A says the flywheel should be replaced so that a longer release bearing can be installed.

   Technician B says the problems may be corrected by installing a precision shim between the flywheel and the crankshaft.

   Who is correct?
   - A. A only
   - B. B only
   - C. Both A and B
   - D. Neither A nor B

7. When discussing the causes of rapid clutch disc wear,

   Technician A says that driving habits influence how long a clutch assembly will last.

   Technician B says that insufficient pedal free travel will cause the disc to wear rapidly.

   Who is correct?
   - A. A only
   - B. B only
   - C. Both A and B
   - D. Neither A nor B

8. When reviewing the procedure for installing a clutch disc,

   Technician A says that the disc should always be installed with its hub and springs toward the engine.

   Technician B says that the splines of the disc should be clean, dry, and never oiled or greased.

   Who is correct?
   - A. A only
   - B. B only
   - C. Both A and B
   - D. Neither A nor B

9. When discussing ways to determine if a pilot bushing bearing is faulty,

   Technician A says that a faulty pilot bearing will squeal constantly when the engine is running.

   Technician B says that a careful inspection of the transmission input shaft can determine the condition of the pilot bushing bearing.

   Who is correct?
   - A. A only
   - B. B only
   - C. Both A and B
   - D. Neither A nor B

10. When installing a new release bearing,

    Technician A says that the outer surface of the transmission's front bearing retainer should be lubricated.

    Technician B says that the release lever and pivot should be lubricated.

    Who is correct?
    - A. A only
    - B. B only
    - C. Both A and B
    - D. Neither A nor B

ASE CHALLENGE QUESTIONS

1. A bearing-type noise is heard immediately when the clutch pedal is depressed. As the pedal is pushed to the floor, the noise remains about the same. When the clutch pedal is released, the noise disappears.

   Technician A says that the clutch pilot transmission input bearing could be at fault.

   Technician B says that the clutch release bearing could be at fault.

   Who is correct?

   A. Technician A
   B. Technician B
   C. Both A and B
   D. Neither A nor B

2. When discussing clutch problems,

   Technician A says that a stretched clutch cable could cause gear clash when shifting.

   Technician B says that a leaking slave cylinder could cause premature release bearing wear.

   Who is correct?

   A. Technician A
   B. Technician B
   C. Both A and B
   D. Neither A nor B

3. A customer says that occasionally when he is at a stoplight with the transmission in first gear and the clutch pedal depressed the vehicle will begin to move by itself.

   Technician A says that there may be insufficient clutch pedal free-play.

   Technician B says that the clutch master cylinder may be leaking internally.

   Who is correct?

   A. Technician A
   B. Technician B
   C. Both A and B
   D. Neither A nor B

4. A clutch replacement has just been performed on a vehicle. The transmission shifts perfectly when the engine is turned off. However, when the engine is started, severe gear clashing occurs whenever shifting into reverse gear is attempted. Shifting into the forward ranges is very difficult.

   Technician A says that there may be excessive clutch pedal free-play.

   Technician B says that there may be air in the clutch hydraulic system.

   Who is correct?

   A. Technician A
   B. Technician B
   C. Both A and B
   D. Neither A nor B

5. A brief squeal comes from the clutch/transmission area of a vehicle just as the clutch is being engaged. There is no noise when the clutch is fully engaged.

   Technician A says that the clutch pilot bearing may be worn out.

   Technician B says that the transmission input shaft bearing may be faulty.

   Who is correct?

   A. Technician A
   B. Technician B
   C. Both A and B
   D. Neither A nor B

Name _____ Date _____

HYDRAULIC CLUTCH SERVICE

Upon completion of this sheet you will be able to bleed a hydraulic clutch system, refill the clutch master cylinder fluid level, and check for leaks.

NATEF MAST Task Correlation

Clutch Diagnosis and Repair

Task #4, Bleed clutch hydraulic system

Task #5, Check and adjust clutch master cylinder fluid level; check for leaks

    a. *Inspect, adjust, replace, and bleed hydraulic clutch slave cylinder/actuator, master cylinder, lines, and hoses.*

    b. *Inspect fluid condition and type; clean and flush hydraulic system; refill with proper fluid.*

Tools and Materials

Clean fluid for the system

A rule

Basic hand tools

Protective Clothing

Goggles or safety glasses with side shields

Describe the Vehicle Being Worked On

Year _____ Make _____ VIN _____

Model _____ Engine type and size _____

Procedure

_____
Task Completed

1. With your foot, slowly work the clutch pedal up and down. Pay attention to its entire travel. Checking for any binding of the pedal linkage and/or a defective return spring. Describe your findings.

    _____

    _____

2. When working the pedal, does it feel like the hydraulic system is working properly?

    _____

    _____

3. Check the level and the condition of the fluid in the reservoir of the master cylinder. Describe your findings.

    _____

    _____

4. If the fluid level is low, fill the reservoir to the proper level.

5. If the fluid was contaminated, the entire hydraulic system must be flushed and bled after a thorough inspection of all of the parts in the system.

Task Completed _____

6. Carefully check the master cylinder, slave cylinder, hydraulic lines, and hoses for evidence of leaks. Describe your findings.

_____

_____

Task Completed _____

7. If there was any evidence of a leak, find and repair the leak by replacing the leaking part.

8. Anytime the system has been opened to make a repair, the system should be bled. Bleeding also may be necessary if the system was allowed to be very low on fluid. Before bleeding, double-check the system for evidence of leaks and the reservoir for proper fluid level.

9. Check all mounting points for the master and slave cylinders. Make sure these do not move or flex when the pedal is depressed. Describe your findings.

_____

_____

10. Loosen the bleed screw on the slave cylinder approximately one-half turn.

11. Fully depress the clutch pedal, and then move the pedal through three quick and short strokes.

12. Close the bleeder screw immediately after the last downward movement of the pedal. Then release the pedal rapidly.

13. Recheck and correct the fluid level in the reservoir.

14. Repeat steps 10, 11, and 12 until no air is evident in the fluid leaving the bleeder screw.

15. Recheck and correct the fluid level in the reservoir.

16. Measure clutch pedal free travel according to the manufacturer's recommendations. What were the results of this check?

_____

_____

17. Adjust the slave cylinder rod to set the clutch release bearing in its proper location and to set proper pedal free travel.

18. With your foot, work the clutch pedal and describe how it feels now.

_____

_____

Instructor's Response: _____

_____

_____

Name _____ **Date** _____

SERVICING CLUTCHES

Upon completion of this job sheet, you should be able to identify vehicles that need clutch service.

NATEF MAST Task Correlation

General: Drivetrain Diagnosis

Task #2, Research applicable vehicle and service information, fluid type, vehicle service history, service precautions, and technical service bulletins.

Clutch Diagnosis and Repair

Task #1, Diagnose clutch noise, binding, slippage, pulsation, and chatter; determine necessary action.

This job sheet is related to the ASE Manual Drivetrain and Axles Test's content area: *Clutch Diagnosis and Repair, Task: Diagnose clutch noise, binding, slippage, pulsation, and chatter problems; determine needed repairs.*

Describe the Vehicle Being Worked On

Year _____ Make _____ VIN _____

Model _____

Procedure

_____

Task Completed

1. Bring the vehicle into the shop area.

2. List the type of vehicle.

   _____

3. Locate and record the VIN #.

   _____

4. From the tag or sticker under the hood, determine engine size and identify the manufacturer.

   _____

5. Obtain a service manual for that year and make of vehicle.

6. Locate the section that identifies the material covering the vehicle and then find the information needed; locate and identify the following information. If you need to raise the vehicle up, use safety rules for lifting.

   Engine size _____

   Transmission tag # _____

   Manufacturer of the components: _____

7. Find the pages that cover the vehicle you are working on, then find the pages that cover clutches.

Instructor's Response: _____

_____

_____

Name _____ **Date** _____

Clutch Linkage Inspection

Upon completion of this job sheet, you should be able to inspect the clutch linkage and clutch components and determine needed service.

NATEF MAST Task Correlation

Clutch Diagnosis and Repair

Task #2, Inspect clutch pedal linkage, cables, automatic adjuster mechanisms, brackets, bushings, pivots, and springs; perform necessary action.

Task #3, Inspect and replace clutch pressure plate assembly, clutch disc, release (throw-out) bearing and linkage, and pilot bearing/bushing (as applicable).

Task #6, Inspect flywheel and ring gear for wear and cracks; determine necessary action.

This job sheet is related to the ASE Manual Drivetrain and Axles Test's content area: *Clutch Diagnosis and Repair, Task: Inspect, adjust, and replace clutch pedal linkage, cables and automatic adjuster mechanisms, brackets, bushings, pivots, and springs.*

Describe the Vehicle Being Worked On

Year _____ Make _____ VIN _____

Model _____

Procedure

Inspect the following, indicating whether or not inspected parts are serviceable.

	SERVICEABLE	NONSERVICEABLE
Clutch linkage	_____	_____
Clutch cable	_____	_____
Clutch cylinders	_____	_____
Hydraulic systems	_____	_____
Release bearing	_____	_____
Release lever and pivot	_____	_____
Flywheel	_____	_____
Flywheel ring gear	_____	_____
Pilot bearing	_____	_____
Pressure plate and cover assembly	_____	_____
Clutch disc	_____	_____
Power train mounts	_____	_____

Instructor's Response: _____

_____

_____

Name _____ Date _____

CLUTCH IN-CAR SERVICE

Upon completion of this job sheet, you should be able to service and inspect a vehicle's clutch and linkage.

NATEF MAST Task Correlation

General: Drivetrain Diagnosis:

Task #1, Identify and interpret drivetrain concerns; determine necessary action.
Clutch Diagnosis and Repair

Task #2, Inspect clutch pedal linkage, cables, automatic adjuster mechanisms, brackets, bushings, pivots, and springs; perform necessary action.

Task #5, Check and adjust clutch master cylinder fluid level; check for leaks.

This job sheet is related to the ASE Manual Drivetrain and Axles Test's content area: *Clutch Diagnosis and Repair, Tasks:*

    a. *Inspect, adjust, and replace clutch pedal linkage, cables, and automatic adjuster mechanisms, brackets, bushings, pivots, and springs.*

    b. *Inspect, adjust, repair, and replace hydraulic clutch slave and master cylinders, lines, and hoses.*

    c. *Inspect, adjust, and replace release (throw-out) bearing, lever, and pivot.*

Tools and Materials

A vehicle with a standard transmission that you may road-test.

A jack and safety stands, or a hoist for raising the vehicle.

Describe the Vehicle Being Worked On

Year _____ Make _____ VIN _____

Model _____ _____

Procedure

  Task Completed _____

  1. List the problems found in your diagnosis of this clutch.

PROBLEM	POSSIBLE CAUSES
a. _____	a. _____
b. _____	b. _____
c. _____	c. _____

  2. List the parts that need to be replaced.

a. _____	a. _____
b. _____	b. _____
c. _____	c. _____

3. Determine the type of clutch linkage in the automobile.

4. List the specification given by the manufacturer.
   a. Pedal free-play _____
   b. Disengagement (travel) _____

   (If not available from manufacturer, use general specification in text.)
   c. Clearance between cable and lever _____
   d. Clearance between slave cylinder push rod adjustment nut and wedge _____
   e. Torque for adjusting locknut _____

Task Completed

5. Check bolt tightness
   a. Bell housing to block
   b. Transmission to bell housing
   c. Transmission to mount

6. Did you adjust or tighten to all specifications listed? ☐ Yes ☐ No
   If no, why not? _____

7. Road-test results_____
   _____

Instructor's Response: _____
_____
_____

Name _____　Date _____

ROAD CHECK FOR SLIPPAGE

Upon completion of this job sheet, you should be able to road-check a vehicle and determine whether the clutch is slipping.

NATEF MAST Task Correlation

General: Drivetrain Diagnosis

Task #1, Identify and interpret drivetrain concerns; determine necessary action.
Clutch Diagnosis and Repair

Task #1, Diagnose clutch noise, binding, slippage, pulsation, and chatter: determine necessary action.

This job sheet is related to the ASE Manual Drivetrain and Axles Test's content area: *Clutch Diagnosis and Repair, Tasks: Diagnose clutch noise, binding, slippage, pulsation, and chatter problems; determine needed repairs.*

Describe the Vehicle Being Worked On

Year _____ Make _____ VIN _____

Model _____

Procedure

_____
Task Completed

1. Obtain a vehicle that the owner has given permission to road-check.

2. In town, drive the vehicle at 35 mph, then quickly accelerate the vehicle to 45 mph. Did the engine rev up without accelerating the vehicle accelerate?　☐ Yes　☐ No

3. On the road, drive 45–50 mph in high gear, then accelerate to about 60 mph. Again, did you feel the engine rev up?　☐ Yes　☐ No

4. Start and stop a few times. Does the clutch feel as if it is holding when you start and stop?　☐ Yes　☐ No

5. Upon completing the road test, determine the condition of the clutch. Is it slipping?　☐ Yes　☐ No

6. Explain why you believe it may be slipping.

_____

_____

_____

_____

_____

_____

_____

_____

Instructor's Response: _____

_____

_____

Name _____ Date _____

CLUTCH INSPECTION

Upon completion of this job sheet, you should be able to inspect the clutch assembly and determine needed service.

NATEF MAST Task Correlation

Clutch Diagnosis and Repair

Task #3, Inspect and replace clutch pressure plate assembly, clutch disc, release (throw-out) bearing and linkage, and pilot bearing/bushing (as applicable).

Task #6, Inspect flywheel and ring gear for wear and cracks; determine necessary action.

Task #7, Measure flywheel runout and crankshaft end play; determine necessary action.

This job sheet is related to the ASE Manual Drivetrain and Axles Test's content area: *Clutch Diagnosis and Repair, Tasks:*

a. *Inspect and replace clutch pressure plate.*
b. *Inspect and replace clutch disc assembly.*
c. *Inspect, repair, or replace flywheel and/or ring gear.*
d. *Inspect engine block, clutch (bell) housing, and transmission case mating surfaces; determine needed repairs.*
e. *Measure flywheel-to-block runout and crankshaft end play; determine needed repairs.*
f. *Measure clutch (bell) housing bore-to-crankshaft runout and face squareness; determine needed repairs.*

Tools and Materials

Appropriate hand tools, mechanic's basic set

Hoist or jack and jack stands

Transmission jack

Appropriate special tools (listed in vehicle service manual)

Safety glasses with side shield

Service manual for vehicle you are working on

Describe the Vehicle Being Worked On

Year _____ Make _____ VIN _____

Model _____

Procedure

Measure the following, recording both the manufacturer's specification and your own measurements.

	MANUFACTURER'S SPEC.	YOUR MEASUREMENT
Clutch linkage	_____	_____
Flywheel runout	_____	_____
Flywheel warpage	_____	_____
Crankshaft end play	_____	_____
Clutch face thickness	_____	_____
Clutch housing bore runout	_____	_____
Clutch housing to engine block runout	_____	_____
Pressure plate face	_____	_____

Instructor's Response: _____

_____

_____

Chapter 4

SERVICING TRANSMISSIONS/ TRANSAXLES

BASIC TOOLS
Basic mechanic's tool set
Transmission jack
Snapring plier set
Bearing service set
Press
Feeler gauge
Dial indicator

UPON COMPLETION AND REVIEW OF THIS CHAPTER, YOU SHOULD BE ABLE TO:

- Diagnose transmission/transaxle noise, hard shifting, jumping out of gear, and fluid leakage problems; determine needed repairs.

- Inspect and replace transmission/transaxle gaskets, seals, and sealants; inspect sealing surfaces.

- Inspect, adjust, and replace transmission/transaxle shift linkages, brackets, bushings, cables, pivots, and levers.

- Remove and replace transmission/transaxle.

- Disassemble and clean transmission/transaxle components.

- Inspect, adjust, and/or replace transmission/transaxle shift cover, forks, grommets, levers, shafts, sleeves, detent mechanisms, interlocks, and springs.

- Inspect and replace transmission/transaxle input shaft and bearings.

- Inspect and replace transmission/transaxle main shaft, gears, thrust washers, bearings, and retainers.

- Inspect and replace transmission/transaxle synchronizer hub, sleeve, inserts, springs, and blocking rings.

- Inspect and replace transmission counter gear, shaft, bearings, thrust washers, and retainers.

- Inspect, repair, and replace transmission extension housing and transmission/transaxle case, including mating surfaces, bores, bushings, and vents.

- Inspect and replace transmission/transaxle speed sensor drive gear, driven gear, and retainers.

- Inspect transmission/transaxle lubrication devices.

- Measure transmission/transaxle gear and shaft end play/ preload (shim/spacer selection procedure).

- Diagnose transaxle differential case assembly noise and vibration problems; determine needed repairs.

- Remove and replace transaxle differential case assembly.

- Inspect, measure, adjust, and replace transaxle differential pinion gears, shaft, side gears, thrust washers, and case.

- Inspect and replace transaxle differential side bearings.

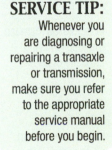

Manual transmissions are strong and typically trouble-free units requiring little maintenance. Normal maintenance may include linkage adjustments and oil changes. However, manual transmissions do break, and this chapter covers the servicing and repair of typical transmissions and transaxles.

Prior to beginning any service or repair work, be sure you know exactly which transmission you are working on. This will ensure that you are following the correct procedures and specifications and are installing the correct parts. Proper identification can be difficult because transmissions cannot be accurately identified by the way they look. The same transmission case may be used to house wide or close ratio gears. The only positive way to identify the exact design of the transmission is by its identification numbers.

SERVICE TIP:
Whenever you are diagnosing or repairing a transaxle or transmission, make sure you refer to the appropriate service manual before you begin.

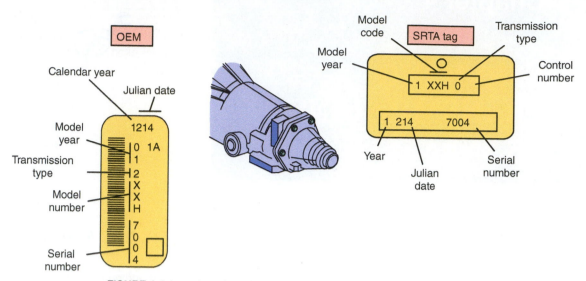

FIGURE 4-1 Location of and information found on GM transaxle ID decals.

Transmission identification numbers are found either as stamped numbers in the case, on a decal attached to the case, or on a metal tag held by a bolt head (Figure 4-1). Use a service manual to decipher the identification number. Most identification numbers include the model, gear ratios, manufacturer, and assembly date. Whenever you work on a transmission with a metal ID tag, make sure the tag is put back on the transmission so that the next technician will be able to properly identify it.

If the transmission does not have an ID tag, it must be identified by comparing it with those in the vehicle's service manual. Some transmissions are identified by the distance, in millimeters, between the centerlines of the main shaft and the countershaft (such as 77 mm).

Transmission and Transaxle Maintenance

Transmissions and transaxles require little maintenance if they are operated properly. Like all gearboxes, transmissions and transaxles require clean lubricant to function properly and to be durable. The moving metal parts must not touch each other and should be continuously separated by a thin film of lubricant to prevent excessive wear. Bearings rely on lubricant to allow for smooth and low-friction rotation of the shafts they support. Lubricant is continuously being wiped away as meshed gears rotate. This action causes friction and high heat, and without a constant supply of lubricant, the gears would rapidly wear. The lubricant also limits the amount of power loss due to friction by reducing the friction in the transmission/transaxle. It also protects the gears and related parts from rust and corrosion, keeps the parts cool, and keeps the internal parts clean.

The type and amount of lubricant varies with the different designs. Most newer transmissions and transaxles use automatic transmission fluid (ATF) or a special oil. OEM manufacturers often specify lubricants that contain special additives to ensure proper synchronizer operation. Other special oils are a mineral oil mixed with an extreme pressure (EP) additive. Many older units used heavy gear oil.

In some cases, EP additives are not recommended because the synchronizers may not be effective with the additive. EP additives reduce friction, which may stop the synchros from working.

Thinner lubricants, such as 30-weight oil and AFT, are more commonly recommended because these lubricants are better at lubricating the bearings inside the constantly meshing gears and they also absorb less engine power, which increases power output and increases fuel economy.

EP additives increase the load-carrying capacity of an oil.

The higher the numerical classification of an oil, the thicker the oil is.

⚠️
CAUTION:
Using incorrect oil will cause transmission failure.

DIAGNOSTICS

A transmission receives power from the engine through the clutch (Figure 4-2). Before beginning to rebuild or repair a transmission, make sure it has a problem and requires repairs. Many problems that seem to be transmission problems are actually clutch problems. Common clutch problems include slipping, grabbing, difficult gear selection, or operating noises. Some of these problems may be caused by either the clutch or the transmission. Gathering as much information as you can to describe the problem will help you decide whether it is a clutch malfunction or a problem in the transmission.

Visual Inspection

Begin your diagnostics by conducting a quick and careful visual inspection. Check all drivetrain parts for looseness and leaks. Carefully check the transmission/transaxle housing for gasket or porosity leaks that may show up as seepage of lubricant. Then, from the driver's seat, operate the clutch with the engine off. Check the free play and smoothness of the clutch. When working the clutch pedal, listen for noises. Carefully identify the source of any noise. If the noise is from the linkage, repair the linkage as is necessary and readjust the free play to specifications.

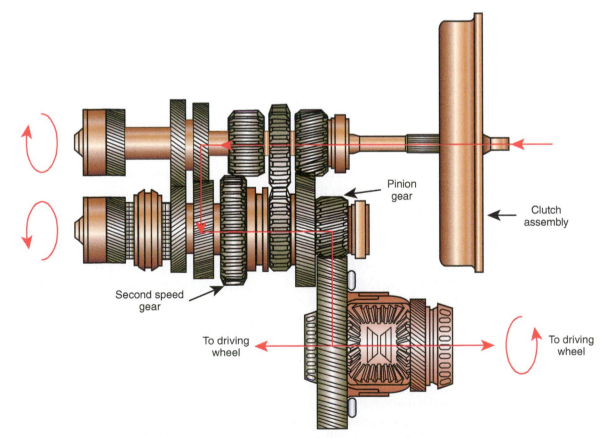

FIGURE 4-2 A transmission receives engine power through the clutch assembly and sends it out to the drive wheels.

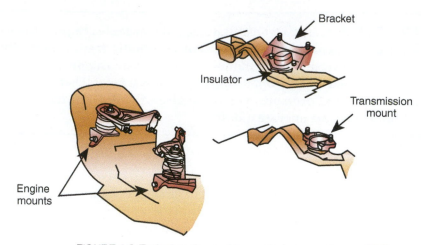

FIGURE 4-3 **Typical engine and transmission mounts on a RWD car.**

If the noise seems to come from the clutch, begin your diagnosis with a check for broken or worn drivetrain mounts (Figure 4-3). If the mounts are good, troubleshoot the clutch as described in the previous chapter.

The engine mounts on FWD cars are important to the operation of the clutch and transaxle (Figure 4-4). Any engine movement may change the effective length of the shift and clutch control cables and therefore may affect the engagement of the clutch or gears. A clutch may slip due to clutch linkage changes as the engine pivots on its mounts. A broken clutch cable may be caused by worn mounts and improper cable routing. Inspect all clutch and transaxle linkages and cables for kinks or stretching. Often transaxle problems can be corrected by replacing or repairing the clutch or gearshift cables and linkage.

Shift Linkage

Check the shift linkage for smooth gear changes and full travel. If the linkage cannot move enough to fully engage a gear, the transmission/transaxle will jump out of gear while it is under a load. Some FWD cars have experienced the problem of jumping out of second or fourth gear. Two causes have been identified with this problem: the upshift light interferes with the shifter or there are improper shifter-to-shifter boot clearances. Both conditions prevent the transaxle's shift forks from moving enough to fully engage the synchronizer collars to their mating gears. If correcting these problems does not solve the complaint, the cause may be the engine mounts or an internal problem in the transaxle.

Lubricant Leaks

Fluid leaks from the seal of the extension housing can be corrected with the transmission in the car. Often the cause for the leakage is a worn extension housing bushing that supports the sliding yoke. When the drive shaft is installed, the clearance between the sliding yoke and the bushing should be minimal. If the clearance is satisfactory, a new oil seal will correct the leak. If the clearance is excessive, the repair requires that a new seal and a new bushing be installed. If the seal is faulty, the transmission vent should be checked for blockage. If the vent is plugged, the oil will be under high pressure when the transmission is hot, and this pressure can cause seal leakage (Figure 4-5).

An oil leak at the speed sensor (Figure 4-6) can be corrected by replacing the O-ring seal. An oil leak stemming from the mating surfaces of the extension housing and the transmission case may be caused by loose bolts. To correct this problem, tighten the bolts to the specified torque.

A leak from the input shaft seal will require transmission/transaxle removal. Closely inspect the clutch assembly for oil contamination.

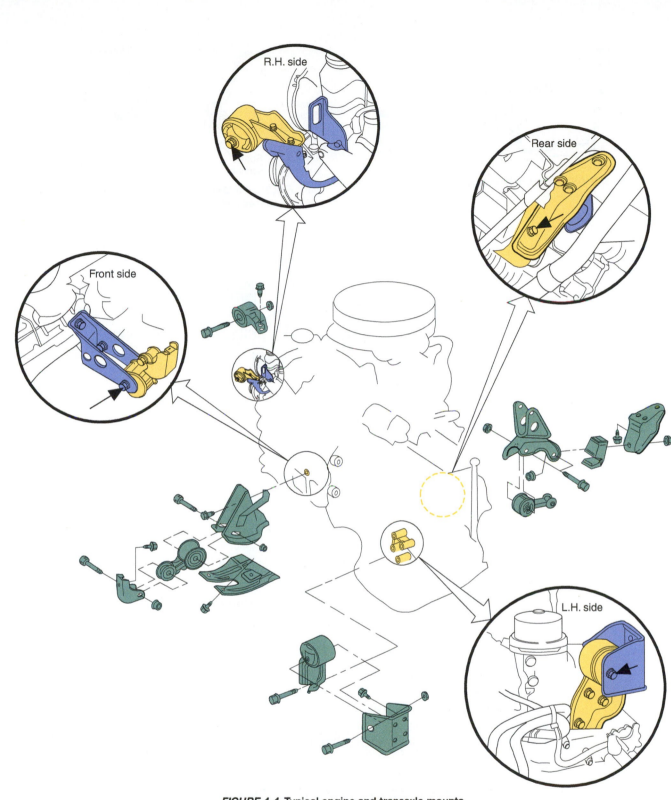

FIGURE 4-4 Typical engine and transaxle mounts.

The entire transmission should be inspected for leaks. The transmission or transaxle is normally comprised of several main castings, each of which is sealed to another (Figure 4-7). In most cases, these gaskets and seals only can be replaced with the transmission on a bench or workstand. Although new gaskets and seals are installed during the reassembly of the unit, it is wise to pay attention to the source of fluid leaks. The mating surfaces or attaching threads could be damaged and the unit will leak even with the new gaskets or specified sealer.

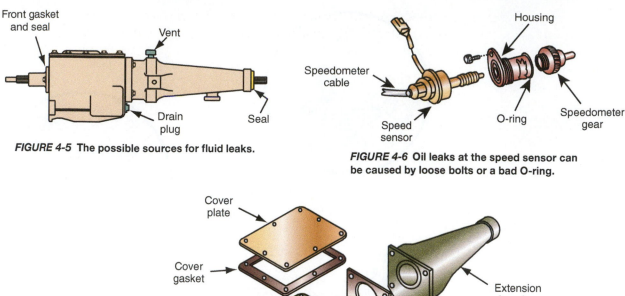

FIGURE 4-5 The possible sources for fluid leaks.

FIGURE 4-6 Oil leaks at the speed sensor can be caused by loose bolts or a bad O-ring.

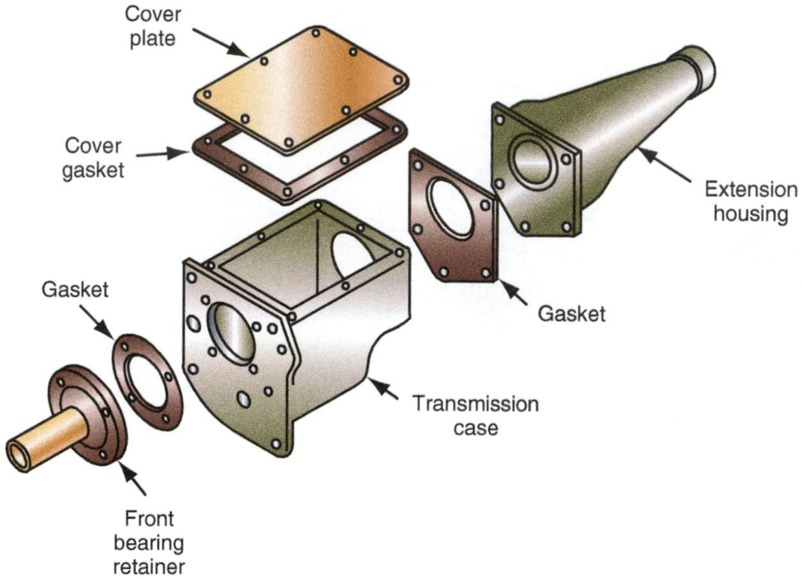

FIGURE 4-7 Various sealing points in a typical transmission.

Normally, the location and cause of a gear oil leak can be quickly identified by a visual inspection. Common causes for fluid leakage are:

1. An excessive amount of lubricant in the transmission or transaxle.
2. The use of the wrong type of gear oil, allowing the oil to foam excessively and leave through the vent.
3. A loose or broken input shaft bearing retainer.
4. A damaged input shaft bearing retainer O-ring or lip seal.
5. Loose or missing case bolts.
6. Case is cracked or has a porosity problem.
7. A leaking shift lever seal.
8. Gaskets or seals are damaged or missing.
9. The drain plug is loose.

TROUBLESHOOTING TRANSMISSION/TRANSAXLE PROBLEMS

All transmission complaints should be verified by test-driving the vehicle and duplicating the complaint. A knowledge of the exact conditions that cause the problem and a thorough understanding of transmissions will allow you to accurately diagnose problems. Many problems that appear to be transmission problems may be caused by problems in the engine, clutch, drive shaft, U- or CV joints, wheel bearings, wheel/tire imbalance, or other conditions.

Make sure these are not the cause of the problem before you begin to diagnose and repair a transmission. Transmission complaints usually concern either a noise or shifting problem. Diagnosis becomes easy if you think about what is happening in the transmission when the

Classroom Manual
Chapter 4, page 97

problem occurs. If there is a shifting problem, think about the parts being moved and what these parts are attempting to do. If the transmission will not engage in second gear but engages into the other gears without effort, the cause of the problem is something unique to second gear. It is very unlikely that a defective clutch would cause a problem only in second gear because the clutch is involved in all gear changes.

Noise

Most transmission noises are caused by gears or bearings (Figure 4-8). When you are diagnosing a noise problem, run the engine in neutral and then shift through all the gears. Pay attention to when the noise gets louder and softer and think about what is happening inside the transmission. Gear noise tends to increase with an increase in speed. Therefore, listen for an increase in noise as each gear is selected. Bearing noise also increases under load and is usually described as a growl that gets louder with speed. By thinking of the power-flow for each selected gear, you can determine the most likely causes of the problem. Remember that transmission components come to a stop when the clutch is disengaged. If the noise stops when you push the clutch in, the cause is most likely to be in the transmission.

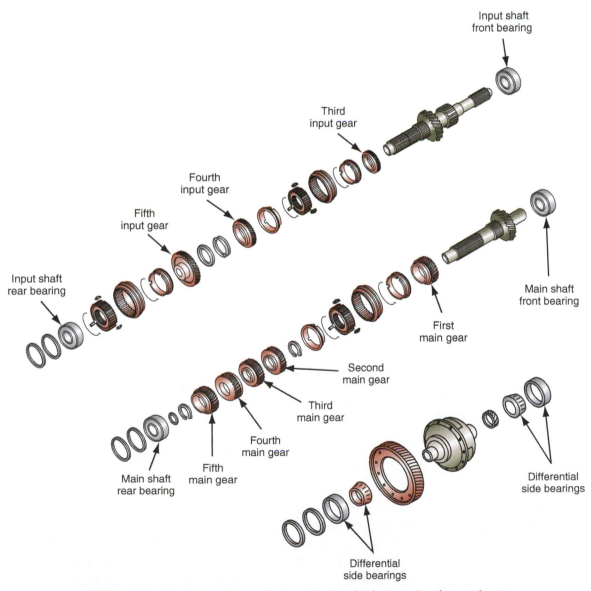

FIGURE 4-8 Exploded internal view of a transaxle showing its many bearings and gears.

Lugging is a term used to describe an operating condition in which the engine is operating at too low an engine speed for the selected gear

Certain terms are normally used to describe transmission and transaxle noises. **Gear rattle** is a repetitive metallic impact or rapping noise that occurs when the vehicle is **lugging** in gear. The intensity of the noise increases with operating temperature and engine torque, and decreases with increasing vehicle speed. Because gear ratios have been designed to achieve maximum fuel economy, there may be instances when gear rattle is distinctly noticeable under lugging conditions. This, however, is not detrimental to the engine or transaxle, provided the appropriate gear is selected for the vehicle speed.

Neutral rollover rattle has the same characteristics as gear rattle except that rollover noise occurs with the engine idling, the transmission or transaxle in neutral, and the clutch engaged. The intensity of rollover noise increases with operating temperature and engine loading, which is caused by the engagement of engine-driven accessories, such as the air conditioning compressor and alternator. Neutral rollover noise is inherent in manual transmissions and transaxles and is not detrimental to the engine or transmission. Neutral gear rattle or rollover noise is most commonly caused by "out of time" engines, or engine harmonic vibrations that cause the gears to rattle while in neutral because of improper clutch damping or worn-out dual-mass flywheels. To diagnose this condition, check the fluid in the transmission. Then, start the engine, place the transmission in neutral, and engage the clutch. Raise the idle slowly in increments to 2,000–2,500 rpm. If the noise decreases with the increase in engine speed, the cause is not the transmission but the engine. However, in vehicles in which the engine idle speed is below specification and/or rough, the rollover noise can increase to a level in which it makes a harsh clattering sound similar to loose marbles in a metal housing. An overhaul of the transmission will not correct this condition. The engine's idle speed must be set or the cause of the improper idle corrected.

> **CUSTOMER CARE:** You should become knowledgeable about the normal noises of a transmission or transaxle. Not only will you avoid wasting time trying to stop a normal noise, but you can explain the cause of these noises to the customer. Make sure you understand why they make certain sounds and be able to explain them to someone who does not share your technical background.

Gear rollover noise, caused by engine torsional vibrations, and clutch throw-out bearing noise are sometimes mistaken for transmission/transaxle bearing noise. Gear rollover noise will disappear when the transmission is engaged in gear; throw-out bearing noise will not. Transaxle repairs will not eliminate gear rollover noise or clutch throw-out bearing noise.

Gear knock is caused by a broken tooth or teeth. The noise may only be heard when the transmission is in that gear and therefore loaded. Broken teeth on the input shaft or counter gear (Figure 4-9) is sometimes mistakenly diagnosed as an engine knock, but will disappear as soon as the clutch is disengaged.

Noise Diagnosis

Diagnosing noise problems can be difficult. However, if you know when the noise occurs, the task becomes easier. Often the cause of the noise is not related to the transmission or transaxle. Hard, imbalanced, or out-of-round tires will cause a noise of vibration whenever the vehicle is moving and seems to worsen with an increase in speed. This is also true of wheel bearing problems. A transmission problem that causes a noise evident only when the vehicle is moving is related to the output shaft or the differential area in a transaxle.

If the noise is evident with the engine running while the vehicle is stopped and also when driving, the most likely problem sources are the engine, transmission or transaxle, or exhaust.

If the noise is evident when the clutch is engaged and the transmission is in neutral, clutch problems or internal transmission problems could be the cause. Probable causes are neutral

SERVICE TIP:
Always refer to your service manual to identify the components of the transmission you are diagnosing. It is also helpful to check for any technical service bulletins that may be related to the customer's complaint.

FIGURE 4-9 This broken front counter gear caused a knocking sound.

rollover rattle, damaged input gear bearings, misalignment of transmission with engine, worn or damaged bearings, low-fluid levels, worn or broken gears, excessive countershaft end play, or loose main shaft gears.

If the noise is only evident when the transmission is in gear, check for worn or broken shaft bearings, a defective engine vibration dampener, worn speed sensor drive gear, insufficient lubricant, worn or damaged input and/or output shaft, loose transmission/transaxle to engine block bolts, worn or damaged gear teeth, and the items listed for noise in neutral.

If the noise only occurs when the transmission/transaxle is in a particular gear, suspect only those components that are engaged in that gear. For example, if the noise is only evident in first gear, suspect damaged or worn first-speed gears or the first gear synchronizer. If the noise is evident in reverse only, suspect a worn or damaged reverse idler gear, shaft or shaft bushing, a damaged synchronizer sleeve that also is the reverse driven gear, or a worn or broken reverse gear. Remember that many transmissions have spur cut reverse gears, and a whine when driving in reverse may be a normal condition. Many racing transmissions use spur gears for every speed, and are very noisy by their nature (Figure 4-10).

Transaxles have unique noises associated with them. These noises can result from problems in the differential or drive axles. More details on the diagnosis of these noises are found in Chapters 5 and 7 of this manual. Use the following as a guide.

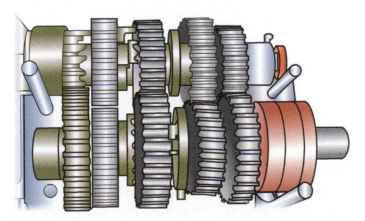

FIGURE 4-10 Spur cut racing gears whine in all ranges.

A knock at low speeds:

- Worn drive axle CV joints
- Worn differential side gear shoulder counterbore

Noise most pronounced on turns:

- Differential gear noise

Clunk on acceleration or deceleration:

- Loose engine mounts
- Worn differential pinion shaft or side gear shoulder counterbore in the case
- Worn or damaged drive axle inboard CV joints or worn axle splines

Clicking noise in turns:

- Worn or damaged outboard CV joint

Classroom Manual
Chapter 3, page 62

DIAGNOSIS BY SYMPTOM

Proper diagnosis involves locating the exact source of the problem. Table 4-1 is a troubleshooting chart for common transmission and transaxle problems. Remember to begin all diagnostics with an interview of the customer. Then verify the customer's complaint or concern. Check the service history of the vehicle and research any relevant technical service bulletins. Also, after repairs are made, make sure you verify the repair.

Many transmission and transaxle problems are related to clutch problems. Either the problem is actually a clutch problem and not a transmission problem, or a clutch problem has caused a transmission problem. For these reasons, the clutch should be thoroughly and carefully inspected and tested prior to suspecting a transmission problem. Transmission and transaxle problems can be grouped into the following basic categories:

1. The unit shifts hard or gears clash while shifting.
2. The unit will not shift into a certain gear.
3. The unit jumps out of gear.
4. The unit is locked in one gear.
5. Excessive vibration is coming from the unit.

The Unit Shifts Hard or Gears Clash While Shifting

If the transmission or transaxle shifts hard or the gears clash while shifting, a common cause of the problem is the clutch. Check the clutch pedal free-travel adjustment if applicable. Make sure the clutch releases completely. Also check for worn clutch parts (Figure 4-11) or a binding input shaft pilot bearing.

Shift linkage problems can also cause this problem. If the shift lever is worn, binding, or out of adjustment, proper engagement of the gears is impossible. A dry (no lubricant) linkage will also cause shifting difficulties. Worn or damaged shift levers, rails, or forks can also cause this problem.

Low-lubricant levels or the use of the incorrect type of gear oil may cause this problem. If the shifting problem mainly occurs when the transmission is cold, the gear oil is too thick. Gear clashing can also be caused by internal transmission or transaxle problems including broken, binding, or worn blocking rings, excessive output shaft end play, weak or damaged detent springs, worn or damaged shift forks, or misaligned housings and/or shafts.

The Unit Will Not Shift into a Certain Gear

If the transmission/transaxle shifts into all gears except one, the problem must be related to a part or system unique to that gear. This could also result from a linkage problem, such as a damaged shift mechanism, interference between the floor shift handle and the console or floor

TABLE 4-1 TRANSMISSION/TRANSAXLE TROUBLESHOOTING CHART

Problem	Possible Cause	Remedy
Gear clash when shifting from one gear to another	1. Clutch adjustment incorrect 2. Clutch linkage or cable binding 3. Clutch housing misaligned 4. Lubricant level low or incorrect lubricant 5. Gearshift components or synchronizer blocking rings worn or damaged	1. Adjust the clutch. 2. Lubricate or repair as necessary. 3. Check runout at the rear face of the clutch housing. Correct runout. Replace as needed. 4. Drain and refill the transmission/transaxle and check for lubricant leaks if the level was low. Repair as necessary. 5. Remove, disassemble, and inspect the transmission/transaxle. Replace worn or damaged components as necessary.
Clicking noise in any one gear range	1. Damaged teeth on input or intermediate shaft gears (transaxles) or damaged teeth on the countergear, cluster gear assembly, or output shaft gears (transmissions)	1. Remove, disassemble, and inspect the unit. Replace worn or damaged components as necessary.
Do not shift into one gear	1. Gearshift internal linkage or shift rail assembly worn, damaged, or incorrectly assembled 2. Shift rail detent plunger worn, spring broken, or plug loose 3. Gearshift lever worn or damaged 4. Synchronizer sleeves or hubs damaged or worn	1. Remove, disassemble, and inspect the transmission/transaxle cover assembly. Repair or replace components as necessary. 2. Tighten the plug or replace worn or damaged components as necessary. 3. Replace the gearshift lever. 4. Remove, disassemble, and inspect the unit. Replace worn or damaged components.
Locked in one gear or cannot be shifted out of that gear	1. Shift rails worn or broken, shifter fork bent, setscrew loose, center detent plug missing or worn 2. Broken gear teeth on countershaft gear input shaft, or reverse idler gear 3. Gearshift lever broken or worn, shift mechanism in cover incorrectly assembled or broken, worn or damaged gear train components	1. Inspect and replace worn or damaged parts. 2. Inspect and replace worn or damaged parts. 3. Disassemble the transmission/transaxle. Replace the damaged parts of the assembly correctly.
Slips out of gear	1. Clutch housing misaligned 2. Gearshift offset lever nylon insert worn or lever attachment nut loose 3. Gearshift mechanisms, shift forks, shift rall rail, detent plugs, springs or shift cover worn or damaged 4. Clutch shaft or roller bearings worn or damaged 5. Gear teeth worn or tapered, synchronizer assemblies worn or damaged, excessive and play caused by worn thrust washers or output shaft gears 6. Pilot bushing worn	1. Check runout at the rear face of the clutch housing. 2. Remove the gearshift lever and check for the loose offset lever nut or worn insert. Repair or replace as necessary. 3. Remove disassemble and inspect the transmission cover assembly. Remove worn or damaged components as necessary. 4. Replace the clutch shaft or roller bearings as necessary. 5. Remove, disassemble, and inspect the transmission/transaxle. Replace worn or damaged components as necessary. 6. Replace the pilot bushing.

(Continued)

TABLE 4-1 TRANSMISSION/TRANSAXLE TROUBLESHOOTING CHART (*Continued*)

Problem	Possible Cause	Remedy
Rough-growing noise when vehicle moving isolated in transmission/transaxle and heard in all gears	1. Intermediate shaft front or rear bearings worn or damaged (transaxle) or output shaft rear bearing worn or damaged (transmission)	1. Remove, disassemble, and inspect the transmission/transaxle. Replace damaged components as necessary.
Rough growing noise when engine operating with transmission/transaxle in neutral	1. Input shaft front or rear bearings worn or damaged (transaxle) or input shaft bearing, countergear, or countershaft bearings worn or damaged (transmission)	1. Remove, disassemble, and inspect the transmission/transaxle. Replace damaged components as necessary.
Vehicle moving–rough growing noise in transmission–noise heard in all gears except direct drive	1. Output shaft pilot roller bearings	1. Remove, disassemble, and inspect the transmission. Replace damaged components as needed.
Transmission/transaxle shifts hard	1. Clutch adjustment incorrect 2. Clutch linkage binding 3. Shift rail binding 4. Internal bind in transmission/transaxle caused by shift forks, selector plates, or synchronizer assemblies 5. Clutch housing misalignment 6. Incorrect lubricant	1. Adjust the clutch. 2. Lubricate or repair as necessary. 3. Check for mispositioned roll pin, loose cover bolts, worn shift roll bores, worn shift rail, distorted oil seal, or extension housing not aligned with the case. Repair as necessary. 4. Remove, disassemble, and inspect the unit. Replace worn or damaged components as necessary. 5. Check runout at the rear of the clutch housing. Correct runout. 6. Drain and refill.

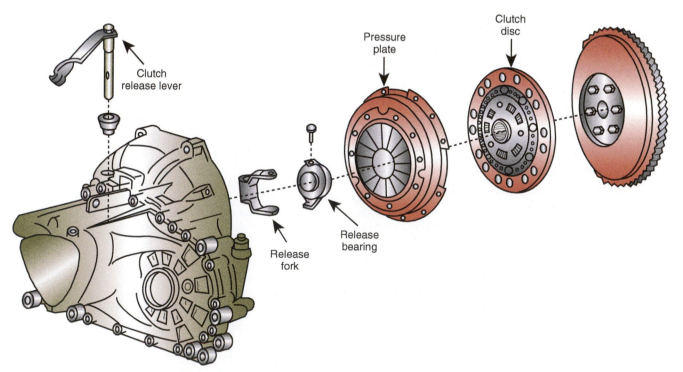

FIGURE 4-11 The typical arrangement of clutch components on a FWD car.

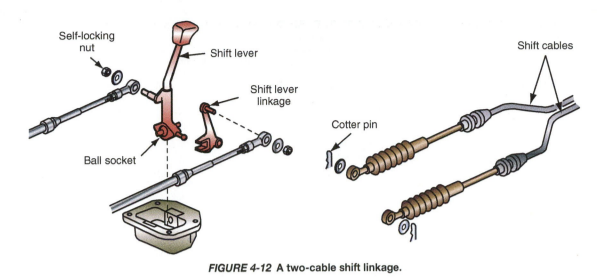

FIGURE 4-12 A two-cable shift linkage.

cutout, or restricted travel of the shift linkage and/or rails due to misalignment. This problem may also be caused by damage to the selected gear or the synchronizer for that gear. Suspect a faulty reverse light switch if the unit will not shift into reverse.

The Unit Jumps out of Gear

If the transmission jumps out of gear, inspect external shift components first. Check the adjustment of the shift linkage or the cable. Make sure the cable assembly is not bent or damaged. Also check to make sure the cables or linkage move freely and do not bind or have excessive movement (Figure 4-12). If the linkage moves too much or seems loose, check for worn pivots, grommets, and levers. Check the gear shifter to make sure there is no interference with it and its boot on the floor or at the transmission. A misaligned shift boot may pull a shift lever out of gear more often when it is cold and stiff. Also check the shifter stabilizer bar for looseness. On models where the shifter goes directly into the transmission, check to see if a plastic isolator at the bottom of the stick is worn or broken.

A misaligned or loose transmission case and/or clutch housing may also be the cause of this problem, especially if a rear-wheel-drive unit is jumping out of direct drive. Anything that causes the input shaft to be out of alignment with the output shaft on a rear-wheel-drive transmission can cause the front synchro ring to slip off the dog teeth of the input gear. Check the mounting of the transmission to the engine block for looseness. Look for anything trapped between the block and transmission. A missing engine dowel or dowels can cause this problem. Check the condition of the front bearing and pilot bearing.

Gear slip-out may also be caused by bent or worn shift forks, levers, or shafts. Other internal problems may cause this condition, such as a bent output shaft, worn or broken synchronizer, worn dog teeth on a speed gear or synchronizer sleeve, excessive input shaft end play, badly worn transmission bearings, or a broken or loose input shaft bearing retainer.

The Unit Is Locked in One Gear

If the transmission or transaxle tends to stick or lock in a gear, low or contaminated gear oil could cause the problem. Clutch release problems may also be the cause of this condition.

Shift linkage problems, both internal and external, will also cause a transmission or transaxle to lock in a gear. Check the adjustment of the linkage. Make sure the linkage or cable is not binding. Misassembled or stuck interlock or detent pins or balls will also prevent the disengagement of a gear (Figure 4-13). So too will binding or damaged synchronizer parts—which prevent the sliding action of the sleeve and hub—such as battered teeth on the sleeves, blocking rings frozen to the gear cones, or burrs on the sliding surfaces.

Classroom Manual
Chapter 4, page 109

Classroom Manual
Chapter 4, page 108

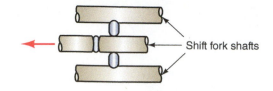

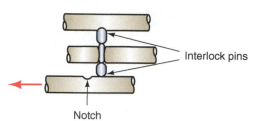

FIGURE 4-13 If the interlock pins or detents are not free to move out of the notches in the shift fork shafts (rails), the transmission will stick in a gear.

Excessive Vibration Is Coming from the Unit

Often the cause of excessive vibration is created by components not directly related to the transmission or transaxle. Worn, scored, or misadjusted wheel bearings, imbalanced or out-of-round tires, damaged or loose axle shafts, worn U- or CV joints, or incorrect drive axle angles will cause a vibration that will seem to increase with an increase in speed. Most transmission or transaxle problems that would result in a vibration will also cause abnormal noises and should be diagnosed by the type and frequency of the noise. Typically, transmission internal components do not have enough mass to cause noticeable vibration. The vibration may come from a faulty crankshaft damper assembly or clutch assembly. A neutral engine run-up test done while using an electronic vibration analyzer (EVA) can help isolate these types of vibration.

IN-CAR SERVICE

Much service and maintenance work can be done to transmissions while they are in the car. Only when an internal repair, complete overhaul, or clutch service is necessary does the transmission need to be removed from the car. Some simple repairs can be made to the transmission assembly while it is still in the car. The following paragraphs summarize procedures for performing common service operations: fluid changes and checks, the replacement of a rear oil seal and bushing, linkage adjustments, and replacement of the back-up light switch and the speed sensor retainer and drive gear.

Fluid Checks

The transmission/transaxle gear oil level should be checked at the intervals specified in the service information. Normally, the intervals range from every 7,500 to 30,000 miles (12,000 to 48,000 km). Fluid-level checks should be made with the vehicle on a level surface and the engine shut off. If the engine has been running, wait 2–3 minutes before checking the gear oil level.

Some transaxles are equipped with a dipstick to check the fluid level (Figure 4-14), while others allow you to only visually check the level. If the transmission/transaxle is equipped with a dipstick, clean the area around the dipstick, remove it, and check the fluid level. Use a funnel and add oil as needed.

On vehicles without a dipstick, raise the vehicle on a lift, and check the oil level through the **fill plug** opening on the side of the unit. The fill plug typically has a square bore in the center designed for a special wrench; however, most of these plugs will loosen with a

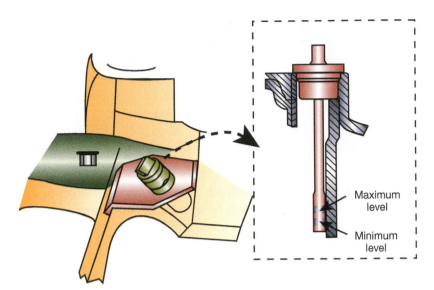

FIGURE 4-14 Example of a dipstick used in manual transaxles.

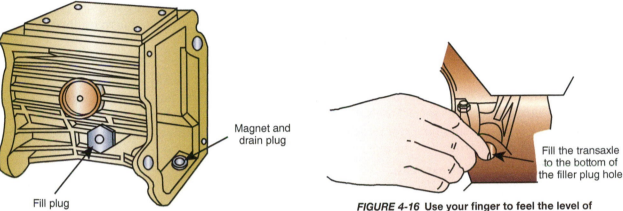

FIGURE 4-15 Typical location of a transmission fill plug.

FIGURE 4-16 Use your finger to feel the level of fluid in the case.

½-inch socket extension and ratchet assembly. Others use a raised square, which can be loosened with an open-end or adjustable wrench (Figure 4-15).

Clean the area around the plug before loosening and removing it. Normally, lubricant should be level with, or not more than, ½ inch (12.7 mm) below the fill hole (Figure 4-16). Always check the service information for the proper level. If the oil is low, add the recommended lubricant to bring it to its proper level. Add the proper grade lubricant as needed using a filler pump.

Fluid Changes

The manufacturers of most modern transmissions do not recommend that the oil be changed at any scheduled time unless there is a need for oil removal, such as during a transmission overhaul. Older transmissions typically had 20,000-mile oil change intervals. When a car has been operated under severe conditions, such as in high heat or dusty road conditions, the fluid may need to be periodically changed. Check the service manual for the manufacturer's recommendations.

To change the transmission fluid, raise the car and safely support it on jack stands. Locate the oil drain plug in the bottom of the transmission case or extension housing. Make sure the car is level so that all of the fluid can drain out. Remove the drain plug with a catch pan positioned below the hole, and let the oil drain into the pan. Let the transmission drain completely—the fluid may be thick and it takes some time to drain out.

SPECIAL TOOLS

Slide hammer
Seal remover tool (Figure 4-17)
Bushing removing tool
Seal driver
Bushing driver
Seal installer (Figure 4-18)

Classroom Manual
Chapter 4, page 97

Inspect the drained oil for gold-color metallic and other particles. The gold-color particles come from the brass blocking rings of the synchronizers. Metal shavings are typically from the wearing of gears or bearings. After the oil has drained out, take a small magnet and insert it into the drain hole, then sweep it around the inside to remove all metal particles. Because brass is not magnetic, it will not show on the magnet. Some metal particles on the magnet may come from normal wear of the reverse idler or reverse driven gear, which must slide into each other when being engaged. An excess of metal particles indicates severe wear in the transmission and the unit is undoubtedly in need of an overhaul.

⚠️ **WARNING:** **Before opening the drain plug of a transmission or transaxle, allow the assembly to cool. Gear oil will tend to adhere to your skin. If the oil is hot, the sticking oil can cause severe burns.**

Before refilling the transmission, reinstall the drain plug. Remove the filler plug, which is normally located above the drain plug. Check your service manual to identify the location of the fill plug and the proper grade and quantity of oil for that transmission. Fill the transmission case until the oil just starts to run out the filler hole or until it is at the bottom of the filler plug hole. Reinstall the plug. You should check the case's vent to make sure it is not blocked with dirt. If the case is not properly vented, the fluid can easily break down due to trapped moisture.

Rear Oil Seal and Bushing Replacement

There are many reasons why the rear oil seal and/or bushing may need to be replaced. The common reason is fluid leakage. On most transmissions, the procedure is quite straightforward but requires special tools to remove and install the seal and bushing. The procedures for the replacement of the rear oil seal and bushing vary little with each car model. The following is a general procedure for replacing oil seals and bushings.

To replace the rear seal:

1. Mark the position of and remove the drive shaft.
2. Remove the old seal from the extension housing.
3. Lubricate the lip of the seal, then install the new seal in the extension housing (Figure 4-19).
4. Install the drive shaft in its original position.

To replace the rear bushing and seal:

1. Mark the position of and remove the drive shaft from the car. Remove the rear seal.
2. Insert the appropriate puller tool into the extension housing until it grips the front side of the bushing.

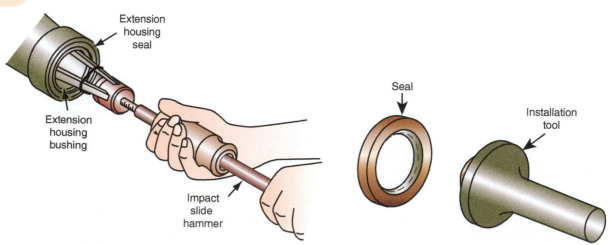

FIGURE 4-17 **A typical seal remover.**

FIGURE 4-18 **A typical seal installer.**

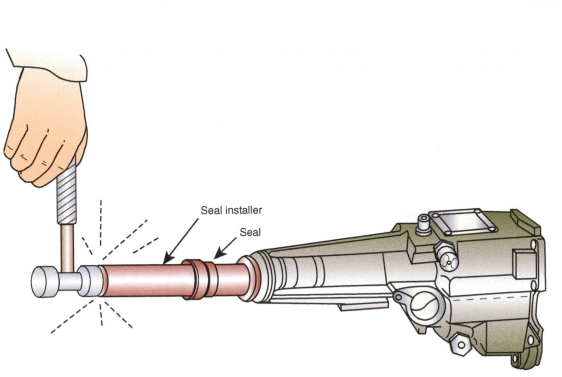

FIGURE 4-19 Drive the new seal in place with a hammer and seal driver.

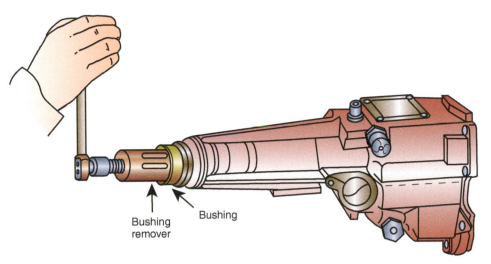

FIGURE 4-20 Removing the extension housing's bushing with a special tool.

3. Pull the bushing from the housing (Figure 4-20).
4. Drive a new bushing into the extension housing.
5. Install a new seal in the housing.
6. Install the drive shaft in its original position.

Linkage Adjustment

Transmissions with internal linkage have no provision for adjustments. However, many external linkages can be adjusted. Linkages are adjusted at the factory but worn parts may make adjustments necessary. Also after a transmission has been disassembled, the shift lever and other controls may need adjustment.

⚠️ **WARNING:** Because you must work under the car to adjust most shift linkages, make sure you properly raise and support the car before working under it. Also wear safety glasses or goggles while working under the car.

Classroom Manual
Chapter 4, page 107

SPECIAL TOOLS
Various-sized rods or drill bits
Trouble light
Torque wrench

Only externally controlled gearshift levers and linkages can be adjusted. To begin the adjustment procedure, raise the car and support it on jack stands. Then follow the procedure given in your service manual. The following is a typical adjustment procedure for an externally mounted shifter:

1. Place the shift lever in the neutral position.
2. Insert a ¼-inch rod into the alignment hole in the shifter assembly. If the rod will not enter, check for bent shift rods. If the shift rods are straight, check the lever locknuts at the rod ends for looseness.
3. Reset the linkage by loosening the rod-retaining locknuts and moving the levers until the ¼-inch rod can enter the alignment holes (Figure 4-21).
4. Double-check the position of the shift lever to ensure that it is in neutral. Ensure the transmission is in neutral by rotating the drive shaft.
5. Tighten the locknuts and operate the shift levers to make sure the detents are engaging.
6. Lower the car and check for smooth shifter operation.

Rod-type linkages are adjusted in much the same way as the lever type. After the alignment pin is in place, the shift rod is adjusted to allow free removal of the pin (Figure 4-22).

Shift cables are normally adjusted by inserting a particular-sized pin into alignment holes in the control assembly when the transaxle is in the specified gear position. These pins keep the transaxle in position while the cables are adjusted. Both cables are adjusted to remove any slack while the transaxle is in gear. After the slack is adjusted out, remove the alignment pins and test the adjustment. The shifter should feel solid with definite gear positions.

Speed Sensor Service

Begin to remove the speed sensor retainer and drive gear by cleaning off the top of the retainer. Then remove the hold-down screw that keeps the retainer in its bore. Carefully pull up on the sensor, pulling the retainer and drive gear assembly from its bore. Unplug the speed sensor wires or unscrew the speedometer cable, if so equipped, from the retainer.

To reinstall the retainer, lightly grease the O-ring on the retainer and gently tap the retainer and gear assembly into its bore while aligning the groove in the retainer with the screw hole in the side of the clutch housing case. Install the hold-down screw and tighten it in place.

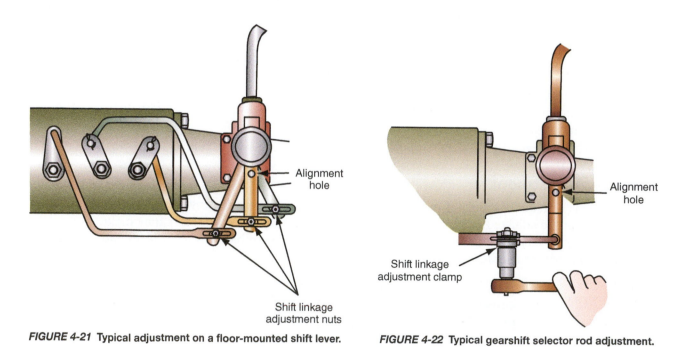

FIGURE 4-21 Typical adjustment on a floor-mounted shift lever.

FIGURE 4-22 Typical gearshift selector rod adjustment.

Back-Up Light Switch Service

Often the back-up or reverse light switch is the source of a fluid leak. To correct this problem may require tightening of the unit or replacement of the switch. To replace the switch, disconnect the electrical lead to the switch. Put the transmission in reverse and remove the switch. Do not shift the transmission until the new switch has been installed. To prevent leaks, wrap the threads of the new switch with Teflon tape in a clockwise direction before installing it. Tighten the switch to the correct torque and reconnect the electrical wire to it.

TRANSMISSION/TRANSAXLE REMOVAL

Removing the transmission from a rear-wheel-drive vehicle is generally more straightforward than removing one from a front-wheel-drive model. Transmissions in FWD cars, because of their limited space, can be more difficult to remove as you may need to disassemble or remove large assemblies such as engine cradles, suspension components, brake components, splash shields, or other pieces that would not usually affect RWD transmission removal.

Once the transmission/transaxle is removed, mark the clutch and flywheel. Disassemble and inspect the clutch assembly. Check the flywheel for scoring, wear, and heat damage. Check the surface of the flywheel for taper wear with a straightedge and a feeler gauge.

RWD Vehicles

The following is a list of components that are typically removed or disconnected while removing a transmission from a RWD vehicle. This list is typical and is arranged in a suggested order of events. Some vehicles will require more than this, whereas others will require less.

- Battery ground cable
- Electrical connectors to sensors and switches
- Gearshift linkage
- Starter motor
- Exhaust heat shields
- Exhaust pipes and catalytic converters
- Drive shaft
- Transmission mounts
- Cross member
- Bell housing to engine bolts

The exact procedure for removing a transmission will vary with each year, make, and model of vehicle; always refer to the service information. Normally the procedure begins with placing the vehicle on a hoist so you can easily work under the vehicle and under the hood.

Once the vehicle is in position, disconnect the negative battery cable and place it away from the battery. Check under the hood, carefully, to find and remove anything that may interfere with transmission removal.

FIGURE 4-23 Typical engine support fixture.

CAUTION:
Make sure you are wearing eye protection while working under a vehicle. Rust and dirt can cause serious damage to your eyes.

SPECIAL TOOLS
Trouble light
Ball-peen hammer
Service information
Snapring pliers
Drive axle puller
Transmission jack
Engine support fixture (Figure 4-23)
Pry bar

Then raise the vehicle and disconnect the parts of the exhaust system that may get in the way. Disconnect all electrical connections to the transmission; mark each connector as to where it was connected. Make sure you place these away from the transmission so they are not damaged during transmission removal or installation.

AUTHOR'S NOTE: On some vehicles, it may be necessary to remove the catalytic converter mounting bracket to have room to work. Doing this will save you time and frustration.

Place a drain pan under the transmission and drain the fluid. Once the fluid is out, move the drain pan to the rear of the transmission. Before removing the drive shaft, use chalk to mark the alignment of the rear U-joint and the pinion flange (Figure 4-24). Then remove the drive shaft. Disconnect and remove the shift linkage.

Place a transmission jack under the transmission (Figure 4-25) and secure the transmission to it. Using a chain or other holding fixture, secure the transmission to the jack's pad. Two chains in an "X" pattern around the transmission work well to secure the transmission to the jack. If the transmission begins to slip on the jack while you are removing it, never try to catch it. Let it fall.

Loosen and remove the lower bell housing-to-engine bolts. Then, remove the cross member at the transmission. After the mount is free, lower the transmission slightly so you can access the top transmission-to-engine bolts. Loosen and remove these bolts.

Slowly and carefully move the transmission away from the engine until the input shaft is out of the clutch assembly. Never let the weight of the transmission hang on the clutch disc as the transmission is moved back. Then slowly lower the transmission. Once the transmission is out of the vehicle, carefully move it to the work area and mount it to a stand or bench.

FIGURE 4-25 A typical transmission/transaxle jack.

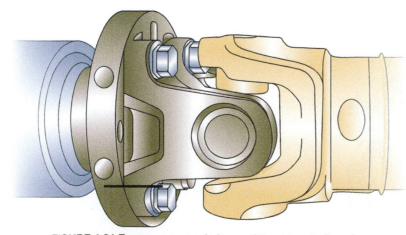

FIGURE 4-24 To ensure proper balance of the drive shaft, make alignment marks on the rear flange and the rear yoke.

On some cars, the engine and transmission must be removed as a unit. The assembly is lifted out of the engine compartment with an engine hoist or lowered underneath the car.

FWD Vehicles

The procedure for removing a transmission from a FWD vehicle is different. The exact items that need to be removed or disconnected will vary from vehicle to vehicle. The following is a list of components that are typically removed or disconnected:

- Battery ground cable
- Underhood electrical connectors to transaxle
- Front wheels
- Electrical connectors at the wheel brake units
- Brake calipers
- Drive axles
- Transaxle to engine brackets
- Electrical connectors to sensors and switches
- Gearshift linkage
- Starter motor
- Exhaust heat shields
- Exhaust pipes and catalytic converters
- Subframe or cross member
- Transaxle to engine mount bolts

On some vehicles, the recommended procedure may include removing the engine with the transaxle. Always refer to the service information before proceeding. You will waste much time and energy if you don't check the information first.

Before removing a transaxle, identify any special tools and precautions that are recommended by the manufacturer. Begin removal by placing the vehicle on a lift. Disconnect the battery before loosening any other components. This is a safety-related precaution to help avoid any electrical surprises. It is also possible to send voltage spikes, which may kill the PCM, if wiring is disconnected while the battery is still connected.

Then, disconnect all electrical connectors to the transaxle. Working under the hood, disconnect the shift linkage or cables and the clutch linkage. Locate the transaxle-to-engine bolts that cannot be removed from under the vehicle and remove them. Install the engine support fixture. Disconnect and remove all items that will interfere with the removal of the transaxle. Drain the fluid from the transaxle.

The drive shafts must be removed next. Loosen the large nut that retains the outer CV joint, which is splined shaft to the hub. It is recommended that this nut be loosened with the vehicle on the floor and the brakes applied.

AUTHOR'S NOTE: Many technicians use their cell phone cameras to help recall the locations of underhood items by taking pictures before work is started. This technique can be quite valuable, considering how complex the underhood systems of current cars have become.

Now raise the vehicle and remove the front wheels. Tap the splined CV joint shaft with a soft-faced hammer. Most will come loose with a few taps. Some FWD cars use an interference fit spline at the hub and require a special puller. The tool pushes the shaft out, and on installation pulls the shaft back into the hub.

The inner CV joint, on some cars, have a flange-type mounting. These must be unbolted to remove the shafts. In some cases, flange-mounted drive shafts may be left attached to the wheel and hub assembly and only unbolted at the transmission. The free end of the shafts should be supported and placed out of the way.

The lower ball joint is now separated from the steering knuckle. The ball joint is either bolted to the lower control arm or held into the knuckle with a pinch bolt. Once the ball joint is loose, the control arm can be pulled down and the knuckle pushed outward to allow the splined CV joint shaft to slide out of the hub. A large pry bar or special control arm tool will bring the control arm down enough to clear the knuckle (Figure 4-26). The inboard joint can then be either pulled or pried out. Some transaxles have retaining clips that must be removed before the inner joint can be removed. A slide hammer with a hook end can be used to remove stubborn inner CV joints (Figure 4-27). Remove the drive axles from the transaxle.

While removing the axles, make sure the brake lines and hoses are not stressed. Suspend the brake calipers with wire to relieve the weight on the hoses and to keep them out of the way.

Now, the shift linkages, electrical connections, and speed sensor wires should be disconnected. The exhaust system may also need to be lowered or partially removed.

Now remove the starter. The starter wiring may be left connected or you can completely remove the starter from the vehicle to get it totally out of the way.

With the transmission jack supporting the transaxle, remove the transaxle mounts. If the car has an engine cradle that will separate, remove the half that allows for transaxle removal.

Then remove all remaining transaxle-to-engine bolts. Slide the transaxle away from the engine. It may be necessary to use a pry bar to separate the two units. Make sure the input shaft is out of the clutch assembly before lowering the transmission.

CAUTION:
The starter should never be left to hang by its wires. The weight of the starter can damage the wires or, worse, break them and allow the motor to fall, possibly on you or someone else. Always support the starter and position it out of the way.

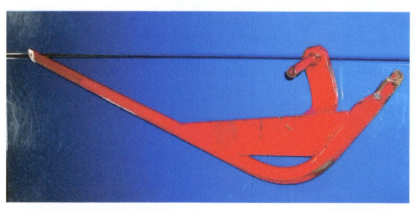

FIGURE 4-26 This specialty tool will safely remove the lower control arm from the steering knuckle.

FIGURE 4-27 A slide hammer attachment to pull the inner CV joint from the transaxle.

DISASSEMBLING A TRANSMISSION

Clean the transmission/transaxle with a steam cleaner, degreaser, or cleaning solvent. During disassembly, pay close attention to the condition of all parts. Plan to replace all worn or damaged parts.

Using a dial indicator, measure and record the end play of the input and main shafts. This will be needed during reassembly for the selection of selective shims and washers.

A brief description of the tear down of a typical five-speed transmission follows; also, Photo Sequence 5 guides you through the disassembly of a typical transmission. Always follow the procedures prescribed by the manufacturer for the transmission being serviced.

Disassembly begins with draining the fluid (if not already drained) from the transmission. Then the clutch or bell housing is removed (Figure 4-28).

With the transmission on a suitable stand or work bench, place the transmission in neutral. Then remove the bolts that attach the shift-lever cover. Lift the cover and shifter out of the extension housing.

Through the shifter housing, remove the retaining pin for the shift rail and linkage. Remove the speed sensor and its retainer. Also remove anything that would prevent the extension housing (sometimes called the **tail shaft**) from being easily removed. Now remove the extension housing attaching bolts.

Slide the extension housing and gasket off. The seal and bushing should be removed from the extension housing. Then, remove any accessible gears from the main shaft.

Remove the shift interlock assembly. Use a pencil-sized magnet to remove the plugs, springs, and balls from the case. Then remove the transmission cover and shift fork assembly.

Using a dial indicator, measure and record the end play of the input and main shafts (Figure 4-29).

Now remove the bearing retainer, its shim, and seal from the case (Figure 4-30). Pull the input gear from the transmission case. Be careful not to lose the roller bearings, the thrust washers, or the thrust bearings.

Remove the rear bearing cup from the transmission case. Tap out the reverse idler shaft to remove the reverse idler gear, remove the roll pin from the reverse idler gear shaft, and then remove the reverse idler gear shaft and gear. Then lift the main shaft out. Remove the bearings, synchronizers, and gears from the shaft. Before you remove a synchronizer, mark the alignment of the sleeve and hub.

SERVICE TIP:
Before disassembling a transmission, observe the effort it takes to rotate the input shaft through all forward gears and reverse. Extreme effort in any or all gears may indicate an end play or preload problem.

SERVICE TIP:
It is good practice to lay the parts on a clean rag as you remove them and to keep them in order as an aid during reassembly. Also, pay close attention to all of the small springs, balls, interlock pieces, and rollers from the transmission's bearings; these parts can easily become lost.

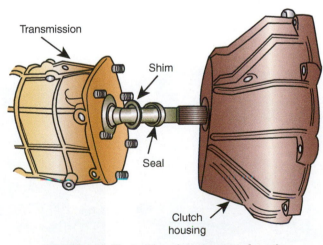

FIGURE 4-28 The bell housing separates from the transmission. Some transmissions use a shim by the seal that helps control end play.

DISASSEMBLY OF A TYPICAL TRANSMISSION

P5-1 Place the transmission in neutral, then remove the bolts that attach the shift-lever cover.

P5-2 Lift the cover and shifter out of the extension housing. As you remove parts, thoroughly inspect them. Replace all worn or damaged parts.

P5-3 Remove the retaining pin for the shift rail and linkage.

P5-4 Remove the extension-housing attaching bolts.

P5-5 Slide the extension housing and gasket off the output shaft.

P5-6 Remove offset lever, roll-pin detent spring and ball from extension-housing detent plate.

P5-7 Remove transmission cover and shift-fork-assembly attaching bolts.

P5-8 Remove cover assembly.

P5-9 Remove the retaining clip for the fifth-reverse arm.

P5-10 Remove fifth-gearshift fork and related parts.

P5-11 Remove the bolts that attach the input-shaft bearing retainer to the transmission case.

P5-12 Remove the bearing retainer from the transmission housing.

P5-13 Remove the bearing retainer, the shim, and the seal from the case.

P5-14 Pull the input gear from the transmission case. Do not lose the roller bearings, the thrust washers, or the thrust bearing.

P5-15 Lift the output-shaft assembly from the case. Remove the countershaft rear bearing cup from the case.

P5-16 Remove the bearing retainer, the gasket, the shim, and the front bearing cup from the case and lift the countershaft assembly through the case.

P5-17 Remove the roll pin from the reverse idler gear shaft.

P5-18 Remove the reverse idler gear shaft and gear.

FIGURE 4-29 Use a dial indicator to measure the end play of the shafts before disassembling the unit.

FIGURE 4-30 Unbolt and pull the bearing retainer away from the transmission case. A worn front bearing retainer.

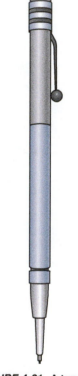

FIGURE 4-31 A tungsten carbide scriber used to index transmission parts.

> **AUTHOR'S NOTE:** Transmission parts are hard . . . really hard. One of the best tools for indexing transmission components is a tungsten carbide scriber, or "pen" (Figure 4-31). Many technicians simply scribe a small "X" on a part to index it with a similarly marked matching component. The technique is also used to index parts from the back or front of the transmission, depending on which way the part is removed from a shaft.

Remove the countershaft bearing retainer, gasket, shim, and cup from the case. Now drive the counter gear's shaft out of the case and lift out the counter gear assembly. Carefully remove the bearings from the countershaft. Inspect the case for any loose small parts. If there is a case magnet, remove it and thoroughly clean it.

DISASSEMBLING A TRANSAXLE

Each model of transaxle has some unique features and one procedure will not work for all models. Much time and frustration can be saved by following the correct procedures. It is especially important to maintain the recommended sequence of steps during disassembly.

Clean the transmission/transaxle with a steam cleaner, degreaser, or cleaning solvent. As the unit is being disassembled, pay close attention to the condition of its parts. It is good practice to lay the parts on a clean rag as you remove them, and to keep them in order to aid you during reassembly. Also, pay close attention to all of the detent springs and balls, interlock components, and rollers from the transaxle's bearings; these parts can easily become lost.

⚠️ **WARNING: Whenever you must press a part on or off with a hydraulic press, make sure you fit a safety collar around the items and wear safety glasses. The parts can easily explode from the pressure exerted on them.**

A brief description of the teardown of a typical five-speed transaxle follows. Also, the disassembly of a typical transaxle is shown in Photo Sequence 6. Always follow the procedures prescribed by the manufacturer.

Place the transaxle onto a suitable work stand. Loosen and remove all transaxle case-to-clutch housing bolts. Separate the clutch housing from the transaxle case. If the housing is difficult to loosen, tap it with a soft mallet. Do not drive a chisel or screwdriver between the case halves, as this could damage the sealing surface. Typically, a gasket is not used to seal these two cases; instead, anaerobic sealant is used.

Now, remove the shift levers, shift interlock assembly, and reverse idler gear. Then remove the shift-lever shaft by gently pulling on it. Some transaxles use a lock tab to secure the bolt that attaches the shift shaft. If a lock tab is used, bend it back before loosening the bolt.

Grasp the input and main shafts and lift them out as an assembly. Then remove the fifth-gear shaft and the fifth-gear fork. Note or photograph the position of the shift forks as an aid when reinstalling them. Then carefully separate the shift rail and forks from the gears on the main shaft.

Lift the differential assembly out of the case and set it aside. Remove the bearing at the end of the main shaft and slide the last gear from the shaft. Now, remove the synchronizer blocking ring from the assembly. This will allow access to the third- and fourth-gear synchronizer retaining ring.

Remove the retaining ring and with a press or puller remove the synchronizer assembly off the shaft. Then remove the remaining gears and synchronizer assembly as a unit.

⚠️ **WARNING: When removing snaprings, take great care not to overstretch the ring to avoid possible personal injury.**

Separate the synchronizer's hub, sleeve, and keys (Figure 4-32), noting their relative positions and scribing their location on the hub and the sleeve prior to separation to make sure it is assembled with the same index. Also make a mark to denote the direction the hub faces the shaft.

Classroom Manual Chapter 4, page 105

DISASSEMBLY OF A TYPICAL TRANSAXLE

P6-1 Remove reverse idler shaft retaining bolt and detent-plunger retaining screw. Then loosen and remove all transaxle case-to-clutch housing attaching bolts.

P6-2 Separate the housing from the case. If the housing is difficult to loosen, tap it with a soft mallet.

P6-3 Remove the "C-clip" retaining ring from the fifth-gear shift-relay-lever pivot pin.

P6-4 Remove the fifth-gear shift-relay lever, reverse idler shaft, and the reverse idler gear from the case.

P6-5 Use a punch to drive the roll pin from the shift-lever shaft.

P6-6 Remove the shift-lever shaft by gently pulling on it.

P6-7 Remove the reverse bias spring assembly.

P6-8 Grasp the input and main shafts and lift them as an assembly from the case. Note the position of the shift forks as an aid when reinstalling them.

P6-9 Remove the fifth-gear shaft assembly and fifth-gear fork assembly.

P6-10 Remove the differential assembly from the case.

P6-11 Remove the bolts for the shift-relay-lever support-bracket assembly, then remove the assembly.

P6-12 Carefully separate the shift rail and forks from the gears on the main shaft.

P6-13 Remove the bearing at the fourth-gear end of the main shaft.

P6-14 Slide the fourth gear from the shaft.

P6-15 Remove the synchronizer blocker ring from the assembly.

P6-16 Remove the third- and fourth-gear synchronizer retaining ring.

P6-17 Lift the assembly off the shaft. Then remove the remaining gears and synchronizer assembly as a unit.

P6-18 Separate the synchronizer's hub, sleeve, and keys, noting their relative positions and scribing their location on the hub and the sleeve prior to separation. Also make a mark to denote the direction the hub faces the shaft.

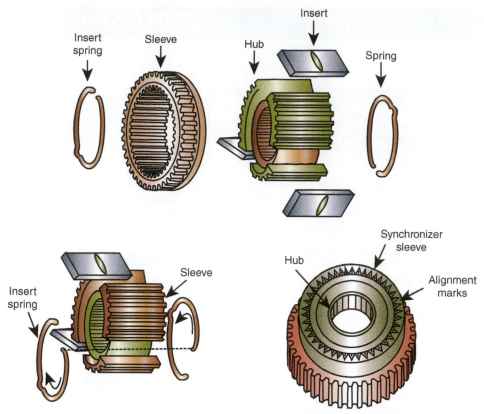

FIGURE 4-32 Synchronizer assembly and location of alignment marks.

INSPECTION OF TRANSMISSION/TRANSAXLE PARTS

The inspection of parts during disassembly should be used to confirm your diagnosis and to identify any and all parts that may be worn or damaged. Your inspection is not complete until you can directly relate what you find to what the customer's complaint was about the transmission.

During and after disassembly, you should conduct a thorough inspection of all of the components of the transmission/transaxle. What follows are some guidelines for the inspection of individual components of a transmission/transaxle. Some defective parts will be obvious, whereas others will require a close inspection to be detected.

Transmission Case

The transmission and clutch housings should be inspected for cracks, worn or damaged bearing bores, damaged threads, or any other damage that could affect the operation of the transmission/transaxle (Figure 4-33). Normally, the housings should be replaced if they are cracked or badly damaged. If any of the threaded holes are stripped out, repair them by using a thread repair kit. Inspect the transmission case and clutch housing mating surfaces for small nicks or burrs that could cause an oil leak or a misalignment of the two halves. These nicks and burrs can be removed with a fine stone or file.

SERVICE TIP:
Care should be taken in assuming a mark is a crack; many casting imperfections look like cracks.

Extension Housing

Check the extension housing for cracks and repair or replace it as needed. Check the mating surfaces of the housing for burrs or gouges and file the surface flat. Check all threaded holes and repair any damaged bores with a thread repair kit. Check the bushing in the rear of the extension housing for excessive wear or damage. Always replace the rear extension housing seal. Remove the bearing from the extension and check it for smooth and quiet rotation. If the bearing is dry, it may sound noisy; if a few drops of clean oil are applied to a good bearing, the noise will cease. Check all bearing cups for excessive wear or damage.

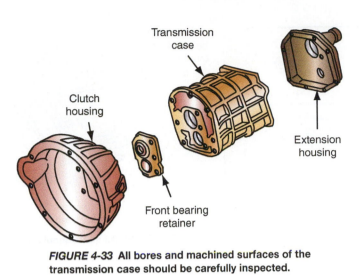

FIGURE 4-33 All bores and machined surfaces of the transmission case should be carefully inspected.

Labels in figure: Transmission case, Clutch housing, Extension housing, Front bearing retainer

Support Plate

Inspect the **center support plate** for cracks. Never attempt to repair this unit; always replace it if it is damaged. Carefully inspect the counter gear bearing for pitting or any other signs of failure, and replace it if it is damaged.

Front Bearing Retainer

Inspect the front bearing retainer for cracks, burrs, or gouges on the case mating surface. Inspect the retainer's snout to see if it is smooth (it should be). Remove any gasket material from the sealing surface and install a new shaft seal before reinstalling the bearing retainer. Make sure the oil return hole in the retainer and the gasket lines up with the oil hole in the case.

Bearings

All bearings should be inspected for signs of **lapping**, **spalling**, or **locking**. Both the inner and outer raceways should be inspected, as should their outer surfaces. Lightly lubricate the bearing and spin the bearing a few times by hand. The bearing should spin evenly and without any roughness. Often roughness is caused by a buildup of metal particles around the ball or roller bearings. Use a magnet to clean the bearing, then apply another thin coat of oil to the bearing and spin it again. If the roughness remains, the bearing should be replaced.

Gears

All gears should show normal wear patterns in the center of their teeth. These wear patterns should appear as a polished finish with little wear on the gear face. Check the gears' teeth carefully for chips, pitting, cracks, and breakage. Often minor damage to the gears' teeth can be filed away. Imperfections can be removed with a file or small high-speed grinder in a process called blend repair (Figure 4-34).

Also inspect the clutch teeth on the speed gears; they should be pointed and not too rounded off (Figure 4-35). If the customer had a complaint about jumping out of gear, carefully inspect the clutch teeth of that gear for back taper wear. Replace the gears if excessive metal is worn from the gear teeth. Inspect the gears' center bores; they should be smooth and free of any imperfections. The cone-shaped area of each gear should be smooth and without nicks, burrs, or gouges that could cause accelerated blocking ring wear or failure. Inspect the cone surface for metal transfer from a failed blocking ring. The surface can be deglazed with emery cloth. Inspect all needle bearings; they should be smooth and shiny.

A **center support plate** is used in some physically large transmissions to help support the weight of the shaft and gears. It is located between the case and the extension housing. It is also called an intermediate bearing support.

Classroom Manual
Chapter 4, page 99

Lapping is the result of small particles of dirt that grind away on the bearings and their races.

Spalling is caused by overloading the bearing and is evidenced by pits on the bearings or their races.

Locking is caused by large particles of dirt that become trapped between a bearing and its race.

1). Grind chip/nick from corner face of gear—drive side. (Approved)	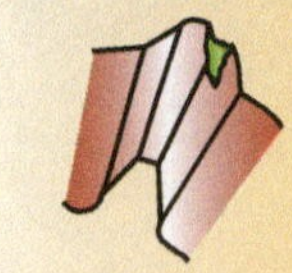
2). Grind chip/nick from edge of gear O.D. — may extend slightly into gear pattern on tooth face. (Approved)	
3). Grind chip/nick from corner face of gear—coast side. (Approved)	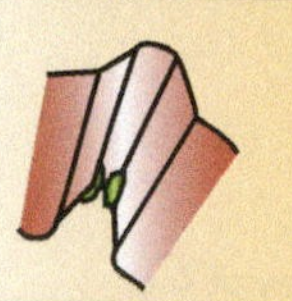
4). Grind chip/nick from gear edge—may extend slightly into gear face. (Approved)	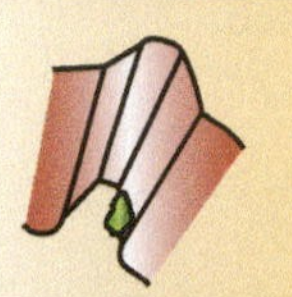
5). Grind chip/nick from gear face—in gear pattern area. (Not approved)	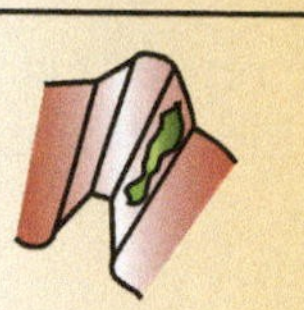

FIGURE 4-34 Samples of gear damage. Those identified as "approved" can be filed down, whereas those marked "not approved" cannot be and the gear should be replaced.

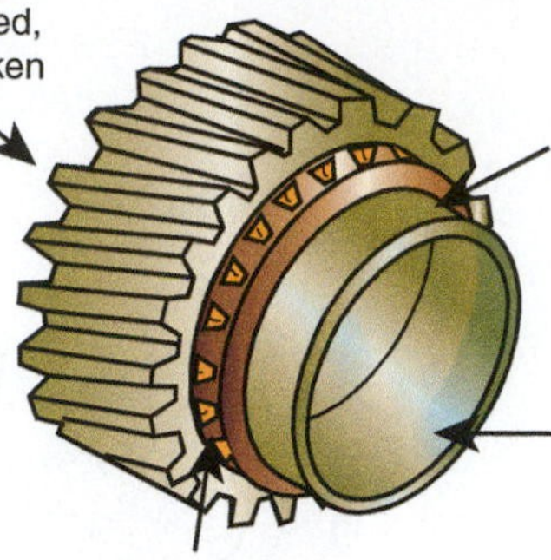

Teeth must not be pitted, cracked, chipped, or broken

Tapered area must be smooth

Bore must be smooth

Inspect clutch teeth

FIGURE 4-35 All gears should be carefully inspected.

AUTHOR'S NOTE: Racing transmission gears face much more severe use than that experienced by road vehicles. Preventing wear and damage requires special manufacturing techniques. Some racing gears go through cryogenic treatment, where the gear is frozen to negative 300° F and brought slowly back up to temperature. The gear becomes tougher as a result of molecular changes in the material. Another process involves tumbling the gear in a special medium for many hours. This process superpolishes the gear surface, reducing friction, heat, and power-robbing resistance.

Interpreting Gear Wear. All new gear teeth have slight imperfections, but these normally disappear during break-in as the teeth are oiled and become polished. The teeth of a normally worn gear will have a polished surface that extends the full length of the tooth from the bottom to the tip of the tooth. Gears that are manufactured properly, well lubricated, and not overloaded or improperly installed will look this way after many hours of service. Common gear conditions and their causes follow.

- Excessive wear or grooves on the teeth are normally caused by fine particles carried in the lubricant or embedded in the tooth surfaces. The usual sources of these fine particles are metal particles from gear teeth, abrasives left in the gear case, or sand and scale from the housing casting.
- Scratching on the surface of the teeth also can be caused by suspended or embedded particles. However, scratching of these surfaces normally occurs when the particles are larger than the fine abrasives that cause excessive wear. These particles normally are metal pieces from the gears themselves, resulting from the gear being subjected to loads for which it was not designed.

- If the contact surface of the teeth is worn smoothly, the gears have been overloaded and metal has been removed by the sliding pressure of the meshing gears. This causes a worn depression along the length of the teeth. Continuous use under this condition will result in backlash and severe peening, which can be misleading as to the real cause of the wear.
- **Rolling** is the result of overload and sliding, which leaves a burr on the tooth edge. Insufficient bearing support results in rolling of the metal because of the sliding pressure.
- **Peening** is the result of backlash, which causes the teeth to hammer on others with tremendous impact so that they look as if they were beaten with a hammer. The impact forces the lubricant out of the teeth so that the gears mesh without a protecting layer of oil, causing heavy metal-to-metal contact.
- A wavy surface or "fish scales" on the teeth at right angles to the direction of the sliding action is referred to as "rippling." This may be caused by a lack of lubrication, heavy loads, or vibration.
- **Scoring** is caused by a temperature rise and thinning or rupture of the lubricant's film because of extreme loads. Pressure and sliding action heats the gear and permits the metal surface to transfer from one tooth to the face of another. As the process continues, chunks of metal loosen and gouge or scratch the teeth in the direction of the sliding motion.
- **Pitting** is a condition normally associated with a thin oil film, possibly because of high oil temperatures. A very small amount of pitting gives the surface of the gear teeth a gray appearance.
- Spalling is a common wear condition that starts with fine surface cracks and eventually results in large flakes or chips coming off the tooth face. Improperly hardened teeth are most often subject to this kind of damage due to the brittle nature of the metal. Spalling may occur on only one or two teeth, but the chips may then damage the other teeth.
- Corrosive wear results in an erosion of the tooth surfaces by acid. The acid is the result of moisture combining with impurities in the lubricant and contaminants in the air. Normally, the surfaces initially become pitted and then become chipped or spalled.
- Evidence of high heat or burning is usually caused by failure of the lubricant or by a lack of lubrication. During high stress and sliding motion, friction causes excessive heat and the temperature limits of the metal are exceeded. Burned gear teeth become extremely brittle and are easily broken.
- A misalignment of the gears that places heavy loads on small portions of a gear will result in interference wear. It also can be the result of meshed gears with teeth that are not designed to work together.
- Ridges are scratches appearing near one end of a tooth, especially on a hypoid pinion gear. These are caused by excessive loads or a lack of lubrication.
- Broken teeth can be the result of many defects: high-impact forces, overloading, fatigue, or defective manufacturing processes. Examine the break carefully. If the break shows fresh metal all over the break, an impact overload was probably the cause. If the break shows an area in the center of fresh metal with the edges dark and old-looking, the breakage was because of fatigue that started with a fine surface crack.
- Cracks are normally caused by improper heat treating during manufacturing or improperly machined tooth dimensions. Most cracks resulting from improper heat treating are extremely fine and will not show up until a gear has been used for some time.

Input Shaft

Inspect the tip that rides in the pilot bearing for smoothness (Figure 4-36). Check the splines of the shaft for any wear that could prevent the clutch disc from sliding evenly and smoothly. Inspect the area of the shaft that the oil seal rides on. This area should be free of all imperfections; a scratch or nick could cause the oil seal to leak. Rotate the bearing; it should rotate smoothly and quietly. Check the gear for cracks, pitting, and chipped or broken teeth. It is

Classroom Manual
Chapter 4, page 98

FIGURE 4-36 The input shaft assembly should be carefully inspected. This input shaft shows damage from a failed pilot bearing.

normal for the front of the gear teeth to show some signs of wear. The dog teeth should be pointed and not cracked or broken. Remove the main shaft pilot roller bearings from the shaft and inspect the bore for burrs or pits. Also check the bearing's rollers for roughness, pitting, or any other signs of failure. All bearings should be examined for signs of wear or overheating. Any damaged component should be replaced. Most transmission rebuilders replace all tapered roller bearings and races when rebuilding a transmission. If the old bearings are good and will be reused, keep the bearing with its race.

> **AUTHOR'S NOTE:** Significant engine performance modifications can result in transmission component failure. The knowledgeable technician will be able to advise the customer what transmission modifications should be made to keep up with increased engine torque. The input shaft is often the first part to fail. An aftermarket billet steel input shaft might be the answer. In making a billet shaft, a steel ingot is heated to near-melting point, then placed under extreme pressure to compact and realign the structure of the metal. This process results in denser and stronger material from which to machine the input shaft.

Reverse Gear

Reverse gear should be inspected like any other gear, except for sliding reverse gears. These gears do not have all the features of forward gears, so inspection is limited to the gear teeth and the center bore and bearings. Since reverse is accomplished by sliding either an idler or reverse driven gear into position, some wear and small chips on the leading edge of the teeth can be acceptable. If reverse gear is noticeably damaged, it should be replaced.

Reverse Idler Gear

Inspect the reverse idler gear for pitted, cracked, nicked, or broken teeth. Check its center bore for a smooth and mar-free surface. Carefully inspect the needle bearing that the idler gear rides on for wear, burrs, and other defects. Also, inspect the reverse idler gear shaft's surface for scoring, wear, and other imperfections (Figure 4-37). Replace any part that is defective or excessively worn.

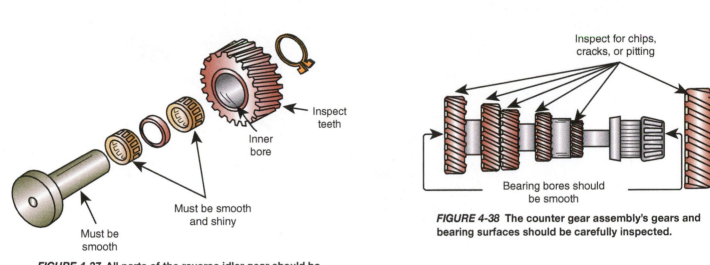

FIGURE 4-38 The counter gear assembly's gears and bearing surfaces should be carefully inspected.

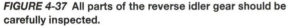

FIGURE 4-37 All parts of the reverse idler gear should be carefully inspected.

Counter Gear

All gears should show normal wear patterns in the center of their teeth. These wear patterns should appear as a polished finish with little wear on the gear face. Check the gears' teeth carefully for chips, pitting, cracks, or breakage (Figure 4-38). Also inspect the bearing surfaces to be sure they are smooth. Any damage to the assembly requires that it be replaced.

Main Shaft

Inspect the bearing surfaces of the main shaft; they should be smooth and show no signs of overheating (Figure 4-39). Also inspect the gear journal areas on the shaft for roughness, scoring, and other defects. Check the shaft's splines for wear, burrs, and other conditions that would interfere with the slip yoke being able to slide smoothly on the splines. Carefully inspect the main shaft pilot where it rides in the bearing set into the back of the input shaft. Finally, check the output bearing for smooth and quiet rotation. Replace any damaged part.

When inspecting the main and other shafts, make sure to check the machined splines. These are critical. If the splines of the shafts, synchronizer hubs, or sleeves are damaged, the synchronizers will not be able to slide into position. Also, check all bearing shoulders for wear or damage (Figure 4-40). If these areas are worn or damaged, the bearings or gears cannot be installed in their correct position on the shaft.

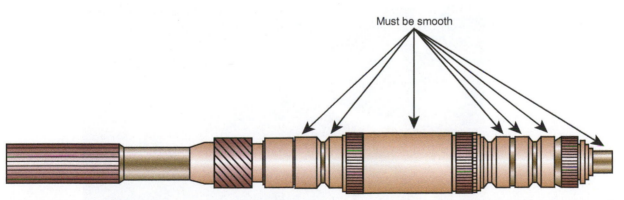

FIGURE 4-39 The bearing surfaces of the main shaft should be inspected after the gears and bearings have been removed.

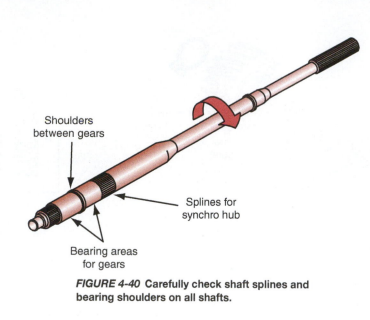

FIGURE 4-40 Carefully check shaft splines and bearing shoulders on all shafts.

Shoulders between gears

Splines for synchro hub

Bearing areas for gears

Classroom Manual

Chapter 4, page 90

Synchronizer Assembly

Because they are relatively inexpensive, blocking rings are often replaced whenever the transmission is overhauled to ensure precise shifting. Inspect the rest of the synchronizer units for smooth operation. Disassemble the units and inspect each part as follows:

1. Carefully inspect the assemblies for worn teeth and damaged cone surfaces. The dog teeth of the synchro sleeve may appear to be slightly rounded; this is normal if it is only slightly rounded and will not interfere with normal operation. However, because any rounding can cause hard shifting, synchronizers with round teeth should be replaced.
2. Inspect the sides of the coupling teeth for wear on the back taper angles of the speed gear dog teeth (Figure 4-41) and the synchro sleeve that can cause gear popout (Figure 4-42).
3. Move the synchronizer sleeves on their hubs and feel for any imperfections that may suggest the sleeve is not able to move freely on the shaft. The sleeve's inner splines should be pointed (Figure 4-43). The sleeve should also be carefully inspected for signs of fatigue, such as cracks or chips.

FIGURE 4-41 Excessive back taper wear on the dog teeth of a speed gear.

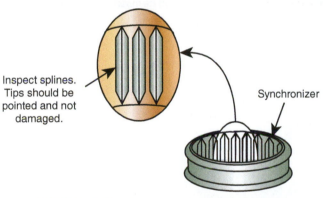

FIGURE 4-42 Excessive back taper wear on a synchronizer sleeve.

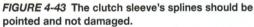

Inspect splines. Tips should be pointed and not damaged.

Synchronizer

FIGURE 4-43 The clutch sleeve's splines should be pointed and not damaged.

4. Carefully clean any fine metal particles from the inside diameters of the sleeve and hub. Centrifugal force will pack metal particles in to these areas, and often they can only be scraped off with a sharp object or stiff wire brush.

5. If there are no alignment marks on the sleeve and hub, make some before disassembling the unit.

6. Examine the position of the insert springs. Also the springs should not be bent, distorted, or broken (Figure 4-44).

7. Inspect the blocking rings for wear marks on the face of the splined end, which may indicate that the ring was bottoming on the gear face because of wear of the blocker ring. The blocking ring also can be checked for wear by placing a feeler gauge between the ring and the speed gear face (Figure 4-45). Compare your measurement to the manufacturer's specifications (Figure 4-46).

8. The rings should be inspected for cracks, breakage, and flatness. To check the flatness of a ring, place it on a flat surface and check how it sits on it. The dog teeth should be pointed and have smooth surfaces for the inner splines of the synchronizer sleeve to ride on. Check the cone; its threads should be well defined and sharp (Figure 4-47).

9. Check for metal contamination in the oil-cutting grooves on the inside of the blocking ring. Any trapped metallic debris will cause the ring to slip on the speed gear cone and cause gear clash.

10. Make sure the hub is not bent, cracked, or has signs of possible failure.

SERVICE TIP:
Many transmissions and transaxles contain press-fit synchronizer hubs that require a puller to remove them from the shaft (Figure 4-48) and a press to install them (Figure 4-49). Other manufacturers may use press-fit synchronizer hubs that should not be removed from the main shaft. The removal of the hub will damage it and the main shaft will need to be replaced. Always check the service manual before removing or disassembling a synchronizer hub.

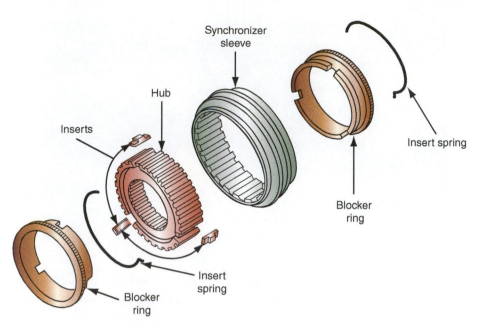

FIGURE 4-44 A synchronizer assembly with its insert springs, which must be placed in their proper position.

FIGURE 4-45 A feeler gauge in position to measure blocking ring clearance.

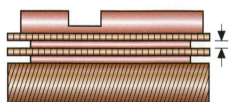

FIGURE 4-46 By measuring the clearance between the blocking ring and the gear, and then comparing this measurement to specifications, you can determine the wear of the ring.

Cone Synchronizers. Many modern synchronizer rings are lined with composite materials. A procedure is often given to measure the reserve of the lining. This is typically done with a feeler gauge.

Clean the cones of the speed gears with solvent and allow them to air dry. Paint the cones with machinist dye and twist the ring onto the cone of the speed gear cone. Check the contact area. If the contact is just at the top or bottom of the cone, gear clash can result. To check for taper problems, do the same thing but with a known good ring. If the contact area is not in the center of the ring, the speed gear cone is distorted and the unit should be replaced.

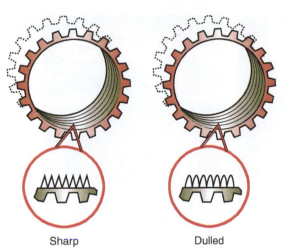

Sharp   Dulled

FIGURE 4-47 The grooves on the internal surfaces of the blocking rings must be sharp and free of debris.

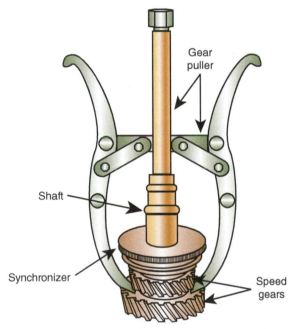

FIGURE 4-48 Using a puller to remove a synchronizer from the shaft.

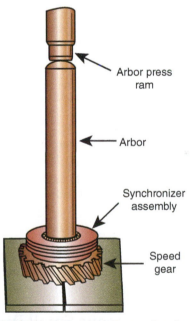

FIGURE 4-49 Pressing the synchronizer hub onto the main shaft.

Shift Forks

The shift forks should be inspected to make sure they are not worn, bent, cracked, or broken (Figure 4-50). If an iron or steel shift fork has ground into the synchro sleeve, most manufacturers recommend the sleeve be replaced as well. If equipped, examine plastic shift fork pads for wear. The pads may be available separately from the fork assembly if they need to be replaced. Slide the forks over the shift rail; they should slide easily but without much play or wobble. Fit each shift fork into its synchronizer sleeve. Check the clearance between the fork and sleeve with a feeler gauge (Figure 4-51). Compare your measurement against specifications.

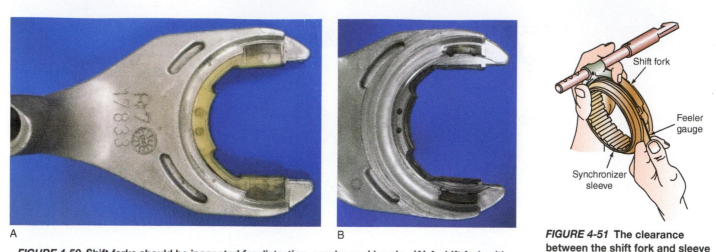

A   B

FIGURE 4-50 Shift forks should be inspected for distortion, cracks, and breaks. (A) A shift fork with plastic tips showing minimal wear. (B) A severely worn shift fork.

FIGURE 4-51 The clearance between the shift fork and sleeve should be checked with a feeler gauge.

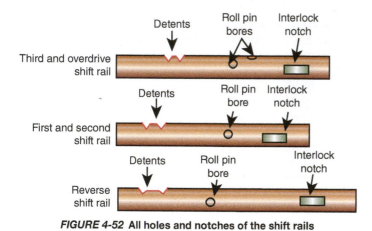

FIGURE 4-52 All holes and notches of the shift rails should be inspected for signs of wear and damage.

FIGURE 4-53 All thrust bearings, races, and washers should be carefully inspected.

Shift Rails

Inspect the shift rails to be sure they are not worn, bent, cracked, or broken (Figure 4-52). Place the rails into their bores in the case or extension housing and check the bore wear by noting the movement of the rail in the bores.

Miscellaneous Small Parts

Inspect all small parts in the transmission for wear. Your service manual may list specifications for the thickness of parts, such as the thrust washers (Figure 4-53). If specifications are available, measure the parts and compare them to the specifications. If specifications are not available, inspect each part for signs of wear or breakage. Some scoring is normal on most thrust washers; however, if the scoring has reduced the thickness of the washer, it should be replaced. Replace any part that is worn or defective. Normally, all the snaprings, roller bearings, washers, and spacers are replaced during a transmission overhaul. Most manufacturers sell a small parts kit that includes all of these parts.

SERVICING A TRANSAXLE'S DIFFERENTIAL ASSEMBLY

Although part of the transaxle, the differential is often kept together while servicing the transmission part of the transaxle. The differential case is normally removed when the transaxle case is separated. It may be the source of a problem and be the only part of the transaxle that needs service.

The following are the basic steps for disassembling and reassembling a differential (Figure 4-54).

1. Separate the ring gear from the differential case if replacement is necessary. Some ring gears are retained by rivets that must be drilled out to remove the gear. Replacement of the unit may be required.
2. Remove the pinion shaft lock bolt.
3. Remove the pinion shaft, then remove the gears and thrust washers.
4. If the differential side bearings are to be replaced, use a puller to remove the bearings. Use the correct installer for reinstallation of the side bearings.
5. Clean and inspect all parts. Replace parts as required.
6. Install the gears and thrust washers into the case and install the pinion shaft and lock bolt. Tighten the bolt to the specified torque.
7. If removed, attach the ring gear to the differential case.

Shim Selection

Keep side bearing shims and bearing races together and identified for reinstallation in their original location. Carefully inspect the bearings for wear and/or damage and determine if a bearing should be replaced. Some replacement tapered roller bearings are available with a nominal-thickness service shim.

When it is necessary to replace a bearing, race, or housing, refer to the manufacturer's recommendation for nominal shim thickness. If only other parts of the differential are

> ⚠️ **CAUTION:** Be careful when unbolting the ring gear. Some transaxles use left-handed bolts to retain the gear.

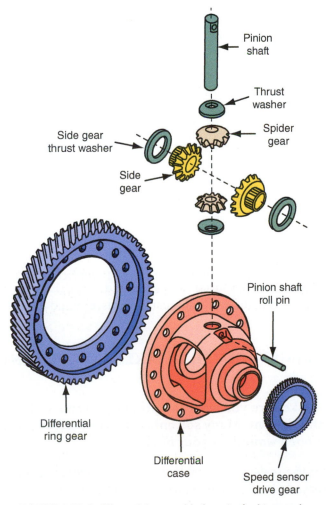

Pinion shaft

Thrust washer

Side gear thrust washer

Spider gear

Side gear

Pinion shaft roll pin

Differential ring gear

Differential case

Speed sensor drive gear

FIGURE 4-54 A differential assembly for a typical transaxle.

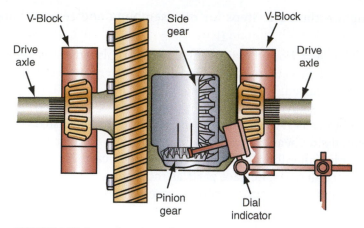

FIGURE 4-55 Setup for measuring the backlash of the pinion gears in a transaxle's differential.

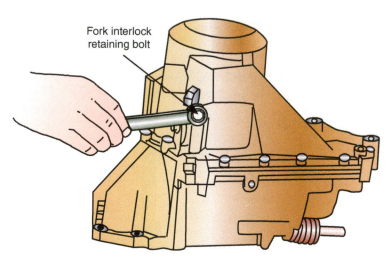

FIGURE 4-56 Location of the shift fork interlock retaining pin.

replaced, reuse the original shims. When repairs require the use of a service shim, discard the original shim. Never use the original shim together with the service shim. Details of the procedures for checking and adjusting bearing preload, as well as other differential services, are given in Chapter 7 of this manual. A common procedure for checking a FWD differential involves mounting the differential, with its drive axles installed, on V-blocks. The backlash of the pinion gears is measured with a dial indicator (Figure 4-55). If the backlash is not within specifications, the differential must be disassembled and the correct size thrust washers installed.

CLEANING TRANSMISSION/TRANSAXLE COMPONENTS

WARNING: Make sure you wear the proper protective clothing when cleaning the parts with a solvent. Many solvents can cause serious injury if they get into your eyes, an open wound, or your mouth.

Wash all parts, except the seals, in a suitable degreaser or cleaning solvent. Brush or scrape all heavy dirt and grease from the parts. When doing this, be careful not to mar the mating surfaces of any part. After the parts have been cleaned, carefully use compressed air to dry them. Rotate the bearings in the cleaning solvent until all built-up lubricant is removed. Then, hold the bearing firmly to prevent it from rotating and dry it with compressed air.

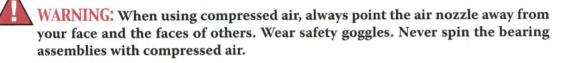 **WARNING: When using compressed air, always point the air nozzle away from your face and the faces of others. Wear safety goggles. Never spin the bearing assemblies with compressed air.**

Immediately lubricate the bearings with the recommended gear oil, then wrap them in a clean, lint-free cloth or paper until you are ready to reinstall them. Clean all metal shavings and grime from the case and its magnet.

Aluminum Case Repair

Normally, the case is replaced if it is cracked or damaged. However, some manufacturers recommend the use of an epoxy-based sealer on some types of leaks in some locations on the transmission (Figure 4-57). Refer to the manufacturer's recommendations before attempting to repair a crack or correct for porosity leaks.

If a threaded area in an aluminum housing is damaged, helicoil-type service kits can be used to insert new threads in the bore. Some threads should never be repaired; check the service manual to identify which ones can be repaired. To replace threads (Figure 4-58), these procedures should be followed carefully.

1. Drill out the damaged threads using the drill size indicated for the insert by the manufacturer.
2. Select the proper tap from the insert toolkit and tap the bore. The taps are marked for the size of the thread being repaired, not for resultant thread size.
3. Select the correct insert installation tool.
4. Place the insert on the tool and adjust the sleeve to the length of the insert being used.
5. Press the insert against the face of the tapped hole, turn the tool clockwise, and wind the insert into the hole until the insert is one-half turn below the surface.
6. Bend the insert tang until it breaks off at the notch.
7. Remove the loose tang.

SERVICE TIP: Improperly installed thread inserts can be removed with the extractor tool included in the kit. Place the extractor tool in the insert with the blade resting against the top coil one-quarter to one-half turn away from the end of the coil. Tap the tool sharply with a hammer to allow the blade to cut the insert. Push down on the tool and turn it counterclockwise until the insert is removed.

A   B

FIGURE 4-57 (A) A crack around the filler plug opening. (B) The crack was repaired with epoxy.

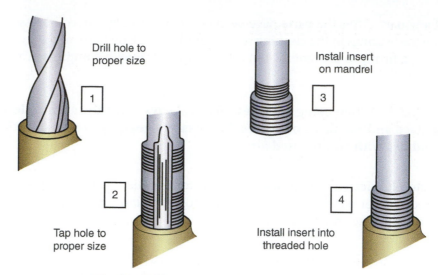

FIGURE 4-58 The procedure for installing a threaded insert to restore damage threads.

SPECIAL TOOLS

Dial indicator and mounting fixtures

Torque wrench

Feeler gauge

Large screwdrivers

Many driveline technicians use "Trans Gel" rather than petroleum jelly during assembly.

GUIDELINES FOR REASSEMBLING A MANUAL TRANSMISSION/TRANSAXLE

Specific repair and assembly instructions depend on the model of transaxle or transmission. Therefore, before reassembling the unit, gather the specific information about the unit you are working on. Also verify that all replacement parts are for the unit being worked on.

Lightly oil the parts of the synchronizers and gear assemblies. Assemble the synchronizer assemblies (Figure 4-59); make sure the parts are aligned according to the scribed marks. Then, carefully install one insert spring and pry the spring back and insert the keys one at a time, being sure to position the wire spring so it is "captured" by the keys. Install the spring on the opposite side, with the open segment of the spring out of phase with the open segment on the other side. Some synchronizer units use balls and small coil springs to lock the shift sleeve in place. It is critical that these are properly positioned (Figure 4-60). To help position the sleeve over the balls, depress them with a small screwdriver while moving the sleeve over them. If the unit is fitted with paper-type blocking rings, soak them in ATF prior to installing them.

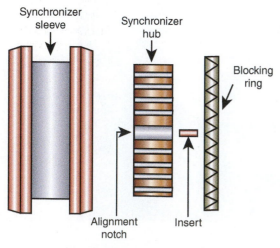

FIGURE 4-59 Align the synchronizer insert with the notch in the blocking ring.

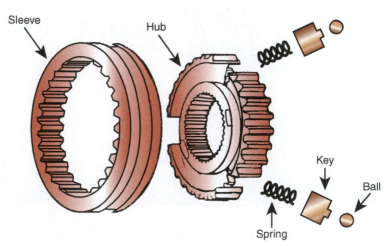

FIGURE 4-60 Typical ball and spring synchronizer assembly.

REASSEMBLE A TRANSMISSION

The following are the steps involved with assembling a typical transmission. This is presented as an example, as is Photo Sequence 7. Always follow the procedures given for a specific transmission.

Install the reverse idler gear and shaft into the case. Then tap the roll pin into place to secure the reverse idler shaft.

Install the countershaft front bearing. Install the shaft's front washer and use a coat of petroleum jelly to hold it in place. Then, lower the countershaft and bearings into the case. Using the proper-sized installer tool, gently tap the countershaft's rear bearing cup into the case.

Some older countergear/shaft assemblies are fitted with small rollers (Figure 4-61) and it is difficult to retain the rollers in the counter gear while installing the shaft. To hold them in place during assembly, coat the bore with petroleum jelly. Press the rollers into the grease; this should hold them.

Installing countershafts with small roller bearings is easier if a "dummy shaft" (Figure 4-62) is used. A **dummy shaft** is like the countershaft, except it is only long enough to hold the rollers and any thrust washers in place. This allows for the lowering of the counter gear, its bearings, and washers into the case. The actual shaft is tapped through the rear bore of the case and through the counter gear to put it in position and drive out the dummy shaft.

Install all gears and bearings to the output shaft. Then, lower the shaft into the case by feeding it through the bore at the rear of the case until the gears are in mesh with the countershaft gears. Now install the bearing cup at the rear of the output shaft.

Apply a liberal coat of Lubriplate7 or equivalent to the thrust washers and thrust bearing of the input shaft, and then install them.

Install the input shaft onto the main shaft and into the case. The main shaft pilot bearing rollers can be held in place by petroleum jelly on the inner bearing surface of the gear. When inserting the input shaft, do not disturb the roller bearings in the input shaft or the thrust washers.

Install a new oil seal in the front bearing retainer with the lip facing toward the transmission case mounting surface. Coat the lip of the seal with Lubriplate7 or equivalent. Place the original shim and bearing cup into the recess of the front bearing retainer. If the retainer uses an O-ring for sealing, lubricate it and position it in the bearing retainer groove.

Classroom Manual
Chapter 4, page 100

A **dummy shaft** can be made from a broom handle cut to the length of the counter gear.

SERVICE TIP: Some synchronizer hubs and some bearings have a significant interference fit on the shaft they will be installed on. Manufacturers may recommend that the hub or bearing be placed on a hot plate to expand the component slightly. Some technicians will place the shaft in a freezer for a few hours to shrink it just enough to help during assembly.

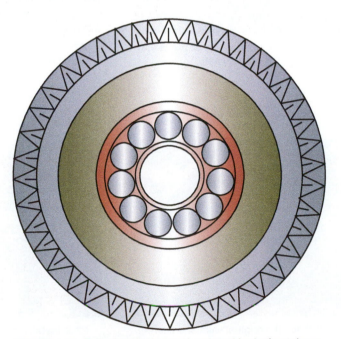

FIGURE 4-61 Roller bearings must be placed in the inner bore of the input shaft counter gear.

PHOTO SEQUENCE 7

REASSEMBLY OF A TYPICAL TRANSMISSION

P7-1 Install the reverse idler gear and shaft into the case. Then tap the roll pin into place to secure the reverse idler shaft.

P7-2 Install the countershaft front bearing. Using a coat of petroleum jelly or trans gel set the countershaft front washer in place. Then, lower the countershaft and bearings into place.

P7-3 Gently tap the countershaft's rear-bearing cup into the case. Do not strike directly onto the bearing; use the proper-sized installer tool.

P7-4 Lower the main shaft into the case by feeding the output shaft through the bore at the rear of the case until the shaft is in mesh with the countershaft gears.

P7-5 Install the bearing cup at the rear of the output shaft.

P7-6 Lubricate the needle and thrust bearings of the input shaft.

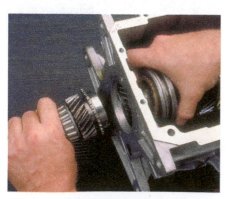

P7-7 Install the input shaft onto the main shaft and into the case.

P7-8 Place the original shim and bearing cup into the recess of the front-bearing retainer.

P7-9 Install the reverse gearshift fork and related parts, if not already installed.

P7-10 Position the front-bearing cup, the shim, and the bearing retainer to the front of the transmission case. Torque the cap screws while rotating the gear. If the gear rotating effort increases while tightening the bearing retainer, replace the shim with a thinner one.

P7-11 With the shift rails and forks in place, apply a thin coat of sealer to the mating surface of the transmission cover.

P7-12 Carefully lower the cover in place, making sure the shift forks are aligned with the appropriate synchronizer sleeve slots.

P7-13 Tighten the cover bolts to the specified amount.

P7-14 Position the extension housing onto the transmission case. Rock the housing to clear the shift mechanisms while moving it toward the transmission case.

P7-15 Install the extension-housing bolts and tighten them to specification.

P7-16 Install the retaining pin for the shift rail and linkage.

P7-17 Apply a thin coat of sealant onto a new shift-lever cover and install the cover onto the transmission case.

P7-18 Install the attaching bolts and tighten them to specifications. Check the operation of the shifter prior to reinstalling the transmission in the vehicle.

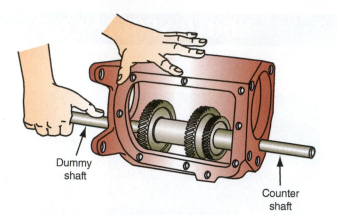

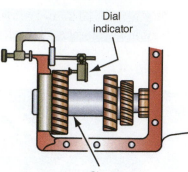

FIGURE 4-62 Push the dummy shaft in to push the countershaft out and push the countershaft in to push out the dummy shaft, when removing and installing the counter gear assembly.

FIGURE 4-63 The typical setup for checking counter gear end play.

Position the front bearing cup, shim, new gasket, and bearing retainer to the front of the transmission case. Make sure the oil hole in the retainer lines up with the oil feed hole in the case. As an aid, mark the position of the oil hole in the retainer on the outside of the retainer with chalk. Torque the cap screws while rotating the gear. If the gear's rotating effort increases while tightening the retainer, replace the shim with a thinner one.

Most transmissions and transaxles have specifications for end play, backlash, and preload. Follow the correct procedures for measuring and adjusting these (Figure 4-63). These clearances are normally set by shims under the bearing caps. If possible, reuse the original shims.

With the shift rails and forks in place, apply a thin coat of sealer to the mating surface of the transmission cover. Then, carefully install the cover, making sure the shift forks are aligned with the synchronizer sleeve slots. Tighten the cover bolts to the specified torque.

Install reverse speed gear and the speed sensor gears onto the main shaft. Install a new seal and bushing into the extension housing, a new extension housing gasket, and the extension housing. Rock the housing to clear the shift mechanisms while moving it toward the transmission case. Tighten the bolts to specification.

Install the retaining pin for the shift rail and linkage. Then apply a thin coat of sealant on the shift lever cover and install it onto the transmission case. Tighten the attaching bolts to specifications. Check the action of the shifter prior to installing the transmission in the vehicle.

AUTHOR'S NOTE: The important final step in transmission/transaxle assembly is to "bench shift" the unit. Rotating effort and shaft speed changes are noted as each gear is selected. This step is more often forgotten with rear-wheel-drive units, probably because the shift lever is still inside the car or truck, which is outside of the shop. A successful bench shift test certainly reduces any chance of having to pull the unit back out because of an assembly problem.

CAUTION: Whenever installing a snapring, make sure it is fully seated in its groove before proceeding with other service.

REASSEMBLE A TRANSAXLE

The following procedure and Photo Sequence 8 are presented as examples and demonstrate the reassembly of a typical transaxle.

Install the synchronizer assemblies onto the main shaft. Now install fourth gear with its bearing. The bearing may need to be lightly pressed into position. Then, install and tighten the shift-relay-lever support-bracket assembly.

The differential unit can now be installed into the transaxle case. At times, a dummy axle stub or some wire must be used to hold the side gears while installing the differential.

Install the fifth-gear shaft assembly and fork shaft into the case. Then place the output shaft control shaft assembly on the main shaft so that the shift forks engage in their respective slots in the synchronizer sleeves. Now install the output shaft assembly.

Most transaxles have specifications for end play, backlash, and preload; make sure these specifications are met. For most transaxles, there are specifications for the end play or preload of the input shaft, output shaft, and the differential. These are usually set by shims under the bearing races. These races can be difficult to remove. Some manufacturers specify special pullers to remove the races. Some technicians will weld a bead around the inside diameter of the race, causing it to expand then shrink, making removal easy (Figure 4-64). The procedure for setting the desired end play or preload on these shafts varies by manufacturer. A special shim selector toolkit may be specified for some transaxles (Figure 4-65). An alternative procedure is to place two sections of electrical solder in place of the shim under each of the three bearing races (Figure 4-66). The transaxle is bolted together with the input shaft, output shaft, and differential in position, squeezing the solder between the race and case. The solder sections are measured, then shims are installed that are typically .002–.003-inch thicker to provide the required preload for tapered roller bearings. If the shaft is supported by a roller or ball bearing, shim size is reduced from the solder thickness to provide proper end play.

Properly position the shift lever, bias, and kickdown springs. Then install the inhibitor spring and ball. Now, depress the inhibitor ball and spring using a drift and slide the shift lever shaft through the shift lever. Tap the shaft into its bore in the clutch housing.

Install the reverse idler gear shaft and gear. Position the shaft so that the retaining bolt bore lines up with the hole in the case. Then, install the fifth-gear relay shaft and align it with the fifth-gear fork slot and interlock spring.

Apply a thin bead of sealer on the case's mating surface for the clutch housing. Always use the sealant recommended by the manufacturer. Install the housing to the case. Make sure the

SERVICE TIP:
It is a good idea to coat all bolts and plugs used in the transmission/transaxle case, except where they are used in blind holes, with a suitable sealant to prevent leakage.

FIGURE 4-64 A transaxle bearing race that has been welded to assist in removal.

FIGURE 4-65 A transaxle preload shim selector toolkit.

REASSEMBLY OF A TYPICAL TRANSAXLE

P8-1 Lightly oil parts of the synchronizer and gear assemblies.

P8-2 Assemble the synchronizer assemblies, being careful to align the index marks made during disassembly.

P8-3 Install the synchronizer assemblies onto the main shaft.

P8-4 Install fourth gear and its bearing. The bearing may need to be lightly pressed into position.

P8-5 Install and tighten the shift-relay-lever support-bracket assembly.

P8-6 Install the differential assembly into the transmission case. Use a "dummy" axle stud or wire to hold the side gears in place.

P8-7 Install the fifth-gear shaft assembly and fork shaft into the case.

P8-8 Place the main-shaft control-shaft assembly on the main shaft so that the shift forks engage in their respective slots in the synchronizer sleeves. Then install the main-shaft assembly.

P8-9 Properly position shift lever, bias, and kickdown springs.

P8-10 Install the inhibitor spring and ball in the fifth and reverse-inhibitor shaft hole.

P8-11 Depress the inhibitor ball and spring using a drift and slide the shift-lever shaft through the shift lever. Then tap the shaft into its bore in the clutch housing.

P8-12 Install reverse idler gear shaft and gear into the appropriate bore in the case. Position the shaft so that the retaining bolt bore lines up with the hole in the case.

P8-13 Install the fifth-gear relay shaft and align it with the fifth-gear fork slot and interlock spring.

P8-14 Install the retaining clip onto the fifth-gear relay shaft.

P8-15 Apply a thin bead of sealer on the case's mating surface for the clutch housing.

P8-16 Install the clutch housing to the case. Be careful that the main shaft, shift-control shaft, and fifth-gear shaft align with their bores in the case.

P8-17 After the housing and case are fit snugly together, tighten the attaching bolts to the specified torque.

P8-18 Install and tighten the reverse idler shaft retaining bolt.

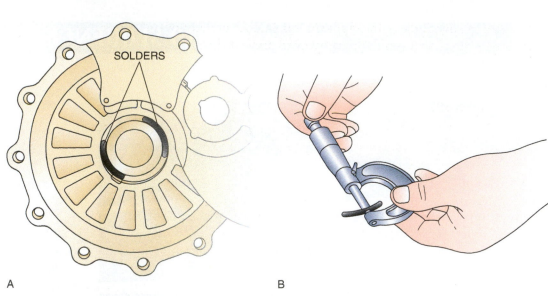

FIGURE 4-66 (A) Solder pieces can be used in the shim selection process by substituting pieces for the original shim and bolting the case halves together. (B) Measuring the flattened solder section.

main shaft, shift control shaft, and fifth-gear shaft align with their bores in the case. After the housing and case are fit snugly together, tighten the attaching bolts to the specified torque. Then install and tighten the reverse idler shaft-retaining bolt. Finally, bench-shift the transaxle to ensure engagement through all the gears.

AUTHOR'S NOTE: Transaxles and transmissions typically use sealers instead of gaskets for most applications. The two most common types are anaerobic sealers and RTV sealers. RTV (room temperature vulcanizing) sealers are silicone based, and come in a variety of formulations. Anaerobic sealers will only cure in the absence of oxygen. Anaerobic sealers are typically specified for surfaces with a low RMS (root, mean square) finish. RMS numbers measure the relative roughness of a surface, with low numbers being smoother. Always use the type of sealant specified by the manufacturer.

TRANSMISSION/TRANSAXLE INSTALLATION

Transmission/transaxle installation is generally a reverse of the removal procedure. A quick check of the following will greatly simplify your installation and reduce the chances of destroying something during installation.

- Always check for free rotation of the transmission in all gears before installing it. If the shafts do not rotate freely, identify the cause and correct it.
- After the unit is together, install the clutch assembly and a new throw-out bearing.
- Make sure the block alignment pins (dowels) are in their appropriate bores and are in good shape and that the alignment holes in the bell housing are not damaged. Use the dowels to locate and support the transmission during installation.
- The input shaft should be lightly coated with grease. This will aid installation and serve as a lubricant for the pilot bearing. Avoid putting too much grease on the shaft, since the excess may fly off and get on the clutch disc, causing it to slip and/or burn.

- Make sure the pilot hole in the crankshaft is smooth and not out-of-round.
- Secure all wiring harnesses out of the way to prevent their being pinched between the bell housing and engine.

Basic Procedure

Set the transmission on a transmission jack, secure it with chain or straps, and move it under the car. Raise the transmission to get close alignment to the engine block. Once the input shaft is seated in the splines of the clutch disc, check the dowel pin alignment with the bell housing and the engine. After all is aligned, push the transmission against the engine. If the transmission/transaxle does not fit snugly against the engine or if you cannot move the transmission into place, Do *not* force it. Pull the transmission back and lower it. Inspect the input shaft splines for dirt or damage. Also, check the mating surfaces for dirt or obstructions. If you try to force the transmission into place by tightening the bolts, you may break the case.

Once the transmission is properly seated, start two bell housing bolts across from each other and slowly tighten them. Then, install the other bolts and torque them to specifications.

On FWD cars, if the engine is held in place with a support bar, the transmission jack may be removed at this time. If the car has a split cradle-type subframe, it should be installed now.

On RWD vehicles, it may be necessary to leave the transmission jack in place while the exhaust crossovers or frame cross members are installed. Don't connect the gearshift linkage until the transmission is mounted on the cross member and the transmission jack is removed. Once the transmission mount is attached to the cross member, the jack may then be removed.

While connecting the shift linkage, pay attention to the condition of the plastic bushings. If these are worn or missing, replace the bushings.

Some other areas that require special attention during installation are the grounding straps and any rubber tubes or vents. These are often overlooked and can cause problems if not checked. New parts should be used if necessary. The ground straps must be in good condition and free of corrosion as these provide an electrical ground path to the body of the vehicle. Failure to clean or attach these straps or cables can also result in poor signals to the PCM, voltage spikes that can damage the PCM, electrolysis through the fluid that welds internal transmission components, or even fused cables as the electrical current looks for a path to flow.

On RWD vehicles, lightly coat the drive shaft's slip joint and carefully insert it into the extension housing. This will prevent possible damage to the rear oil seal. Make sure you install the drive shaft according to the marks you made during removal.

The drive shafts of FWD cars are installed in the reverse manner in which they were removed and the related components should be installed. Use a new nut on the outer CV joint stud, if the bolt is the self-locking type. The torque of this nut is critical. Torquing the nut should be done with the car's tires on the ground and/or with the brakes held. Air-type impact tools should not be used, as they can damage the wheel bearing or hub.

Connect the battery cables. Then fill the transmission with the correct type and amount of fluid.

Road-test the vehicle to check the operation of the transmission and anything else that may have been affected by your work. If any adjustments are necessary, make them. After the road test, visually inspect the transmission for signs of leakage. Also carefully look around the engine and transmission to see if any wires, hoses, or cables are disconnected or positioned in an undesirable spot.

SERVICE TIP:
When installing a rear-wheel-drive transmission, make sure the transmission is in gear when you put it up to the engine. If the input shaft splines don't quite line up with the clutch, you can turn the output shaft slightly and the input shaft will rotate into position.

TERMS TO KNOW

Center support plate
Dummy shaft
Fill plug
Gear knock
Gear rattle

CASE STUDY

A customer brought his 2008 Toyota into the shop with a complaint of the transaxle jumping out of fifth gear. The technician verified the complaint, then proceeded to check the adjustment of the shift linkage. No problem was found after she checked the adjustment and inspected the linkage. She proceeded to check the alignment of the transaxle to the engine, and again the cause of the problem was not found.

The transaxle was removed and disassembled; again no cause for the problem was identified. The technician carefully reassembled the transaxle and installed it back in the car. During the road test, she again experienced the problem. Because she could not find a cause for the problem, she returned it to the customer and told him the problem could not be fixed. Taking this approach is a good way to lose customers and your job. True, she did check all of the right things. However, she probably did not check the internal parts of the transaxle carefully enough. Wear on the back taper of the dog clutch teeth of the speed gear and/or slider will cause this problem. Only a careful inspection will reveal this.

ASE-STYLE REVIEW QUESTIONS

1. When discussing shift problems,

   Technician A says that broken or worn engine and transaxle mounts can cause a transaxle to have shifting problems.

   Technician B says that poor shift boot alignment can cause a transaxle to jump out of gear.

   Who is correct?

   A. A only
   B. B only
   C. Both A and B
   D. Neither A nor B

2. When inspecting a transaxle's gears,

   Technician A says that a wear pattern on the center of the gear teeth is normal.

   Technician B says that it is normal for a reverse idler gear to have small chips on its engagement side but these do not affect operation.

   Who is correct?

   A. A only
   B. B only
   C. Both A and B
   D. Neither A nor B

3. When inspecting the synchronizers of a transaxle,

   Technician A says that if the dog teeth of the synchronizer are slightly rounded, the synchronizer assembly must be replaced.

   Technician B says that the movement of the synchronizer sleeve on the shaft should be checked.

   Who is correct?

   A. A only
   B. B only
   C. Both A and B
   D. Neither A nor B

4. When discussing transmission disassembly,

   Technician A says that the alignment of the parts of each synchronizer must be marked prior to disassembling the synchronizer assembly.

   Technician B says that most synchronizer hubs are splined to the shaft and are pressed on and off the shaft.

   Who is correct?

   A. A only
   B. B only
   C. Both A and B
   D. Neither A nor B

5. When disassembling a transaxle, a severely worn second gear synchronizer blocker ring is found.

   Technician A says that this should have caused the transaxle to jump out of gear.

   Technician B says that this should have caused difficult shifting into second gear.

   Who is correct?

   A. A only
   B. B only
   C. Both A and B
   D. Neither A nor B

6. Transmission fluid level and type are being discussed.

   Technician A says that the proper fluid level for a transmission is normally to the bottom of the filler hole.

   Technician B says that most manual transmissions require heavy gear lube for proper operation.

   Who is correct?

   A. A only
   B. B only
   C. Both A and B
   D. Neither A nor B

7. When discussion transmission repair,

    Technician A says that every part of the transmission should be carefully inspected, regardless of the problem.

    Technician B says that if the lubricant had traces of gold in it, it is very likely that the gears will be severely worn.

    Who is correct?

    A. A only

    B. B only

    C. Both A and B

    D. Neither A nor B

8. When installing the shafts into the transmission case,

    Technician A says that the end play of the output shaft should be checked.

    Technician B says that the thrust washers and bearings should be lubricated prior to installing the shafts.

    Who is correct?

    A. A only

    B. B only

    C. Both A and B

    D. Neither A nor B

9. When inspecting the gears of a transmission, a shiny wear pattern is found on the center of each gear.

    Technician A says that this was caused by worn shaft bearings.

    Technician B says that this was caused by worn synchronizers.

    Who is correct?

    A. A only

    B. B only

    C. Both A and B

    D. Neither A nor B

10. While assembling a transmission:

    Technician A says end play, backlash, and preload cannot be adjusted on newer transmissions and transaxles.

    Technician B says shims can be used to set the end play of the input shaft.

    Who is correct?

    A. A only

    B. B only

    C. Both A and B

    D. Neither A nor B

ASE CHALLENGE QUESTIONS

1. A bearing-type noise is coming from inside a manual transmission when the vehicle is stationary. The noise disappears when the clutch is disengaged.

    Technician A says that the input shaft bearing could be faulty.

    Technician B says that the rear main shaft support bearing could be faulty.

    Who is correct?

    A. Technician A

    B. Technician B

    C. Both A and B

    D. Neither A nor B

2. A vehicle equipped with a five-speed manual transmission sometimes jumps out of high (fourth) gear.

    Technician A says that this may be caused by worn cluster gear teeth.

    Technician B says that a worn main shaft pilot bearing could cause this problem.

    Who is correct?

    A. Technician A

    B. Technician B

    C. Both A and B

    D. Neither A nor B

3. Which of the following would not cause a harsh 3–4 upshift on a vehicle equipped with a five-speed manual transmission?

    A. Worn synchronizer sleeve

    B. Broken synchronizer sleeve

    C. Misaligned shift rails

    D. Worn teeth on fourth gear of the main shaft

4. When discussing shifter linkages,

    Technician A says that occasionally the internal shifter linkage of some manual transaxles may need to be adjusted.

    Technician B says that worn external shifter linkage may cause excessive clutch pedal free play.

    Who is correct?

    A. Technician A

    B. Technician B

    C. Both A and B

    D. Neither A nor B

5. Which of the following is *not* something that should be done prior to installing a transmission into a vehicle?

    A. Check for free rotation of the transmission in all gears before installing in a vehicle.

    B. Clean the input shaft and remove all grease so that it will not fly off and get on the clutch assembly.

    C. Make sure the block alignment pins (dowels) are in their appropriate bores and are in good shape.

    D. Make sure the pilot bushing in the crankshaft is smooth and not out-of-round.

Name _____ Date _____

DRIVETRAIN FLUID SERVICE

Upon completion of this job sheet, you will be able to diagnose the cause of excessive fluid loss and contaminated fluid. You will also be able to drain and fill a manual transmission/transaxle and final drive unit with the correct fluid.

NATEF MAST Task Correlation

General: Drivetrain Diagnosis

Task #3 Check fluid condition; check for leaks; determine necessary action.

Task #4 Drain and refill manual transmission/transaxle and final drive unit.

Tools and Materials

Appropriate service information   Drain pan

Lift   Clean rags

Protective Clothing

Goggles or safety glasses with side shields

Describe the Vehicle Being Worked On

Year _____ Make _____ VIN _____

Model _____

Procedure

1. Raise the vehicle and visually inspect the vehicle's drivetrain for signs of fluid leakage. Summarize your results.

   _____

   _____

   _____

   _____

2. Based on the above, what service do you recommend?

   _____

   _____

3. If fluid is leaking from around the seal at the transmission's extension housing, what should you check before replacing the seal?

   _____

   _____

4. List at least five possible sources for a fluid leak in the drivetrain of this vehicle.

_____

_____

_____

_____

_____

5. Explain why a fluid leak can also cause the fluid to become contaminated with water or dirt.

_____

_____

_____

_____

6. Locate the preventive maintenance schedule for the drivetrain in the service information or the vehicle's owners' manual. Summarize what should be done and when it should be completed.

_____

_____

_____

_____

7. What type of fluids should be used in this drivetrain? Be specific.

_____

8. Locate the drain plug for the transmission/transaxle. Describe what it looks like and where you found it.

_____

_____

_____

☐ 9. Position the drain pan so it can catch all fluid from the drain plug.

☐ 10. Allow all of the fluid to drain from the unit.

11. Inspect the fluid in the drain pan for contamination and metal and other particles. Describe what you found.

_____

_____

12. What would cause the fluid to be contaminated with gold-color metal particles?

_____

_____

☐ 13. Reinstall the drain plug.

14. Locate the filler plug and describe where it is located.

_____

_____

15. Remove the filler plug and fill the housing with the fluid until it reaches the level described by the manufacturer. How much fluid does the transmission/transaxle normally hold and how do you know when there is enough fluid in the housing?

_____

_____

_____

_____

_____

16. Reinstall the filler plug after the correct fluid level is reached.   ☐

17. Does the final drive unit require its own fluid or is the fluid shared with the transmission?

_____

_____

18. If the final drive has its own fluid reservoir, locate the drain plug and describe what it looks like and where you found it.

_____

_____

_____

19. Position the drain pan so it can catch all fluid from the drain plug.   ☐

20. Allow all of the fluid to drain from the unit.   ☐

21. Inspect the fluid in the drain pan for contamination and metal and other particles. Describe what you found.

_____

_____

22. Install the drain plug and then locate the filler plug and describe where it is located.

_____

_____

23. Remove the filler plug and fill the housing with the fluid until it reaches the level described by the manufacturer. How much fluid does the housing normally hold and how do you know when there is enough fluid in the housing?

_____

_____

_____

_____

_____

☐ **24.** Reinstall the filler plug after the correct fluid level is reached.

Instructor's Response: _____
_____
_____

Name _____ **Date** _____

ROAD-CHECK VEHICLE FOR TRANSMISSION PROBLEMS

Upon completion of this job sheet, you should be able to road-check a vehicle for transmission problems.

NATEF MAST Task Correlation

Transmission/Transaxle Diagnosis and Repair

Task #3 Diagnose noise concerns through the application of transmission/transaxle power-flow principles.

Task #4 Diagnose hard shifting and jumping out of gear concerns; determine necessary action.

Task #5 Diagnose transaxle final drive assembly noise and vibration concerns; determine necessary action.

Tools and Materials

A vehicle for which the owner has given permission to road-check

Describe the Vehicle Being Worked On

Year _____ Make _____ VIN _____

Model _____

Procedure

1. Driving the vehicle in town, obey the speed laws. Shift the transmission through the gears and identify any problem you might feel.

   _____

   _____

2. Did it easily shift into first? ☐ Yes ☐ No

3. Did the gears feel smooth? ☐ Yes ☐ No

4. Shift the vehicle into second gear. Did it feel smooth? ☐ Yes ☐ No

5. Did the shift seem to move smoothly? ☐ Yes ☐ No

6. Shift into third gear. Did it shift smoothly? ☐ Yes ☐ No

7. Did it feel as if the gears slipped in smoothly? ☐ Yes ☐ No

8. Shift into fourth gear (if it has one). Did it shift smoothly? ☐ Yes ☐ No

9. Did it feel as if the gears slipped in smoothly? ☐ Yes ☐ No

10. Shift into fifth gear (if it has one). Did it shift smoothly? ☐ Yes ☐ No

11. Did it feel as if the gears slipped in smoothly?   ☐ Yes   ☐ No

12. Describe what you noticed about the shift on this vehicle.

Does it need service?   ☐ Yes   ☐ No

13. What would you tell the customer about this vehicle?

_____

_____

_____

_____

_____

_____

Instructor's Response: _____

_____

_____

Name _____ **Date** _____

CHECKING TRANSAXLE MOUNTS

Upon completion of this job sheet, you should be able to inspect, replace, and align powertrain mounts.

NATEF MAST Task Correlation

General: Drivetrain Diagnosis

Task #1 Identify and interpret drivetrain concerns; determine necessary action.

Tools and Materials

Engine support fixture

Engine hoist

Basic hand tools

Protective Clothing

Goggles or safety glasses with side shields

Describe the Vehicle Being Worked On

Year _____ Make _____ VIN _____

Model _____ Engine type and size _____

Transmission type and model _____

Procedure

1. Many shifting and vibration problems can be caused by worn, loose, or broken engine and transmission mounts. Visually inspect the mounts for looseness and cracks. Give a summary of your visual inspection.

   _____

   _____

   _____

2. Pull up and push down on the transaxle case while watching the mount. If the mount's rubber separates from the metal plate or if the case moves up but not down, replace the mount. If there is movement between the metal plate and its attaching point on the frame, tighten the attaching bolts to an appropriate torque. Describe the results of doing this.

   _____

   _____

3. From the driver's seat, apply the foot brake, set the parking brake, and start the engine. If the vehicle has an automatic transmission/transaxle, put the transmission into a forward gear and gradually increase the engine speed to about 1,500 to 2,000 rpm. If the vehicle has a manual transaxle, put it in first gear, raise the engine speed to about 1,500 to 2,000 rpm, and gradually release the clutch pedal. Watch the torque reaction of the engine on its mounts. If the engine's reaction to the torque appears to be excessive, broken or worn drivetrain mounts may be the cause. Describe the results of doing this.

_____

_____

4. If it is necessary to replace the transaxle mount, make sure you follow the manufacturer's recommendations for maintaining the alignment of the driveline. Describe the recommended alignment procedures.

_____

_____

☐ 5. When removing the transaxle mount, begin by disconnecting the battery's negative cable.

☐ 6. Disconnect any electrical connectors that may be located around the mount. Be sure to label any wires you remove to facilitate reassembly.

☐ 7. It may be necessary to move some accessories, such as the horn, in order to service the mount without damaging some other assembly.

☐ 8. Install the engine support fixture and attach it to an engine hoist.

☐ 9. Lift the engine just enough to take the pressure off the mounts.

☐ 10. Remove the bolts attaching the transaxle mount to the frame and the mounting bracket, and then remove the mount.

11. To install the new mount, position the transaxle mount in its correct location on the frame and tighten its attaching bolts to the proper torque. What is the torque specification?

_____

_____

☐ 12. Install the bolts that attach the mount to the transaxle bracket. Before tightening these bolts, check the alignment of the mount.

☐ 13. Once you have confirmed that the alignment is correct, tighten all loosened bolts to their specified torque.

☐ 14. Remove the engine hoist fixture from the engine and reinstall all accessories and wires that may have been removed earlier.

Instructor's Response: _____

_____

_____

JOB SHEET

16

Name _____ **Date** _____

INSPECT TRANSMISSION LINKAGE

Upon completion of this job sheet, you should be able to inspect a vehicle for transmission linkage problems.

NATEF MAST Task Correlation

Transmission/Transaxle Diagnosis and Repair

Task #1 Inspect, adjust, and reinstall shift linkages, brackets, bushings, cables, pivots, and levers.

Tools and Materials

A vehicle for which the owner has given permission to inspect shift linkage

A hoist or a jack with some safety stands

A creeper if you have no hoist

Describe the Vehicle Being Worked On

Year _____ Make _____ VIN _____

Model _____

Procedure

Task Completed

1. Follow the safety rules and raise the vehicle in the air.   ☐

2. Get under the vehicle and inspect the shift linkage.   ☐

3. Is the linkage loose?   ☐ Yes   ☐ No

4. Are all the bushings tight, or do they need replacing?   ☐ Yes   ☐ No

5. Move the linkage with your hand. Does it move freely or does it bind at different points?   ☐ Yes   ☐ No

6. What is your diagnosis of the linkage: Does it need service or is it okay?   ☐ Yes   ☐ No

7. What would you tell the customer?

_____

_____

_____

_____

_____

Instructor's Response: _____

_____

Name _____ **Date** _____

TRANSAXLE POWER-FLOW

Upon completion of this job sheet, you should be able to trace the power-flow of all the gear ranges of a transmission, calculate the ratios of the individual gear set with the transaxle, and calculate the overall gear ratio for each of the transmission.

Tools and Materials

Appropriate hand tools, mechanic's basic set

Transmission holding fixture

Appropriate special tools listed in service manual

Service manual to follow proper procedure

Protective Clothing

Goggles or safety glasses with side shields

Describe the Vehicle Being Worked On

Year _____ Make _____ VIN _____

Model _____

Procedure

Task Completed

1. Reassemble transmission main shaft on the work bench outside of the transaxle case. ☐

2. Lay out transmission input shaft, assembled main shaft, countershaft, reverse idler gear, and overdrive gears, if separate, so you can easily see the relationship the gears have to each other. ☐

3. Count and record the teeth on each gear.

GEAR	NUMBER OF TEETH
Input shaft gear	_____
Cluster shaft input gear	_____
First gear on cluster shaft	_____
First gear on the main shaft	_____
Second gear on cluster shaft	_____
Second gear on the main shaft	_____
Third gear on the cluster shaft	_____
Third gear on the main shaft	_____
Fourth gear on the cluster shaft (skip if N/A)	_____
Fourth gear on the main shaft (skip if N/A)	_____

GEAR	NUMBER OF TEETH
Fifth gear on the main gear shaft	_____
Fifth gear on cluster shaft	_____
Reverse gear on the main shaft	_____
Reverse idler gear	_____
Reverse gear on the main shaft	_____

4. Use the gear tooth count for each gear to calculate the gear ratios of each gear set.

Input gear gear set: Input shaft/cluster shaft _____

First gear gear set: First gear cluster shaft/first gear main shaft _____

Second gear gear set: Second gear cluster shaft/second gear main shaft _____

Third gear gear set: Third gear cluster shaft/third gear main shaft _____

Fourth gear gear set: Fourth gear cluster shaft/first gear main shaft _____

Fifth gear gear set: Fifth gear cluster shaft/fifth gear main shaft _____

Reverse idler gear set: reverse gear cluster shaft/reverse idle gear _____

Reverse main shaft gear gear set: Reverse gear idler shaft/reverse gear main shaft
_____

5. Using the above figures, state the final gear ratio for each gear.

GEAR	FINAL GEAR RATIO
First	_____
Second	_____
Third	_____
Fourth	_____
Fifth (O/D)	_____
Reverse	_____

6. Using the parts that you laid out in step 1, explain power-flow through the transmission to your instructor. Your description should include:

   a. Rotational speed of each gear—1,000 rpm engine speed
   b. Direction of rotation of each gear—clockwise input
   c. Relative torque of each gear—100 ft.-lb. of input torque
   d. Synchronizer action for entering each gear
   e. Shift fork position for each gear

_____

_____

_____

_____

_____

Instructor's Response: _____

_____

_____

REMOVE AND INSTALL A TRANSMISSION/TRANSAXLE

Name _____ Date _____

Upon completion of this job sheet, you should be able to remove and install a manual transmission or transaxle.

Tools and Materials

Lift

Engine hoist

Engine support

Transmission/transaxle jack

Drain pan

Basic hand tools

Protective Clothing

Goggles or safety glasses with side shields

Describe the Vehicle Being Worked On

Year _____ Make _____ Model _____

VIN _____ Engine type and size _____

Transmission or transaxle _____

Procedure

For removing transmissions:

_____

Task Completed

1. Disconnect the negative cable of the battery. ☐

2. Disconnect the gearshift lever or linkage from the transmission. Describe the procedure followed to do this:

   _____

   _____

   _____

   _____

3. Using chalk, mark the drive shaft at the rear U-joint and the rear axle flange so that the   ☐
   drive shaft can be installed correctly. Remove the drive shaft bolts at the flange of the
   rear axle. Then remove the drive shaft.

4. Disconnect all wiring to the transmission. Mark the wires so you can easily reconnect
   them to the correct terminals. How did you mark them?

   _____   _____

   _____   Task Completed

☐ 5. Disconnect the speedometer cable (if equipped).

☐ 6. Use the transmission jack to support the transmission.

7. Loosen and remove the starter motor. If there is enough battery cable to the starter, you may unbolt the starter and move it out of the way without disconnecting the cables. When doing this, use heavy wire to support the starter; do not allow the starter's weight to pull on the electrical wiring and cables. Describe what you needed to do with the starter.

_____

_____

☐ 8. Loosen and remove the bell housing-to-engine bolts.

☐ 9. Slide the transmission away from the engine until the input shaft is away from the clutch assembly.

☐ 10. Lower the transmission jack and place the transmission in a suitable work place.

11. Problems encountered:

_____

_____

_____

☐ **For removing transaxles:**

☐ 1. Disconnect the negative cable of the battery.

2. Disconnect the gearshift lever or linkage from the transmission. Describe the procedure followed to do this:

_____

_____

_____

3. Disconnect all wiring to the transmission. Mark the wires so you can easily reconnect them to the correct terminals. How did you mark them?

_____

_____

☐ 4. Disconnect the speedometer cable (if equipped).

5. Install the engine support fixture.

☐ 6. Use the transmission jack to support the transmission.

7. Loosen and remove the starter motor. If there is enough battery cable to the starter, you may unbolt the starter and move it out of the way without disconnecting the cables. When doing this, use heavy wire to support the starter; do not allow the starter's weight to pull on the electrical wiring and cables. Describe what you needed to do with the starter.

_____

_____

8. Identify the transaxle-to-engine block bolts that cannot be removed from under the vehicle and remove them with the vehicle at ground level. Disconnect anything that may interfere with the movement of the transaxle. Then install the engine support fixture.

9. Raise the vehicle and place the transmission jack under the transaxle. ☐

10. Remove the suspension parts that will interfere with the removal of the transaxle. Name the parts you needed to remove:

_____

_____

_____

_____

11. Disconnect and remove the drive axles from the transaxle. Describe the process you followed to do this:

_____

_____

_____

_____

12. Remove the remaining transaxle-to-engine bolts. ☐

13. Slide the transaxle away from the engine until the input shaft is away from the clutch assembly. ☐

14. Problems encountered:

_____

_____

_____

_____

For installing transmissions:

1. Place the transmission on the transmission jack, secure the transmission with a chain or straps, and place the jack into position for installing the transmission into the vehicle. ☐

2. Coat the input shaft with a thin layer of grease. Slowly slide the transmission into position. Make sure the splines of the input shaft line up with the internal splines of the clutch disc. If necessary, rotate the input shaft slightly so the splines line up. Do not force the shaft into the disc. Describe any problems you encountered and what adjustments you needed to make to the procedure.

_____

_____

_____

_____

3. Make sure the transmission is fully seated against the engine. Then tighten the bell housing-to-engine bolts. Is there a torque specification for these bolts?

☐ Yes ☐ No How much? _____

4. Install the starter motor and its wires and cables. Tighten the bolts evenly. Do not use the bolts to fully seat the starter; rather, hold it in place while tightening the bolts. Some starters use shims to ensure proper tooth contact with the flywheel. If shims are used, use the ones that were removed. Describe what you needed to do with the starter.

_____

_____

☐ 5. Remove the transmission jack.

☐ 6. Connect the speedometer cable (if equipped).

☐ 7. Connect all wiring to the transmission according to the marks you made during removal.

☐ 8. Install the drive shaft so that the index marks made during removal are aligned. Install and tighten the drive shaft bolts at the rear flange.

☐ 9. Connect the gearshift lever or linkage to the transmission.

☐ 10. Connect the negative cable of the battery.

11. Problems encountered:

_____

_____

_____

_____

For installing transaxles:

1. Place the transaxle on the transmission jack and place the jack into position for installing the transaxle into the vehicle.

2. Coat the input shaft with a thin layer of grease. Slowly slide the transaxle into position. Make sure the splines of the input shaft line up with the internal splines of the clutch disc. If necessary, rotate the input shaft slightly so the splines line up. Do not force the shaft into the disc. Describe any problems you encountered and what adjustments you needed to make to the procedure.

_____

_____

_____

_____

3. Make sure the transaxle is fully seated against the engine. Then tighten the bell housing-to-engine bolts. Is there a torque specification for these bolts?
   ☐ Yes ☐ No   How much? _____

4. Carefully inspect the drive axle boots and install the axles into the transaxle. Describe the process you followed to do this:

   _____

   _____

   _____

   _____

5. Reinstall the suspension parts that were removed during transaxle removal. Name the parts you needed to reinstall:

   _____

   _____

   _____

   _____

6. Remove the transmission jack and lower the vehicle. ☐

7. Reinstall and tighten the transaxle-to-engine bolts that remain out. Remove the engine support fixture. During transaxle removal you may have needed to remove some additional components to allow access to the transaxle or to allow the transaxle to come out. Make sure you reconnect them now. Describe what parts you needed to reinstall:

   _____

   _____

   _____

   _____

8. Install the starter motor and wires. Make sure you tighten the bolts evenly and that the starter is fully seated against the engine or transaxle housing. ☐

9. Reconnect the speedometer cable (if equipped). ☐

10. Reconnect all wiring to the transaxle according to the marks you made during removal. ☐

11. Reconnect the gearshift lever or linkage from the transmission. ☐

12. Reconnect the negative cable of the battery.

13. Problems encountered:

    _____

    _____

    _____

Instructor's Response: _____

_____

_____

Chapter 5

SERVICING FRONT DRIVE AXLES

BASIC TOOLS

Basic mechanic's tool set

Long-handled breaker bar

Torque wrench

Lift

Jack stands

UPON COMPLETION AND REVIEW OF THIS CHAPTER, YOU SHOULD BE ABLE TO:

- Conduct a safe and effective road test to identify axle and joint problems.

- Diagnose axle shaft and CV joint noise and vibration problems; determine needed repairs.

- Inspect, service, and replace shaft, boots, and CV joints.

- Inspect, service, and replace FWD front wheel bearings and their hubs.

FRONT DRIVE AXLE SERVICE

A complete front drive axle service normally includes a CV joint inspection, front hub bearing inspection and possible replacement, and CV joint boot replacement (Figure 5-1). CV joint boots should be replaced whenever the drive axle is serviced and excessively worn parts, such as CV joints and hub bearings, must be replaced.

MAINTENANCE

No periodic service to the drive axles or CV joints is required. However, if a CV joint boot is punctured or damaged, the boot or the axle should be replaced immediately; loss of lubricant and entry of dirt and moisture will quickly damage the joint.

Most FWD cars have permanently sealed front-wheel bearings. These do not require periodic adjustment or lubrication. These units are replaced if they are defective.

DIAGNOSING FWD AXLE PROBLEMS

Classroom Manual
Chapter 5, page 121

Diagnostics of drive axle and CV joint problems should include talking to the customer to gather as much information as possible about the problem, road-testing the car, and thoroughly inspecting all components. The technician should examine the past service history of the vehicle, and check for any applicable technical service bulletins.

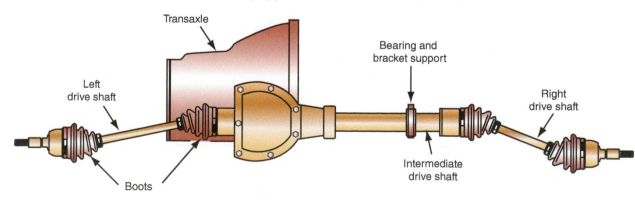

FIGURE 5-1 Typical FWD components.

Talk to the Customer

Although your customer probably doesn't know the cause of the problem, what he or she does know will be of great help to you while you are diagnosing the problem. Your task while talking to the customer is to find out as much as you can about the problem.

> **CUSTOMER CARE: When you are questioning customers, make sure you always treat them in a nonthreatening way. Customers know less about their cars than you do, but they know more about something else than you do. Never belittle them because they lack knowledge about cars. Show them respect.**

Good questions to ask the customer are:

- What is the problem? Is it a shake, vibration, or noise?
- Can you feel and hear the symptom?
- If the problem is a noise, what kind of noise is it?
- If you feel a vibration or shake, where do you feel it?
- When does the noise or vibration occur?
- When did you first notice it?
- Were you going straight? Turning left? Turning right? Braking the car?
- What work has been done on the vehicle lately?

It will be very helpful if you write a summary of the customer's responses on the repair order or in a notebook. This will give you a quick reference to the problem during the road test or other diagnostic work.

Another thing to determine is whether any other repairs have been done that could have contributed to the customer's complaints or to the failure of a CV joint component. Examples of such repairs are body repair that could indicate half shaft misalignment; hub-and-bearing replacement during which an impact wrench was used to tighten the CV joint nut; or engine, transaxle, or cradle removal and replacement.

Road Test

Because many times a customer may misstate the problem or not communicate it clearly, a road test is a must. A road test can confirm the customer's complaint and also show related problems that the customer may not have noticed. Drive the car in an attempt to duplicate the symptoms. You should be listening and feeling for key indicators of CV joint wear.

Drive the car in a safe area where you can weave the vehicle from side to side. During the road test, a good method of identifying a problem is to take full right and left turns while going in reverse. This forces the joints to operate out of their normal positions and makes noise and vibration more noticeable. A common complaint is a clicking or popping noise when the vehicle is turning. This sound is normally caused by a worn or damaged outer CV joint (Figure 5-2). Drive the car through full turns. If only a faint click is heard, put the car in reverse and repeat the test. Sometimes the noise may be more noticeable in reverse. The source of this noise can be pinpointed by paying attention to the noise as the vehicle is turned to the right and left. Normally this noise is caused by the outboard CV joint that is on the inside of the turn when the noise is the most noticeable (Figure 5-3). The noise on turns is caused as the balls of the CV joint go back and forth over a worn divot, usually in a spot on the outer race where the balls ride in the straight-ahead position.

If clicking was not heard, quickly accelerate then decelerate the car. A clunking noise during torque output changes indicates an inner CV joint wear problem. Also note any **wheel shimmy**, shudder, or shaking during acceleration, which could also indicate a bad inner CV joint. The problem occurs under load as the components in the inner joint move out of their normal ride-height location. The shimmy or shudder typically disappears on deceleration. Other possible causes of this noise are loose inner CV joint flange bolts, a worn half shaft damper, or excessive backlash in the transaxle and differential gears.

Wheel shimmy is best described as the wobble of a tire.

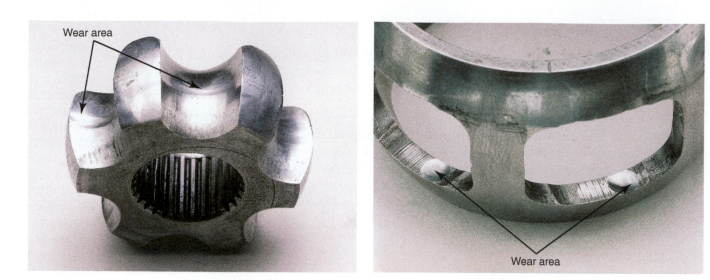

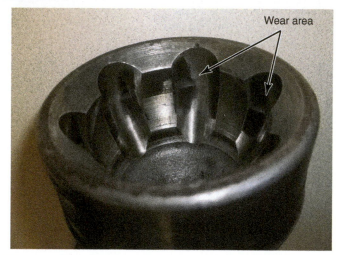

FIGURE 5-2 A worn cage, race, or housing can cause a clicking sound during a turn.

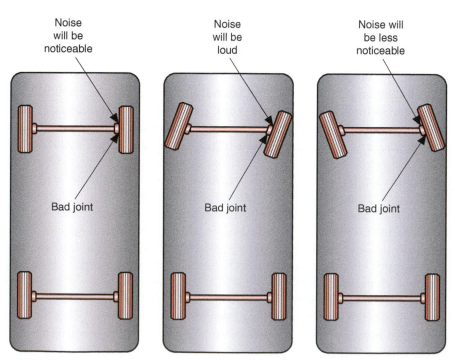

Noise will be noticeable

Bad joint

Noise will be loud

Bad joint

Noise will be less noticeable

Bad joint

FIGURE 5-3 A bad outboard joint on the inside of a turn will cause a noticeable noise.

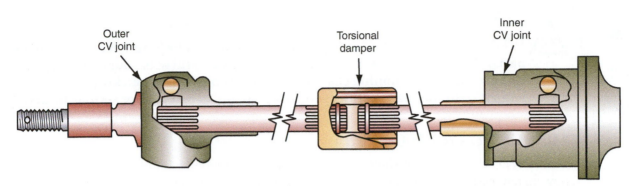

Outer CV joint

Torsional damper

Inner CV joint

FIGURE 5-4 Shudder can be caused by a stuck joint or worn, damaged, or missing torsional damper.

Next, drive at a constant speed and begin to coast lightly. If a **shudder** or steering wheel shimmy exists, a bad outboard joint is the possible cause. Coast slowly through a sharp turn. Clicking noises are usually a sign of a bad outer joint or defective wheel bearings. If you feel the driveline shudder at the same time you hear a clunk, suspect a faulty inner CV joint. This might mean a stuck tripod or a bad intermediate shaft bearing. If the shudder or vibration is intermittent or occurs only at certain speeds, it may be caused by a bent or twisted half shaft. A defective or missing damper can also cause this problem (Figure 5-4); however, although it may feel like a shudder, the problem will actually be a vibration. Shudder during acceleration can be caused by excessive wear on the race surface of an inboard joint.

When you reach a safe area for higher speeds, raise the vehicle speed up to about 50–55 mph. If a vibration becomes obvious at about 45 mph and increases with an increase in speed, a worn transaxle bushing is a possible cause. If the vibration pulses and appears at 45–60 mph, the cause may be an out-of-balance wheel or a worn inner tripod joint. The vibration may be caused by **dimpling** or roughness of the tripod roller tracks on the inner joint. The vibration seems to pulsate because when the roller hits the dimple or rough patch on the bad track, small jerks in the driveline are created.

If there is a continuous humming or growling sound, suspect the source of the noise to be worn transaxle gears, worn wheel bearings, a faulty intermediate shaft bearing, or poorly lubricated CV joint. If the noise is felt through the steering wheel, suspect worn hub bearings or disc brakes.

While diagnosing the cause of a noise, make sure you consider the front wheel bearings. Often wheel bearing noises are misdiagnosed as drive axle, transaxle, or differential problems. Weaving the car left and right in a safe area will usually cause front wheel bearing noise to change. The humming may become louder in one direction, then be quieter in the other. Since there are two rows of bearings in each wheel bearing assembly, one row is loaded and the other unloaded during turns. The technician must determine which side is bad unless the noise is obvious.

> **CUSTOMER CARE:** Wheel bearing noise often develops gradually, and the customer slowly gets used to it. The technician may test-drive with the customer and immediately hear what sounds like a large airplane sound coming from the front. The technician should resist the urge to make a comment such as "How can you possibly drive this thing?" Remember, the customer has feelings!

If the source of the noise is found to be in the drive axles, transaxle, or wheel bearings, inspect and test the electrical ground for the vehicle. These components can fail because of the electrolysis that occurs when the components become the current path to ground.

If the road test did not verify the customer's complaint, raise the car, set the jack stands beneath the lower control arms, and carefully lower the suspension down on the stands. All-wheel-drive vehicles must have all four wheels off the ground. Now start the engine with the brakes applied. Place the transaxle in gear and release the brakes. Apply power and listen to the drive axles for noises that may not have been evident during the test drive. The maximum speed of the tires during this test should be 25–30 mph.

A **shudder** is a shake or shiver movement.

When a car has a shimmy, its front wheels have an abnormal shake or vibration that transmits through the suspension to the steering wheel.

Dimpling is brinelling or the presence of indentations in a normally smooth surface.

⚠ **WARNING:** Don't allow any part of your body or clothing near any rotating parts. Never grab a rotating wheel. Be especially careful with your hair. Long hair can get close to a spinning shaft and static electricity can quickly pull it in and wrap it onto the shaft, pulling your scalp off and causing other serious injury.

TORQUE STEER DIAGNOSIS

Most FWD vehicles exhibit some torque steer. When equipped with high torque and horsepower engines, torque steer tends to be more noticeable. This is a normal trait of these vehicles. However, the cause of abnormal torque steer should be found and corrected. The best way to determine if the amount of torque steer exhibited by a vehicle is normal or excessive is to drive a similar vehicle and observe the amount of normal torque steer. Torque steer should not be confused with pulling or wandering. If the vehicle had a tendency to drift or pull to one side while driving at normal speeds or during braking, the cause is most likely to be a brake, tire, or alignment problem. If the vehicle drove normally, but pulled to the right only during acceleration, there is torque steer. Unusually severe torque steer can be the result of bad control arm bushings.

AUTHOR'S NOTE: If the vehicle has been modified to improve its handling, torque steer may be more evident. This is especially true if the vehicle was fitted with lower profile tires, stiffer springs, stiffer shocks, and/or urethane control arm bushings. When this is the case, make sure the customer understands this, as the only way to reduce torque steer on these vehicles would be to remove those items and replace them with OEM parts. However, some of the torque steer can be reduced with a good wheel alignment.

DIAGNOSIS BY SYMPTOM

The results of the road test and the concerns noted can lead to the proper diagnosis of the problem. Table 5-1 includes the most common symptoms and their causes.

TABLE 5-1 PROBLEM DIAGNOSIS AND SERVICE FOR FWD DRIVELINES

Problem	Possible Cause	Corrective Remedy
Vibrations in steering wheel at highway speeds	Front-wheel balance	Front-wheel unbalance is felt in the steering wheel. Front wheels must be balanced.
Vibrations throughout vehicle	Worn inner CV joints	Worn parts of the inner CV joint are not operating smoothly.
Vibrations throughout vehicle at low speed	Bent axle shaft	Axle shaft does not operate on the center of the axis; thus, vibration develops.
Vibrations during acceleration	Worn or damaged inner CV joints Fatigued front springs	CV joints are not operating smoothly due to damage or wear. Sagged front springs are causing the inner CV joint to operate at too great an angle, causing vibrations.
Grease dripping on ground or sprayed on chassis parts	Ripped or torn CV joint boots	Front-wheel-drive CV joints are immersed in lubricant. If the CV joint boot has a rip or is torn, lubricant leaks out. The condition must be corrected as soon as possible.
Clicking or snapping noise heard when tuning curves and corners	Worn or damaged outer CV joint Bent axle shaft	Worn parts are clicking and noisy as loading and unloading on the CV joint takes place. Irregular rotation of the axle shaft is causing a snapping, clicking noise.

VISUAL INSPECTION

After the test drive, return to the shop and prepare the vehicle for a visual inspection. Before assuming that a noise or vibration is caused by a faulty CV joint or axle, check all other possible problem areas. These should be checked and corrected before paying full attention to the drive axles and their joints.

1. Check the tires and wheels for damage and proper inflation (Figure 5-5). Check tire runout and examine each tire's wear pattern for signs of steering and suspension problems.
2. Check the control arm, sway bar, ball joints, and shock bushings for damage or wear.
3. Check engine and transaxle mounts for wear and deterioration, which can cause clunking during power applications.
4. Check the strut, brake hardware, and steering components for wear or damage. These can create noise sometimes thought to be CV joint related.
5. Check all engine and chassis electrical grounds. If these are bad, the electrical system may find a ground through the wheel bearing and/or drive axles. When this happens, these components can be damaged.

With the engine off, raise the vehicle and look for torn CV joint boots; missing, loose, or incorrect boot clamps; and evidence of grease having been slung or thrown in the area of the CV joint boot. Boots should be airtight and have secure clamps on both ends. Check both the outer and the inner joint on both sides of the vehicle (Figure 5-6). Remember, CV joint boots not only keep the lubricant in but also keep moisture and dirt out of the joint. Defective boots are the most common cause of CV joint failure.

Boots can fail for the following reasons:

1. Cuts or tears from foreign objects.
2. Sheet-metal damage resulting from an accident.
3. Improper hookup during towing. The tow truck operator may have connected to the drive axle instead of the frame or control arm.

Classroom Manual
Chapter 5, page 116

Classroom Manual
Chapter 5, page 121

CONDITION		PROBABLE CAUSE	CORRECTIVE ACTION
Shoulder wear 		• Underinflation (both sides wear) • Incorrect wheel camber (one side wear) • Hard cornering • Lack of rotation	• Measure and adjust pressure • Repair or replace axle and suspension parts • Reduce speed • Rotate tires
Center wear 		• Overinflation • Lack of rotation	• Measure and adjust pressure • Rotate tires
Toe-in or toe-out wear 	Feathered edge	• Incorrect toe	• Adjust toe
Uneven wear 		• Incorrect camber or camber • Malfunctioning suspension • Unbalanced wheel • Out-of-round brake drum • Other mechanical conditions • Lack of rotation	• Repair or replace axle and suspension parts • Repair, replace, or if necessary, reinstall • Balance or replace • Correct or replace • Correct or replace • Rotate tires

FIGURE 5-5 Tire wear problems and needed repairs.

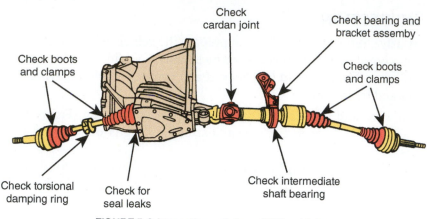

FIGURE 5-6 Inspection points on FWD vehicles.

Check boots and clamps

Check cardan joint

Check bearing and bracket assemby

Check boots and clamps

Check torsional damping ring

Check for seal leaks

Check intermediate shaft bearing

4. Ice forming around or near the boot and cutting it when the engine's torque breaks the ice away and turns the axle.
5. Deterioration caused by the environment and fluid leaks.
6. Boot clamp failure.
7. Improper service on a suspension component.
8. Improper removal of the half shaft.

Although the condition of the boot is a good indicator of drive axle and joint condition, it is possible to have a good boot on a defective joint or shaft. Continue your inspection by checking the half shafts for signs of damage or being bent. Also check for sloppy transaxle output shaft bushings, which are evident by fluid leaks at the output shaft. Worn transaxle output shaft bushings often produce excessive side-to-side play on the inboard joints. Also check for signs of leakage where the inner joints fit into the transaxle case. Leakage may occur from the seal through which the joint passes and the seal should be replaced (Figure 5-7).

If there is any indication of torn boots, failed joints, bad clamps, slung grease from the joints, or worn transaxle carrier assemblies, the shaft should be removed from the vehicle and the problem fixed.

SERVICE TIP:
Never attempt to diagnose CV joint problems on a lift with the wheels hanging. The weight of the wheels will make a loose joint seem tight.

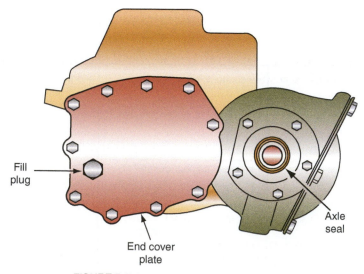

Fill plug

End cover plate

Axle seal

FIGURE 5-7 Axle seal in a transaxle housing.

Drive Axle Inspection

Rotate the wheels slowly and examine the shaft for damage and excessive **runout**. Some runout is allowed on certain cars (as much as ¼ inch) and will not cause vibration problems. Always refer to the vehicle's specifications before coming to a conclusion about the shaft.

The drive axles should also be checked for signs of contact or rubbing against the chassis. Rubbing can be a symptom of a weak or broken spring or engine mount as well as chassis misalignment. Also check the half shaft mounting flanges and bolts at the transaxle and the wheel spindles for looseness. On FWD transaxles with equal-length half shafts, inspect the intermediate shaft U-joint, bearing, and support bracket for looseness by rocking the wheel back and forth and watching for any movement. Oil leakage around the inner CV joints indicates a faulty transaxle shaft seal. To replace the seal, the half shaft must be removed.

Support the tires so their weight is not on the half shafts. Then, grasp the shaft near where it fits into the transaxle and check for excessive play or movement. The output shaft should not wobble. Various models of vehicles have had problems in this area. You cannot accurately check a CV joint by moving it to see if it is loose. CV joints will have some physical looseness because of their design. Also, when the car is raised, the inner CV joint will be extended beyond the point of normal wear and any excessive looseness may not be evident. The best way to check a joint is to observe its operation.

DRIVE AXLE REMOVAL

The exact procedure for removing and installing drive axles varies with car models. Always refer to your service manual for the special procedures pertaining to the model you are working on. A typical procedure for removing and installing drive axles is represented in Photo Sequence 9. Additional guidelines to follow when removing and installing a half shaft follow.

1. Loosen and tighten the axle hub nut with the vehicle on the ground. It is advisable not to use an impact. Wheel bearings and CV joint races can be damaged by an air impact. Use a large breaker bar or a long-handled ratchet instead. Use the weight of the car and the traction of the wheel on the shop floor to manually break loose the shaft nut. Have a helper apply the brakes while you loosen or tighten the axle nut.
2. In most cases the easiest way to disconnect the suspension at the wheel is to separate the lower ball joint from the steering knuckle. After you pull the bolt out of the steering knuckle, the stud should slide out. Remove the **pinch bolt** and use a pry bar or special control arm tool to separate the ball joint from the steering knuckle. Some Ford models with aluminum lower control arms require a special puller to separate the ball joint from the control arm (Figure 5-9).
3. Disconnect and separate the tie-rod end from the steering knuckle. If you remove a pinch bolt, be sure to replace it when reassembling the suspension and tighten the nut to the proper specifications. If an adjustable suspension connection was separated (Figure 5-10), a front-end alignment must be performed on the vehicle before it is returned to the customer.
4. If possible, disconnect the sway bar bolts to release tension on the lower control arm. This will make the job of loosening the control arm easier. On some vehicles, it may be necessary to remove the brake caliper. If this is done, make sure to suspend it with wire.

FIGURE 5-8 Typical drive axle remover.

SERVICE TIP:
You cannot accurately check a CV joint by moving it to see if it is loose. CV joints will have some physical looseness because of their design. Also, when the car is raised, the inner CV joint will be extended beyond the point of normal wear and any excessive looseness may not be evident. The best way to check a joint is to observe its operation.

SPECIAL TOOLS
Hub puller
Drive axle remover (Figure 5-8)
Snapring pliers
Brass drift
Boot protector
Staking punch
Soft-faced hammer
Pry bars
Boot clamp pliers

A **pinch clamp** and **pinch bolt** are used on many front suspensions to hold the lower control arm's ball joint to the steering knuckle.

A **pinch bolt** is sometimes referred to as a clamp bolt.

TYPICAL PROCEDURE FOR REMOVING AND INSTALLING DRIVE AXLES

P9-1 With the tire on the ground, remove the axle hub nut and loosen the wheel nuts.

P9-2 Raise the vehicle and remove the tire and wheel assembly.

P9-3 Remove the brake caliper and rotor. Be sure to support the caliper so that it does not hang by its weight.

P9-4 Remove the bolts that attach the lower ball joint and tie-rod end to the steering knuckle. Then pull the steering-knuckle assembly from the ball joint and tie-rod.

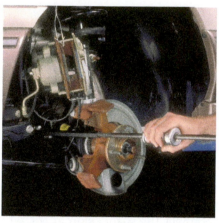

P9-5 Remove the drive axle from the transaxle by using the special tool.

P9-6 Remove the drive axle from the hub and bearing assembly using a spindle remover tool.

P9-7 Remove the drive axle from the car.

P9-8 Install joint in a soft-jawed vise to make necessary repairs to the shaft and joints and prepare to reinstall the shaft.

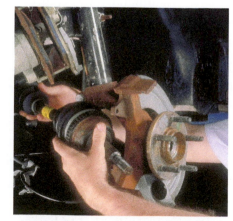

P9-9 Loosely install the half shaft into the steering knuckle and transaxle.

P9-10 Tap the splined end into the transaxle or install the flange bolts and tighten them to specifications.

P9-11 Pull the steering knuckle back and slide the splined outer joint into the knuckle.

P9-12 Fit the ball joint to the steering knuckle assembly. Install the pinch bolt and tighten it.

P9-13 Install the axle hub nut and tighten it by hand.

P9-14 Install the rotor and brake caliper.

P9-15 Torque the caliper bolts to specifications.

P9-16 Install the wheel and tire assembly.

P9-17 Lower the car.

P9-18 Torque the hub nut to the specified amount.

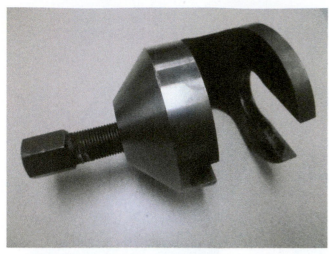

FIGURE 5-9 Ball joint separator for some Ford models.

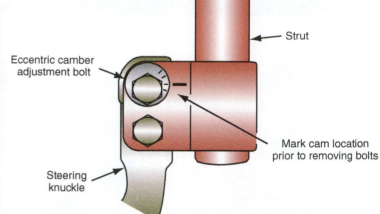

FIGURE 5-10 If an adjustable suspension connection, such as one with an eccentric washer, is disconnected or loosened, the wheels must be aligned before the vehicle is returned to the customer.

SERVICE TIP:
If it is necessary to drive the axle through the hub, use a soft-tip mallet or a block of wood between the hammer and the axle.

5. When removing a press-fit joint from the steering knuckle, push the half shaft back through the knuckle with a two- or three-leg puller (Figure 5-11). Make sure the adapter's attaching nuts or bolts are fully threaded onto the hub. Once the tool is in place, hold the tool stationary while tightening the tool's nut. The use of a hammer and a drift punch to remove a press-fit joint from the steering knuckle can cause damage to the transaxle's differential gear teeth. Always use the proper pullers and recommended special tools to remove the shaft from a wheel bearing to prevent damage to the bearings and their races.

6. Most Ford applications require 5–7 tons of pressure to remove or install the outer CV joint in the wheel hub.

7. On vehicles equipped with an antilock brake system, be careful not to damage the sensor, sensor cable, or the trigger wheel mounted on the CV joint housing. The air gap between

FIGURE 5-11 Using a hub puller to push the axle out of the steering knuckle.

the sensor and the trigger wheel is critical to proper ABS operation (Figure 5-12). The gap should be measured with a nonmetallic feeler gauge. If the air gap is wrong or if the sensor, sensor cable, or trigger wheel is damaged, the ABS control unit will shut down the system. This will allow normal braking without ABS and will cause the amber ABS light to be lit.

8. Never allow the half shaft to dangle unsupported. Severe damage to the inner joint/boot could occur. Do not stretch the brake hose. Remove and support the caliper if necessary (Figure 5-13).

9. Always follow the correct service procedures when removing the inner CV joint's outer housing from the transaxle; damage to the internal components of the inner CV joint can occur.

Most cars with ABS have labels or warning lights that announce the fact that the car has an ABS.

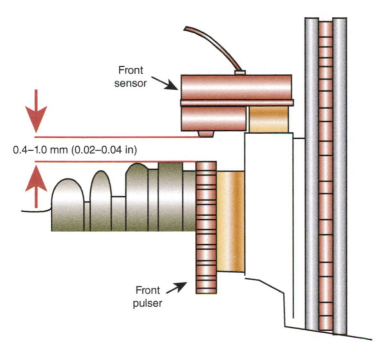

FIGURE 5-12 Air gap between the wheel speed sensor and the trigger wheel (pulser) on a front CV joint assembly.

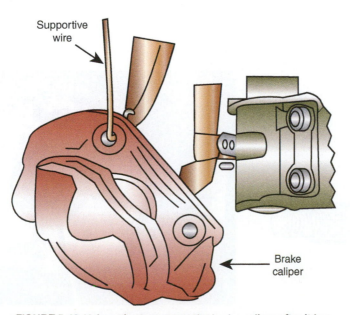

FIGURE 5-13 Using wire to support the brake caliper after it has been unbolted from the steering knuckle.

10. Mark the mounting positions on flange-type CV joints to be sure of retaining balance when they are reassembled.

11. Mark the positions of the boots on the shafts and joint housings. There will not always be a ridge to show you where to put it back. Also use a boot protector on each axle boot to prevent damaging them while removing the axle (Figure 5-14).

12. Some transaxles require that a spacer be placed in the differential to keep the differential side gears from falling loose if you pull both half shafts out at the same time. If this center piece is not available or if you are not sure which transaxles require this piece, remove one half shaft at a time.

13. When disconnecting the inner joint from a stub-shaft flange, four-square, hex, or allen-head bolts might need to be removed. Always clean the inside of the bolt heads to make sure your tool can fully grip the bolts. Spline-type joint housings can be pried or pulled from the transaxle. Be careful not to damage the transaxle case and seal. Some Chrysler half shafts are spring loaded and can be pulled out by hand.

14. Only use new axle hub nuts. Never reuse torque prevailing nuts; they will not hold specifications the second time around.

15. Before installing the half shaft, make sure all surfaces of the shaft are clean. Put a slight film of lube on the inner joint where it passes through the transaxle seal.

16. On half shafts that are press-fit to the steering knuckle, use the correct special tool to pull the CV joint into the knuckle.

17. Overtorquing the control arms' pinch bolts can lead to bolt failure. Always tighten to specifications and never use an impact wrench on these bolts.

18. When installing the washer and axle hub nut, use a torque wrench and tighten it to specifications. Make sure the threads of the shaft are clean. Use a staking punch to lock the nut into position (Figure 5-15).

> A staking punch is a chisel-like punch used to create a large dimple in metal. This dimple prevents the nut from loosening.

AUTHOR'S NOTE: Sometimes the inner CV joint is really stuck into the side gear of the differential. A slide hammer tool might not work, and the next step is a big pry bar. Patience and a professional attitude can be important here. More than one transaxle case has been cracked when a technician gets too rough trying to pry an inner joint out.

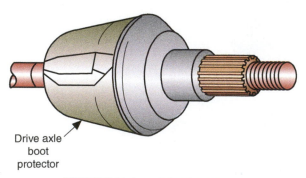

Drive axle boot protector

FIGURE 5-14 An axle boot protector.

FIGURE 5-15 Most axle hub nuts are staked after they are tightened to specifications in order to lock them in place.

BENCH INSPECTION

Mount the half shaft in a vise. Be careful not to tighten the vise too far, as overclamping could damage the shaft. Mark the shaft where the inner end of the boot sits, unless the boot seats in a locating groove (Figure 5-16).

Remove the outer joint boot and wipe any excess grease from around the inner race and cage. Carefully inspect the grease wiped from the assembly and the remaining grease in the joint for contamination (Figure 5-17). Water inside the joint will cause emulsification, which gives the grease a milky appearance. If the grease is gritty from sand or dirt contamination and the vehicle has been run for some time with a torn boot or other condition that allowed the contamination, the joint must be replaced. If the grease is not contaminated with water or dirt, only a replacement boot may be required.

Look for cracks, chips, pits, or rust on any interior component, including the balls, cage, inner race, or housing. A discolored outer CV joint housing is normal, and is caused by the heat treatment it received during manufacturing. Although a shiny surface on the races of the joint does not mean the joint must be replaced, pitting or brinelling do.

A test that may be performed to get the feel of any internal wear on inner CV joints is to tilt the joint to its maximum angle. While it is tilted, attempt to plunge the joint inward and outward. If the joint catches, stops, or sticks along its path of travel, one or more of the bearings or rollers have wear spots on them and the joint should be replaced.

On Rzeppa and cross-groove joints check for signs of cracks, wear, or dimpling in the cage windows, making sure that each ball fits snugly in its proper place (Figure 5-18). Also check for wear, cracks, and dimpling or rough grooves in the inner race or outer housing of the joint.

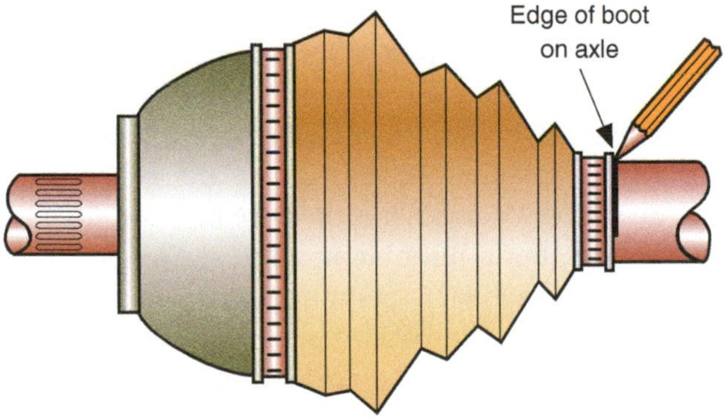

Classroom Manual
Chapter 5, page 124

Emulsification forces water droplets to mix with the grease, resulting in a thicker solution than normal grease.

Brinelling is evident by dents in the surface.

An outer CV joint will tend to wear ten times faster than an inboard joint.

Edge of boot on axle

FIGURE 5-16 Mark the position of the boot on the axle shaft.

Check lubricant for contamination by rubbing between two fingers

Any gritty feeling indicates a contaminated CV joint

FIGURE 5-17 Carefully check the joint's grease.

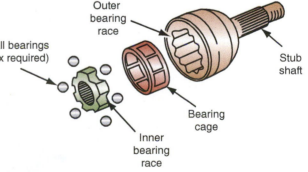

Outer bearing race

Ball bearings (six required)

Stub shaft

Bearing cage

Inner bearing race

FIGURE 5-18 Carefully inspect the balls, races, and housings of a Rzeppa joint.

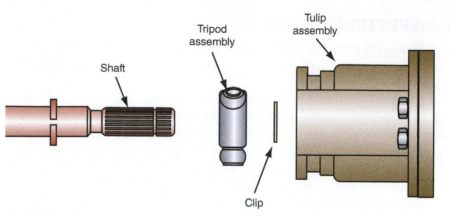

FIGURE 5-19 The tripod and tulip assembly should be carefully inspected for cracks, wear, and damage.

If the inner joint has become separated, determine the cause of the problem. Check for a bent or rusted frame member, worn suspension components, and bent, rusted, or worn engine/transaxle mounts.

With tripod-design CV joints (Figure 5-19), look for wear, cracks in rollers, needle bearings, and roller tracks. The balls, because they usually have a chrome or high-alloy finish, are not likely to fail, rust, or discolor unless they have gotten very hot or have been exposed to too much moisture.

OBTAINING CV JOINT AND DRIVE AXLE REPLACEMENT PARTS

Whenever you replace an axle, CV joint, or boot, make sure the parts are correct for that vehicle. Be careful. Often manufacturers change the type and style of joint during production.

If the axle needs to be replaced, install a complete shaft. Most part suppliers offer a complete line of original equipment drive shafts for FWD vehicles. These shafts come fully assembled and ready for installation.

If only the CV joints need service, a CV joint service kit should be installed. Joint service kits typically include a CV joint, boot, boot clamps and seals, special grease for lubrication (the correct quantity is packed in each kit), retaining rings, and all other attachment parts.

Part manufacturers also produce a line of complete boot sets for each application, including new clamps and the appropriate type and amount of grease for the joint (Figure 5-20). CV joints require a special high-temperature, high-pressure grease. Substituting any other type of grease may lead to premature failure of the joint. Be sure to use all the grease supplied in the joint or boot kit. The same rule applies to the clamps. Use only those clamps supplied with the replacement boot. Follow the directions for positioning and securing them.

CAUTION:

Old boots should never be reused when replacing a CV joint. In most cases, failure of the old joint is caused by some deterioration of the old boot. Reusing an old boot on a new joint usually leads to the quick destruction of the joint.

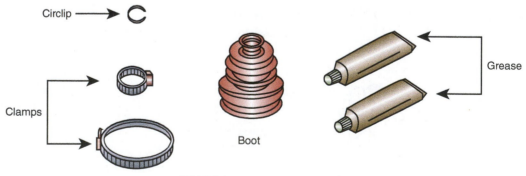

FIGURE 5-20 A complete boot kit.

GENERAL SERVICE PROCEDURES

Replacing Joint Boots

Photo Sequence 10 guides you through the replacement of a typical CV joint boot. The exact procedure for doing this varies with the various makes and models of cars, therefore always refer to the appropriate service manual when replacing a CV joint boot.

Here are some basic guidelines to follow whenever you are replacing a boot. Whenever a CV joint is serviced, always install a new boot. Never reuse a boot.

Mark on the drive axle the location of the small end of the old boot before you remove it. This will help you to properly position the new boot on the axle. If possible, the axle should be mounted in a vise when removing or installing boots. To remove the boot, remove the clamps (Figure 5-21) and slip the boot off the shaft.

When replacing a boot, always inspect the grease in the joint for contamination. If the grease is clean, the joint is probably working properly. Clean as much grease away from the joint as possible, and then apply the grease that comes with the boot kit (Figure 5-22). Next, install the new boot and new clamps onto the shaft. Spread the clamps just enough so that it can be slipped over the axle. The clamp should be positioned so that the end with the holes points in the direction of axle rotation.

Now slide the clamp from the axle to the desired end of the boot. Position the clamp so that the end with the holes fits over the locking hooks (Figure 5-23). Using moderate hand

Classroom Manual
Chapter 5, page 121

SPECIAL TOOLS
Clean rags
Diagonal cutters
Boot clamp tool
Vise with soft jaw inserts
Dulled screwdriver

SERVICE TIP:
Make sure the new grease is evenly distributed in the boot. Also, use all of the grease provided in the kit.

⚠️

CAUTION:
Never mix greases when packing the CV joint. Incorrect matching of greases may lead to boot deterioration.

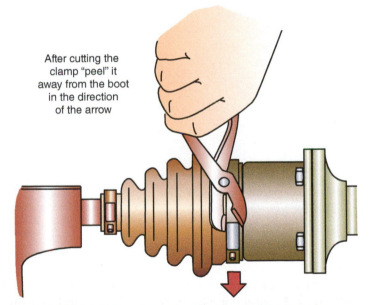

After cutting the clamp "peel" it away from the boot in the direction of the arrow

FIGURE 5-21 Use diagonal cutters to cut the boot clamp.

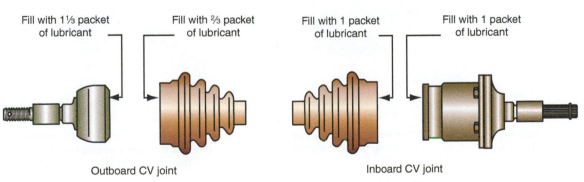

Fill with 1⅓ packet of lubricant

Fill with ⅔ packet of lubricant

Fill with 1 packet of lubricant

Fill with 1 packet of lubricant

Outboard CV joint

Inboard CV joint

FIGURE 5-22 Use all of the grease supplied with the kit.

REMOVING AND REPLACING A CV JOINT BOOT

P10-1 Removing the axle from the car begins with the removal of the wheel cover and the wheel hub cover. The hub nut should be loosened before raising the car and removing the wheel and tire assembly.

P10-2 After the car is raised and the wheel is removed, the hub nut can be unscrewed from the shaft.

P10-3 The brake line holding clamp must be loosened from the suspension.

P10-4 The ball joint must be separated from the steering knuckle assembly. To do this, first remove the ball joint retaining bolt. Then, pry down on the control arm until the ball joint is free.

P10-5 The inboard joint can be pulled free from the transaxle.

P10-6 A special tool is normally needed to separate the axle shaft from the hub. This allows the axle to be removed from the car.

P10-7 The axle shaft should be mounted in a soft-jawed vise for work on the joint. Pieces of wood on either side of the axle work well to secure the axle without damaging it.

P10-8 Begin boot removal by cutting and discarding the boot clamps.

P10-9 Scribe a mark on the axle to indicate the boot's position on the shaft. Then, move the boot off the joint.

P10-10 Remove the circlip and separate the joint from the shaft.

P10-11 Slide the old boot off the shaft.

P10-12 Wipe the axle shaft clean and install the new boot onto the shaft.

P10-13 Place the boot into its proper location on the shaft and install a new clamp.

P10-14 Using a new circlip, reinstall the joint onto the shaft. Pack joint grease into the joint and boot. The entire packet of grease that comes with a new boot needs to be forced into the boot and joint.

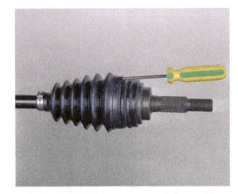

P10-15 Pull the boot over the joint and into its proper position. Use a dull screwdriver to lift the boot up, to purge it of any air that may be trapped, and to correct any collapsing of the boot.

P10-16 Install the new large boot clamp and reinstall the axle into the car. Tighten the hub nut after the wheels have been reinstalled and the car is sitting on the ground.

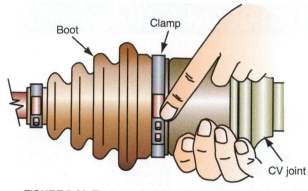

FIGURE 5-23 Fit the end of the clamp into the clamp's hooks.

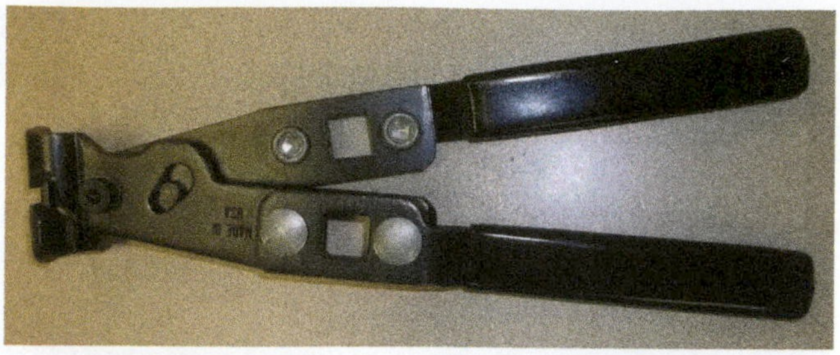

FIGURE 5-24 Clamp pliers designed to be used with a torque wrench and breaker bar.

Classroom Manual
Chapter 5, page 122

pressure, engage the hook at the tightest possible position. Squeeze the ear of the clamp firmly with the clamp pliers. If not crimped properly, the clamp will not form a good seal. Hard plastic boots, such as DuPont's Hytrel boots, require high clamping forces to provide the proper pressure needed to keep the boot from coming loose. A typical clamp tool with cut-outs for a breaker bar and torque wrench is shown (Figure 5-24). Torque specifications can be as high as 100 ft./lbs.

If the boot is dimpled or collapsed after you have installed it, carefully insert a dulled screwdriver between the boot and the shaft to allow air to enter the boot. The air should allow the boot to return to its normal shape (Figure 5-25).

Replacing CV Joints

The typical procedure for replacing CV joints and other FWD components begins with the removal of the drive axle. Etch a mark where the rubber boot ends on the half shaft. Remove the clamps, then the boots. Because the boots will be replaced, you can cut off the old ones to save some time. Clean all excess grease from the shaft and joint. Using your service manual, determine how the joints are retained to the shaft (Figure 5-26). Check for a snapring inside the joint, they are usually buried in the old grease. Replace all snaprings you remove during service. Snaprings lose their tension with time and also when you pinch them open or closed to remove them.

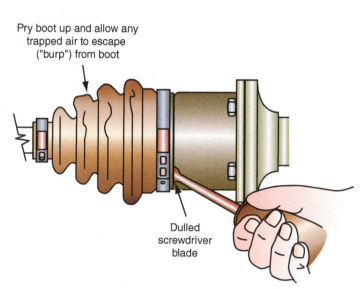

FIGURE 5-25 Use a dulled screwdriver to release trapped air from the boot.

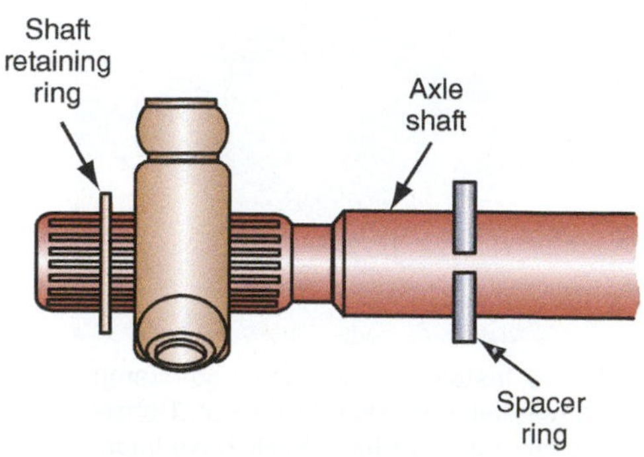

FIGURE 5-26 Normally, a circlip or snapring is used to retain the joint on the shaft.

The Rzeppa joint is the most commonly used fixed CV joint. This type of joint is held to the half shaft or the intermediate shaft in one of two ways. There may be an external retaining ring that fits into a bearing race. This clip must be expanded to separate the joint from the shaft. The other retention method is the use of an internal retaining clip. This clip fits into a groove in the shaft and expands into the inner race of the joint. This type of joint can typically be driven off the shaft with a brass punch and a hammer. Three types of plunging joints are most commonly used today: the cross groove, double offset, and tripod tulip. The cross-groove joint is found primarily on German vehicles. The joint consists of an inner race held to the shaft by a retaining ring and six balls held in place by a cage in the outer race. The outer race is usually a flange bolted to the transaxle.

Double-offset joints also have six balls and a cage to retain them. However, this joint is normally splined to the transaxle. Most often the shaft and joint are held in position by an internal clip or retaining ring. To remove this type of joint, it must be disassembled. Other double-offset joints use a circle clip. This type of joint can be driven off the shaft.

Tripod tulip joints use a three-point spider with rollers (called the tripod) and an outer race (called the tulip). The housing can be splined to the transaxle or bolted to a flange.

If it is a tripod-type joint, index the housing and inner tulip. Be sure to check your service manual to determine how the joint is retained. These joints are retained in various ways and failure to identify how they are retained may result in joint damage and frustration. Remove the retainer or tabs from the outer race, and slide the race from the joint while holding the balls and needle bearings in place. Remove the circlip nearest the end of the half shaft and discard it. Do not remove the heavy inner spacer ring. It needs to remain on the shaft to help position the joint properly when you put the assembly back together. To remove the spider from the housing, lift the retaining ring out of its groove. While removing an inboard tripod-type joint's spider from its housing, hold the three balls in place to prevent them and their needle bearings from falling off the joint. Once the spider is free from the housing, wipe away all excess grease and wrap tape or a rubber band around the balls and spider to prevent them from separating while the spider assembly is outside of the housing. Use a brass punch and drive the spider off the shaft (Figure 5-27).

If it is a Rzeppa joint, index the parts then use a drift punch to tap the CV joint assembly off the half shaft. Hit the joint near the spline (Figure 5-28).

Thoroughly clean the shaft, inspect it for wear, and check the grease for evidence of contaminants such as dirt and water. Wipe any remaining grease from the splines and shaft. Clean the surface where the boot clamps on the half shaft with solvent and dry thoroughly. Slide the new clamp and boot onto the half shaft. Then, position the new joint on the splines and

SPECIAL TOOLS
Needle-nose pliers
Snapring pliers
Appropriate pullers
Drift punch
Tape

SERVICE TIP:
When servicing CV joints on vehicles equipped with ABS, make sure you protect the wheel-speed sensors and tone rings, located on the outer-joint housings, to avoid ABS problems.

The internal retaining ring may be called a circle clip or circlip.

Joints that are retained by an internal clip are sometimes called knock-off joints.

SERVICE TIP:
If a joint is retained by a snapring or circlip, use needle-nose pliers or snapring pliers to compress the ring. When you are doing this, use a screwdriver to pry into the side gear. The screwdriver will prevent the ring from falling back into its groove.

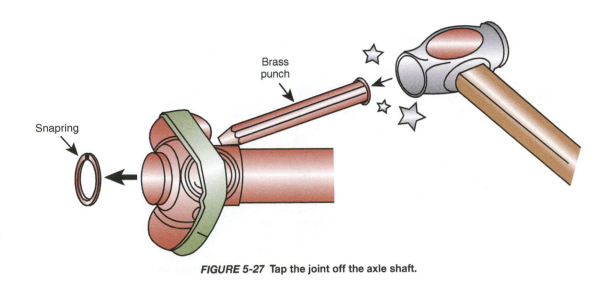

Snapring

Brass punch

FIGURE 5-27 Tap the joint off the axle shaft.

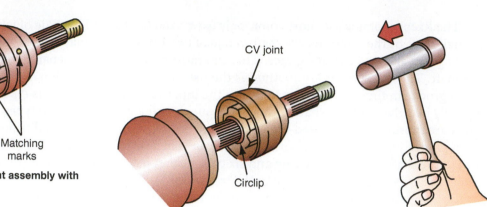

FIGURE 5-28 Index the joint assembly with the outer housing.

FIGURE 5-29 Use a soft-faced mallet to drive the joint into position on the axle shaft.

tap the joint on until you hear a click, which indicates that the circlip has seated in the groove inside the joint (Figure 5-29). Pull on the joint to check that it is properly locked on the shaft.

Install a new circlip. Do not overexpand or twist the circlip during installation. Thoroughly pack all of the supplied grease into the assembly or joint.

Line up the boot to the etch mark and tighten the small clamp. Squeeze the remaining grease into the boot over the joint housing. Make sure all of the grease provided in the kit is applied to the joint. Make sure the boot is not twisted or collapsed. If it is collapsed, allow air into the boot by venting it with a screwdriver.

Pull the large end of the boot over the joint housing. Where applicable, distribute the second packet of grease into the boot. Reposition the large end of the boot, install the clamp, and tighten it. The axle can now be reinstalled in the car.

Many FWD vehicles use an intermediate shaft to allow for equal-length right and left half shafts. Intermediate shafts have support bearings and seals (Figure 5-30) that are serviceable. To provide noise-free operation, the bearings must be well lubricated. The seals retain the

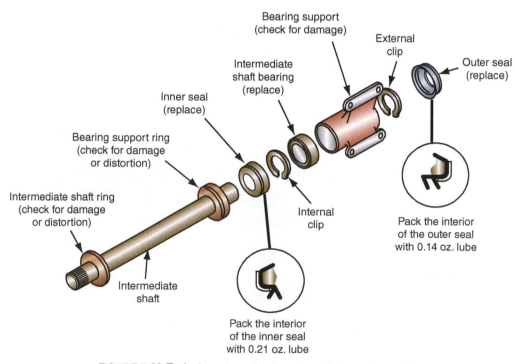

FIGURE 5-30 Typical service guidelines for an intermediate shaft.

lubricant in the bearings as well as seal out contaminants. If the intermediate shaft is wet or covered with grease, carefully check the support bearing and replace the seals.

FWD Front Wheel Bearings

Servicing a front wheel bearing on a FWD car is different than servicing one on a RWD car. Most FWD cars don't use a separate inner and outer tapered roller bearing like the front wheels of RWD cars. They are usually equipped with a bearing assembly per wheel (Figure 5-31). This bearing may have two rows of balls, and is permanently sealed so that it requires no periodic lubrication. The double-row ball bearings have races that allow the bearing to handle radial and side loads. Other FWD vehicles use inner and outer tapered roller bearings contained in a one-piece sealed unit.

Wheel Bearing Inspection

A hub-and-bearing assembly will become noisy when it is excessively worn, if spalling or brinelling of the bearings has occurred, or if the unit has a cracked bearing race. However, other signs of failure may be evident before the unit becomes noisy.

Bearing wear can be checked by grasping a front tire at the 12 o'clock and 6 o'clock positions, with the vehicle supported under its lower control arms. Pull outward on the top of the tire while pushing inward at the bottom. Alternate pushing and pulling while you check for play. This is the same procedure used for checking front wheel bearing play on RWD cars that use tapered roller bearings in the front hubs. This check can also be conducted on the rear wheels of GM and other cars that have sealed hubs.

Normally, any play indicates a defective wheel bearing assembly. Any movement that is felt while conducting the aforementioned check, without forcing the tire to move, will indicate more than the allowable play. Movement of less than 0.010 inch will not be felt.

Replacement is necessary when a front hub-and-bearing assembly has excessive play or is making noise. Not replacing a unit that is failing may cause the assembly to overheat and lock up while the vehicle is driven.

WARNING: Failure to replace a defective wheel bearing can result in the driver losing all braking and steering control. Locked-up front hub-and-bearing assemblies can cause, and have caused, serious accidents.

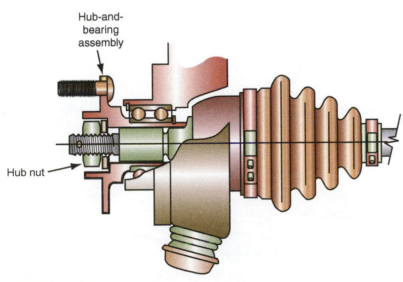

FIGURE 5-31 FWD front wheel and hub assembly.

CAUTION:
Whenever a new CV joint is installed, always use a new boot. Never reuse the old boot because it was probably the reason for joint failure.

Classroom Manual
Chapter 5, page 131

SPECIAL TOOLS
Two- or three-leg gear puller
Appropriate seal installer
Pry bar
Brass drift
Soft-faced hammer
CV joint boot protector

To check for a binding hub-and-bearing assembly, simply rotate the tire and wheel. If the tire does not turn freely, suspect dragging brakes or a binding hub-and-bearing assembly.

Replacing Front Wheel Bearings

The wheel bearings on most FWD vehicles are one-piece, unitized assemblies that do not require periodic lubrication, but must be replaced if they are worn or damaged. The procedure for removing and replacing a front wheel bearing is different for different makes of vehicles. Photo Sequence 11 shows the procedure for replacing wheel bearings on GM and some Chrysler vehicles. This design is the easiest to replace. The hub assembly, with the bearing assembly, bolts onto the steering knuckle.

Photo Sequence 12 shows the procedure for replacing a front wheel bearing on nearly all FWD Ford and Asian vehicles, as well as some Chrysler models. These use a bearing that is pressed into the steering knuckle. The steering knuckle must be completely removed to service the bearings.

Have someone apply the brakes with the vehicle's weight on its tires. Loosen the axle hub nut and wheel lug nuts with the appropriate tools. Then remove the axle hub nut.

Raise the car and remove the wheels. At this point it's a good idea to install a boot protector over the CV joint boot to protect it while you are working. Next, unbolt the brake caliper from the steering knuckle and support it on the strut with a piece of wire. Make sure the flexible brake hose doesn't support the caliper's weight.

Remove the brake rotor, which is normally attached to the knuckle with one or two bolts. Then, loosen and remove the pinch bolt that holds the lower ball joint to the steering knuckle. Remove the cotter pin and castellated nut from the tie-rod end and pull the tie-rod end from the knuckle. The tie-rod end can usually be freed from its bore in the knuckle by using a small gear puller or by tapping on the metal surrounding the bore in the knuckle. On some cars, an **eccentric washer** is located behind one or both of the bolts that connect the spindle to the strut. These eccentric washers are used to adjust camber. Mark the exact location and position of this washer on the strut before you remove it. This allows you to reinstall the washer and the assembly without drastically disturbing the car's camber angle.

Once the washer is marked, remove the hub-and-bearing to knuckle retaining bolts. With a soft-faced hammer, tap on the hub assembly to free it from the knuckle. The brake splash guard may come off with the hub assembly. Install the appropriate puller and press the hub-and-bearing assembly from the half shaft. Pull the half shaft away from the steering knuckle (Figure 5-32). Do not allow the CV joint to drop downward when you remove the hub assembly. Carefully check the external splines on the axle stub of the half shaft and the internal

SERVICE TIP:
If no one is available to apply the brakes for you, use a brake-pedal depressor from a front-end alignment machine to depress and hold the brake pedal down.

An **eccentric washer** is a normal-looking washer except that its hole is offset from the center.

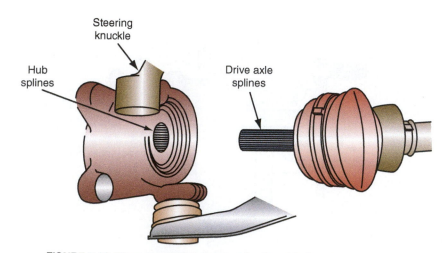

Hub splines

Steering knuckle

Drive axle splines

FIGURE 5-32 When separating the steering knuckle from the axle, check the external and internal splines.

REMOVING AND REPLACING A FWD GM FRONT WHEEL BEARING

P11-1 Loosen the hub nut and wheel nuts.

P11-2 Jack up the vehicle and remove the tire and wheel assembly.

P11-3 Unbolt the front brake caliper and move it out of the way.

P11-4 Suspend the caliper with wire.

P11-5 Remove the rotor.

P11-6 Remove the hub retaining bolts.

P11-7 Remove the air deflector.

P11-8 Remove the center hub nut.

P11-9 Using a puller, remove the hub assembly from the knuckle.

P11-10 Remove the seal from the inside of the knuckle. Inspect and clean the hub's mounting surface on the knuckle.

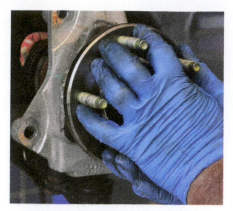

P11-11 Place the new seal and hub assembly onto the knuckle.

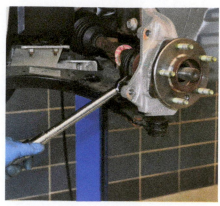

P11-12 Tighten the retaining bolts to the specified torque.

P11-13 Install the rotor and caliper assembly.

P11-14 Loosely install the hub nut.

P11-15 Install the tire and wheel assembly.

P11-16 Lower the vehicle and tighten the hub and lug nuts to specifications.

REMOVING AND REPLACING A FWD FORD FRONT WHEEL BEARING

P12-1 Loosen the wheel nuts.

P12-2 Loosen the hub nuts.

P12-3 Jack up the vehicle and remove the tire and wheel assembly.

P12-4 Unbolt the front brake caliper.

P12-5 Suspend the caliper with wire.

P12-6 Remove the rotor.

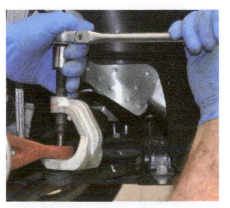

P12-7 Separate the lower ball joint and tie-rod end from the knuckle.

P12-8 Index the camber eccentric bolt to ensure proper camber adjustment during reassembly.

P12-9 Remove the knuckle-to-strut bolts.

P12-10 Remove the steering knuckle and support the drive axle with wire.

P12-11 Using a press with the proper mandrel, press the hub out of the knuckle.

P12-12 With a puller, remove the inner bearing race from the hub.

P12-13 Unbolt the retaining plate.

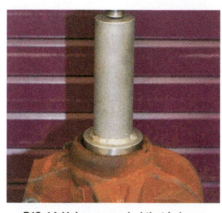

P12-14 Using a mandrel that is large enough, press the bearing out of the knuckle.

P12-15 Inspect the knuckle. Lightly lubricate the outer surface of the new bearing assembly and the inner bore of the knuckle.

P12-16 Press the new bearing into the knuckle using a mandrel that is large enough to touch the outer race of the bearing.

P12-17 Install the retaining plate.

P12-18 Using a hollow mandrel, press the hub into the bearing assembly. The mandrel must only touch the inner race of the bearing.

P12-19 Slide the knuckle assembly over the drive axle.

P12-20 Mount the knuckle to the strut and tighten the bolts.

P12-21 Make sure the camber marks to the strut are positioned properly.

P12-22 Reinstall the lower control arm ball joint and tie-rod end to the steering knuckle. Install the rotor, then mount the brake caliper. Torque all bolts to specifications.

P12-23 Install the tire and wheel assembly, then install the wheel lug nuts and the new hub nut.

P12-24 Lower the vehicle. Tighten all bolts to specifications. Stake or use a cotter pin to keep the hub nut in place after it is tightened.

splines in the hub. If either is damaged, they must be replaced. Support the half shaft and the CV joint from the back side of the knuckle (Figure 5-33).

Remove the pinch bolt holding the steering knuckle to the lower ball joint, and then remove the knuckle from the car (Figure 5-34).

With the steering knuckle off, remove the snapring or collar that retains the wheel bearing. Mount the knuckle on a hydraulic press so that it rests on the base of the press with the hub free. Using a mandrel slightly smaller than the inside diameter of the bearing, press the hub out of the bearing. On some cars, this will cause half of the inner race to break out. Because half of the inner race is still on the hub, it must be removed. To do this, use a small gear puller (Figure 5-35). If the proper puller is available, mount the steering knuckle in a vise. Insert the tool through and around the bearing and hub assembly (Figure 5-36). Tighten the tool. This should press the bearing out into the tool.

Remove the snaprings that hold the bearing in the knuckle. Then, using a mandrel large enough to touch the outer race of the bearing, press the bearing out of the knuckle. Inspect

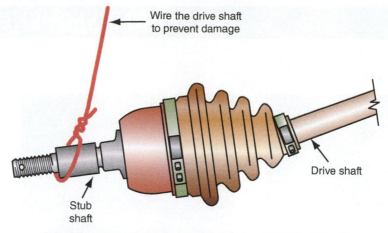

Wire the drive shaft
to prevent damage

Drive shaft

Stub
shaft

FIGURE 5-33 The outer end of the axle should not be allowed
to hang freely; suspend it with wire.

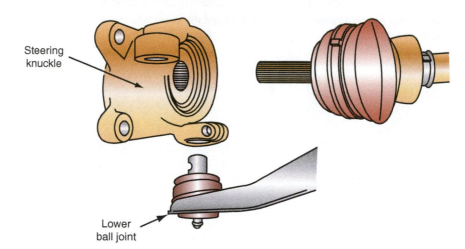

Steering
knuckle

Lower
ball joint

FIGURE 5-34 After the axle has been pulled and the strut disconnected,
the steering knuckle can be removed from the car.

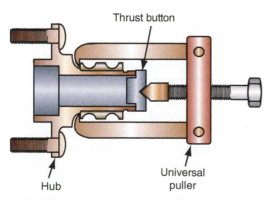

Thrust button

Hub

Universal
puller

FIGURE 5-35 Using a puller to remove bearing
race from the hub.

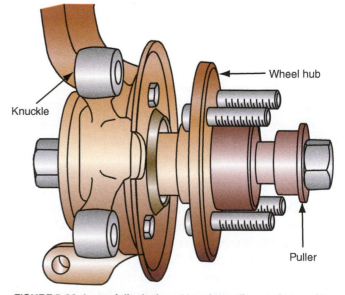

Knuckle

Wheel hub

Puller

FIGURE 5-36 A specially designed bearing puller can be used to
remove the bearing from the steering knuckle.

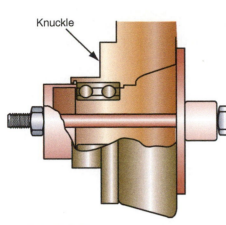

Knuckle

FIGURE 5-37 Use the correct tools to press the bearing into the knuckle.

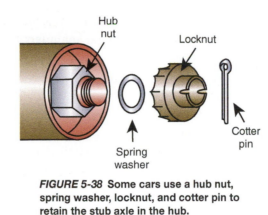

Hub nut

Locknut

Cotter pin

Spring washer

FIGURE 5-38 Some cars use a hub nut, spring washer, locknut, and cotter pin to retain the stub axle in the hub.

the bearing's bore in the knuckle for burrs, score marks, cracks, and other damage. If no defects are found, install a new snapring in the hub side, lightly coat the outer surface of the new bearing and the inner surface of the knuckle bore with moly lube, then press the bearing into the knuckle. Make sure the mandrel applies pressure only to the outer race of the bearing (Figure 5-37). Applying pressure to any other part of the bearing will ruin it.

Using the proper tool, press the new steering knuckle seal into place. Lubricate the lip seal and coat the inside of the knuckle with a thin coat of grease. With the bearing and seal fully seated against the hub-side snapring, install the snapring on the other side of the knuckle. Then, lubricate and install a new O-ring around the rear bearing housing. Install the splash guard and hub assembly on the knuckle. Then reattach the brake splash shield.

To complete the assembly of the steering knuckle, the hub must be pressed into the inner race. Use a hollow mandrel that contacts only the inner race of the bearing and that has an inner circumference large enough to allow the hub to fully seat in the bearing. The steering knuckle is now ready to be reinstalled in the car. Push the new hub/bearing assembly onto the half shaft, while holding the CV joint for support. Be careful to line up the half shaft splines with those in the hub. Tighten all bolts to specifications.

Tighten the axle hub nut to pretorque specifications. This ensures that the bearing assembly is properly seated. Then install the brake caliper and rotor. Now, install the tire/wheel assembly and lower the vehicle. Once the car is set on the ground, tighten the new axle hub nut and wheel lug nuts to their specified torque (Figure 5-38). Road-test the car and recheck the torque on the hub nut.

GUIDELINES FOR SERVICING CV JOINTS AND DRIVE AXLES

Always refer to the manufacturer's recommended procedures when servicing CV joints, drive axles, and wheel bearings. Although the exact procedures vary for servicing the different drive axle designs on the many different models of cars, there are some important rules that should always be followed.

1. FWD vehicles that have been towed should be examined for drive shaft problems. If the tow truck operator inadvertently used the drive shaft to support the hook, the drive shaft could be damaged.
2. Never reuse ball joint pinch bolts and nuts on Chrysler or Ford vehicles when reconnecting the lower ball joints.
3. Always torque the axle hub nuts to specifications in order to properly preload the wheel bearings.
4. Be especially careful when servicing CV joints on vehicles equipped with ABS. You must protect the wheel-speed sensors and tone rings, located on the outer-joint housings, to avoid damage and ABS problems.

SERVICE TIP:
After you have replaced one or more CV joints or removed a steering knuckle, always check the vehicle's wheel alignment. Although indexing the alignment during disassembly will get the alignment angles close, they won't be at the desired settings. Camber can be off by as much as ¾ degree due to differences between the size of the camber holes and the bolt.

CAUTION:
Hub nut torque specification is critical. If the nut is loose, the bearing preload will be wrong and the bearing will become noisy and fail in a short time.

A dummy CV joint is normally an old joint and stub axle assembly.

5. Whenever you replace a CV joint or boot, make sure the replacement parts are correct for that model car. Often car manufacturers change the type and style of joint during production, so make sure you have the correct parts.

6. Never reuse staked or previously tightened axle hub nuts. Always use new ones. These were designed to be used only once. This is also true of torqued ball joint nuts and bolts that are removed.

7. Never lower a FWD car onto its front wheels with an axle shaft removed unless a dummy CV joint and stub axle are installed and the axle nut tightened. The dummy shaft will keep the bearings in their races, thereby preventing damage to them. Never move a vehicle, even a few feet, while one or more CV joints are removed. This can damage the wheel bearings.

8. Never hammer on the threaded end of an axle. If you need to loosen the axle, use a punch centered on the end of the axle or a pusher-type tool that bolts to the wheel studs.

9. Check for a retaining or snapring on the inner race of an outboard CV joint before attempting to tap the joint off the shaft.

10. Never mix parts from different types of CV joints. Although they may look the same, their fit may not be exactly right.

11. Never allow a drive axle to hang unsupported. Often it may be necessary to wire the axle to the strut or frame. An unsupported axle can damage its joint or boot. Always support the drive axle at its ends and the axle horizontal.

12. Never pull on the axle shaft when attempting to remove it or when checking the engagement of differential side gears. Applying pressure to the shaft can cause the inner joint to come apart. Pressure should only be applied to the housing of the inboard CV joint.

13. Whenever you remove the spider of a tripod joint from the shaft, mark or note where the spider was installed on the axle so that you can reinstall it the same way.

14. After you have removed a Rzeppa joint from the shaft, mark the housing, cage, and inner race so that you can reassemble these components in the same position.

15. Never pry the bearing cage out of a Rzeppa joint. Tilt the cage by pressing it down on one side of the inner race. Some joints may use a retaining ring inside the outer race, which must be removed so the cage can be tipped. Tip the cage far enough to remove a ball (Figure 5-39), then tilt it further until you can pull the cage out of the outer race and remove the remaining balls. With the cage removed, rotate the inner race inside the cage and lift it out.

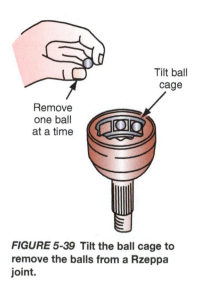

FIGURE 5-39 Tilt the ball cage to remove the balls from a Rzeppa joint.

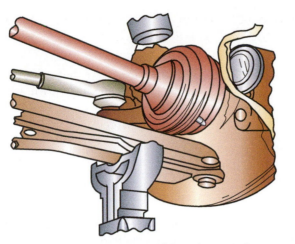

FIGURE 5-40 Support the vehicle on its control arm while spin balancing wheels on a FWD vehicle.

16. Never use the plunging action of an inboard CV joint as a slide hammer to disengage the joint from the transaxle. To free the joint, only use a pry bar or a crow's foot on a slide hammer. Never jerk or pull on a half shaft with inner tripod joints. The joint may pull apart, dropping its needle bearings from the rollers. Pull on the inner housing, supporting the outer end of the shaft until it is completely removed.

17. Never overtighten an axle in a vise while servicing it. Some axles are hollow and the vise can easily distort them.

18. When spin balancing the front wheels of a FWD car, support the car's weight at the control arms with a jack or stands to prevent damage to the joints and axles. Also position the jack stands so that the car's axle shafts are at their proper operating angle (Figure 5-40).

CASE STUDY

A customer brings in her car stating the car vibrates and makes a clicking noise. After the service writer asks the right questions, it is discovered that the only time it makes the noise is when the car is turning to the right. But it vibrates at all speeds and directions, and its intensity increases as the speed increases.

The technician takes the car for a road test and confirms both the clicking sound and the vibration. Knowing that this noise is normally caused by a faulty outer CV joint and that the joint on the inside of the turn is normally the cause, the technician recommends that a CV joint service kit be installed on the right side. The customer gives permission for the repair and the technician installs the kit.

After the kit was installed, the technician took the car out for a short test drive to verify the repair. While doing this, she notices that the clicking noise has disappeared but the vibration remains. She brings the car back to the shop and informs the customer that she overlooked something and will need to do more diagnostics. The customer appreciates her honesty and gives her permission to continue testing. She does, however, request that the technician call her back with a new estimate.

The technician begins to check those things she assumed were okay during her first diagnosis, such as the tires, suspension and steering parts, the transaxle, and the drive axles. While doing this, she finds traces of a fluid leak near the output shaft of the transaxle. Knowing this is an indication of a worn transaxle bushing and that a worn bushing will cause a vibration, she believes she has found the cause of the problem. However, to save the possible embarrassment of having to call the customer for permission to make a third attempt to determine the cause of the problem, she continues her visual inspection. Finding nothing else wrong, she informs the customer and makes the repair. Now both problems are solved.

TERMS TO KNOW

Dimpling
Eccentric washer
Pinch bolt
Runout
Shudder
Wheel shimmy

ASE-STYLE REVIEW QUESTIONS

1. When discussing the possible causes for a shudder during acceleration,

   Technician A says that a worn inner CV joint could be the cause.

   Technician B says that insufficient differential gear backlash could be the cause of the problem.

   Who is correct?
   A. A only
   B. B only
   C. Both A and B
   D. Neither A nor B

2. When discussing axle hub nuts,

   Technician A says that they are never to be reused after they have been staked.

   Technician B says that they should be chiseled off the stub axle if they are difficult to remove.

   Who is correct?
   A. A only
   B. B only
   C. Both A and B
   D. Neither A nor B

3. When installing a new CV joint boot,

   Technician A disassembles and cleans the joint to remove all old grease.

   Technician B packs all of the lubricant contained in the service kit into the joint and the boot.

   Who is correct?
   A. A only
   B. B only
   C. Both A and B
   D. Neither A nor B

4. When attempting to diagnose a vibration that is present only when the car is cruising at highway speeds,

   Technician A says that the problem may be caused by out-of-balance tires.

   Technician B says that the problem may be caused by a worn wheel bearing.

   Who is correct?
   A. A only
   B. B only
   C. Both A and B
   D. Neither A nor B

5. *Technician A* says a sticking inboard joint can cause the car to shimmy during acceleration.

   Technician B says a worn inboard joint can cause a clunking during acceleration.

   Who is correct?
   A. A only
   B. B only
   C. Both A and B
   D. Neither A nor B

6. When reviewing the service procedures for CV joints and drive axles,

   Technician A says that the axle can be used as a slide hammer to help remove a stubborn inboard joint from the transaxle.

   Technician B says that a drive axle should never be allowed to hang by its own weight during removal or installation.

   Who is correct?
   A. A only
   B. B only
   C. Both A and B
   D. Neither A nor B

7. When reviewing the procedure for replacing an axle boot,

   Technician A says that the position of the old boot should be marked on the shaft before removing it.

   Technician B says that if the boot is dimpled or collapsed after the new boot is installed, a dulled screwdriver should be used to allow air to enter the boot.

   Who is correct?
   A. A only
   B. B only
   C. Both A and B
   D. Neither A nor B

8. While discussing axle hub nuts,

   Technician A says that all axle nuts are torqued to the same specification.

   Technician B says the best way to install a hub nut is with an air impact.

   Who is correct?
   A. A only
   B. B only
   C. Both A and B
   D. Neither A nor B

9. When discussing the procedure for servicing a tripod-type CV joint,

   Technician A says that this type of joint is always retained with a snapring, which must be removed before the joint can be serviced.

   Technician B says that the location of the spider should be marked on the shaft before it is removed.

   Who is correct?
   A. A only
   B. B only
   C. Both A and B
   D. Neither A nor B

10. When diagnosing the cause of a clicking noise heard only when the car is turning,

    Technician A says that this can be caused by a worn axle torsional damper.

    Technician B says that this can be caused by a worn outboard CV joint.

    Who is correct?
    A. A only
    B. B only
    C. Both A and B
    D. Neither A nor B

ASE CHALLENGE QUESTIONS

1. A vehicle makes a loud clicking sound during a left turn; during a right turn the clicking noise is still apparent but is less pronounced.

   Technician A says that the left inner CV joint may be the source of the noise.

   Technician B says that the left outer CV joint may be causing the noise.

   Who is correct?

   A. A only   C. Both A and B

   B. B only   D. Neither A nor B

2. A vehicle vibrates when under acceleration at all speeds above 10 mph; it does not vibrate during coast down.

   Technician A says that a worn inner CV joint could cause this problem.

   Technician B says that an out-of-balance tire is the likely problem.

   Who is correct?

   A. A only   C. Both A and B

   B. B only   D. Neither A nor B

3. A discussion concerning CV joint testing is being held.

   Technician A says that a recommended method to check CV joints is to raise the vehicle so that the lower control arms are hanging freely; at this point the engine can be accelerated while noise checks can be performed.

   Technician B says that the vehicle should be road tested so that the CV joints can be checked under load.

   Who is correct?

   A. A only   C. Both A and B

   B. B only   D. Neither A nor B

4. A half shaft is being removed from a vehicle. The outer CV joint appears to be pressed into the hub.

   Technician A says that a brass punch or hammer can be used to remove the CV joint from the hub.

   Technician B says that a puller should be used to remove the CV joint from the hub.

   Who is correct?

   A. A only   C. Both A and B

   B. B only   D. Neither A nor B

5. The CV joint boots on the left side of a vehicle appear to be squeezed together; the boots on the right side of the vehicle appear to be stretched apart.

   Technician A says that one or more of the engine/ transaxle's mounts may be broken.

   Technician B says that one or more of the engine/ transaxle's subframe mounts may be broken.

   Who is correct?

   A. A only   C. Both A and B

   B. B only   D. Neither A nor B

Name _____ Date _____

DRIVE AXLE INSPECTION AND DIAGNOSIS

Upon completion of this job sheet, you should be able to inspect and diagnose problems with drive axles.

NATEF MAST Task Correlation

Drive Shaft and Half Shaft, Universal and Constant-Velocity (CV) Joint Diagnosis and Repair

Task #1 Diagnose constant velocity (CV) joint noise and vibration concerns; determine necessary action.

Describe the Vehicle Being Worked On

Year _____ Make _____ VIN _____
Model _____

Procedure

1. Describe any noise heard in the drive axle.

   _____

   _____

   _____

   _____

   _____

2. Explain what causes the noise.

   _____

   _____

   _____

   _____

   _____

3. Describe any looseness found in the drive axle.

   _____

   _____

   _____

   _____

   _____

4. Was there any abnormal vibration found in the drive axle?   ☐ Yes   ☐ No
If yes, describe the cause of the vibration. _____

_____

_____

_____

5. List your recommendations for correction of problems found in the drive axle.

_____

_____

_____

_____

Instructor's Response: _____

_____

_____

Name _____ Date _____

SERVICING DRIVE AXLE CV JOINTS

Upon completion of this job sheet, you should be able to service problems with drive axle CV joints.

NATEF MAST Task Correlation

Drive Shaft and Half Shaft, Universal and Constant-Velocity (CV) Joint Diagnosis and Repair

Task #1 Diagnose constant-velocity (CV) joint noise and vibration concerns; determine necessary action.

Task #4 Inspect, Service, and replace shafts, yokes, boots, and universal/CV joints.

Tools and Materials

Basic hand tools
Hydraulic floor jack
Safety stands
Special tools for the vehicle

Describe the Vehicle Being Worked On

Year _____ Make _____ VIN _____

Model _____

Procedure

1. List all problems found in drive assembly before removal. _____

   a. _____

   b. _____

   c. _____

   d. _____

2. Were the CV joints repacked?   ☐ Yes   ☐ No

3. Were the CV joints replaced with new ones?   ☐ Yes   ☐ No
   If yes, why? _____

   _____

   _____

   _____

4. Was there any noise or vibration in the drive axles after new or repacked joints were installed?   ☐ Yes   ☐ No

If yes, what caused it? _____

_____

_____

_____

5. Did you correct all problems dealing with vibration and noise in the drive axles?
☐ Yes   ☐ No

If yes, how? _____

_____

_____

_____

_____

_____

Instructor's Response: _____

_____

_____

Name _____ Date _____

IDENTIFYING TYPES OF CV JOINTS ON THE AXLE

Upon completion of this job sheet, you should be able to demonstrate the ability to identify CV joints.

NATEF MAST Task Correlation

Drive Shaft and Half Shaft, Universal and Constant-Velocity (CV) Joint Diagnosis and Repair

Task #1 Diagnose constant-velocity (CV) joint noise and vibration concerns; determine necessary action.

Tools and Materials

A vehicle

A hoist or jack and safety stands

A creeper if you don't have a hoist

Describe the Vehicle Being Worked On

Year _____ Make _____ VIN _____

Model _____

Procedure

Task Completed

1. Lift the vehicle and turn the vehicle's wheels so you can inspect them. ☐

2. Turn the wheels and watch for broken boots. Do you see any grease working out of any place?　☐ Yes　☐ No

3. Identify the inner and outer CV joints on the driver's side. ☐

4. Identify the inner and outer CV joints on the passenger's side. ☐

5. List the name of each kind.

_____

_____

_____

_____

_____

_____

Instructor's Response: _____

_____

_____

Name _____ Date _____

Servicing FWD Wheel Bearings

Upon completion of this job sheet, you will be able to replace front-wheel-drive front wheel bearings.

NATEF MAST Task Correlation

Drive Shaft and Half Shaft, Universal and Constant-Velocity (CV) Joint Diagnosis and Repair Task #3 Inspect, remove, and replace front-wheel-drive (FWD) bearings, hubs, and seals.

Tools and Materials

Lift   Brass drift
Torque wrench   Soft-faced hammer
Hydraulic press   CV joint boot protector
Various drivers and pullers   Service information

Protective Clothing

Goggles or safety glasses with side shields

Describe the Vehicle Being Worked On

Year _____ Make _____ VIN _____
Model _____

Procedure

NOTE: *This is a typical procedure. Check with the service information for the exact procedure you should follow and mark the differences in procedure as you progress through the steps below.*

Task Completed

1. Loosen the hub nut and wheel lug nuts. Have someone apply the brakes with the vehicle's weight on its tires. Loosen the axle hub nut and wheel lug nuts with the appropriate tools. Then, remove the axle hub nut. ☐

2. Jack up the vehicle and remove the tire and wheel assembly. At this point, it's a good idea to install a boot protector over the CV joint boot to protect it while you are working. ☐

3. Unbolt the front brake caliper, move it out of the way, and suspend it with wire. Make sure the flexible brake hose doesn't support the caliper's weight. ☐

4. Remove the brake rotor. ☐

5. Then, loosen and remove the pinch bolt that holds the lower control arm to the steering knuckle and separate the lower ball joint and tie rod end from the knuckle. ☐

☐ 6. On some cars, an eccentric washer is located behind one or both of the bolts that connect the spindle to the strut. These eccentric washers are used to adjust camber. Mark the exact location and position of this washer on the strut before you remove it. This allows you to reinstall the washer and the assembly without drastically disturbing the car's camber angle.

☐ 7. Once the washer is marked, remove the hub-and-bearing to knuckle retaining bolts.

☐ 8. Remove the cotter pin and castellated nut from the tie rod end and pull the tie rod end from the knuckle.

☐ 9. The tie rod end can usually be freed from its bore in the knuckle by using a small gear puller or by tapping on the metal surrounding the bore in the knuckle.

☐ 10. Remove the pinch bolts holding the steering knuckle to the strut, and then remove the knuckle from the car.

☐ 11. With a soft-faced hammer, tap on the hub assembly to free it from the knuckle. The brake backing plate may come off with the hub assembly.

☐ 12. Install the appropriate puller and press the hub-and-bearing assembly from the half shaft.

☐ 13. Pull the half shaft away from the steering knuckle. Do not allow the CV joint to drop downward while you remove the hub assembly. Support the half shaft and the CV joint from the back side of the knuckle.

☐ 14. With the steering knuckle off, remove the snapring or collar that retains the wheel bearing.

☐ 15. Mount the knuckle on a hydraulic press so that it rests on the base of the press with the hub free. Using a driver slightly smaller than the inside diameter of the bearing, press the bearing out of the hub. On some cars, this will cause half of the inner race to break out. Since half of the inner race is still on the hub, it must be removed. To do this, use a small gear puller.

☐ 16. Inspect the bearing's bore in the knuckle for burrs, score marks, cracks, and other damage.

☐ 17. Lightly lubricate the outer surface of the new bearing assembly and the inner bore of the knuckle.

☐ 18. Press the new bearing into the knuckle. Make sure all inner snaprings are in place.

☐ 19. Install the outer retaining ring and bolt the brake splash shield onto the knuckle.

☐ 20. Using the proper tool, press the new steering knuckle seal into place. Lubricate the lip seal and coat the inside of the knuckle with a thin coat of grease.

☐ 21. Slide the knuckle assembly over the drive axle.

☐ 22. Mount the knuckle to the strut and tighten the bolts. Make sure the camber marks to the strut are positioned properly.

☐ 23. Reinstall the lower control arm ball joint and tie rod end to the steering knuckle. Then mount the rotor and brake caliper. Torque all bolts to specifications. What are the specifications?

_____

24. Install the tire/wheel assembly and lower the vehicle.

25. Once the car is set on the ground, tighten the new axle hub nut and wheel lug nuts to their specified torque. Install the wheel lug nuts and the new hub nut. What are the specifications? □

_____

26. Lower the vehicle. Tighten all bolts to specifications. Stake or use a cotter pin to keep the hub nut in place after it is tightened. What are the specifications? □

27. Road-test the car and recheck the torque on the hub nut. Summarize the results of the road test.

_____

_____

28. Check the vehicle's wheel alignment. Although indexing the alignment during disassembly will get the alignment angles close, they won't be at the desired settings. Camber can be off by as much as ¾ degree due to differences between the size of the camber holes and the bolt. □

Instructor's Response: _____

_____

_____

Chapter 6

SERVICING DRIVE SHAFTS AND UNIVERSAL JOINTS

BASIC TOOLS
Basic mechanic's tool set
Jack stands
Drain pan
Dial indicator
Angle gauge

UPON COMPLETION AND REVIEW OF THIS CHAPTER, YOU SHOULD BE ABLE TO:

- Diagnose shaft and Universal joint noise and vibration problems and determine needed repairs.

- Inspect, service, and replace shaft, yokes, boots, and Universal joints.

- Inspect, service, and replace shaft center support bearings.

- Check and correct shaft balance.

- Measure shaft runout.

- Measure and adjust shaft angles.

DIAGNOSING DRIVE SHAFT PROBLEMS

Diagnostics of drive shafts, U-joints, and drive axles are normally focused on finding the cause of a noise or vibration. Begin by clearly defining the symptom or the customer's complaint. To do this, talk to the customer, conduct a thorough road test, and complete a careful inspection. Always review the service history of the vehicle, and check for relevant technical service bulletins.

The differential assembly is an integral part of the rear drive axle assembly (Figure 6-1). Before proceeding to correct any problem in the driveline, especially with differentials, refer to the appropriate service manual for the correct procedures.

Talking to the Customer

The diagnosis of any problem should begin with a conversation with the customer. You should find out as much information as you can about the problem. Ask the customer to carefully describe the problem. This description should include any noises or vibrations, when the problem is evident, and when the problem first became noticeable. Find out where it is felt or heard. Ask the customer whether engine speed or vehicle load changes the vibration. Find out if the noise or vibration seems related to engine or vehicle speed. This information will not only help you pinpoint the cause of the problem, but will also allow you to determine if the problem is in the driveline or if it is caused by other systems. Many noise and vibration problems are caused by damaged or worn suspension, brake, and exhaust components.

Find out if the problem resulted from an event or mishap, such as running into a curb or pothole. Ask the customer about the service history of the car; often driveline problems can be caused by mistakes made by technicians. Never jump to conclusions based on the information gathered from the customer. Your diagnostic procedure should continue, using the customer's information as guidelines. Always road-test the vehicle to verify the customer's complaint and to identify any other problems.

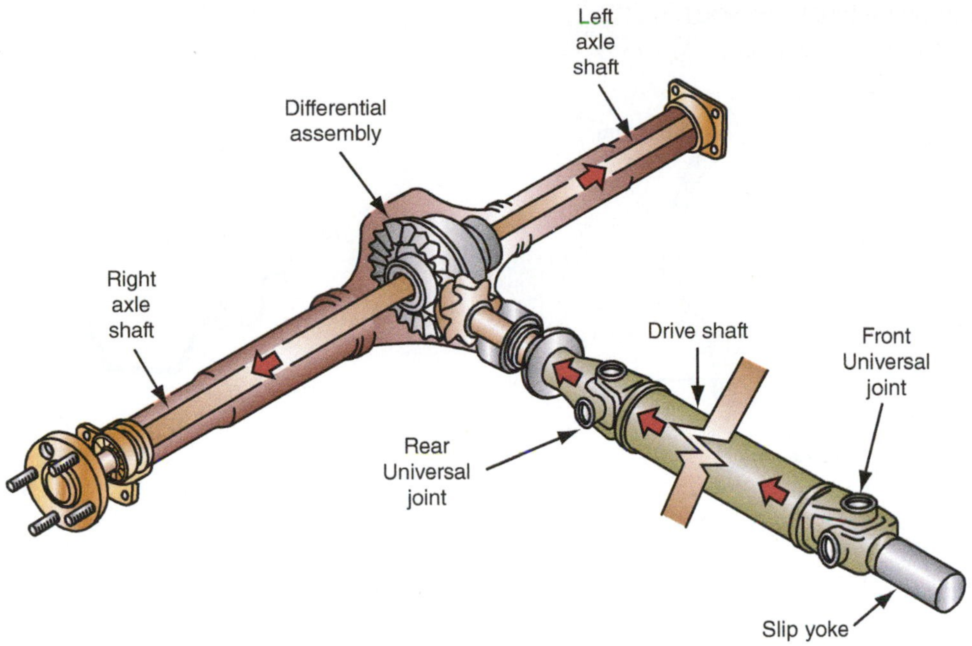

FIGURE 6-1 The differential unit is an integral part of the rear drive axle.

Road Test

If it is possible, take the customer along with you on the road test. Attempt to duplicate the customer's complaint by operating the vehicle under the conditions in which the problem occurs. Pay careful attention to the vehicle during those and all operating conditions. Keep notes of the vehicle's behavior while driving under various conditions. Note the engine and vehicle speeds at which the problem is most evident.

During the road test, the car should be slowly accelerated to highway speed on a smooth road. If a vibration is noted, accelerate to a speed slightly higher and then shift the vehicle into neutral and coast back through the vibration range. If the vibration is still noticed, it is related to vehicle speed. From highway speed, note the engine rpm when the vibration is felt, then safely downshift to the next lower gear. If the vibration is felt again at the same rpm, it is related to the engine or its accessories, or to the clutch, flex plate, or torque converter. All noises and vibrations should be noted. The car should also be driven at various constant speeds. It is also helpful to note how a change in speed affects the noise or vibration.

As a general rule, a worn U-joint is most noticeable during acceleration or deceleration and is less speed-sensitive than an unbalanced tire or bad wheel bearing. After the road test, the chassis should be carefully checked on a hoist. Exhaust system components positioned too close to the frame or underbody may be the cause of the noise. Loose or bent wheels (Figure 6-2), bad tire treads, worn U-joints, and damaged engine mounts all are capable of creating noise and/or vibration that appear to be coming from the axle.

Vibrations

In an automobile, there are many parts that vibrate as they operate. Through careful engineering, these vibrations can be isolated or insulated, thereby reducing the chances of the vibrations moving through the vehicle. Vibration control also is important for the reliability of components. If the vibrations are not controlled, the object could shake itself to destruction. Vibration control is the best justification for always mounting components in the way they were designed to be mounted.

Unwanted and uncontrolled vibrations are typically the result of one component vibrating at a different frequency than another part. When two waves or vibrations meet, they add up

An unbalanced tire will normally cause a vibration while driving 30–60 mph.

Bad wheel bearings will cause a vibration that increases with an increase in speed.

Classroom Manual
Chapter 6, page 143

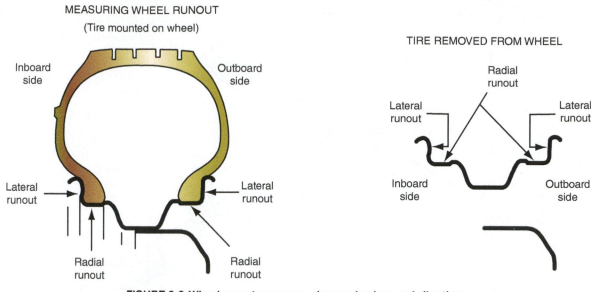

MEASURING WHEEL RUNOUT
(Tire mounted on wheel)

Inboard side

Outboard side

Lateral runout

Lateral runout

Radial runout

Radial runout

TIRE REMOVED FROM WHEEL

Radial runout

Lateral runout

Lateral runout

Inboard side

Outboard side

FIGURE 6-2 Wheel runout can cause abnormal noises and vibrations.

or interfere. Making unwanted vibrations tolerable may involve canceling out each with an equal and opposite vibration. This approach to vibration reduction is best illustrated by the construction and setup of the drive shafts used in rear-wheel-drive (RWD) and four-wheel-drive (4WD) vehicles.

> **AUTHOR'S NOTE:** The automotive industry long ago adopted standard terminology for categorizing issues related to noise and vibration. Noise, vibration, harshness (NVH) concerns are sometimes the most difficult issues to address in the automotive repair field. A more recent development in diagnostic terminology relates to less serious, but annoying, concerns. Buzz, squeak, rattle (BSR) diagnosis plays an important part in maintaining customer trust and satisfaction.
>
> NVH and BSR diagnosis is often left up to the more experienced technicians in a shop.

Vibration Diagnosis. Vibrations can be classified by whether they can be felt or heard. All vibrations are related to **frequency**, measured in **hertz**. In drivetrain diagnosis the frequency corresponds to the rotational speed of various components. Vibrations that can be felt can be broken out into different levels including shake, roughness, buzz, or tingling. Vibrations that can be heard include a low-frequency boom, a moan or drone, or higher-frequency howl or whine.

Electronic vibration analyzers (Figure 6-3 and Figure 6-4) monitor vibration frequency and **amplitude** (the strength of the vibration) and use an accelerometer to pick up vibrations and present the information on the tool's screen. The technician often has to enter relevant vehicle information such as tire size, transmission and differential gear ratios, and drive shaft configuration. Software programs can assist the technician in diagnosing the source of a vibration or noise.

Many driveline vibrations are caused by tires or driveline angle. Vibrations that seem to increase with speed can be caused by the drive shaft. The shaft may be damaged or dented, have undercoating or other heavy materials built up on it, or the balance pad has fallen off the shaft. Carefully inspect the entire length of the shaft (Figure 6-5). This problem can also be caused by worn or defective transmission and/or engine mounts, worn or loose wheel bearings, or unbalanced tire/wheel assemblies.

If the vibration is only evident at high speeds, the most probable cause is an unbalanced drive shaft. If the vibration is most noticeable while accelerating to the testing speed, suspect a worn or defective centering ball-and-socket of a double Cardan U-joint.

Classroom Manual
Chapter 6, page 138

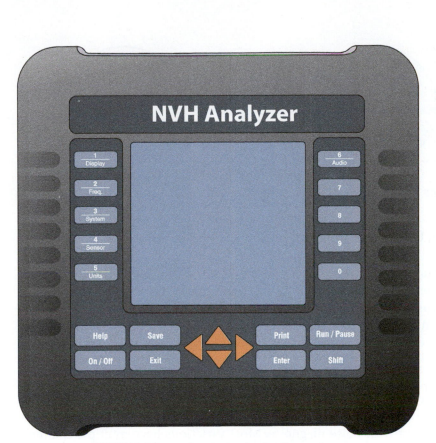

FIGURE 6-3 A popular electronic vibration analyzer.

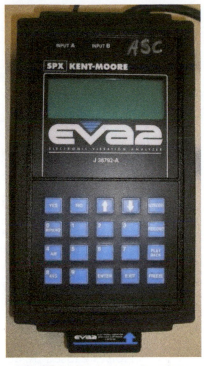

FIGURE 6-4 Revised version of one of the original electronic vibration analyzers.

If the vibration is most evident when the vehicle is traveling at low speeds, suspect incorrect shaft and joint installation angles, worn and sagging rear springs, or excessive vehicle load (Figure 6-6). If the vehicle is weighted down, the suspension will sag, which changes the operating angle of the Universal joints. A broken engine mount can allow the engine to lift under hard acceleration, lowering the back of the transmission and causing a vibration due to

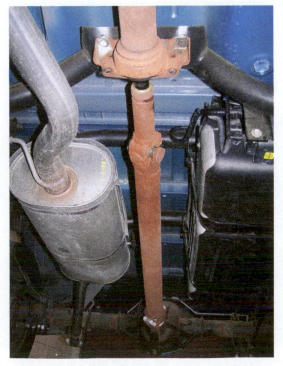

FIGURE 6-5 The entire length of the drive shaft should be inspected.

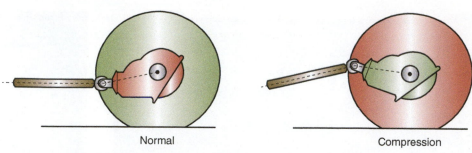

Normal	Compression

FIGURE 6-6 Spring compression can cause excessive joint angles.

the temporary change in U-joint operating angles. Binding U-joints can also cause low-speed vibration.

A vibration that always occurs while driving is often caused by a faulty Universal joint, bent drive pinion shaft, or damaged pinion flange. All of these cause the drive shaft to run off center. Also, check the runout of the pinion flange. Install a dial indicator with its base on the carrier and its plunger on the flange. Rotate the wheels while observing the movement of the indicator. Any reading indicates some runout. Compare the reading to specifications.

Noises

Noises caused by defective drive shafts or U-joints are normally felt, as well as heard. By describing the noise, you can usually identify the most probable problem areas. Many shops use "electronic ears" to assist with noise diagnosis. Most variations have multiple channels that correspond to individual pick-ups placed at different suspected components on the car or truck (Figure 6-7).

Clunk. The most noticeable noise to a customer is clunk. Typically this noise is noticed when the vehicle is shifted from forward to reverse, when accelerated from a dead stop, or when gears are changed. It is caused by metal-to-metal contact and tends to disappear when the driveline load is constant. The likely causes for this problem are worn or damaged U-joints or

FIGURE 6-7 The electronic ear tool can assist in noise diagnosis.

excessive backlash in the driveline. However, excessive clearance between the slip joint and the extension housing bushing, a loose companion flange, and loose upper or lower control bushing bolts can also cause clunking.

A noise similar to clunk but sharper is snap. This sound is often heard only once as the vehicle leaves a stop or comes to a complete stop. The sound is usually caused by interference between the inner splines of the slip yoke and the output shaft. Manufacturers may specify a special lubricant to eliminate the noise.

Squeak. Squeaking can be a very dull or a loud noise that occurs when two surfaces rub against each other. A worn or damaged Universal joint may cause a squeaking noise whose frequency increases with vehicle speed. Obviously the U-joint will not be as worn or loose as one that causes a clunking sound. A poorly lubricated joint or one that has its retaining U-bolts tightened too far can also be the cause of squeaking.

Bearing Noise. Bearing rumble sounds like marbles tumbling in a container. This noise is usually caused by faulty wheel bearings. Since wheel bearings rotate at approximately one-third the speed as the driveline, the noise has a much lower tone than other faulty driveline bearings.

Bearing "whine" is a high-pitched sound normally caused by faulty pinion bearings that rotate at drive shaft speed and occur at all speeds. Although they travel at lower speeds, dry or poorly lubricated roller-type wheel bearings may make the same noise. Pinion bearings make a high-pitched, whistling sound. However, if only one pinion bearing is faulty, the noise may change with changes in driving modes.

> **CUSTOMER CARE:** If a customer's vehicle has a worn U-joint, make sure he or she knows the joint should be replaced before the drive shaft or yoke is damaged. Both of these problems are expensive to repair and can be avoided. More importantly, if a U-joint breaks, the loose shaft can penetrate the floor or truck box and cause a possible loss of control.

Causes of Failure

There are many causes for U-joint failure. The most common cause is a lack of lubrication. Without proper lubrication, the needle bearings in the joint will overheat; this can totally destroy a U-joint. The beginning of wear due to a lack of lubrication is evidenced by fine grooves worn in the bearing areas of the joint.

Other causes of failure include abnormal operating conditions, including overloading the joints by pushing another car or towing a trailer; abruptly changing gears; going from forward to reverse or reverse to forward at high speeds; or carrying heavy loads that change the ride height of the car and cause the joints to operate at angles greater than they were designed for.

Lubricating Universal Joints

Original equipment U-joints are usually sealed units and require no periodic lubrication; in fact, they cannot be lubricated. However, the U-joints on some model vehicles do require lubrication. Also, grease fittings are included on most replacement U-joints (Figure 6-8). Look for the grease fittings, commonly called **zerk fittings**, at each of the U-joints, including the center joints if the drive shaft is so equipped. Proper assembly of a drive shaft with original equipment grease fittings demands the correct placement of grease fittings in relationship to each other (Figure 6-9). Refer to the lubrication chart given in the service manual for the proper procedures and precautions. Some U-joints may have plugs that can be removed to install a grease fitting. A sealed joint must be replaced if it shows signs of low lubrication.

Classroom Manual
Chapter 6, page 147

The beginning of wear due to a lack of lubrication is called brinelling.

Zerk fittings are very small check valves that allow grease to be injected into a part, but keep grease from squirting out.

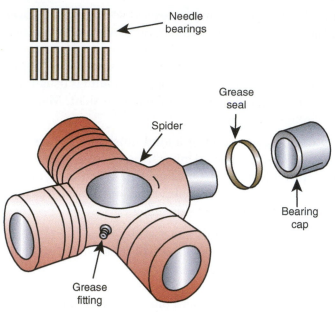

Needle bearings

Grease seal

Spider

Bearing cap

Grease fitting

FIGURE 6-8 Location of grease fittings on a U-joint.

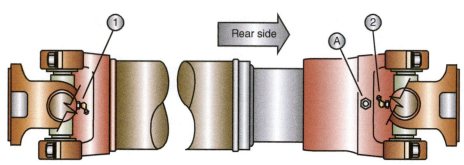

Rear side

FIGURE 6-9 When assembling this drive shaft, the grease fittings on the U-joints (1 and 2) and the grease fitting at the sliding yoke (A) must be positioned properly.

SPECIAL TOOLS

Hand grease gun
Clean rags

SERVICE TIP:
Conduct all in-shop diagnostic work with the vehicle at normal curb height and weight. Always refer to the service manual before raising the vehicle on a lift to identify the proper lift points.

Classroom Manual

Chapter 6, page 137

To lubricate the joint, wipe off the nozzle of the fitting. Then attach the hose fitting of the grease gun to the nozzle. Pump grease slowly into the grease fitting. When grease begins to appear at all four bearing cups, stop pumping the grease gun. Too much pressure can damage the seals of the joint. If grease is not evident at one or more of the bearing cups, the grease passages inside the joint are clogged. To unclog the passages, the joint must be removed and thoroughly cleaned or replaced. Wipe off any excess grease at the caps or fitting with a rag. Grease all of the joints in the same way.

If grease cannot be pumped through a grease fitting, the fitting is clogged with dirt or other materials. Remove the fitting and replace it with a new one. If the fitting still doesn't accept grease, the joint is clogged and should be replaced.

DRIVE SHAFT INSPECTION

It is often very difficult to accurately pinpoint driveline problems with only a test drive. Therefore a complete undercar inspection is necessary. Place the vehicle on a lift and raise it up.

⚠ **WARNING:** Whenever working under a vehicle, make sure it is supported on safety stands or the lift's lock set. Also be sure to wear safety glasses or goggles because dirt and/or rust can fall off the car and cause serious injury to your eyes.

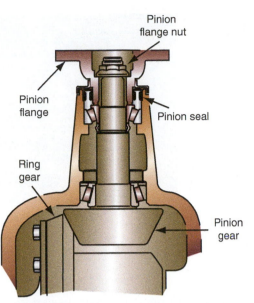

FIGURE 6-10 A common source of rear axle oil leaks is a damaged pinion seal.

Begin your inspection by looking for fluid leaks. Leaks from related components can indicate the actual cause of apparent driveline problems. If there is evidence of a leak around the rear axle's pinion shaft, the noise or vibration could be caused by a bad pinion bearing. To determine if this is the source of the problem, start the engine and put the transmission in gear. While engaging the clutch, pay close attention to the noises from the rear axle housing. If any bearings are noisy, they should be replaced with the seal. If the bearings sound normal, replace only the seal. On some vehicles, pinion seal replacement is a simple procedure in which the pinion flange is removed and the seal replaced (Figure 6-10). However, always refer to the service manual before attempting to replace the seal.

⚠️ **WARNING: Hot exhaust system parts can cause severe burns. Be careful when working around the exhaust system. It is best to let it cool before working around it.**

The transmission end of the driveline should also be examined for evidence of fluid leaks. If the extension housing seal is leaking, it can be easily replaced. However, before replacing the seal, check the extension housing bushing and replace it with the seal if it is worn (Figure 6-11). If the outside surface of the slip yoke is rough or damaged, it will destroy the seal. Inspect the slip yoke's surface if the seal was leaking.

After looking for signs of fluid leaks, inspect the U-joint's grease seals for signs of rust, leakage, or contamination of the lubricant. Universal joints can be quickly checked for wear or damage. To do this, grasp the U-joint's yoke and the drive shaft. Carefully watch the two parts as you turn them in opposite directions; there should be no noticeable movement. Then, attempt to move the shaft up and down in the yoke. If any movement is possible, the joint is worn. Naturally, the amount of movement indicates the amount of wear.

Check the entire length of the drive shaft for dents, missing weights, undercoating or dirt on the shaft, and other damage. If the vehicle is equipped with a two-piece drive shaft, inspect the center bearing assembly for signs of wear or other damage (Figure 6-12). All of these could cause the drive shaft to be unbalanced, which would cause a vibration. Any buildup of dirt or undercoating should be removed with a scraper and wire brush.

The runout of the drive shaft should also be checked. If there is excessive runout, determine the cause and make the necessary repairs. If the runout is fine, check the phasing of the

The drive pinion shaft is located near the center of the rear axle housing and its flange is the part that the rear U-joint attaches to.

The grease seals of a Universal joint are located in the body of the trunnion or sometimes in the end of a bearing cap.

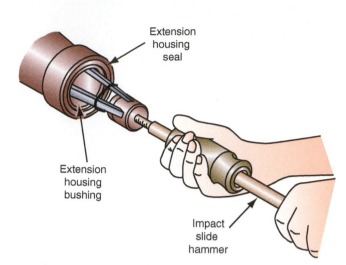

FIGURE 6-11 Using a seal remover tool and slide hammer to remove the extension housing's seal.

Extension housing seal

Extension housing bushing

Impact slide hammer

FIGURE 6-12 Center bearing assembly.

joints and their angle. To check their operating angle, use an inclinometer. This instrument, when attached to the drive shaft, displays the angle of the drive shaft along any point. Your finding from this test should be compared to specifications. Normally, if the angles are wrong, the rear axle has moved in its mounting or the vehicle's ride height is not correct.

Sometimes the only way to check a U-joint is to remove the drive shaft and move the joint back and forth to check for binding.

REMOVING AND INSTALLING A DRIVE SHAFT

Before removing a drive shaft from a car, always place an indexing mark on the driving and driven yokes of the rear U-joint using a center punch, chalk, or paint stick (Figure 6-13). If the vehicle is equipped with a two-piece drive shaft, index each of the separate parts of the center support bearing, including the ends of the drive shaft that attach to the support housing. Be careful not to remove the marks when you are servicing the drive shaft. The index marks will allow you to reinstall the drive shaft in the correct position and prevent the possibility of a vibration that originally was not there. Many two-piece shafts are master-splined, or keyed, to ensure correct U-joint phasing during assembly.

CAUTION:

Be careful, hot exhaust system parts can cause severe burns. It is best to let it cool before working around it.

CAUTION:

Be careful not to overtighten the U-bolt retaining nuts, as doing so can distort the roundness of the bearing caps. This can cause the trunnion and bearing cap to bind, which contributes to short U-joint life and drive-line vibrations.

Marks

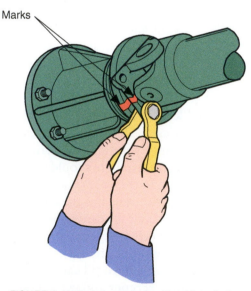

FIGURE 6-13 Before removing the drive shaft, make an alignment mark on the flange yoke and the end yoke.

To remove a drive shaft, raise and support the vehicle safely on a hoist. Use a box-end wrench and remove the two U-bolts or the nuts or bolts that attach the shaft to the differential's pinion flange. Sometimes the nuts are locked in place with metal tabs that must be pried away from each nut before a wrench can be placed over them. If some of the bolts are difficult to reach with the wrench, put the transmission in neutral and rotate the shaft until you are able to reach all of the bolts or nuts. If the joint is retained by U-bolts, remove the bolts from the pinion flange after the nuts have been removed. Often, after removing the bolts in the flange, the drive shaft and joint must be pried away from the flange with a screwdriver or pry bar.

If the drive shaft is a two-section shaft, unbolt the center support bearing assembly before you attempt to remove the shaft from the car. A scribe mark around the assembly before removal will assist in proper placement during installation.

WARNING: Support the front section of the drive shaft as you unbolt the center support. Use a jack stand or have an assistant hold the shaft in place. If the shaft falls, it could cause injury or be damaged.

Slide the drive shaft toward the transmission, lower the drive shaft, and pull it to the rear of the car until its slip joint is out of the transmission. Then remove the drive shaft.

WARNING: Always place an oil drain pan below the transmission's extension housing seal to catch transmission fluid as the drive shaft is removed. This will prevent the fluid from dripping onto the shop's floor, which could make the floor slippery and very unsafe. Also install a special transmission rear plug (Figure 6-14) or spare slip yoke into the extension housing to prevent the transmission from leaking when you are servicing the drive shaft.

Securely mount the shaft in a bench vise, taking care not to damage the shaft by over-tightening the vise. To help prevent damage the vise should have soft brass or lead jaws. Never clamp the shaft in the vise on a balance pad or on the shank portion of a slip yoke.

To remove a torque-tube-type drive shaft, the rear axle assembly must be unbolted in order to make room for the drive shaft to move back and disengage from the transmission.

Drive Shaft Installation

To reinstall the drive shaft, match up all index marks, then insert the slip joint into the extension housing. Push the joint to its most forward position.

CAUTION:
Remember to check the fluid in the transmission after the drive shaft has been installed. Fluid was lost through the rear of the extension housing when the slip joint was removed. Running the transmission while it is low on fluid can cause serious damage.

SERVICE TIP:
To make the task of installing the slip yoke into the transmission easier and with less chance of damaging the transmission's rear seal, apply a thin coat of white grease or specified lubricant to the internal splines. Apply a film of transmission fluid to the outside surface of the slip yoke (Figure 6-15).

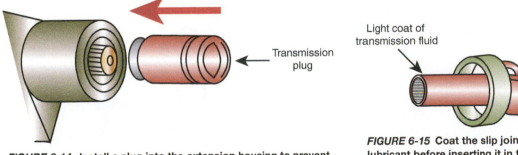

FIGURE 6-14 Install a plug into the extension housing to prevent fluid from leaking out when the drive shaft has been removed.

Transmission plug

Light coat of transmission fluid

FIGURE 6-15 Coat the slip joint with lubricant before inserting it in the extension housing.

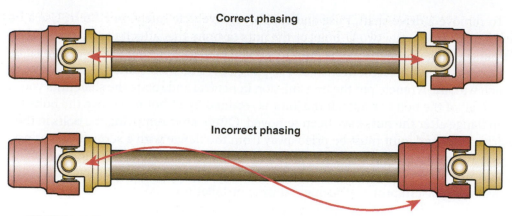

Correct phasing

Incorrect phasing

FIGURE 6-16 Two-piece drive shafts must be assembled so that the U-joints are in the same plane, or phased.

If the drive shaft assembly has a center support, loosely bolt the housing in place. Then carefully raise the other end of the shaft so that it can mate with the pinion flange. Make sure the index lines are aligned and tighten the attaching bolts to the recommended torque. Now tighten the center support bearing attaching bolts to specifications. Remember that two-piece drive shafts must be properly phased when the sections are assembled at the slip joint (Figure 6-16).

After the drive shaft is installed, check the level of the transmission fluid, lubricate all joints that are equipped with a grease fitting, and then road-test the vehicle.

DISASSEMBLING AND ASSEMBLING UNIVERSAL JOINTS

Universal joints should be replaced whenever they are defective. Never use parts of another joint to rebuild a U-joint. To replace them, they must be separated from the drive shaft and this requires disassembly of the joint. Although there are different styles of single U-joints, the procedure for disassembling them is very similar. Photo Sequence 13 shows how to remove and install a single Cardan-type U-joint (Figure 6-17) using a special Universal joint tool. Photo Sequence 14 shows the same procedure but uses a bench vise and sockets. Using the universal tool is the preferred method.

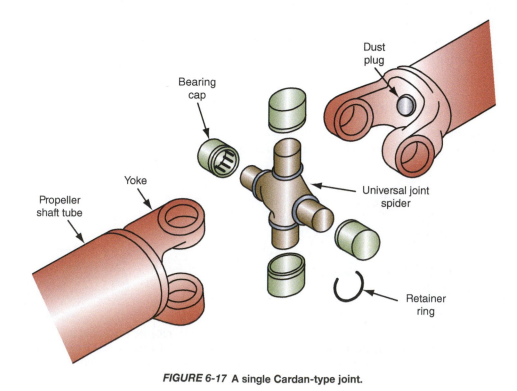

FIGURE 6-17 A single Cardan-type joint.

U-Joint Service with Special Tool

P13-1 A Universal joint tool.

P13-2 Clamp the drive shaft in a vise so that a U-joint freely extends from the vise. Be sure to support the other end of the drive shaft.

P13-3 Then, remove the lock rings on the tops of the bearing caps.

P13-4 Install the tool on the two ends of the U-joint and carefully tighten the tool to press the bearing cap out of the yoke.

P13-5 Install one of the bearing caps of the new U-joint.

P13-6 Lubricate and install the new U-joint into the yoke and install the bearing cap.

P13-7 Install the other bearing cap over the spider and onto the yoke.

P13-8 With the special tool, press the U-joint into place in the yoke.

P13-9 Install new locking rings.

P13-10 Install and tighten a new zerk fitting into the joint. Then lubricate the joint.

Classroom Manual
Chapter 6, page 147

SPECIAL TOOLS
Bench vise
1¼-inch socket
⁹⁄₁₆-inch socket
Vise grips
Soft-faced hammer
Brass drift

U-Joint Service Guidelines

■ Before pressing on the joint's spiders, make sure the locking rings are removed (Figure 6-18). U-joints with plastic retainers will break away when you press the bearing caps out of the shaft and yoke and will be replaced with a new U-joint assembly that uses inside snap-rings. Plastic retainers can also be removed by melting them with a propane torch.

⚠ **WARNING:** Pressure can build up behind the bearing cup of the U-joint while being heated. The bearing cap can explode out of the drive shaft or yoke. Never heat a plastic retainer with the bearing cup pointed at you.

FIGURE 6-18 Remove all retaining rings from the joint before attempting to disassemble the joint.

DIASSEMBLING AND ASSEMBLING A SINGLE U-JOINT WITH A BENCH VISE

P14-1 Clamp the slip yoke in a vise and support the other end of the drive shaft.

P14-2 Then, remove the lock rings on the tops of the bearing caps. Make index marks in the yokes so that the joint can be assembled with the correct phasing.

P14-3 Select a socket that has an inside diameter large enough for the bearing cap to fit into; usually a 1- to 1¼-inch socket will suffice.

P14-4 Select a second socket that can slide into the shaft's bearing cap bore, usually a 9/16-inch socket.

P14-5 Open the vise and place the large socket against one vise jaw. Position the drive shaft yoke so that the socket is around a bearing cap.

P14-6 Position the other socket to the center of the bearing cap opposite to the one with the large socket.

P14-7 Carefully tighten the vise to press the bearing cap out of the yoke and into the large socket.

P14-8 Turn the shaft over and drive the spider and remaining bearing cap down through the yoke with a brass drift punch and hammer.

P14-9 Use a drift and hammer to drive the joint out of the other yokes.

P14-10 Clean any dirt from the yoke and the retaining ring grooves.

P14-11 Carefully remove the bearing caps from the new U-joint.

P14-12 Place the new spider inside the yoke and push it to one side.

P14-13 Start one cap into the yoke's ear and over the spider's trunnion.

P14-14 Carefully place the assembly in a vise and press the cap partially through the ear.

P14-15 Remove the shaft from the vise and push the spider toward the other side of the yoke.

P14-16 Start a cap into the yoke's ear and over the trunnion.

P14-17 Place the shaft in the vise and tighten the jaws to press the bearing cap into the ear and over the trunnion. Install the locking ring, making sure they are seated in their grooves.

P14-18 Rotate the drive shaft and position the joint's spider in the drive shaft yoke and install the two remaining bearing caps locking rings.

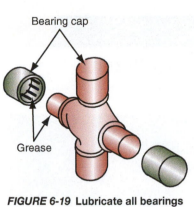

Bearing cap

Grease

FIGURE 6-19 Lubricate all bearings before installing them into the drive shaft yoke.

Color	Thickness mm (in.)
Green	1.384 (0.0545)
Red	1.435 (0.0565)
Black	1.486 (0.0585)
Copper	1.511 (0.0595)
Silver	1.537 (0.0605)
Yellow	1.588 (0.0625)
Blue	1.638 (0.0645)

FIGURE 6-20 Some snaprings are selective to ensure the proper fit of the joint in the yoke and are often color-coded to indicate their thickness.

- If a bearing cap cannot be pressed all the way out of the yoke, remove the shaft from the vise and pull the cap out with a pair of vise grips or drive it out with a brass drift.
- After removing the U-joint, check the drive shaft yokes for any burrs and rough spots. Remove all burrs with a fine file. Also inspect the yokes for signs of breakage or cracks. If there is any evidence of damage, the shaft should be replaced or a new yoke installed.
- Before installing a new joint, clean any dirt from the yoke and the retaining ring grooves. Dirt will prevent the retaining snapring from completely seating and the joint will come apart as the shaft is rotating.
- Carefully remove the bearing caps from the new U-joint. Make sure the needle bearings are well lubricated (Figure 6-19) and in the correct position against the inside of the cap.
- As you tighten the tool or vise to seat the trunnions in the bearing caps, frequently turn the spider to make sure the trunnions move freely on the bearings. If the movement becomes difficult, check to see if the needle bearings have tipped in their cage.
- It may be necessary to move the bearing caps to allow room for the retaining rings. If you do this, install the snapring on one side, then press on the other side to install the second ring.
- Make sure not to damage the seal on the bearing caps during installation. The seal is important to the life of the joint. It keeps lubricant in and dirt out.
- After installing the snaprings, use a brass drift punch and hammer to gently tap the spider out against the new caps.
- Most new U-joints are equipped with a grease fitting. If it has one, fill the joint with grease until the seals of the bearing cap begin to bulge.

Some U-joint assemblies have selective size snaprings. These snaprings ensure the proper fit of the joint in the yoke and are often color-coded to indicate their thickness (Figure 6-20). To determine the correct size snapring, the width of the yoke is measured from the bottom of one snapring groove to the bottom of the other. The total width of the Universal joint is also measured (Figure 6-21). The difference between the two is divided by two, and this figure determines the correct width of the snaprings. When doing this, it is important that the bearing caps are firmly positioned on the spider. This can be done by tightening the spider with its caps in a vise when taking the measurement.

DOUBLE CARDAN U-JOINTS

Double Cardan U-joints are simply two Cardan joints mounted together by a ball-and-socket unit or centering ball joint. The procedures for disassembling and assembling this type of U-joint is much the same as for a single Cardan joint. However, there are some unique steps that must be followed in order to prevent damage to the centering ball joint. Photo Sequence 15 shows those procedures. The following are some guidelines for disassembling and reassembling a double Cardan-type joint.

CAUTION:
The plastic retainers are flammable; always wear the appropriate protective gear.

SPECIAL TOOLS
Bench vise
1¼-inch socket
9/16-inch socket
Vise grips
Soft-faced hammer
Brass drift

Classroom Manual
Chapter 6, page 149

TYPICAL PROCEDURE FOR DISASSEMBLING AND REASSEMBLING A DOUBLE CARDAN UNIVERSAL JOINT

P15-1 Index all parts of the joint assembly by marking each part's location in relationship to the other parts.

P15-2 Remove the bearing caps according to the recommended sequence.

P15-3 Disengage the flange yoke and trunnion from the centering ball.

P15-4 Pry the seal from the ball socket and remove the washers, spring, and ball seats.

P15-5 Inspect the ball and socket carefully for wear, cracks, and other defects.

P15-6 Use the grease furnished with the kit to lubricate all parts.

P15-7 Install the new centering ball seal and retainer by placing a large socket over it and tapping it gently until it is fully seated.

P15-8 Install the flange yoke on the centering ball.

P15-9 Install the trunnions and bearing caps. Be sure all parts are properly aligned.

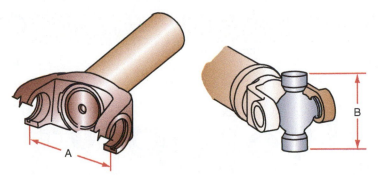

FIGURE 6-21 To determine the correct size snapring, measure the total width of the yoke and U-joint. Half of that difference is the correct size for the snapring on both sides of the joint.

1. Disassembly should begin with the indexing of all parts of the joint assembly by marking each part's location in relationship to the other parts.

2. There is a fixed sequence by which the bearing caps of this type joint should be removed (Figure 6-23). Always follow this sequence. Remove the bearing cups in the same way they were removed from a single joint.

3. Don't attempt to remove the centering ball unless it needs to be replaced. It is pressed onto a stud and is part of the ball-stud yoke. To replace the centering ball, use a special puller. The fingers of the puller are placed under the ball. Then the collar is put on the tool and a nut is tightened on the puller screw threads, which pulls the ball off the ball stud. Then a new ball can be driven onto the stud.

4. Inspect the ball and socket carefully for wear, cracks, and other defects. Replace the ball and socket unit if it is damaged.

5. Install all parts included in the service kit into the ball-seal cavity in this order: spring, small washer, three ball seats with the largest opening outward to receive the balls, large washer, and seal (Figure 6-24).

6. A seal tool is needed to seat the cavity's seal; make sure it is flush with its bore.

7. When installing the flange yoke onto the centering ball, make sure the alignment marks line up.

SPECIAL TOOLS
Small pry bar
Seal installer
Small gear puller

SERVICE TIP:
Some center support bearings are made in one piece with the rubber support assembly and must be replaced as a unit (Figure 6-22).

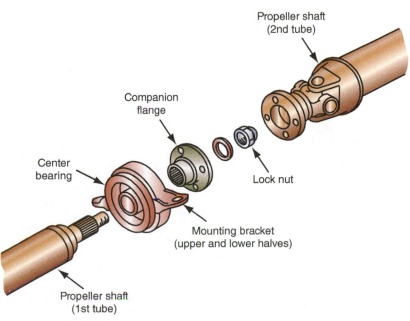

FIGURE 6-22 Always refer to the service manual when servicing drive shafts with center bearings.

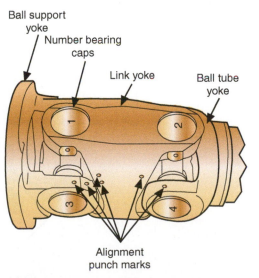

FIGURE 6-23 Make alignment marks as reference points and remove the bearing caps according to the specified sequence when disassembling a double Cardan joint.

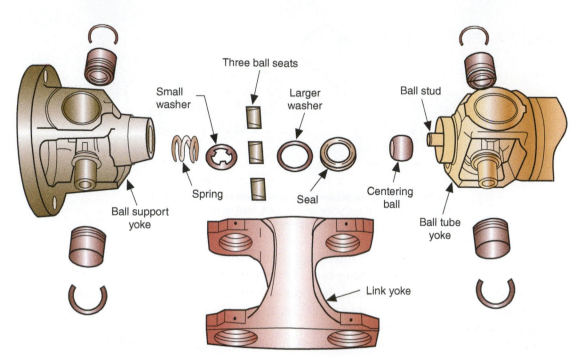

FIGURE 6-24 Components of a double Cardan joint.

Three ball seats

Small washer

Larger washer

Ball stud

Spring

Seal

Centering ball

Ball support yoke

Ball tube yoke

Link yoke

⚠️

CAUTION:
When installing a new ball, make sure it is fully seated against the shoulder at the base of the stud. If it is not fully seated, the entire unit can come apart when the vehicle is operating.

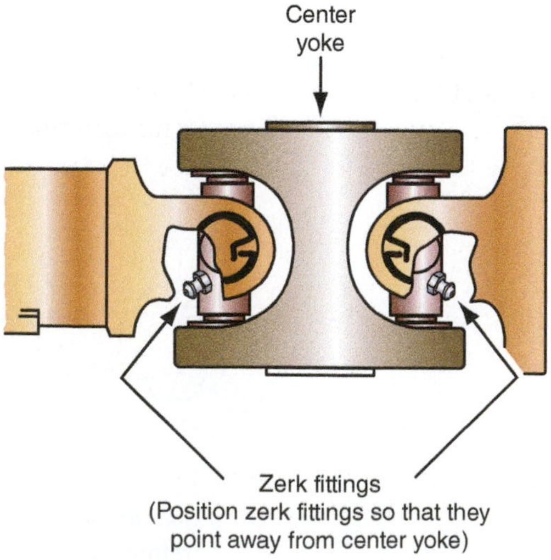

8. Install the trunnions and bearing caps according to the procedure used on single joints, except install the bearing caps in the reverse order of their removal sequence. Make sure the grease fittings are accessible by pointing them away from the center yoke (Figure 6-25).

9. When installing a two-piece drive shaft, make sure the center support bearing is set to keep the drive shaft straight. The support bearing should be positioned exactly 90 degrees from the centerline of the shaft (Figure 6-26). Once this angle is set, torque the center bearing mounting bolts to specifications.

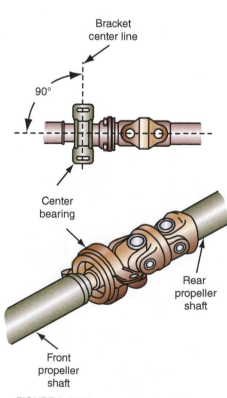

Bracket center line

90°

Center bearing

Rear propeller shaft

Front propeller shaft

FIGURE 6-26 The center support bearing should be positioned exactly 90 degrees from the centerline of the shaft.

Center yoke

Zerk fittings
(Position zerk fittings so that they point away from center yoke)

FIGURE 6-25 Proper position of the grease fittings allow the joint to be lubricated after assembly.

DRIVE SHAFT FLEX DISCS

A few RWD vehicles, mostly European, use flexible coupler discs instead of U-joints. The flex discs serve as the attachment points for the drive shaft to the transmission and rear axle. These discs are designed to isolate engine vibrations from the rest of the driveline. They are made of reinforced rubber reinforced by internal rope cords. The discs are also called guibos and are quite durable. However, they can only handle a moderate range of drive shaft angles and if they operate at extreme angles, they will fail.

These should be inspected about every year. As the discs are exposed to the environment, they tend to develop cracks and begin to disintegrate. Oil contamination from a leaking seal can deteriorate the disc. The discs should be replaced whenever there is evidence of cracking or wear.

The typical procedure for replacing a flex disc follows:

1. Raise the vehicle on a lift or hydraulic jack. Secure the vehicle in position by the locking mechanism on the lift or jack stands.
2. In some cases, the exhaust system should be removed. Disconnect all oxygen sensor wires attached to the exhaust pipes.
3. Remove any shields, brackets, and fixtures that may interfere with removing the drive shaft.
4. Make index marks on the transmission's output flange and the drive shaft.
5. Remove the bolts from the flex disc. Doing this may require rotating the drive shaft to gain access to all of the bolts.
6. Remove the bolts that secure the center bearing and move the assembly just enough to disconnect the front drive shaft from the center bearing.
7. Remove the front drive shaft.
8. Remove the flex disc from the drive shaft.
9. Install the new flex disc with new bolts and locking nuts to the drive shaft. There may be arrows on the disc to indicate which way the bolts should be pushed through the disc; make sure to install them in the correct direction.
10. Torque the bolts onto the drive shaft.
11. Place the disc onto the transmission's output shaft and install the mounting bolts. Make sure the drive shaft is aligned with the index marks made during removal.
12. Reattach the center drive shaft bearing. Push it forward as you tighten it. This removes any play in driveline.
13. Tighten the bolts from the flex disc to the transmission.
14. Reinstall all shields, brackets, and mounts that were removed during removal.
15. Reinstall the exhaust system and connect all oxygen sensors.

DRIVE SHAFT BALANCE

It is possible for drive shafts to be damaged by rocks or other heavy debris. A drive shaft that is dented or has lost its balance pad(s) will be out of balance. Any out-of-balance condition may cause a driveline vibration at various speeds. If the drive shaft has lost its balance weights, it should be removed and rebalanced by a shop equipped to do so.

A dented or damaged drive shaft cannot be rebalanced; it must be replaced. However, minor balance problems can be isolated by installing a pair of worm drive hose clamps. The clamps are installed on the shaft with their heavy ends opposite the heavy side of the shaft. This balances the drive shaft much like the weights used to balance tire and wheel assemblies.

Checking Shaft Balance

Drive shaft balance, while the shaft is in the car, can be checked with a timing light and EVA or can be mechanically checked. Specialty shops have special machines that rotate the shaft to precisely balance them. The following procedure is considered the mechanical check:

1. Place the car on a lift with the rear axle supported and the rear wheels free to rotate. Make sure the transmission is in neutral and the parking brakes are not applied.

CAUTION:
Wear eye protection while working under the vehicle.

Classroom Manual
Chapter 6, page 138

SPECIAL TOOLS
Frame contact or twin post lift
Piece of white chalk
Two screw-type hose clamps

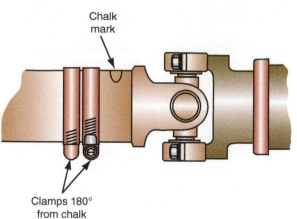

FIGURE 6-27 Install large hose clamps around the drive shaft with their screw assembly 180 degrees away from the chalk mark.

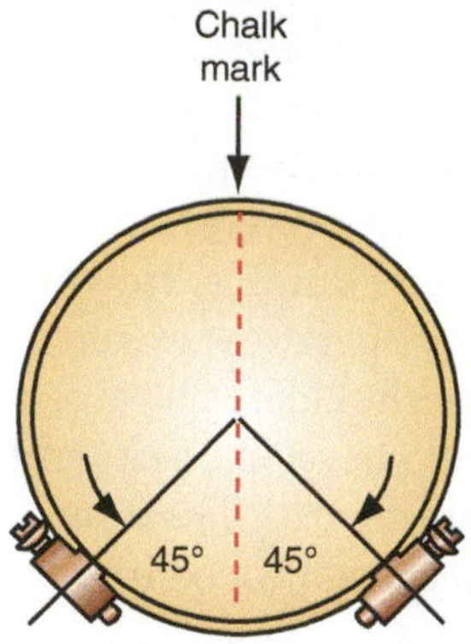

FIGURE 6-28 Move the clamps in equal amounts to achieve drive shaft balance.

⚠️

CAUTION:

Don't run the engine faster than 55 mph and don't let it run for a long period of time. Prolonged running or high speeds without a load can cause damage to the engine and/or transmission.

⚠️

CAUTION:

Because of tight clearances, you cannot install hose clamps around the drive shaft on some cars. Although the clamps may fit, they may hit the underbody when the suspension is compressed. This could lead to severe damage to the vehicle. Always check the service manual for information regarding drive shaft balance before installing hose clamps to the shaft.

Classroom Manual

Chapter 6, page 138

2. Remove the rear wheels and install the lug nuts with their flat side against the brake drums.
3. Seat an assistant in the vehicle and raise it up. Be sure to set the lift's lock.
4. Have your assistant start the engine and run it in high gear to a speedometer speed of 40–50 mph.
5. Carefully raise the piece of chalk close to the rear of the drive shaft. Use a stand to support and steady your hand, or use a runout fixture.

⚠️ **WARNING: Be very careful when conducting this check. Avoid any physical contact with the rotating parts. Especially be careful not to work near the balance pads on the drive shaft. These can severely cut you. Make sure you are wearing safety glasses or goggles.**

6. Continue to raise the chalk until it contacts the rotating shaft. It will leave a mark on the shaft, which will indicate the location of a heavy spot of the shaft.
7. Have the assistant gently apply the brakes and turn off the engine.
8. Install the two hose clamps on the drive shaft so that their screw assembly is directly opposite the chalk mark (Figure 6-27).
9. Tighten the clamps and run the engine to the desired speed again.
10. If no vibration is felt, install the wheels, lower the vehicle, and road-test it.
11. If a vibration is felt, rotate the two clamps away from each other in equal amounts, tighten them, and repeat the test (Figure 6-28). Repeat this procedure until no vibration is felt. Then install the wheels, lower the vehicle, and road-test it.

Correcting Shaft Balance

The installation of hose clamps on the shaft may be a permanent fix to a balance problem. However, some manufacturers recommend that the clamps be treated as a temporary fix and the shaft should be sent out for balancing. On other cars, if the drive shaft becomes unbalanced and vibrates, the drive shaft should be replaced.

In some cases the drive shaft can be corrected for an unbalanced condition by disconnecting the shaft at the rear and rotating it to another position on the companion flange. If the vibration is still evident, the shaft should be replaced.

If undercoating on the shaft is the primary suspect for an unbalanced condition, remove the undercoating from the shaft and road-test the vehicle. If the vibration still exists, remove the drive shaft and rotate it 180 degrees. Repeat the road test. If this does not correct the problem, return the shaft to its original position and continue testing to determine the true cause of the problem.

DRIVE SHAFT RUNOUT

A bent or damaged drive shaft can cause vibration and noise. A good drive shaft will have very little runout during rotation. Excessive runout can cause noise and vibrations.

Measuring Runout

Drive shaft runout is measured with a dial indicator. Runout readings are taken at the center and near the two ends of the shaft. To prepare for measuring runout, put the transmission in neutral and make sure the parking brake is off. Raise the vehicle on a lift and place large jack stands under the axle so that the shaft is at its normal operating angle. Clean all dirt from the surfaces of the drive shaft where the measurements will be taken. Figure 6-29 shows the locations at which runout measurements should be taken.

Mount the indicator on the frame or underbody of the vehicle (Figure 6-30). Position the plunger of the indicator so that it contacts a smooth and clean area of the drive shaft. Zero the dial indicator, then slowly rotate the drive shaft by hand. The plunger will move in or out as the drive shaft moves toward or away from the indicator. As the plunger moves, the dial indicator will show the amount of runout. Acceptable **total runout** is given in the service information for the vehicle. If the runout is greater than the allowable amount, the drive shaft must be replaced. A typical maximum runout specification for one major manufacturer is .040-inch..Check for a bent U-joint flange or slip yoke before replacing a drive shaft with a runout problem.

SERVICE TIP:
Make sure the dial indicator's plunger is not riding over a ridge or flat on the drive shaft while attempting to measure shaft runout.

SPECIAL TOOLS
Dial indicator
Magnetic mounting bracket for indicator

Runout is the distance that the surface of a rotating shaft moves in or out.

Total runout is the sum of the maximum readings below and above the zero line on the indicator.

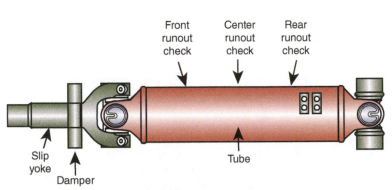

FIGURE 6-29 Drive shaft runout checkpoints.

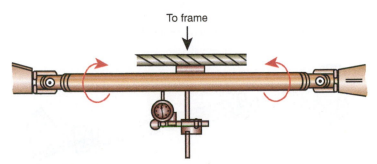

FIGURE 6-30 Checking drive shaft runout with a dial indicator.

An **inclinometer** can be either a special protractor and bubble gauge tool or a degree gauge with a weighted cord.

If an inclinometer is not available, a carpenter's protractor can be used.

SERVICE TIP: On some coil-spring rear suspensions, the distance between the top of the axle housing and the frame must be measured and corrected before measuring the Universal joints' operating angles.

UNIVERSAL JOINT AND SHAFT ANGLES

The angle at which the transmission is mounted to the frame does not change during vehicle operation. However, the angle at which the rear axle pinion is mounted to the frame changes constantly while the car is moving down the road. When the car is not in operation, the drive shaft is set to a specific angle (Figure 6-31). This angle is referred to as the installation angle.

If the input yoke of a U-joint is rotating at a constant speed, the output yoke speeds up and slows down twice during each revolution. This speeding up and slowing down can cause vibration in the driveline. When the two Universal joints are properly assembled and are at equal angles, the vibration caused by the speed changes of the front Universal joint is canceled by the speed changes of the second Universal joint. This is a condition referred to as canceling angles (Figure 6-32).

In order for the angles of the Universal joints to have a canceling effect, their yokes must be on the same plane. When they are, the Universal joints are said to be in-phase.

Normally the drive shaft's operating angle will be correct if the car is at its manufactured ride height and the rear axle housing and transmission have not moved on their mounts. Typically, the maximum allowable drive shaft angle is 3 or 4 degrees from the car's horizontal. The drive shaft angle is checked with an **inclinometer**. Specifications are typically given for a single joint and for the allowable difference between the joints.

Measuring the Angles

Before checking the angles of the U-joints, make sure the vehicle is empty and has a full or near-full fuel tank. Raise the car and support it under the axle housing. Rotate the drive shaft so that the bearing caps in the rear axle's pinion flange and the transmission's slip yoke are straight up and down. Then, clean the bearing caps of each joint.

Some shops use an electronic level to check the angle of the drive shaft. These are accurate and have a locking function that allows the measured angle to be observed without fear of changing the reading while working (Figure 6-33). A less expensive device measures driveline angles with a dial readout (Figure 6-34). The below procedure discusses the use of an inclinometer, which is a bit more difficult to use.

Install the inclinometer with the magnet on the downward-facing bearing cap (Figure 6-35). Using the adjusting knob, center the weighted cord. Remove the inclinometer. Make sure not to bump the adjustment knob.

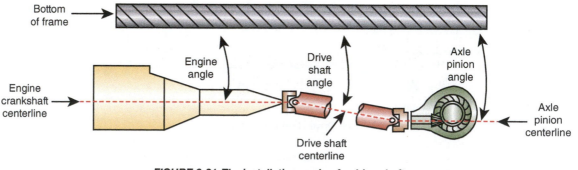

FIGURE 6-31 The installation angle of a drive shaft.

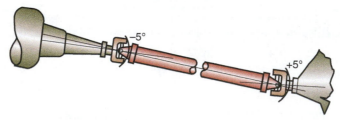

FIGURE 6-32 A drive shaft set with canceling angles.

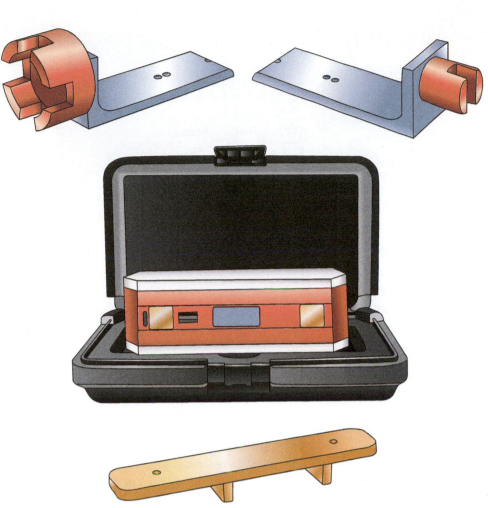

FIGURE 6-33 A professional-grade electronic level used to check driveline angles.

FIGURE 6-34 A magnetic angle locator.

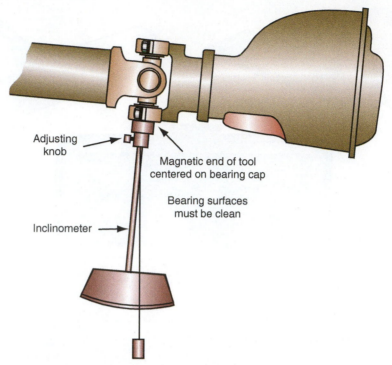

FIGURE 6-35 Preparation for checking the rear joint's operating angle.

SERVICE TIP:
Do not use force when positioning the inclinometer. Hold the instrument firmly but gently to obtain accurate readings.

Now, rotate the drive shaft 90 degrees and attach inclinometer to the bearing cap now facing down and record the reading (Figure 6-36). Remove the inclinometer and rotate the drive shaft 90 degrees. Then attach the inclinometer to the bearing cap, on the front joint, that is now facing downward. Using the adjusting knob, center the cord in its scale (Figure 6-37).

Remove the inclinometer. Make sure not to bump the adjusting knob. Again, rotate the drive shaft 90 degrees. Attach inclinometer to the bearing cap now facing down, and record the reading (Figure 6-38). Compare rear and front angle readings. The difference is the operating angle.

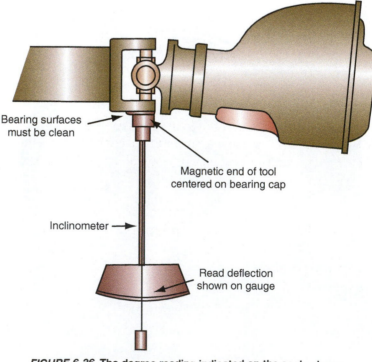

FIGURE 6-36 The degree reading indicated on the scale shows the operating angle of the joint.

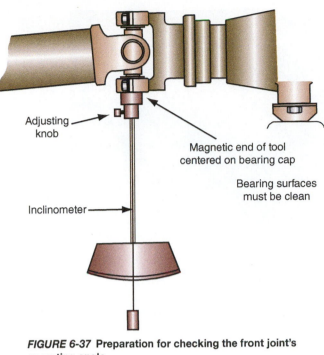

FIGURE 6-37 Preparation for checking the front joint's operating angle.

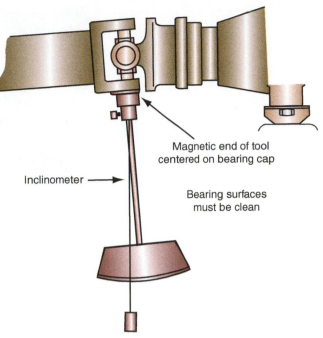

FIGURE 6-38 The degree reading indicated on the scale shows the operating angle of the joint.

Adjusting the Angles

If the installation angles of the two joints are not equal, normally the rear angle is adjusted so that it equals the front angle. When doing this, refer to your service manual for the recommended procedure and angles. On vehicles with leaf springs, the rear angle can be changed by rotating the rear axle assembly on the spring pads (Figure 6-39) or by installing tapered shims between the springs and the spring pads (Figure 6-40).

> **CUSTOMER CARE:** Lift kits installed on four-wheel-drive trucks can have a large effect on U-joint operating angles. In addition to tapered shims installed at the leaf spring pads, many kits provide spacers used to drop the cross member that supports the transmission and transfer case.

If the vehicle uses rear coil springs, the joint angle can be changed by inserting wedge-shaped shims between the rear axle assembly and rear control arms, adjusting an eccentric washer at the control arm, or by replacing the control arm with one of a different length. Enthusiasts who require a specific pinion angle for racing purposes often install aftermarket adjustable control arms (Figure 6-41).

If it is necessary to change the angle of the front U-joint, shims should be installed between the transmission's extension housing and the transmission mount.

⚠️ **CAUTION:** Whenever adjusting a joint's angle with the inclinometer in place, be careful not to jar or bump the tool. It can easily come loose, fall, and break.

SERVICE TIP: While tire size can change the angle that the vehicle sits relative to the road, tire size has no effect on the working angles of the driveline.

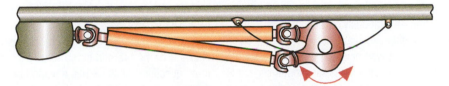

FIGURE 6-39 The rear joint angle can be changed by rotating the axle assembly on its spring pads.

FIGURE 6-40 This tapered shim is labeled for a 4-inch-wide spring and will change the pinion angle by 4 degrees.

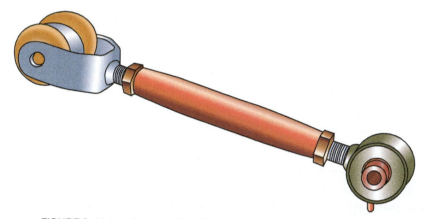

FIGURE 6-41 An aftermarket performance adjustable rear control arm allows fine-tuning of the rear pinion angle.

CASE STUDY

A customer brought his new pickup truck back to the dealership with a complaint of a vibration. This vibration seems to increase with increases of vehicle speed, but is mostly noticeable at highway speeds. The technician is surprised by the complaint because the truck has only 1,200 miles on it and this model has very few vibration problems even after high mileage. So the technician cautiously questions the customer.

The customer explains that the truck rode like a luxury car when it was new, but it started acting up a week ago. The customer was sure something came loose. The technician thought that perhaps a wheel weight came off and a tire was now out of balance. After checking all of the tires, he found this was not the problem. Further inspection located the problem. It seems that the customer liked the truck so much, he wanted it to last forever. He had taken the pick up to a shop to be undercoated and some undercoating got on the shaft. The technician wiped it off, and the pickup's ride was restored.

The technician's day would have been much easier if he had remembered to ask the customer more questions, such as Has anything been done to the pickup since it left the dealership?

ASE-STYLE REVIEW QUESTIONS

1. When discussing the correct procedure for checking drive shaft runout,

   Technician A says that a drive shaft with excessive runout can be straightened by a drive shaft specialty shop.

   Technician B says that runout should be checked at the center of the shaft with the rear wheels hanging freely.

   Who is correct?

   A. A only   C. Both A and B

   B. B only   D. Neither A nor B

2. When attempting to correct the balance of a drive shaft,

   Technician A installs hose clamps around the shaft to offset the heavy spot.

   Technician B rotates the shaft on the pinion flange to another position.

   Who is correct?

   A. A only   C. Both A and B

   B. B only   D. Neither A nor B

3. The transmission extension housing's rear seal leaks.

   Technician A says that this could indicate a worn housing bushing.

   Technician B says that this could be caused by a damaged slip yoke.

   Who is correct?

   A. A only   C. Both A and B

   B. B only   D. Neither A nor B

4. When adjusting the installation angle of a rear Universal joint,

   Technician A installs shims between the differential's pinion shaft and the car's frame.

   Technician B repositions the rear axle housing on its spring pads.

   Who is correct?

   A. A only   C. Both A and B

   B. B only   D. Neither A nor B

5. When diagnosing the cause of a clunking noise each time the speed gears are changed,

   Technician A says that loose rear control arms could be the cause.

   Technician B says that a worn U-joint could be the cause.

   Who is correct?

   A. A only   C. Both A and B

   B. B only   D. Neither A nor B

6. When diagnosing a vibration that becomes greater with vehicle speed,

   Technician A says that the tires may be out of balance.

   Technician B says that the drive shaft may be out of balance.

   Who is correct?

   A. A only   C. Both A and B

   B. B only   D. Neither A nor B

7. When disassembling a double Cardan Universal joint,

   Technician A says that each joint should be removed separately using the same procedures as for a single Cardan joint.

   Technician B says that the location of the parts of the centering ball joint should be indexed.

   Who is correct?

   A. A only   C. Both A and B

   B. B only   D. Neither A nor B

8. While removing a drive shaft from a RWD vehicle,

   Technician A installs a transmission rear plug into the extension housing to prevent the transmission from leaking while repairs are made.

   Technician B places an oil drain pan below the transmission's extension housing seal to catch transmission fluid as the drive shaft is being removed.

   Who is correct?

   A. A only   C. Both A and B

   B. B only   D. Neither A nor B

9. While inspecting a drive shaft,

   Technician A checks the entire length of the drive shaft for undercoating or dirt on the shaft, which may cause the drive shaft to be unbalanced, which would cause a vibration.

   Technician B says if a two-piece drive shaft is in good condition, there is no need to check the center bearing assembly for signs of wear or other damage.

   Who is correct?

   A. A only   C. Both A and B

   B. B only   D. Neither A nor B

10. While diagnosing the cause of a squeaking sound coming from the driveline,

    Technician A checks for a dry or worn U-joint.

    Technician B checks for excessive clearance between the slip joint and the extension housing bushing.

    Who is correct?

    A. A only   C. Both A and B

    B. B only   D. Neither A nor B

ASE CHALLENGE QUESTIONS

1. A RWD vehicle exhibits a vibration that is most evident at 32–34 mph.

   Technician A says that this may be caused by incorrect vehicle height.

   Technician B says that improper pinion angle may be the cause of this problem.

   Who is correct?

   A. A only
   B. B only
   C. Both A and B
   D. Neither A nor B

2. A "clunking" noise is heard when shifting from Park into Reverse on a RWD vehicle.

   Technician A says that this may be caused by a lower than normal engine idle speed adjustment.

   Technician B says that the transmission bushing that supports the drive shaft slip joint could be worn out.

   Who is correct?

   A. A only
   B. B only
   C. Both A and B
   D. Neither A nor B

3. When discussing drive shaft removal and repair,

   Technician A says that when removing a drive shaft it is not necessary to index the slip yoke to the transmission output shaft.

   Technician B says that failure to index the slip yoke or companion flange during Universal joint replacement can result in an incorrect drive pinion angle.

   Who is correct?

   A. A only
   B. B only
   C. Both A and B
   D. Neither A nor B

4. When discussing drive shaft runout,

   Technician A says that a drive shaft with a total runout of .250-inch should be replaced.

   Technician B says that a drive shaft that has zero runout is properly balanced.

   Who is correct?

   A. A only
   B. B only
   C. Both A and B
   D. Neither A nor B

5. When discussing drive shaft inspection and repair,

   Technician A says that drive shaft Universal joint yokes must be on different planes in order to be in-phase.

   Technician B says that when adjusting Universal joint installation angles the rear angle should equal the front angle.

   Who is correct?

   A. A only
   B. B only
   C. Both A and B
   D. Neither A nor B

Name _____ **Date** _____

DRIVE SHAFT AND U-JOINT INSPECTION

Upon completion of this job sheet, you should be able to service problems with drive shaft Universal joints.

NATEF MAST Task Correlation

Drive Shaft and Half Shaft, Universal and Constant-Velocity (CV) Joint Diagnosis and Repair

Task #4 Inspect, service, and replace shafts, yokes, boots, and universal/CV joints.

Task #5 Check shaft balance and phasing; measure shaft runout; measure and adjust drive-line angles.

Tools and Materials

Appropriate hand tools

Lift or jack

Jack stands

Service manual

Protective Clothing

Goggles or safely glasses with side shields

Describe the Vehicle Being Worked On

Year _____ Make _____ VIN _____

Model _____

Procedure

_____

Task Completed

1. Raise vehicle and do the following:
   a. Check transmission seal. (If seal leaks, check rear bushing.) ☐
   b. Check pinion seal. ☐
   c. Check the drive shaft for dents. ☐
   d. Check the drive shaft for missing weights. ☐
   e. Check two-piece drive shaft for proper phasing. ☐
   f. Check bearing caps for lubricant leakage. ☐
   g. Check U-joint for movement. ☐

2. Review the results of your inspection with your instructor. What are your conclusions?

   _____

   _____

Instructor's Response:_____

_____

_____

Name _____ Date _____

DRIVE SHAFT AND U-JOINT ROAD TEST

Upon completion of this job sheet, you should be able to demonstrate the ability to road check a vehicle for drive shaft noises.

NATEF MAST Task Correlation

Drive Shaft and Half Shaft, Universal and Constant-Velocity (CV) Joint Diagnosis and Repair

Task #2 Diagnose Universal joint noise and vibration concerns; perform necessary action.

Tools and Materials

A vehicle for which the owner has given permission to road check.

Describe the Vehicle Being Worked On

Year _____ Make _____ VIN _____

Model _____

Procedure

1. Drive the vehicle at a slow speed. Do you hear any squeaks? ☐ Yes ☐ No

2. Put the vehicle in reverse. Do you hear any squeaks? ☐ Yes ☐ No

3. While holding your foot on the brakes, move the shift lever from forward to reverse a few times. Do you hear a clunk or a click? ☐ Yes ☐ No

4. From what you have heard, is the drive shaft in need of service? ☐ Yes ☐ No

   Explain your findings. _____

   _____

5. What would you tell the customer?

   _____

   _____

   _____

   _____

   _____

Instructor's Response: _____

_____

_____

Name _____ **Date** _____

Check U-Joints

Upon completion of this job sheet, you should be able to check U-joints.

NATEF MAST Task Correlation

Drive Shaft and Half Shaft, Universal and Constant-Velocity (CV) Joint Diagnosis and Repair

Task #4 Inspect, service, and replace shafts, yokes, boots, and Universal/CV joints.

Task #5 Check shaft balance and phasing; measure shaft runout; measure and adjust driveline angles.

Tools and Materials

Appropriate hand tools

Inclinometer

Lift or jack

Jack stands

Service manual

Protective Clothing

Safety glasses with side shields

Describe the Vehicle Being Worked On

Year _____ Make _____ VIN _____

Model _____

Procedure   Task Completed

1. Inspect vehicle to determine whether ride height is correct. (If ride height is not correct, Universal joint operating angle cannot be measured.)   ☐

2. Raise vehicle on twin-post suspension lift so that rear wheels are free to rotate.   ☐

3. Position a cap of the front yoke perpendicular with the ground.   ☐

4. Attach inclinometer and measure the angle; or, if your inclinometer is adjustable, zero the inclinometer. _____   ☐

5. Record your measurement. _____

6. Rotate the drive shaft so that the trailing yoke is now perpendicular to the ground.   ☐

7. Attach inclinometer and measure the angle. _____

8. Record your measurement. The difference between the two measurements is the Universal joint operating angle. _____

☐ 9. Compare your reading to shop manual specifications.

☐ 10. If the angle is not within specification, adjust per shop manual procedures.

Instructor's Response: _____

_____

_____

Chapter 7

SERVICING DIFFERENTIALS AND DRIVE AXLES

BASIC TOOLS

Basic mechanics tool box

Torque wrench

Frame contact lift

Clean rags

UPON COMPLETION AND REVIEW OF THIS CHAPTER, YOU SHOULD BE ABLE TO:

- Diagnose differential and rear-axle noise, vibration, and fluid leakage problems; determine needed repairs.

- Diagnose limited-slip differential noise, slippage, and chatter problems; determine needed repairs.

- Inspect and replace companion flange and pinion seal; measure companion flange runout.

- Inspect and replace ring and pinion gear set.

- Measure ring gear and case runout; determine needed repairs.

- Inspect and replace differential collapsible spacers, sleeves, and bearings.

- Measure and adjust drive pinion depth.

- Measure and adjust drive pinion bearing preload.

- Measure and adjust differential (side) bearing preload and ring and pinion backlash (threaded cup or shim type).

- Measure shaft end play/preload (shim/spacer selection procedure).

- Perform ring and pinion tooth contact pattern checks; determine needed adjustments.

- Remove and replace differential assembly.

- Inspect, measure, adjust, and replace differential pinion gears (spiders), shaft, side gears, thrust washers, and case.

- Inspect and replace differential side bearings.

- Inspect, flush, and refill a limited-slip differential with correct lubricant.

- Inspect, adjust, and replace limited-slip clutch (cone/plate) pack.

- Inspect and replace rear axle shaft wheel studs.

- Remove and replace rear axle shafts.

- Inspect and replace rear axle shaft seals, bearings, and retainers.

- Measure rear axle flange runout and shaft end play; determine needed repairs.

INTRODUCTION

The drive axle assembly serves several important functions: It must secure the drive wheels, transfer power from the transmission to the drive wheels, provide torque to the wheels, and allow the drive wheels to turn at different speeds when the vehicle is turning a corner (Figure 7-1). Even the slightest problem in these units can have a negative effect on the performance, safety, and handling of the vehicle. Minor problems in the drive axle assembly, which may result in noise and vibration, may also become major annoyances for the customer.

This chapter covers the removal, disassembly, inspection, and reassembly of differential units. It begins with general diagnostics and is followed by instructions for the disassembly and assembly of both integral and removable carrier differentials. These procedures are followed by detailed explanations of the critical steps in the procedures. Also included are those special procedures for the repair of limited-slip differentials (LSDs).

Because all cars and trucks have differentials and drive axles, being able to service and repair these units is a must for competent automotive technicians.

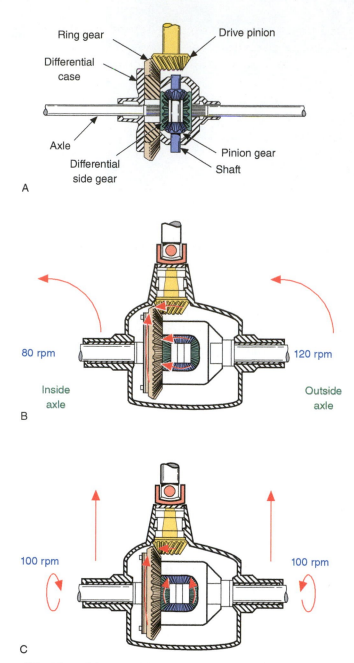

FIGURE 7-1 (A) Basic differential components; (B) differential action while the vehicle is turning left; (C) differential action while the vehicle is moving straight.

DIAGNOSIS OF DIFFERENTIAL AND DRIVE AXLES

Diagnostics of differentials and drive axles are normally centered around finding the cause of a **noise** or vibration. The key to locating the problem is clearly defining the symptom or the customer's complaint. This is done by talking with the customer or service adviser, conducting a thorough road test, and carefully inspecting the components.

Noise is simply defined as an unwanted or annoying sound.

Talking to the Customer

The diagnosis of any problem should begin with a detailed conversation with the customer. The purpose of this discussion is to gather as much information about the problem as possible. Ask the customer to carefully describe the problem. This description should include any noises or vibrations, when the problem is evident, and when the problem first became noticeable. If the customer's complaint is based on an abnormal noise or vibration, find out where it is felt or heard.

You should also find out if the problem resulted from an event or mishap, such as running into a curb or pothole. Ask the customer about the service history of the car; often drive axle problems can be caused by mistakes made by technicians while servicing the vehicle.

Road Test

If it is possible, take the customer along with you on the road test. Attempt to duplicate the customer's complaint by operating the vehicle under the conditions in which the problem occurs. Pay careful attention to the vehicle during those and all operating conditions. Keep notes of the vehicle's behavior while driving under various conditions. Note the engine and vehicle speeds at which the problem is most evident. It can be quite difficult to determine the exact source of a noise or vibration without a thorough road test and without identifying the conditions in which they occur.

> **CUSTOMER CARE:** Never criticize the driver's driving techniques. Use tact to explain why doing something incorrectly may have an adverse effect on the vehicle and may be the cause of the problems that the vehicle is currently experiencing.

Classroom Manual
Chapter 7, page 163

SERVICE TIP:
Never jump to conclusions based on the information gathered from the customer. Your diagnostic procedure should continue, using the customer's information as a guideline. Always road-test the vehicle to verify the customer's complaint and to identify any other problems.

Normally axle noises are more noticeable during one or more of the following operating conditions: drive, cruise, coast, and float. When the vehicle is accelerated with a definite application of the throttle to demand engine torque, the condition is referred to as the drive operation. The cruise operating condition occurs when a constant speed is maintained by a constant throttle opening. The coast operating condition occurs during deceleration with the throttle closed. Float is the operating condition during which there is controlled deceleration by adjusting throttle position carefully to prevent braking or acceleration. Float is considered a "no load" condition for drivetrain components, and requires the technician to develop a practiced technique to achieve this condition.

When diagnosing a drive-axle noise, it is important to operate the vehicle in all four conditions and note any changes in noise during each mode. It is also helpful to note how a change in speed affects the noise or vibration.

Before assuming the drive-axle is the source of a noise or vibration, make certain the tires or exhaust are not the cause. Exhaust systems always make some noise, but at times the exhaust's noise may sound like gear whine or wheel bearing rumble. An exhaust component that is touching the body or frame can resonate through the vehicle and sound like a drivetrain issue. Tires, especially off-road and snow tires, can have a high-pitched tread whine or roar that is similar to gear noise. To determine if the noise is caused by the tires, drive on different road surfaces.

Often noises and vibration appear to be coming from the axle, when actually they are caused by other problems. After the road test, the chassis should be visually checked in the shop while on a hoist. Exhaust system components positioned too close to the frame or underbody may be causing what appears to be axle noise. Loose or bent wheels (Figure 7-2), bad treads on the tires, worn U-joints, and damaged engine mounts are all capable of creating noise and/or vibration that appear to be coming from the axle.

Poor lubrication in the rear axle can also be a source of noise, so the lubricant's level and condition should be checked (Figure 7-3). If the level is low, the axle should be filled with the proper type and amount of lubricant and the car should be road-tested again. If the lubricant is contaminated, it should be drained and the axle refilled with the specified lubricant. Low lubricant in a drive axle indicates a leak in the assembly. Therefore, the assembly should be carefully inspected to locate the source of the leak (Figure 7-4). After locating the leak, check for a plugged vent that can cause high pressure in the housing and result in fluid leaks.

Types of Noises

By describing the noise you hear, you can usually identify the most probable problem areas. Difficult noise issues may require the use of a stethoscope or electronic chassis ears to assist with the diagnosis. The most common noises, a brief description of each, and the most probable causes for each follow.

SPECIAL TOOLS
Stethoscope
Electronic chassis ears

FIGURE 7-2 Check the wheels and the suspension for looseness before proceeding to check the drive-axle assembly.

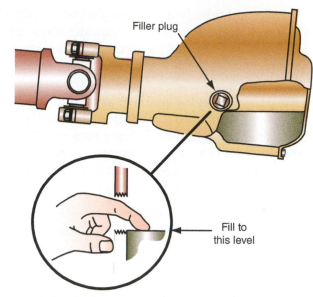

Filler plug

Fill to this level

FIGURE 7-3 Check the final drive's lubricant level.

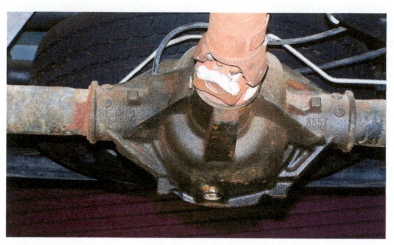

FIGURE 7-4 Wet areas around the differential assembly indicate a fluid leak.

- **Chuckle**—A rattling noise that sounds like a stick rubbing against the spokes of a bicycle wheel. It can be heard only during coasting and is usually most pronounced from about 15 to 25 mph. The frequency varies with the speed of the car. Chuckle is usually caused by side gear shoulder wear and matching wear on the side gear bore in the differential case (Figure 7-5). Any damage to the coast side of a ring or pinion gear can cause a sound similar to chuckle. Often the noise can be eliminated by cleaning up the gear with a small grinding wheel or file. Replace the gear set if the damaged area is larger than 1/8 inch.
- **Knock**—Similar to chuckle, but louder and can occur in all driving modes. Gear tooth damage is a common cause of knocking (Figure 7-6). The technician should drive the vehicle under different load conditions to determine whether the damage is on the drive or coast side of the teeth. Knocking can also be caused by ring gear bolts hitting the inside of the case. This problem may be caused by one or more loose bolts and can be corrected by properly tightening the bolts. A knocking sound on turns only indicates damage to the side gears or differential pinion gears. These "spider gears" do not move when the vehicle is going straight, so any problem will only be noticed when cornering.
- **Clunk**—A metallic noise heard when the throttle is quickly opened or closed while the vehicle is moving. Clunking is caused by excessive backlash in the driveline and is felt or heard in the axle assembly. Clunk can be caused by a bad U-joint, excessive ring and pinion backlash, or by any excessive clearance in the drivetrain.

FIGURE 7-5 A damaged side gear shoulder can cause chuckle.

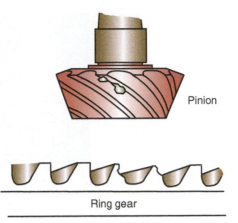

Pinion

Ring gear

FIGURE 7-6 Drive side damage will cause knocking.

- **Gear noise**—A howling or whining of the ring and pinion gears caused by an improperly set gear pattern, gear damage, or improper bearing preload. This whine can occur during all driving conditions and can be continuous or vary with vehicle load. Often the technician can control the throttle to a float, or no load, condition to change the sound of gear noise.
- **Bearing rumble**—A sound that is like marbles tumbling in a container. This noise is usually caused by a faulty wheel bearing. Since wheel bearings rotate at approximately one-third the speed as the drive axle, the noise has a much lower tone than other bearings. Rear axle bearing noise typically changes as the vehicle is weaved side to side.
- **Bearing whine**—A high-pitched sound normally caused by faulty pinion bearings, which rotate at drive shaft speed. Unlike gear noise, which changes with changes in speed, bearing noises occur at all speeds. Pinion bearings make a high-pitched, whistling sound. However, if only one pinion bearing is faulty, the noise may change with changes in driving modes. Pinion bearings are often replaced unnecessarily when correcting gear noise. They should not be replaced unless they are actually scored or damaged, or there is a specific pinion bearing noise. A stethoscope can often pinpoint pinion bearing noise on the hoist.
- **Chatter**—A noise that is evident when the vehicle is making a turn and is accompanied by a vibration. When this happens on vehicles equipped with a limited-slip axle, worn out or incorrect lubricant is the likely cause. In open axles, excessive preloading can cause a condition of partial lockup that creates chatter. If the car is equipped with a limited-slip axle and the wrong type of lubricant was installed, drain the case and refill with the correct type of lubricant, then drive the car around a parking lot, turning it many times in each direction to allow the new lubricant to work into the clutches. If changing the lubricant doesn't fix the chattering, the limited slip clutches are likely worn.

Types of Vibrations

Many vibration problems in the rear end are caused by tires or driveline angle. However, there are problems in the drive axle that can cause vibration, and most often there is a noise associated with the vibration. Diagnosing the noise will usually determine the cause of a vibration. Attention to conditions in which the vibration is felt is a key to accurately diagnosing the cause of the problem. For example, vibration felt during all modes of driving, especially at 35–45 mph, is most often caused by a faulty Universal joint or by a bent-drive pinion shaft.

A vibration that occurs at a constant speed of 35–45 mph may be also caused by a damaged **pinion flange**. A damaged flange should be replaced because it is causing the drive shaft to run off center. The runout of the pinion flange should also be checked. To do this, install a dial indicator with its base on the carrier and its plunger on the flange (Figure 7-7). Rotate the wheels while observing the movement of the indicator. Any reading indicates some runout. Compare the reading to specifications.

Classroom Manual
Chapter 7, page 170

A **pinion flange** is often referred to as the companion flange or yoke.

FIGURE 7-7 Setup for checking the runout of a companion flange.

Limited-Slip Differential Diagnostics

Classroom Manual Chapter 7, page 176

The best check of a limited-slip differential is a summary of its behavior when the vehicle has good traction at both drive wheels and when it has little or no traction at one of the drive wheels. If it works as it should and without noise or chatter, the unit is probably fine. Of course, the best way to check the condition of the unit is to inspect its components.

Some check a limited-slip differential without removing it from the axle housing. Many limited-slip differentials have a designed preload spring load on internal clutches that can be measured. This check is typically called the "breakaway" check and, although it is widely used, many experts say the results may not be dependable. To check breakaway, make sure the transmission is in neutral. Place one drive wheel of the vehicle on the floor with the other off the floor. Specifications in the service manual will list a breakaway torque reading. This is what should be required to start the rotation of the wheel that is raised off the floor. Using a torque wrench, measure the torque required to turn the wheel (Figure 7-8).

The initial breakaway torque reading may be higher than the torque required for continuous turning, but this is normal. The axle shaft should turn with even pressure throughout, without slipping or binding. If the torque reading is less than the specified amount, the differential needs to be checked.

⚠️ **WARNING:** **Never operate a car equipped with a limited-slip differential when only one wheel is raised. Power applied to the wheel remaining on the ground may cause the car to move off the jack or jack stand.**

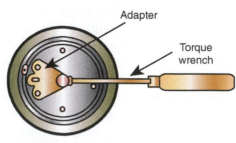

FIGURE 7-8 Checking breakaway torque on a limited-slip differential.

INSPECTION

After the road test, put the vehicle on a hoist and carefully check the entire chassis for potential causes of noise and vibration. Parts of the exhaust system may be positioned too close to the frame or underbody and may be causing the noise. Loose or bent wheels, bad treads on the tires, worn U-joints, and damaged engine mounts all are capable of creating noise and/or vibration that appear to be coming from the drive-axle assembly.

While under the vehicle looking at things, check the fluid in the axle assembly. Poor lubrication in the axle can also be a source of noise, so the lubricant's level and condition should be checked. If the level is low, the axle should be filled with the proper type and amount of lubricant (Figure 7-9), and the car should be road-tested again. If the lubricant is contaminated, it should be drained and the axle refilled with the specified lubricant. Low or contaminated lubricant normally indicates a leak somewhere in the assembly. Therefore, the drive-axle assembly should be carefully inspected to locate the source of the leak.

Checking the Fluid Level

Some vehicle manufacturers recommend periodic differential or final drive fluid changes, and all require fluid-level checks. Be sure to use the proper lubricant whenever you change or top off the fluid. Hypoid final drive gears and limited-slip differentials both require special lubricants. Make sure the lubricant meets the manufacturer's specifications for the component being serviced.

On a RWD differential and final drive assembly, check the fluid level with the vehicle on a level surface and the axle assembly at normal operating temperature. The fluid level should be even with the bottom of the fill plug opening in the axle housing unless otherwise specified. If the fluid level is low, inspect the differential and axle housing for signs of leakage.

On most FWD models with manual transaxles, the transmission and final drive assembly are lubricated with the same fluid and share the same fluid reservoir. Check the fluid level at the transaxle fill plug or dipstick. The fluid should be even with the bottom of the fill plug opening unless otherwise specified.

On FWD cars with automatic transaxles, the transmission may or may not use the same lubricant as the final drive. At times when the same lubricant is used, a separate reservoir is used for the transmission and the final drive. Always refer to the service manual or a lubrication guide to determine the correct way to check or refill the fluid.

FIGURE 7-9 Often rear axles are filled with lubricant using a pump that is graduated so that the amount of fluid entering the unit can be observed.

SERVICE TIP:
There are many different types of transmission fluids specified by manufacturers. The only way you will know that you are refilling a unit with the correct lubricant is to refer to the appropriate service manual. It is very important that transmissions and differentials are lubed with the correct lubricant.

Multiviscosity lubricants maintain their viscosity over a wide range of temperatures.

SPECIAL TOOLS
Pinion flange holding tool
Catch pan
Center punch
Slide hammer
Gasket sealer
Seal driver
Fresh lubricant
Inch-pound torque wrench

If the transmission and final drive use the same lubricant and have a common reservoir, check the level at the transaxle fill plug. If the transmission and final drive have separate fluid reservoirs, check each individually and refill them with the proper lubricant.

Always refer to your shop manual before refilling the axle housing with lubricant. The required types of lubricants and their capacities vary with the type of axle and with the manufacturer. The common transmission/differential lubricant combinations include the following:

1. 90W gear oil in the differential and the transmission.
2. 90W gear oil in the differential and ATF in the transmission.
3. ATF in both the transmission and the differential.

Hypoid gears require hypoid gear lubricant of the extreme pressure type with high viscosity.

Many new models recommend using ATF in manual transaxles because it is a lower viscosity and improves shifting effort and fuel economy. Most RWD vehicles use 90W gear oil in the manual transmission and differential. However, because of the reasons already noted, some manual transmissions should be filled with ATF and the differentials with 90W gear lube.

Multiviscosity lubricants are most commonly required for differentials. These lubricants provide temperature protection to −40° F. 75W–90W and 80W–90W are two of the most common multiviscosity lubricants. These lubricants are compatible with the single-viscosity lubricants used before. Most are also suited for use on axle assemblies equipped with limited-slip or other locking-type differentials. However, most limited-slip differentials require a special oil additive that makes the fluid slicker to allow the unit to slip for cornering.

Before adding friction modifiers to the lubricant of a limited-slip differential, always check its compatibility. Using the wrong additive may cause the axle's lubricant to break down, which would destroy the axle assembly very quickly. Also pay attention to the manufacturer's recommendations. Most modern units require the use of synthetic fluid.

Gear lubricant is circulated throughout the final drive housing by the ring gear. Special troughs or gullies are used to return the lubricant to the ring and the pinion area. The housing is sealed to keep fluid in and dirt out.

Possible Sources of Leaks

To find the source of a leak, carefully inspect the assembly for wet spots. Thoroughly clean the area around the leak so the exact source can be found.

The tail of the transmission should be examined. If the extension housing seal is leaking, it can be easily replaced. However, before replacing the seal, check the extension housing bushing and replace it with the seal if it is worn. Inspect the slip yoke's surface if the seal was leaking. Surface damage on the slip yoke will destroy the seal.

An improperly installed or damaged drive pinion seal will allow lubricant to leak past the outer edge of the seal (Figure 7-10). Any damage to the seal's bore, such as dings, dents, and gouges, will distort the seal casing and allow leakage.

It is also possible for oil to leak past the threads of the drive pinion nut or the pinion retaining bolts. These leaks can be stopped by removing the nut or bolts, applying thread or gasket sealer on the threads, and torquing the nuts or bolts to specifications.

Leakage past the stud nuts that hold the carrier assembly to a removable carrier axle housing can be corrected by installing copper washers under the nuts. This is a recommended procedure whenever the carrier is removed, whether or not copper washers were installed originally. If the assembly was originally equipped with steel washers, replace them with copper. Always make sure there is a copper washer under the axle ID tag.

Most gasket leaks result from poor installation, loose retaining bolts, or a damaged mating surface (Figure 7-11). Many vehicles do not use a gasket on the housing cover; rather, silicone sealer is used. The old sealer should be cleaned off before applying a new coat to the surface. If a housing cover is leaking, inspect the surface for imperfections, such as cracks or nicks. File the surface true and install a new gasket. If the surface cannot be trued, apply some gasket

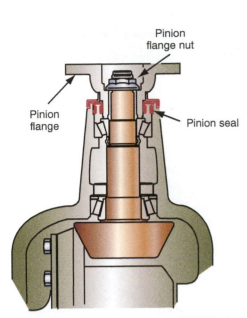

FIGURE 7-10 A common source of fluid leaks is the pinion seal.

FIGURE 7-11 Inspection points for leaks on a removable carrier-type axle housing.

Always use a new gasket with no tears, kinks, or distortion

Check the machined mating surfaces for bumps and nicks

Check surface where copper washers seat for bumps and gouges

sealer to the surface before installing the new gasket (Figure 7-12). Always follow the recommended tightening sequence and torque specifications when tightening the cover.

At times, lubricant will leak through the pores of the housing. There are two recommended ways to repair these leaks, other than replacing the housing. If the porous area is small, force some metallic body filler into the area. After the filler has set, seal the area with an epoxy-type sealer. If the area is large, drill a hole and tap an appropriately sized setscrew into the hole, then cover the area with an epoxy sealer. Minor weld leaks can also be sealed with epoxy sealer. However, if a weld is broken, the housing should be replaced (Figure 7-13).

If the wrong vent was installed in the axle housing or if there is an excessive amount of lubricant or oil turbulence in the axle, lubricant may leak through the axle vent hose. This can only be found by a careful inspection. Check for a crimped or broken vent hose. Replace the hose if it is damaged. If the cause of the leakage is an overfilled axle assembly, drain the housing and refill the unit with the specified amount and type of lubricant.

Axle fluid may be evident on the brakes' backing plate or brake drums. A worn or damaged axle seal usually causes this. However, if the seal's bore in the axle tube is damaged, the

SERVICE TIP:
After servicing a differential, the lubricant may appear to look much like oil mixed with water. This discoloration might be caused by the marking compound that may have been used to examine the gear pattern. Most of the compound should have been wiped off before assembly. The marking compound will eventually disperse in the gear oil and the normal color will return. This will do no harm to the axle and there is no need to change the lube.

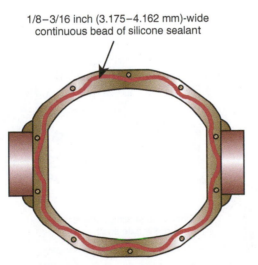

1/8–3/16 inch (3.175–4.162 mm)-wide continuous bead of silicone sealant

FIGURE 7-12 Proper application of sealant to correct axle housing cover leaks.

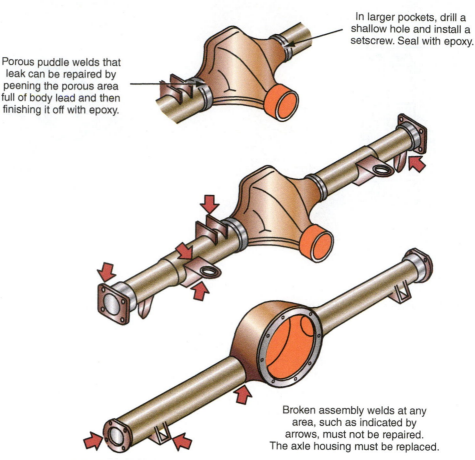

In larger pockets, drill a shallow hole and install a setscrew. Seal with epoxy.

Porous puddle welds that leak can be repaired by peening the porous area full of body lead and then finishing it off with epoxy.

Broken assembly welds at any area, such as indicated by arrows, must not be repaired. The axle housing must be replaced.

FIGURE 7-13 Typical guidelines for correcting axle housing leaks.

seal will be unable to seal properly (Figure 7-14). Whenever fluid is leaking past the axle seal, the bore should be checked and the seal replaced. A worn axle bearing or overheated brakes may be the cause of an axle seal leak. A worn axle bearing on a C-clip retained axle will often damage the axle journal. Axle-saver bearing kits are available for many vehicles that step the bearing out about ¾-inch to a fresh surface of the axle.

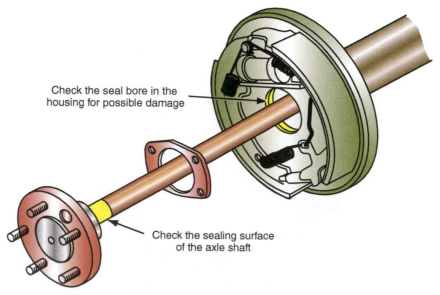

Check the seal bore in the housing for possible damage

Check the sealing surface of the axle shaft

FIGURE 7-14 If fluid leaks around the axle seal, check the seal and the condition of the shaft and the bearing bore.

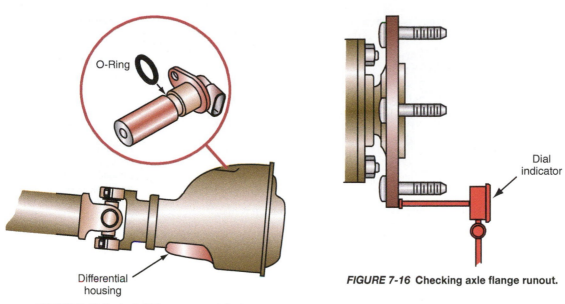

FIGURE 7-15 Typical ABS sensor and O-ring.

FIGURE 7-16 Checking axle flange runout.

Some differential units are fitted with an ABS sensor. Lubricant can leak from around the O-ring (Figure 7-15). To correct his problem, remove the sensor and replace the O-ring.

Driveline Inspection

Check the runout of the axle shaft's flange by positioning a dial indicator against the outer surface of the flange (Figure 7-16). Apply slight pressure to the center of the axle to remove axle end play, and then zero the indicator. Slowly rotate the axle one complete turn and observe the readings on the indicator. The total amount of indicator movement is the total amount of axle flange lateral runout. Compare the measured runout with specifications.

Inspect the wheel studs in the axle flanges. If they are broken or bent, they should be replaced. Also check the condition of the threads. If they have minor distortions, run a die over the stud. If the threads are severely damaged, the stud should be replaced. Studs are normally pressed in and out of the flange. Make sure you install the correct size stud.

Checking Backlash of the Assembly

To check the total backlash of an axle assembly, raise the vehicle on a hoist so that both drive wheels are free to rotate. Clamp a bar between the companion flange and the frame or body. Position this bar so the flange cannot move (Figure 7-17). Lower the left drive wheel onto a chock to keep it from turning. Rotate the right wheel slowly until it feels like it is in the drive condition. Position a piece of chalk onto the side of the tire. While keeping the chalk against the tire, turn the wheel in the opposite direction until you again feel the drive condition. Measure the length of the chalk mark. This equals the total axle backlash. It should be 1 inch or less.

If the backlash is within the proper limit, the cause of the clunking or other noise is not the axle. If the backlash is excessive, the following items should be checked.

1. Check for elongation of the differential pinion shaft holes in the differential case.
2. Check for a missing or damaged differential or side gear washer.
3. Check for galling of the differential pinion shaft and bore.
4. Check the fit of the axle shafts in the splines of the side gears (Figure 7-18).

Checking Bearing Preload

To determine if a vibration or chatter is caused by excessive preload, position the vehicle on a frame-contact lift. Remove the rear wheels, brake drums, and the drive shaft. Install an inch-pound torque wrench on the pinion nut and measure the torque required to turn the pinion.

⚠

CAUTION:
Whenever working under a vehicle, make sure it is supported on safety stands or the lift's lock set. Also be sure to wear safety glasses or goggles; dirt and/or rust can fall off the car and get in your eyes. This can cause serious injury to your eyes.

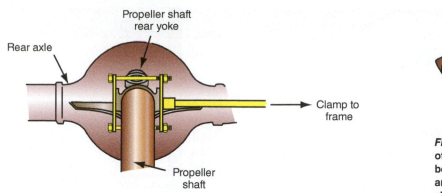

FIGURE 7-17 Using a holding tool to prevent the companion flange from rotating while checking backlash.

FIGURE 7-18 The splines of the side gears should be inspected for wear and their fit on the axle shaft checked.

SERVICE TIP:
Before installing the companion flange into a new oil seal, lubricate the seal area of the flange (Figure 7-19) to prevent damage to the seal.

Compare your reading with the specifications. If the required torque is not within specifications, the unit must be readjusted.

In-Vehicle Service

There is a minimum of service that can be performed on final drives and differentials while they are still in the vehicle. Checking the fluid and changing the pinion seal are the only services that are practical or possible.

Replacing Pinion Seals

If the pinion seal is a source of fluid leakage, it should be replaced. Photo Sequence 16 covers a typical procedure for replacing a pinion seal, as well as checking the runout of the companion flange. To replace a pinion seal, remove the drive shaft from the differential pinion flange, then remove the flange nut. You may need a special wrench to hold the flange while removing the nut. Pull the pinion flange off the drive pinion splines, then inspect the sealing surface of the flange to make sure it is smooth and free of burrs. If the seal has worn a groove on the sealing surface, a repair sleeve can be installed to correct the situation.

With a slide hammer, remove the pinion seal. Before you install a new seal, lubricate the seal's lip and coat the outside diameter of the seal with gasket sealer. Use a seal driver to install the new seal flush against the carrier or axle housing (Figure 7-20).

The tightness of the pinion flange nut is critical because it determines the pinion bearing preload. Methods of tightening the nut vary. Some manufacturers recommend that the nut be tightened a little at a time until the pinion gear turning resistance is within specifications when

FIGURE 7-19 Apply grease to the sealing area of the companion flange.

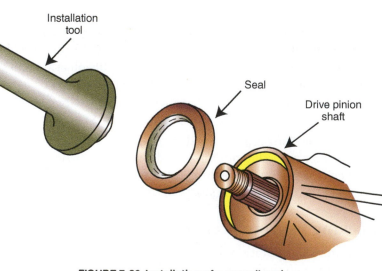

FIGURE 7-20 Installation of a new oil seal.

COMPANION FLANGE AND PINION SEAL SERVICES

P16-1 To measure the runout of the companion flange or to replace the pinion seal, the drive shaft must be removed.

P16-2 Excessive companion flange runout can cause pinion seal wear and driveline vibrations. To check flange runout, mount a dial indicator on the axle housing so that the indicator's plunger rests on the outside face of the flange.

P16-3 Mark the spot where the indicator's plunger contacts the flange.

P16-4 Set the indicator to zero and use the wheels to turn the flange one complete turn.

P16-5 Observe the needle of the indicator when the flange was turning. The total amount of needle deflection on the indicator is the total amount of runout. In this case, it was __0.003__ inches.

P16-6 Compare your findings to the specifications. If there is excessive runout, the companion flange should be replaced.

P16-7 If the pinion seal is a source of fluid leakage, it should be replaced. To replace a pinion seal, remove the flange nut. You may need a special wrench to hold the flange when removing the nut.

P16-8 Place the catch can under the pinion shaft area, and then pull the companion flange off the pinion splines.

P16-9 Inspect the sealing surface of the flange to make sure it is smooth and free of burrs. If it is damaged, it should be replaced.

P16-10 Using a slide hammer–type puller, remove the pinion seal.

P16-11 Before you install a new seal, lubricate the seal's lip and coat the outside diameter of the seal with gasket sealer. Use a seal driver to install the new seal flush against the carrier or axle housing. Make sure the spring around the seal lip remains in place.

P16-12 After the seal is installed, lubricate the seal area of the companion flange, and push it carefully into the seal, making sure the seal is not damaged when doing this.

P16-13 Start a new pinion flange nut onto the threads and set bearing preload. The tightness of the pinion flange nut is critical because it determines the pinion bearing preload. If the flange nut is overtightened, the collapsible collar must be replaced.

P16-14 Once the flange nut is installed and properly tightened, install the drive shaft.

P16-15 Check the fluid in the differential, then road-test the vehicle and again check for leaks.

A **companion flange** is a mounting flange that attaches a drive shaft to another drivetrain component.

measured with an inch-pound torque wrench. Another method is to mark the relative positions of the nut, **companion flange**, and pinion gear with a center punch before taking them apart. Then tighten the nut to its original position plus a couple of degrees. If the flange nut is overtightened, the pinion bearing preload must be readjusted and a new collapsible collar installed.

Once the flange nut is installed and properly tightened, install the drive shaft. Check the fluid in the differential, then road-test the vehicle and again check for leaks.

OUT-OF-VEHICLE SERVICES

Out-of-vehicle service of the differential and final drive is usually required only for major repairs such as ring and pinion gear replacement and adjustments, or differential bearing replacement. FWD final drives are normally an integral part of the transaxle assembly. To service a differential in a FWD vehicle, the transaxle must normally be removed and disassembled. Most of the procedures for servicing a RWD differential apply to FWD differentials.

Precautions

The following should be adhered to preparing to service and during servicing a differential and final drive unit:

1. Before disassembly, clean the outside of the drive axle housing and remove any sand or mud to prevent it from entering the inside of the assembly during disassembly and installation.
2. When removing a part connected to the housing that is made of light alloy, such as a differential carrier cover, tap it off with a plastic hammer. Do not pry it off with a screwdriver.
3. Always arrange the parts as the unit is being disassembled and according to the order of their removal. Also, keep them protected from dirt and dust.
4. Before installation, thoroughly clean and dry each internal part and then apply hypoid gear oil to it.
5. Never use alkaline cleaners on aluminum or rubber parts and ring gear bolts.
6. Coat all sliding surfaces and rotating parts with hypoid gear oil.
7. Be careful not to damage the contact surfaces of the housing; damage may cause oil leakage.
8. Be careful not to damage the surfaces that are assembled with an oil seal, O-ring, or gasket; damage may cause oil leakage.
9. When press-fitting an oil seal, be careful not to damage the oil seal lip and outside casing.
10. When replacing a bearing, replace the inner and outer races as a set.

Check Ring Gear Runout

Before disassembling the differential, measure the runout of the ring gear (Figure 7-21). Excessive runout can be caused by a warped gear, worn differential side bearings, a warped differential case, or particles trapped between the gear and case.

Runout is checked with a dial indicator mounted on the carrier assembly. The indicator stem is loaded slightly against the back face of the ring gear. The plunger should be set at a right angle to the gear. With the dial indicator in position and its dial set to zero, rotate the ring gear and note the highest and lowest readings.

The difference between these two readings indicates the total runout of the ring gear. Normally the maximum permissible runout is 0.003 inch. The cause for the excessive runout must be corrected.

To determine if the runout is caused by a damaged differential case, remove the ring gear from the case. Reassemble the case in the carrier without the ring gear but with good side

Classroom Manual
Chapter 7, page 165

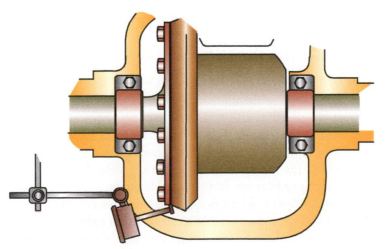

FIGURE 7-21 Position of dial indicator to check ring gear runout.

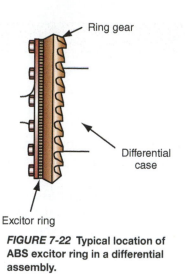

Ring gear

Differential case

Excitor ring

FIGURE 7-22 Typical location of ABS excitor ring in a differential assembly.

Classroom Manual

Chapter 7, page 169

bearings. Set the differential bearing preload. Mount a dial indicator in the same manner as for measuring ring gear runout, except measure the runout of the ring gear mounting face. Runout should not exceed 0.003 inch; if runout is greater than that, the case should be replaced.

If the case's runout was within specifications, the ring gear is probably warped and should be replaced. A ring gear is never replaced without replacing its mating pinion gear.

Some ring gear assemblies are comprised of an excitor ring used in antilock brake systems (Figure 7-22). This ring is normally pressed onto the ring gear hub and can be removed after the ring gear is removed. If the ring gear assembly is equipped with an excitor ring, carefully inspect it and replace it if it is damaged.

Preparing for Removal

Removing the final drive assembly, it is wise to check the adjustments of the ring and pinion gears. The procedures for doing this are explained later in this chapter, as these are extremely critical during reassembly. The gear tooth pattern should be checked with marking compound and any irregular patterns noted and recorded for future reference.

The preload of the pinion bearing and the case bearings should also be measured and recorded for future reference. This is also true for backlash of the gears.

REMOVING AND DISASSEMBLING A FINAL DRIVE UNIT

Photo Sequence 17 shows a typical procedure for removing and disassembling a final drive unit on a removable carrier housing. Photo Sequence 18 shows the same procedure for a typical integral housing. Always refer to the appropriate service material before servicing a unit.

Guidelines for Removing the Differential Assembly

Always follow the correct procedures for removing a differential assembly.

- If the entire rear axle must be removed to service the differential, disconnect the axle housing from all suspension parts and spring mounts and then support it with a jack. Once the housing has been removed, place it on a bench for service.
- Always place a drain pan under the rear of the transmission and rear axle before removing the drive shaft and/or the differential. Once the slip yoke is pulled out, install a plug into the housing's opening to prevent further fluid leakage.
- Always check the shop manual to identify how the axles are retained.
- On non-C-clip axles a slide hammer is used to remove the axles from the housing. This should only be done after the four retainer bolts on the wheel end have been removed.
- Clean the interior of the axle housing and sealing areas on the housing and/or cover after the differential or cover has been removed.

TYPICAL PROCEDURE FOR REMOVAL AND DISASSEMBLY OF A REMOVABLE CARRIER-TYPE FINAL DRIVE UNIT

P17-1 Raise the vehicle and drain lubricant out of the axle housing.

P17-2 Mark the alignment of the drive shaft, then remove it.

P17-3 Remove the tire and wheel assemblies.

P17-4 Remove the brake drums to gain access to the axle shafts.

P17-5 Check and record the end play of the axles with a dial indicator.

P17-6 Using the correct tool, free the axle shafts from the differential unit.

P17-7 Remove the axle assemblies. On some vehicles it may be easier to remove the axle with the brake backing plate and the brakes fully assembled, as shown.

P17-8 Unbolt the carrier to housing attaching bolts.

P17-9 Carefully pull the differential carrier out of the axle housing.

P17-10 Set the carrier on a workbench or in an appropriate fixture.

P17-11 Mark the differential bearing caps and mating bearing supports.

P17-12 With a dial indicator, measure ring-gear backlash and runout.

P17-13 Remove the adjusting nut locks.

P17-14 Remove the bearing caps and adjusting nuts.

P17-15 Lift the differential case from the carrier.

P17-16 Remove the differential side bearings using the appropriate puller.

P17-17 Mark the alignment of the differential case and the ring gear.

P17-18 Remove and discard the ring gear to differential-case bolts. Then separate the ring gear from the case. Ring gear bolts may be left-hand threaded. An "L" on the bolt head indicates that it has left-handed threads.

P17-19 Using a chisel or other tool, flatten the locking tabs for the differential-case bolts.

P17-20 Remove the differential-case bolts.

P17-21 Make alignment marks at the mating surfaces of the differential-case assembly, then separate the case.

P17-22 Remove the side-gear thrust washer and one side gear.

P17-23 Drive out the differential-pinion shaft lock pins. Then remove the positioning block, differential pinions, thrust washers, and the remaining side gear.

P17-24 Loosen and remove the pinion nut.

P17-25 Drive pinion out of the front bearing with a soft-faced hammer and remove it from the rear of the housing.

P17-26 Drive front-bearing cup out of housing and remove the front seal.

P17-27 Using the correct fixture, press the pinion bearing off the shaft. Be sure to remove, measure, and record the thickness of the shim located behind the bearing.

TYPICAL PROCEDURE FOR THE DISASSEMBLY OF AN INTEGRAL CARRIER-TYPE FINAL DRIVE UNIT

P18-1 Raise the vehicle and drain lubricant out of the housing. Then index the drive shaft.

P18-2 Remove the drive shaft.

P18-3 Remove the tires, wheels, and brake drums to gain access to the axle shafts.

P18-4 Remove the rear cover from the housing.

P18-5 Clean the interior of the housing and the gasket sealing area of the housing.

P18-6 Remove the differential-pinion shaft lock bolt.

P18-7 Remove the differential-pinion shaft.

P18-8 Push the axle toward the center of the housing and remove the C-locks.

P18-9 Pull the axle shafts from the housing.

P18-10 Check for evidence of differential-assembly side play.

P18-11 Measure ring-gear runout and backlash.

P18-12 Mark the position of the axle housing and differential bearing caps.

P18-13 Loosen and remove the bearing-cap bolts.

P18-14 Remove the bearing caps.

P18-15 Gently pry the differential case partly out of the housing.

P18-16 Grasping the bearings and shims, pull the differential unit out of the housing.

P18-17 With the differential unit in a suitable fixture, remove and discard the ring-gear bolts. Be careful, some ring-gear bolts have left-handed threads.

P18-18 Using a brass drift, tap the ring gear loose from the differential case and remove it.

P18-19 Rotate the side gears until the pinion gears appear at the case window.

P18-20 Remove the pinion gears, side gears, and thrust washers.

P18-21 Remove the side bearings.

P18-22 Using the correct holding fixture, loosen the pinion nut.

P18-23 Remove the pinion nut and washer.

P18-24 Remove the pinion flange and pinion oil seal.

P18-25 Drive the pinion shaft out of its bearing.

P18-26 Remove the pinion from the rear of the housing. Then remove the front-bearing cup. Remove and discard the collapsible spacer.

P18-27 Use a press and the correct fixtures to remove the pinion bearing. Remove the pinion shim and record its thickness.

- To check differential assembly side play in an integral assembly, position a screwdriver between the left side of the housing and the differential case flange (Figure 7-23). Using a prying motion, determine if side play is present. There should be no side play.
- To check the side play of a removable carrier, attempt to move the differential case with a large screwdriver. Any movement indicates side play.
- Check and record backlash between the ring and pinion before disassembly. Most manufacturers recommend that runout be checked in at least three or four locations. Determine backlash variation between the lowest and highest reading.
- Ring gear runout should be measured before the gear is removed from the housing. Excessive runout can be caused by warped gear, worn differential side bearings, warped differential case, or particles trapped between the gear and case.
- Mount a dial indicator on the case or housing and load the indicator stem slightly with the plunger at a right angle to the back face of the ring gear. Set the dial indicator to zero, rotate the ring gear, and note the highest and lowest readings. The difference between these readings indicates the total runout of the ring gear. Normally the maximum permissible runout is 0.003 inch. Mark the point of maximum runout on the ring gear.
- The axles in most integral carriers are retained by C-locks, which are removed through the opening in the case (Figure 7-24).
- To determine if the runout is caused by a damaged differential case, first mark the bearing caps so that the correct side and direction can be maintained. Next, remove the ring gear from the differential case. Then reinstall the differential case into the housing. Install the bearing caps and lightly tighten the bearing cap bolts. Mount a dial indicator in the same manner as for measuring ring gear runout, except measure the runout of the mounting face for the ring gear. Runout should not exceed 0.003 inch; if runout is greater than that, the case should be replaced. If there is slight (less than 0.003 inch) runout, position the ring gear 180 degrees from the point of maximum runout during reassembly.
- If the case's runout was within specifications, the ring gear is probably warped and should be replaced. A ring gear is never replaced without replacing its mating pinion gear.
- When removing bearing caps and cups, keep them with their respective bearings.

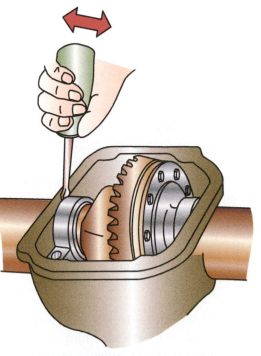

FIGURE 7-23 Check the side play of the differential case before disassembly.

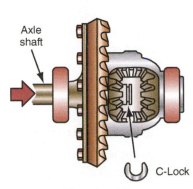

Axle shaft

C-Lock

FIGURE 7-24 Location of axle shaft C-locks.

Removable carrier differentials are sometimes called "chunk" types.

Classroom Manual

Chapter 7, page 173

SERVICE TIP:
Always lubricate the side bearing bores in the axle carrier or housing before installing the differential. This will allow the bearings to move easily during backlash and preload adjustments.

SERVICE TIP:
Some positraction units have two sets of splines that the axle must pass through on each side. When the case is bolted together and the two sets of splines are under spring tension (if they are NOT lined up exactly), the axle will not go all the way in. Stub axles or some sort of special tool will line them up while the case halves are bolted together.

■ After the differential assembly is removed from the housing, mount it in a soft-jawed vise or mount it on a special fixture.

■ Some ring gear assemblies have an excitor ring, used in antilock brake systems. This ring is normally pressed onto the ring gear hub and can be removed after the ring gear is removed. Carefully inspect the excitor ring and replace it if it is damaged.

■ The bolts that attach the ring gear to the differential case may have left-handed threads. Inspect the head of the bolts to see if they are marked with an "L" before attempting to loosen these bolts.

■ Remove and discard the ring gear bolts. Ring gear bolts should never be reused; they stretch when properly torqued and will not be reliable if reused.

■ Make sure you measure and record the thickness of all of the shims during disassembly.

■ If the unit uses selective shims at the side gears, keep them in order when removing the bearings and set them aside for future reference.

INSPECTION OF PARTS

A highly important step in the procedure for disassembling any type of differential assembly is a careful inspection of each part as it is removed. The differential bearings should be looked at and felt to determine if there are any defects or evidence of damage.

If the unit had side play, check the differential bearings. These bearings usually wear from improper preload adjustments. If during the inspection you find the side play to be the result of loose bearing cones on the differential case hubs, the differential case must be replaced.

To inspect a gear set, remove as much lubricant as possible from the gears with clean solvent. Wipe the gears dry or carefully blow them dry with compressed air. Examine each gear for cracks, scoring, or damaged teeth. If either gear is scored or badly damaged, the gear set should be replaced. If the gears are chipped or damaged, the carrier and housing must be thoroughly cleaned to remove all metal shavings or particles.

All other parts of the differential also should be inspected. Cleaning each part individually will aid in this inspection. Any damaged part in the differential must be replaced. While cleaning parts for inspection, use care on limited-slip units to prevent any solvent from entering the clutch pack. The solvent may cause the clutch surfaces to deteriorate. Damaged clutches are the most common cause of differential chatter.

REASSEMBLING A DIFFERENTIAL ASSEMBLY

Removable carrier final drives are installed after they have been serviced on a work bench. The assembly of the unit is performed on the bench. Photo Sequence 19 covers a typical procedure for the assembly and installation of a removable carrier-type final drive unit. Certain checks and measurements must be made during the procedure and are carefully described later in this chapter. Always follow the specific procedures given in a service manual.

Integral carrier final drives are assembled into the axle housing. Photo Sequence 20 covers a typical procedure for the assembly of an integral carrier-type final drive unit. Certain checks and measurements must be made during the procedure and are carefully described later in this chapter. Always follow the specific procedures given in a service manual.

Always thoroughly clean the mounting and sealing surfaces of the axle housing prior to installing the differential or housing cover onto the housing. Seal the opening with a new gasket or sealant and use copper washers under the carrier mounting nuts.

Ring and Pinion Gear Installation

Whenever a ring or pinion gear needs to be replaced, its mating gear should also be replaced. Always install a new gear set. Both the ring and the pinion will have matching numbers to ensure they are installed as a set. As a set, the two gears have been lapped together by the manufacturer and will operate longer and quieter than a set of gears that have not been lapped

Typical Procedure for the Assembly and Installation of a Removable Carrier-Type Final Drive Unit

P19-1 Press differential case bearings onto the case.

P19-2 Place one differential side gear and thrust washer into the case's bore. Always lubricate the side gear shoulder and case bore.

P19-3 Install the pinion-gear assemblies into the case.

P19-4 Install the positioning block into the case.

P19-5 Place the other side gear into position in the case.

P19-6 Position and align the differential-case halves.

P19-7 Install and tighten the differential-case bolts.

P19-8 Position the ring gear onto the differential case. At times it may be necessary to heat the gear before it will fully seat against the case.

P19-9 Install new ring-gear bolts and tighten them to specifications.

P19-10 Using the correct fixture, press the pinion bearing onto the shaft. Be sure to install the correct size shim behind the bearing.

P19-11 Install the pinion shaft, with spacer, into the housing.

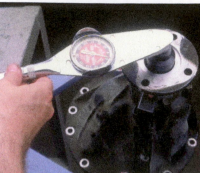

P19-12 Install pinion seal, flange, and nut. Tighten the nut to the specified preload.

P19-13 Measure and set pinion depth with the appropriate gauge set.

P19-14 Lower the gear case carefully into the housing with the bearing cups in position.

P19-15 Make sure the gear set is properly aligned and install the bearing caps.

P19-16 Check backlash and the side-bearing preload with a dial indicator.

P19-17 Turn adjusting nuts to achieve desired readings. Apply marking compound on the ring gear and check the backlash and pinion depth.

P19-18 After the proper pattern is achieved, install and tighten the adjusting nut locks.

P19-19 Apply a thin coat of sealant around the sealing surface of the axle housing.

P19-20 Install the carrier housing using a new gasket.

P19-21 Install carrier bolts and tighten them to the specified torque.

P19-22 Connect all wires and tubes to the carrier assembly.

P19-23 Align and install the drive shaft.

P19-24 Install the axle shafts with seals.

P19-25 Connect and tighten all brake hardware to the axle housing and backing plate.

P19-26 Install the brake drums and wheel assemblies.

P19-27 Fill the axle housing with the correct amount and type of lubricant.

PHOTO SEQUENCE 20

TYPICAL PROCEDURE FOR THE ASSEMBLY OF AN INTEGRAL CARRIER-TYPE FINAL DRIVE UNIT

P20-1 Install the thrust washers on the differential side gears.

P20-2 Position the side gears in the differential case. Always place lubricant on the shoulder and case bore during assembly.

P20-3 Place thrust washers on the pinion gears. Then mesh the pinion gears with the side gears. Install the differential pinion shaft and align and install the lock pin into the pinion shaft.

P20-4 Heat the ring gear with a heat lamp or hot water and install the heated ring gear onto the differential case.

P20-5 Install and tighten the new ring gear bolts.

P20-6 Press the side bearings onto the differential case.

P20-7 Press the pinion bearing onto the pinion shaft.

P20-8 Set pinion depth with appropriate gauge set.

P20-9 Install the pinion gear with a new crush sleeve into the housing.

P20-10 Install the flange and pinion nut and tighten the nut to achieve the proper bearing preload.

P20-11 Install the differential with its bearing cones and proper shims.

P20-12 Install the bearing caps and tighten to specifications.

P20-13 Check the preload and backlash of the gear set.

P20-14 Install the axles into the housing.

P20-15 Insert the C-locks while pushing the axles toward the center of the housing.

P20-16 Install the differential-pinion shaft lock bolt and pinion shaft.

P20-17 Apply a thin bead of sealer onto the rear cover and install on the housing. Tighten the bolts and nuts to the specified torque.

P20-18 Install the brake drums, wheels, and drive shaft. Then refill the housing with the appropriate amount and type of lubricant.

together. Often the gear set will have a prerolled pattern. This serves as a reference for the proper shimming of the pinion.

When installing the ring gear onto the differential case, make sure the bolt holes are aligned before pressing the gear in place (Figure 7-25). Even pressure should be applied when pressing the gear. Likewise, when tightening the bolts, always tighten them in steps and to the specified torque. These steps reduce the chances of distorting the gear. Some manufacturers have lock plates under the ring gear retaining bolts (Figure 7-26). These must be staked to hold the bolts tight. Use a chisel and a hammer to stake all of the lock plates. There may be two claws near each bolt. Stake one claw so that it is flush against the flat surface of the bolt. Stake the other claw against the surface of the bolt head so that it can act as a stopper if the bolt starts to loosen.

Examine the gears to locate any timing marks on the gear set that indicate where the gears were lapped by the manufacturer. Normally, one tooth of the pinion gear will be grooved and painted, and the ring gear will have a notch between two painted teeth. If the paint marks are not evident, locate the notches. Proper timing of the gears is set by placing the grooved pinion tooth between the two marked ring gear teeth. Most gear sets have no timing marks. These gears are hunting gears and do not need to be timed.

When reinstalling the original ring and pinion gear set, use a whetstone to remove any sharp edges on the inside diameter of the ring gear.

Classroom Manual

Chapter 7, page 169

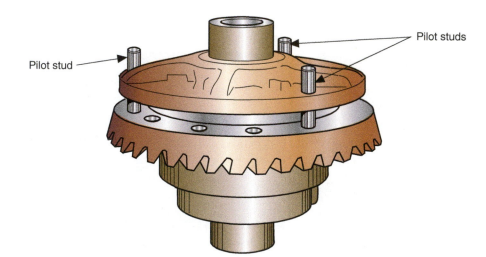

FIGURE 7-25 Pilot studs can be used to properly align the ring gear to the differential case prior to pressing the ring gear onto the case.

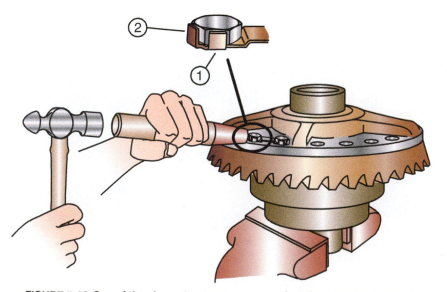

FIGURE 7-26 One of the claws should be staked against the flat side of the bolt head (1) and other against the bolt head (2).

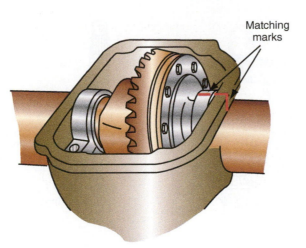

Matching marks

FIGURE 7-27 Align the index marks when reinstalling the bearing caps.

RING AND PINION GEAR ADJUSTMENTS

Whenever the ring and pinion gears or the pinion and/or carrier bearings are replaced, pinion gear depth, bearing preload, and the ring and pinion gear tooth patterns must be checked and adjusted. This holds true for all types of differentials, except FWD differentials which use helical-cut gears and do not require checking of tooth patterns. All hypoid gear-type final drive units must be aligned to ensure a quiet operation.

During reassembly, position the pinion with a slight backlash and tight against the ring gear. Carefully install the adjusting nuts, making sure they are seated on the threads. Tighten the nuts by hand until they contact the bearing cups. Make sure they have engaged into the same amount of threads on each side. Then install the bearing caps; be sure to match the alignment marks made when it was disassembled (Figure 7-27). While holding the caps up, start the bolts into their threads, then lower and snap the caps into place on the cap bolts. Tighten the bolts to the specified torque to fully seat the bearing caps, then loosen each bolt about one-half turn. Now retorque the bolts to 35 foot-pounds to allow the bearings to move during the adjustments. After the adjustments have been made, the bolts should be retorqued to the specified amount.

Pinion Gear Depth

The base pinion depth setting is the distance between the nose of the pinion gear and the centerline of the axles or the differential case bearing bores (Figure 7-28). Pinion gear depth is normally adjusted by changing the thickness of shims that are placed underneath the rear pinion bearing. The thickness of the drive pinion rear bearing shim controls the depth of the mesh between the pinion and ring gear. In addition to the base pinion depth setting, the pinion gear also may be marked with additional adjustment instructions. Check the pinion gear for painted or etched markings on its stem or on its small end (Figure 7-29). These are called checking distance markings. A nominal, or unmarked gear, will have a standard pinion depth shim. For example, a gear with a (–2) marking will have a shim 0.002-inch thicker than that found on a nominal gear to compensate for the thinner pinion head dimension. In this case, if the old gear had a (–2) marking and came with a .035-inch shim, and the new gear had a (+1) marking, the new starting shim would be .032-inch to make up the difference.

Special tools—such as a depth micrometer or a fixed-length arbor, block gauge, and dial indicator—are needed to set the pinion gear depth. Many manufacturers have a dedicated pinion depth shim selector kit for individual axle assemblies. Use the depth micrometer to determine the difference between the actual depth and the specified depth. The difference between the reading and the specifications identifies what should be done to the shims. If the depth is greater than the desired amount, the difference should be subtracted from the shim pack. Likewise, if the depth is less than specifications, the shim pack should be thicker.

Classroom Manual
Chapter 7, page 170

SERVICE TIP:
Don't spend too much time looking for engraved pinion checking distance numbers. Improved manufacturing techniques mean that most drive pinions are made to the nominal, or "perfect," size. In these cases the gear will come with a standard, or nominal, pinion depth shim.

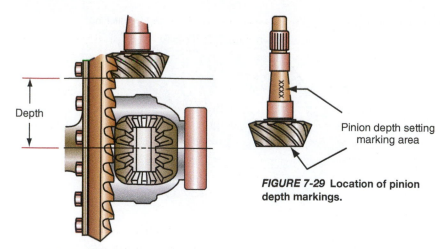

FIGURE 7-29 Location of pinion depth markings.

FIGURE 7-28 Pinion depth.

When using a gauge set, which includes an arbor and block gauge, for a particular differential, a dial indicator or gauge tube can be used to determine the difference between the actual and specified depth. Clamp the arbor into the carrier or axle housing bearing bores and place a gauge block over the pinion gear (Figure 7-30). Measure the difference between the arbor and the gauge block to determine the thickness of the shims required to correct pinion depth.

The procedure for setting pinion depth on various differentials is the same. However, the required measuring tools or gauge kits will vary, as will the desired settings. Therefore always refer to the appropriate service manual when setting pinion depth. The following is a general procedure for measuring and adjusting pinion depth with a gauge set and dial indicator. (*Note:* This procedure is for a removable carrier-type differential and measures the pinion depth from the centerlines of the bearing bores in the differential case and not from the housing.)

1. If the pinion bearing races were removed, install them.
2. Lubricate and install the new pinion bearings.
3. Position the gauge block and rear pinion bearing pilot on the preload stud.
4. Install the gauge block through the rear pinion bearing, front pinion bearing, and pilot.
5. Install the pinion nut until it is snug.
6. Properly seat the bearings by rotating them.
7. Hold the preload stud stationary with a wrench.
8. Tighten the pinion nut until 20 inch-pounds is needed to rotate the bearings.

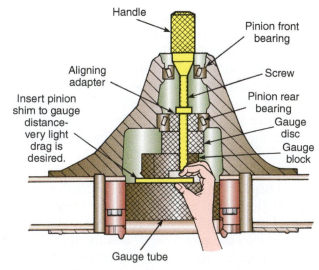

FIGURE 7-30 Typical pinion depth gauge setup.

9. Mount the side bearing gauging discs on the ends of the arbor.
10. Place the arbor into the carrier, making sure the discs are properly seated (Figure 7-31).
11. Install the side bearing caps and tighten the bolts to prevent movement.
12. Position the dial indicator on the mounting post of the arbor.
13. Preload the dial indicator one-half revolution, then tighten the plunger in this position.
14. Place the plunger onto the gauging area of the gauge block.
15. Move the plunger rod slowly back and forth across the gauging area until the dial indicator reads the greatest deflection.
16. Set the dial indicator to zero.
17. Swing the plunger until it is moved off the gauging area.
18. The dial indicator now reads the required pinion shim thickness for nominal pinion depth. Record this reading.
19. Correct the reading by adding or subtracting the amount indicated on the pinion gear. This corrected reading is the required thickness for the shim pack.
20. Remove the bearing caps and gauging tools from the housing.
21. Place the selected shim onto the drive pinion (Figure 7-32).

An Alternate Way to Measure Pinion Depth. Many shops don't have a pinion depth gauge set. The gauge sets are rather expensive and don't cover a wide selection of drive axles. A shop could purchase tool sets for each rear axle they do and may service, or they could make up their own set of tools for each axle.

When a rear axle is being serviced, a technician can use the good bearings that will be replaced with new ones or an additional pair of new bearings as the basis for making a gauge set. With a hone or a grinding stone set in a drill press, lathe, or milling machine, remove enough metal from the bearings' bore to remove the 2 or 3 thousandths of an inch interference fit. Hone the bearing bore just enough so that the bearings become a slip fit and easily removable by hand pressure only. Be careful not to remove too much metal.

Classroom Manual
Chapter 7, page 170

SERVICE TIP: The collapsible sleeve is crushed as the pinion nut is tightened. If you overtighten the nut, the sleeve should be replaced before resetting the preload. To avoid overtightening the nut, tighten the pinion nut to the low side of the torque specification range. Then use an inch-pound torque wrench to measure the force required to turn the pinion shaft. Continue to tighten the pinion nut in small increments and check the preload at each phase until the desired preload is reached. If the desired preload is exceeded, replace the collapsible spacer.

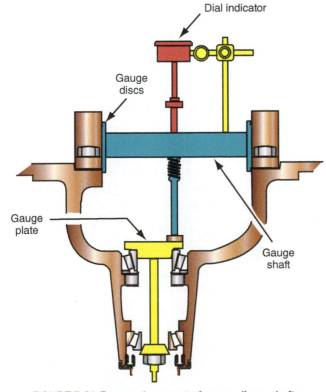

FIGURE 7-31 Proper placement of gauge discs, shafts, and plates for measuring pinion depth.

FIGURE 7-32 Placement of pinion depth shim.

Use these bearings to set up the rear axle. Start with the original shims and make the adjustments necessary to obtain the correct pinion depth and backlash. Verify this by running a contact pattern. Once the axle's gears are set, remove the slip fit bearings. Keeping the shims in place, install the new bearings.

Using this method for adjusting pinion depth avoids damage to new bearings as they are pressed on and off their shafts to change shims.

Once the slip fit bearings are removed, tag them as to what rear axle assembly they were from. These bearings can be used again on the same axle and therefore become a gauge set for that axle.

> **AUTHOR'S NOTE:** The alternative method of measuring pinion depth is especially useful when installing a set of aftermarket gears. These gear sets will typically have an etched pinion depth specification on the small face of the pinion gear (e.g., "2.478"). This figure indicates the precise distance between the face of the pinion gear and the centerline of the ring gear. This distance can be difficult to determine without special tooling, and the pattern adjustment method provides an alternative procedure.

Pinion Bearing Preload

Pinion bearing preload is set by tightening the pinion nut until the desired number of inch-pounds is required to turn the shaft. Tightening the nut crushes the collapsible pinion spacer, which maintains the desired preload. Never overtighten and then loosen the pinion nut to reach the desired torque reading. Tightening and loosening the pinion nut will damage the collapsible spacer, and it must then be replaced. For the exact procedures and specifications for bearing preload, refer to the appropriate service manual.

Incorrect bearing preload can cause differential noise. Driving loads tend to move the teeth of the pinion ring to the outside of the ring gear. It is possible for the pinion gear to move to the edge of the gear's teeth and score them. A collapsible spacer (Figure 7-33) serves as a spring to hold the bearing preload and prevent the small pinion bearing inner race from

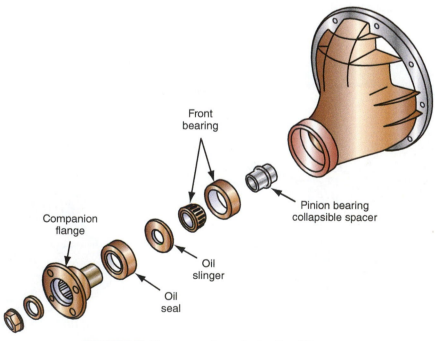

Front
bearing

Companion
flange

Pinion bearing
collapsible spacer

Oil
slinger

Oil
seal

FIGURE 7-33 Placement of a typical collapsible spacer.

spinning on the drive pinion shaft. Some axles are equipped with a solid spacer. The size of the spacer determines the preload on the bearing. Solid spacers may be replaced with a collapsible spacer to avoid the extra work in determining the required size of the spacer. However, if the vehicle is subject to severe use, never replace a solid spacer with a collapsible one.

Pinion Seal Service

It is recommended that a new pinion seal be installed whenever the pinion shaft is removed from the differential. To install a new seal, thoroughly lubricate it and press it in place with an appropriate seal driver (Figure 7-34). Make sure the spring around the seal lip remains in place. After the seal is installed, push the companion flange carefully into the seal, making sure the seal isn't damaged. Start a new pinion flange nut onto the threads and set bearing preload.

A new pinion seal may cause a resistance to the rotation of the pinion shaft. Therefore, a new seal may affect pinion bearing preload readings. Normally, to achieve proper preload after installing a new pinion seal, add 5 or 6 inch-pounds to the specifications for preload. Always refer to the service manual before adding this amount of preload.

Backlash and Side Bearing Preload

Backlash between the pinion gear and ring gear is set whenever the differential is disassembled for bearing or gear replacement. It is adjusted by setting the side-to-side position of the differential case after the pinion depth is properly set. Backlash is normally checked with a dial indicator after the pinion preload and depth has been set and the assembled differential case with the ring gear is in the carrier or axle housing.

To check backlash, position a dial indicator against the face of a tooth on the ring gear (Figure 7-35). Measure the backlash by moving the ring gear back and forth against the teeth of the pinion gear. The total range of movement on the indicator is equal to the amount of backlash. It is advisable to check the backlash at several places around the ring gear.

The side bearing preload limits the amount the differential case is able to move laterally in the axle housing or carrier. Preload is checked with a dial indicator. The differential case is pried to one side of the carrier or housing and its movement recorded on the dial indicator or with a feeler gauge (Figure 7-36).

SPECIAL TOOLS

Seal remover
Seal installer
Fresh lubricant
Inch-pound
torque wrench

Classroom Manual
Chapter 7, page 171

Preload is often referred to as the differential case spread.

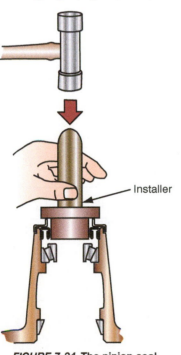

FIGURE 7-34 The pinion seal should be driven in with the correct driving tool.

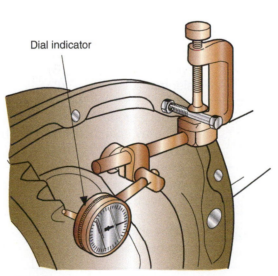

FIGURE 7-35 Position of dial indicator to measure backlash.

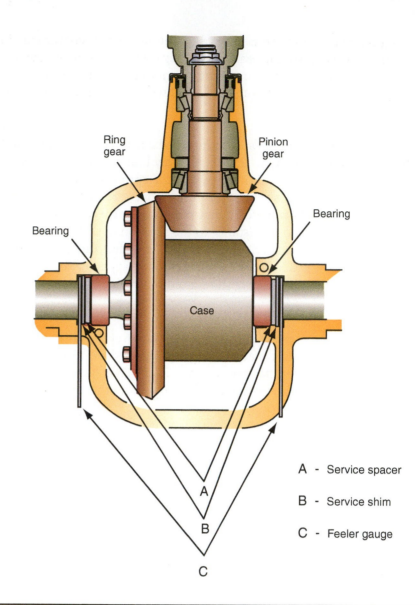

A - Service spacer

B - Service shim

C - Feeler gauge

EXAMPLE		
Ring gear side Combined total of:		**Opposite side** Combined total of:
.250" Service spacer (A) Service shim (B) Feeler gauge (C)	Service spacer (A) Service shim (B) .265" Feeler gauge (C)	
$\dfrac{-.010"}{.240"}$	To maintain proper backlash (.005"–.008") ring gear is moved away from the pinion by subtracting .010" shim from ring gear side and adding .010" to the other side	$\dfrac{+.010"}{.275"}$
+.004"	To obtain proper preload on side bearings, add .004" shim to each side	+.004"
.244"	Shim dimension required for ring gear side	Shim dimension required for opposite side .279"

FIGURE 7-36 **Side bearing preload procedure using a feeler gauge.**

Side bearing preload is adjusted at the same time as the backlash of the ring and pinion gears. Both of these are adjusted by changing shim thicknesses or by turning side bearing adjusting nuts. On differentials with threaded adjusting nuts at the outer ends of the side bearings, turn the adjusting nuts to obtain the specified backlash (Figure 7-37). Then set the preload by tightening both of the nuts a fraction of a turn beyond zero side play, as specified by the manufacturer (Figure 7-38). Some manufacturers specify a certain torque specification on the adjusting nuts to set preload. Others require that a dial indicator be mounted on the housing and measure the outward spread of the bearing caps as the adjusting nuts are tightened.

When shim packs are used to set backlash and side bearing preload, they are typically located behind the outer races of the side bearings. Add or subtract shims to the pack to obtain the desired backlash and zero side play (Figure 7-39). Then set the preload by adding the

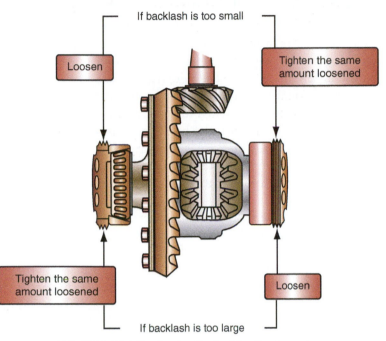

FIGURE 7-37 Adjusting backlash with adjusting nuts.

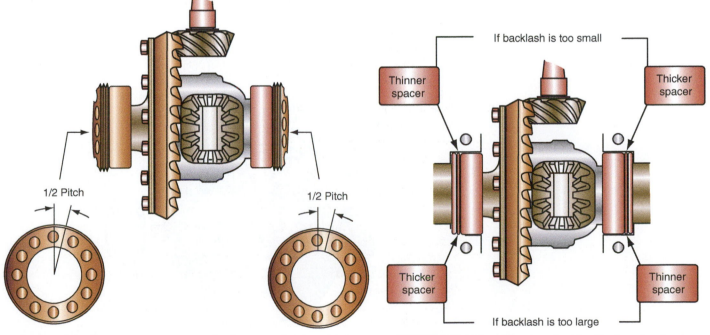

FIGURE 7-38 Setting preload by turning adjusting nuts a fraction of a turn.

FIGURE 7-39 Using shims to adjust backlash and preload.

specified additional amount to both shim packs. On integral carrier axle housings, a special case spreading tool is sometimes needed to add shims between the housing and the side bearings.

Photo Sequence 21 shows a typical procedure for measuring and adjusting backlash and side bearing preload on a differential assembly that uses a shim pack to position the differential case in the axle housing. Always refer to your shop manual before proceeding to adjust side bearing preload and backlash.

Photo Sequence 22 shows a typical procedure for measuring and adjusting backlash and side bearing preload on a differential assembly that uses side bearing adjusting nuts to position the differential case in the axle housing. Always refer to your shop manual before proceeding to adjust side bearing preload and backlash.

Guidelines for Setting Backlash and Side Bearing Preload

Service spacers are used to measure the preload and are not intended to be used in place of shims. Use the service spacers only when measuring preload. If service spacers are not available, the original shims can be used when measuring preload.

After prying the differential case to the side, the indicator reading shows the amount of side play present. The amount indicated is the shim thickness that should be added or the amount the adjusting nuts should be turned to arrive at no preload and no end play.

When installing the new shims into the housing, it may be necessary to gently tap the shims in place (Figure 7-40). Make sure the shims are fully seated and that the differential case turns freely. Be careful when trying to reinstall the original cast iron production shims because they may break when tapped into place. Most manufacturers do not recommend reusing cast spacers.

Classroom Manual
Chapter 7, page 173

Side bearing preload is set by first loosening the right adjusting nut and turning the left nut in until there is zero backlash. The right nut is then used to set the preload. The preload specification may be different for new and used bearings and for different axles. Be sure to set it as specified in the service manual.

The backlash of the gear set should only be checked after the side bearing preload has been set.

To increase backlash, install a thinner shim on the ring gear side and a thicker shim on the opposite side. To decrease backlash, do the opposite. Always recheck the backlash at four points equally spaced around the ring gear. If the variation between those points exceeds 0.002 inch, excessive ring gear runout is indicated.

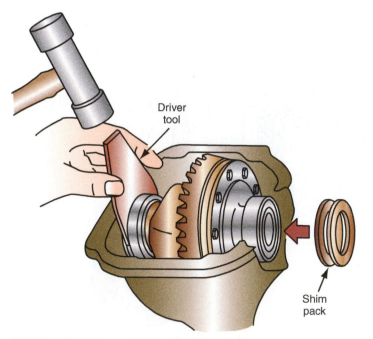

Driver tool

Shim pack

FIGURE 7-40 Driving adjusting shims into place.

TYPICAL PROCEDURE FOR MEASURING AND ADJUSTING BACKLASH AND SIDE BEARING PRELOAD ON A DIFFERENTIAL ASSEMBLY WITH A SHIM PACK

P21-1 Measure the thickness of the original side bearing preload shims.

P21-2 Install the differential case into the housing.

P21-3 Install service spacers the same thickness as the original preload shims between each bearing cup and the housing.

P21-4 Install the bearing caps and finger-tighten the bolts.

P21-5 Mount a dial indicator to the housing so that the button of the indicator touches the face of the ring gear. Using two screwdrivers, pry between the shims and housing. Pry to one side and set the dial indicator to zero, then pry to the opposite side and record the reading.

P21-6 Select two shims with a combined thickness equal to that of the original shims and the indicator reading, then install them.

P21-7 Using the proper tool, drive the shims into position until they are fully seated.

P21-8 Install and tighten the bearing caps to specifications.

P21-9 Check the backlash and preload of the gear set. Check the backlash by rocking the ring gear and noting the movement on the dial indicator. Adjust the shim pack to allow for the specified backlash. Recheck the backlash at four points equally spaced around the ring gear.

TYPICAL PROCEDURE FOR MEASURING AND ADJUSTING BACKLASH AND SIDE BEARING PRELOAD ON A DIFFERENTIAL ASSEMBLY WITH ADJUSTING NUTS

P22-1 Lubricate the differential bearings, cups, and adjusters.

P22-2 Install the differential into the housing.

P22-3 Install the bearing cups and adjusting nuts onto the differential case.

P22-4 Snugly tighten the top bearing cap bolts and finger-tighten the lower bolts.

P22-5 Turn each adjuster until bearing free play is eliminated with little or no backlash present between the ring and pinion gears.

P22-6 Seat the bearings by rotating the pinion several times each time the adjusters are moved.

P22-7 Install a dial indicator and position the plunger against the drive side of the ring gear. Set the dial to zero. Using two screwdrivers, pry between the differential case and the housing. Observe the dial indicator.

P22-8 Set the preload by turning the right adjusting nut.

P22-9 Check the backlash by rocking the ring gear.

P22-10 Adjust the backlash by turning both adjusting nuts the same amount so that the preload adjustment remains unchanged.

P22-11 Install the locks on the adjusting nuts.

P22-12 Tighten the bearing cap bolts to the full specified torque.

Backlash and side bearing preload should be rechecked after the shims have been installed or after the adjusting nuts are secured by using a gear marking compound to check the gear tooth contact pattern.

Gear Tooth Patterns

The pattern of gear teeth determines how quietly two meshed gears will run. The pattern also describes where on the faces of the teeth the two gears mesh. The pattern should be checked during teardown for gear noise diagnosis, after adjusting backlash or side bearing preload, or after replacing the ring and drive pinion or setting up the pinion bearing preload. The following terms are commonly used to describe the possible patterns on a ring gear.

1. **Drive** refers to the convex side of the gear tooth.
2. **Coast** refers to the concave side of the gear tooth.
3. **Heel** refers to the outside diameter of the ring gear.
4. **Toe** refers to the inside diameter of the ring gear.
5. **Face** refers to the area near the top of the tooth.
6. **Flank** refers to the area near the bottom of the tooth.

Gear patterns will appear as markings on both the drive and coast sides of the tooth. Using the terms mentioned previously, patterns are described as being high or low and toward the heel or toe of the gear.

In order to display an accurate tooth pattern, the pinion bearing preload, side bearing preload, and pinion and ring gear backlash must be set to specifications.

Paint several ring gear teeth with nondrying Prussian blue, yellow or white marking compound (Figure 7-41). Use a box-end wrench on the ring gear bolts to rotate the ring gear to allow the painted teeth to contact the pinion gear. Rotate it in both directions until a clearly defined pattern is evident. Maintain a load on the pinion by providing turning resistance, often by wrapping a cloth around the yoke. Then examine the pattern on the ring gear.

Many new gear sets come with a pattern prerolled on the teeth. This pattern will provide the quietest operation for that gear set. Never wipe this pattern off or cover it up during assembly. When checking the pattern on a new gear set, apply the marking compound to half of each tooth that has the prerolled pattern. Rotate the ring gear and compare the new pattern with the prerolled pattern. Examine the drive pattern first to determine if adjustments are necessary. Adjust the preload and backlash to match the prerolled pattern. Normally, the desired pattern is centered on the drive side of the teeth (Figure 7-42).

By examining the displayed pattern, you can determine where adjustments should be made (Figure 7-43). By moving the pinion gear toward the ring gear, the pinion moves deeper

The drive side is designed to handle extreme amounts of force as it is the side in contact with the pinion gear during driving and accelerating conditions.

The coast side contacts the pinion gear during coasting. A tooth pattern on the coast side is only used as a reference to determine if additional adjustment should be made on the drive side.

White or yellow marking compounds are preferred by many technicians because they tend to be more visible than the others.

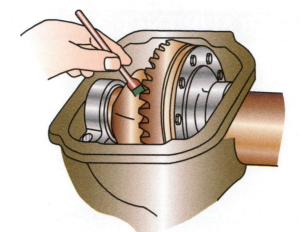

FIGURE 7-41 To check gear tooth patterns, several teeth of the ring gear are coated with a marking compound and the pinion gear is rotated with the ring gear. The resultant pattern shown on the teeth determines how the gear set ought to be adjusted.

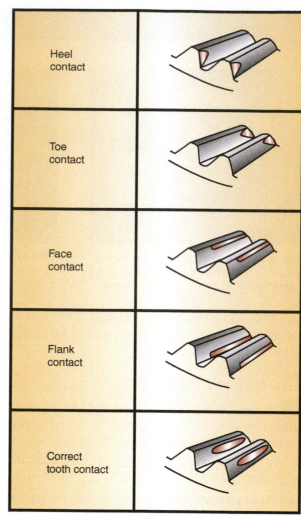

FIGURE 7-42 The desired gear tooth pattern is centered on the drive side of the gear's teeth.

into the ring gear. It moves from a high-heel position to a lower-toe position. By moving the pinion away from the ring gear, the pinion moves from a low-toe position to a higher-heel position. These movements are normally accomplished by changing the pinion shim. On removable carrier axles with a separate pinion housing, thicker shims will pull the pinion away from the ring gear, whereas on integral carrier axles, increasing the shim size will move the pinion closer in. Therefore, if the drive pattern is low, increase the shim thickness on removable carrier axles and decrease it on integral carrier axles. If the pattern is high, reduce shim thickness on removable carrier axles and increase it on integral carrier axles.

Heel characteristics are treated the same as high patterns. Likewise, *toe* characteristics are treated as low patterns. As the drive pattern moves higher or lower, the coast pattern usually will move the same way. However, it doesn't move quite as far, due to the shape of the teeth and their contact area. Therefore, if the drive and coast patterns are high, or both are low, the pattern is correctable. However, if one pattern is high and the other low, the gear set should be replaced. Centering one of the patterns would move the other farther from center, possibly off the tooth.

> **AUTHOR'S NOTE:** A centered drive side pattern usually provides the best contact area for both gear longevity and quiet operation. Drag racing setups are not so concerned about noise, and are usually set with strong toe contact patterns that put load on the strongest part of the gear teeth.

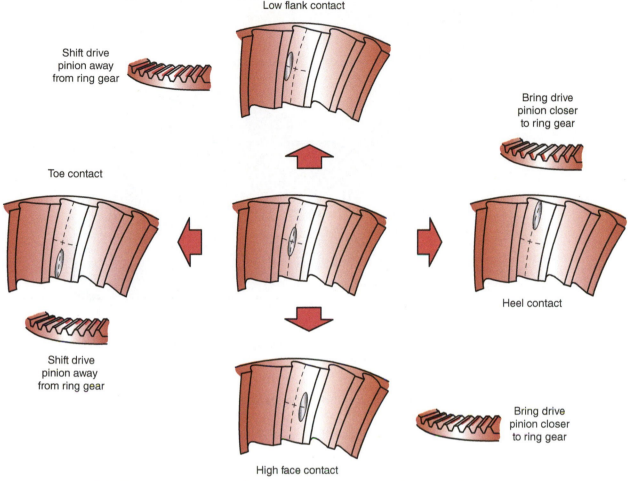

Low flank contact

Shift drive
pinion away
from ring gear

Bring drive
pinion closer
to ring gear

Toe contact

Heel contact

Shift drive
pinion away
from ring gear

Bring drive
pinion closer
to ring gear

High face contact

FIGURE 7-43 The placement of the pattern indicates where adjustments should be made.

INSPECTING FWD FINAL DRIVE UNITS

There are four common configurations used as the final drives on FWD vehicles. Helical, planetary, and chain final drive arrangements are found with transversely mounted engines. Hypoid final drive gear assemblies are normally found in vehicles with a longitudinally placed engine. The hypoid assembly is basically the same unit as would be used on RWD vehicles and are mounted directly to the transmission.

Helical Gear Drives

Helical final drive units should be checked for worn or chipped teeth, overloaded tapered-roller bearings, and excessive differential side gear wear. Excessive play in the differential is a cause of engagement clunk. Be sure to measure the clearance between the side gears and the differential case and check the fit of the gears on the gear shaft. It is possible that the side bearings of some final drive units are preloaded with shims. Select the correct size shim to bring the unit into specifications. With a torque wrench, measure the amount of rotating torque. Compare your readings to specifications.

If bearing preload and end play and the bearings are good, the bearings can be reused. However, new seals should always be installed. The preload on new bearings should be set to the specifications for a new bearing. Used bearings should be set to the amount found during teardown or about one-half of the specification for a new bearing.

To remove and install the transfer shaft gear, a holding tool is used to stop the transfer gear from turning while loosening or tightening the retaining nut (Figure 7-44). The nut is typically

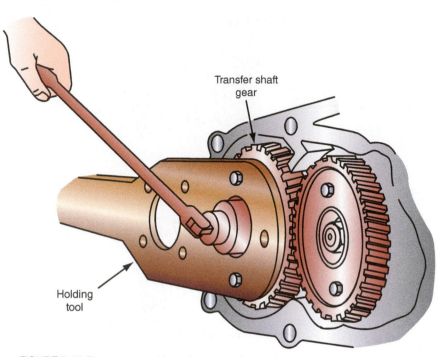

FIGURE 7-44 To remove and install the transfer shaft gear, a holding tool is used to prevent the gear from turning.

tightened to 200 foot-pounds. To remove the transfer shaft from the transaxle case, remove the retaining snapring (Figure 7-45). A puller is used to remove the transfer gear from the output shaft. Then the shaft can be pulled out with its bearing. The shaft's bearing is pressed on and off the shaft.

Behind the transfer shaft gear is a selective shim used to provide for correct meshing of the teeth on the transfer shaft gear and output shaft transfer gear and to control transfer shaft end play. End play is measured with a dial indicator (Figure 7-46). If the end play is not within specifications, install a thrust washer with the correct thickness behind the transfer gear.

The output shaft should also be checked for end play. If the end play is incorrect, a different-size shim should be installed behind the output shaft transfer gear. Since the gear is

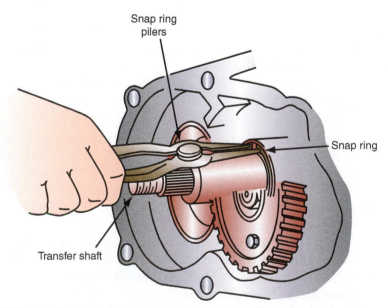

FIGURE 7-45 To remove the transfer gear, the retaining clip must be removed first.

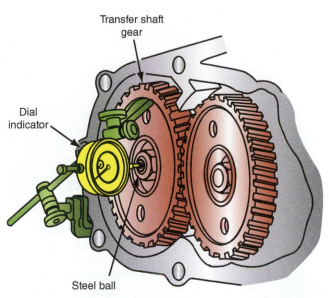

FIGURE 7-46 Use a dial indicator to measure the end play of the transfer shaft. The end play is controlled by a selective shim behind the gear.

splined to the shaft, a thrust bearing is not needed here. After end play has been corrected, the turning torque of the shaft should be measured. If the turning torque is too high, a slightly thicker shim should be installed. If the turning torque is too low, install a slightly smaller shim. The bearings for the transfer and output shafts are pressed on and off the shafts.

It is very important that the transfer gears be tightened to specifications. The torque setting maintains the correct bearing adjustments.

Planetary Gear Assemblies

Planetary final drives (Figure 7-47) are checked for the same differential case problems as the helical type. All planetary gear teeth should be inspected for chips or stripped teeth. Pay particular attention to the planetary carriers. Look for obvious problems like blackened gears or pinion shafts. These conditions indicate severe overloading and require that the carrier be replaced.

Inspect the gears; a bluish tint can be a normal condition as this is part of a heat-treating process used during manufacture. Check the planetary pinion gears for loose bearings. Check each gear individually by rolling it on its shaft to feel for roughness or binding of the bearings. Wiggle the gear to check for looseness on the shaft. Looseness will cause the gear to whine when it is loaded. Inspect the gears' teeth for chips or damage; these also cause whine.

Check the gear teeth around the inside of the planetary ring gear. Look for cracks or other damage to the carrier. Check the carrier gears for end play by placing a feeler gauge between the carrier and the pinion gear (Figure 7-48). Compare the end play to specifications. Replace any abnormal or worn parts.

Drive Chains

The drive chains used in some transaxles should be inspected for side play and stretch. These checks are made during disassembly and should be repeated during reassembly.

Chain deflection is measured between the centers of the two sprockets. Deflect the chain inward on one side until it is tight (Figure 7-49). Mark the housing at the point of maximum deflection. Then, on the same side, deflect the chain outward until it is tight (Figure 7-50). Mark the housing at the point of maximum deflection. Measure the distance between the two marks. If this distance exceeds specifications, replace the drive chain.

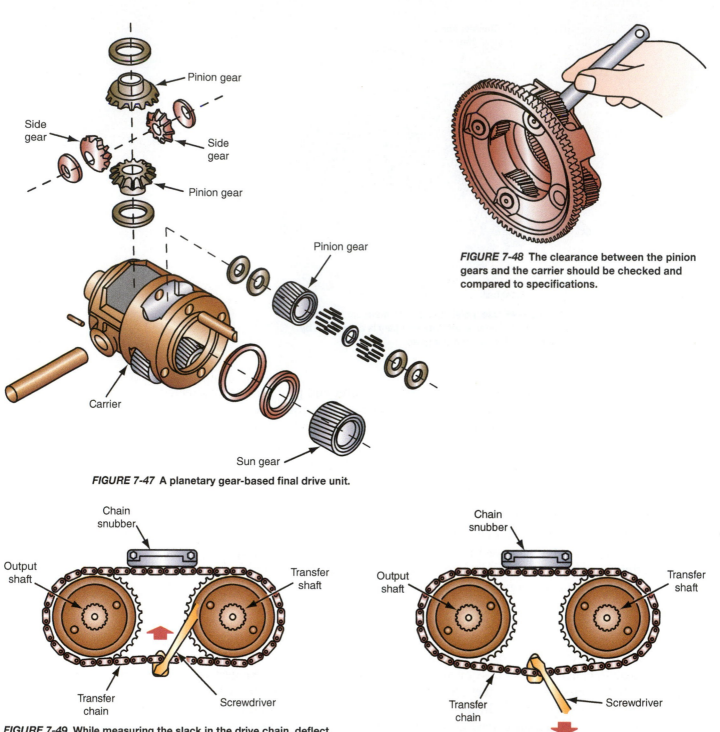

FIGURE 7-47 A planetary gear-based final drive unit.

FIGURE 7-48 The clearance between the pinion gears and the carrier should be checked and compared to specifications.

FIGURE 7-49 While measuring the slack in the drive chain, deflect the chain in one direction and mark the point of deflection.

FIGURE 7-50 Continue measuring the slack in the chain by deflecting it in the opposite direction. Mark the point of deflection. The distance between the two marks is the amount of chain slack.

AUTHOR'S NOTE: According to the major supplier of drive chains, it is impossible to accurately measure chain stretch without expensive gauges. Therefore they suggest that drive chains be replaced every 50,000 miles or during an overhaul of a transaxle or transfer case. Doing this will prevent the damage that can occur when a chain breaks or stretches too far.

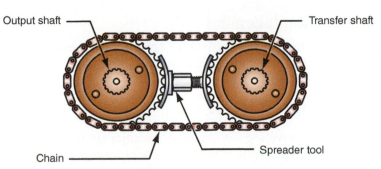

FIGURE 7-51 To remove and install some drive chains and gears, the gears must be spread slightly apart with a special tool.

Check each link of the chain by pushing and pulling the links away from their retaining pins. All of the links should move very little and each move the same amount. It is important to realize that a chain is only as strong as its weakest link. Check each link carefully.

Be sure to check for an identification mark on the chain during disassembly. These can be painted or dark-colored links, and indicate either the top or the bottom of the chain.

The sprockets should be inspected for tooth wear and wear at the points where they ride. If the chain has too much slack, it may have worn the sprockets. A slightly polished appearance on the face of the gears is normal.

Bearings and Bushings. The bearings and bushings used on the sprockets need to be checked. The radial needle thrust bearings are checked for deterioration of the needles and cage. The running surface in the sprocket must also be checked, as the needles may pound into the gear's surface during abusive operation. The bushings should be checked for signs of scoring, flaking, or wear. Replace any defective parts. Typically, the bearings and bushings are removed with a puller and installed with a driver and a press.

The removal or installation of the chain drive assembly of some transaxles requires that the sprockets be spread slightly apart (Figure 7-51). It is important to spread the sprockets just the right amount. If they are spread too far, they will not be easy to install or remove.

The shafts and gears of the assembly have numerous seals and thrust washers. The seals must be replaced whenever the unit is disassembled.

PART REPLACEMENT

Removing and installing parts of a final drive unit require special procedures. The procedures and tools given in the service information should be used.

Side Bearings

The side bearings of most final drive units must be pulled off and pressed onto the differential case.

Final Drive Bearing Replacement

The final drive unit is positioned in the transaxle case or in a separate housing mounted to the transaxle. Regardless of its location, it is supported by bearings. Normally tapered roller bearings are used. These bearings are removed with pullers and installed with a press and driver. Make sure the bearing seats are free from nicks and burrs. These defects will not allow the bearing to seat properly and will give false end play measurements. When new bearings are installed, end play and gear clearances must be adjusted before installing the final drive into the housing.

The ring gear of some transaxles is riveted to the differential case. The rivets must be drilled then driven out to separate the ring gear from the case. When installing the ring gear, nuts and bolts are used in place of the rivets. These nuts and bolts must be of the specified hardness and should be tightened in steps and to the specified torque.

Speed Sensor Gear Replacement

The final drives in a transaxle may be fitted with a speedometer gear or speed sensor trigger wheel pressed onto the differential case under one side bearing. These gears are pulled off and pressed onto the case.

Adjusting Final Drive Units

Some of the procedures for servicing FWD final drive units are the same as for RWD differentials. Because these units typically use the end of the output shaft as the drive pinion gear, all normal pinion shaft adjustments are not needed. Ring gear and side bearing adjustments are still necessary. These adjustments are typically made with the differential case assembled and out of the transaxle case.

The procedure that follows is typical for measuring and adjusting the side gear end play in a helical gear final drive. Always refer to your service manual before proceeding to make these adjustments on a transaxle.

1. Install the correct adapter into the differential bearings (Figure 7-52).
2. Mount the dial indicator to the ring gear with the plunger resting against the adapter.
3. With your fingers or a screwdriver, move the ring gear up and down.
4. Record the measured end play.
5. Measure the old thrust washer with the micrometer.
6. Install the correct-size thrust washer (Figure 7-53).
7. Repeat the procedure for the other side.

The following procedure is typical for the measurement and adjustment of the differential bearing preload in a transaxle. Special tools are required for most transaxles. Don't attempt to this without the proper tools and always refer to the service information before proceeding to make these adjustments on a transaxle.

1. Remove the bearing cup and existing shim from the differential bearing retainer.
2. Select a gauging shim that will allow for 0.001- to 0.010-inch end play.
3. Install the adjusting shim into the differential bearing retainer (Figure 7-54).
4. Press in the bearing cup.
5. Lubricate the bearings and install them into the case.
6. Install the bearing retainer.
7. Tighten the retaining bolts.
8. Mount the dial indicator with its plunger touching the differential case (Figure 7-55).
9. Apply medium pressure in a downward direction while rolling the differential assembly back and forth several times.

SPECIAL TOOLS

Adapter kit for transaxle
Dial indicator
0- to 1-inch micrometer

SPECIAL TOOLS

Set of gauging shims
Dial indicator
Fresh lubricant
Inch-pound
torque wrench

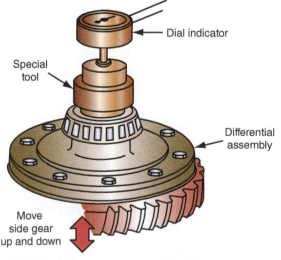

FIGURE 7-52 Tool setup for measuring side gear end play.

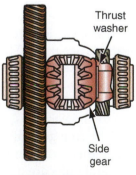

FIGURE 7-53 Location of thrust washer.

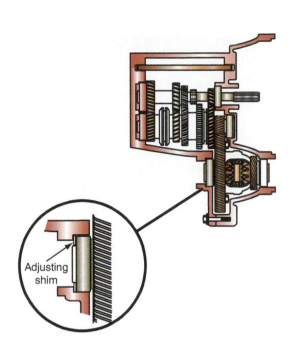

FIGURE 7-54 Typical location of preload shim.

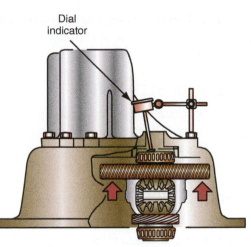

FIGURE 7-55 Setup for determining proper differential bearing preload.

10. Zero the dial indicator.
11. Apply medium pressure in an upward direction while rotating the differential assembly back and forth several times.
12. The required shim to set preload is the thickness of the adjusting shim plus the recorded end play.
13. Remove the bearing retainer, cup, and adjusting shim.
14. Install the required shim.
15. Press the bearing cup into the bearing retainer.
16. Install the bearing retainer and tighten the bolts.
17. Check the rotating torque of the transaxle (Figure 7-56). If this is less than specifications, install a thicker shim. If the torque is too great, install a slightly thinner shim.
18. Repeat the procedure until desired torque is reached.

The side bearings of most final drive units must be pulled off and pressed onto the differential case (Figure 7-57). Always be sure to use the correct tools for removing and installing the bearings.

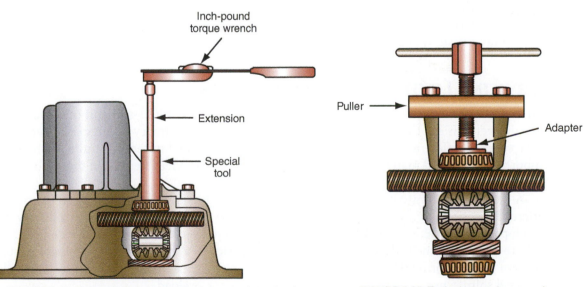

FIGURE 7-56 Using an inch-pound torque wrench to check the turning torque of a transaxle.

FIGURE 7-57 Typical setup for removing side bearings from a differential case.

The final drives of most transaxle differential cases are fitted with a speedometer gear pressed onto the case and under one side bearing. These gears are pulled off and pressed onto the case.

LIMITED-SLIP UNITS

The oil in clutch-based limited-slip units should be changed on a regular basis. This not only provides fresh lubrication but also rids the unit of particles from the friction materials and gears. Normally a special additive is added to the oil. Gear-based units do not require this additive. These units require fluid changes, but not as frequently as clutch-based units.

Limited-Slip Differential Diagnostics

A limited-slip differential is best checked by how it reacts when the vehicle has good traction at both drive wheels and when it has little or no traction at one of the drive wheels. If it works as it should and without noise or chatter, the unit is undoubtedly fine. Of course, the best way to check the unit is to inspect its components.

Clean and inspect all gears, wear plates, clutches, and bearings. Look for signs of galling on the face of the axle gears. If there are slight imperfections, these can be removed by moving the gear over a flat surface covered with 320-grit sandpaper. If this does not remove the imperfections, the gear must be replaced.

In a gear-based unit, the gears and wear plates should be inspected. If the pinion gears or wear plates are worn, it may be reinstalled upside down, as long as they remain on the same side as they originally were. Small chips on the gears can be ground out or the gears replaced.

Some check a limited-slip differential by conducting a "breakaway" check. Although it is widely used, many experts say the results are not always dependable. However, many manufacturers of gear-based LSDs do recommend this check.

To check breakaway, put the transmission in neutral. Place one drive wheel on the floor with the other off the floor. The specifications in the service information will list a breakaway torque reading. This is what should be required to start the rotation of the wheel that is raised off the floor. Using a torque wrench, measure the torque required to turn the wheel.

Normally, the initial breakaway torque is higher than the torque required for continuous turning. However, the axle shaft should turn with even pressure, without slipping or binding. If the torque reading is less than the specified amount, the differential needs to be checked.

> ⚠️ **WARNING:** Never operate a vehicle with a limited-slip differential when only one drive wheel is raised. Power applied to the wheel remaining on the ground may cause the car to move off the jack or jack stand.

SERVICING LIMITED-SLIP DIFFERENTIALS

Most of the procedures for the inspection and servicing of limited-slip and other locking-type differentials are the same as for open differentials. The primary difference in servicing these types of differentials is with the clutch assemblies. Some limited-slip units, such as the Auburn, Torsen, and many with viscous clutches, are not serviceable. When these units fail, they are replaced as a unit.

A clutch-style limited-slip differential uses two sets of multiple disc clutches to control differential action (Figure 7-58). The mounting distance of each side gear is controlled by the clutch pack. The friction and steel plates of the clutch pack are stacked on the side gear hub and are housed in the differential case between the side gears. Positioned with each clutch pack is a preload spring (Figure 7-59) that applies an initial force to the clutch packs. Normally, one steel shim is used on each side to control clutch pack preload.

Classroom Manual
Chapter 7, page 176

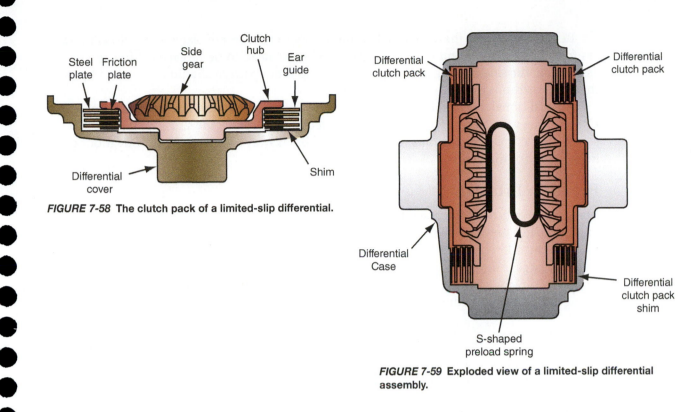

FIGURE 7-58 The clutch pack of a limited-slip differential.

FIGURE 7-59 Exploded view of a limited-slip differential assembly.

The friction plates are splined to the side gear hub, which is splined to the left and right axle shafts (Figure 7-60). The ears of the steel plates fit into the case, causing the clutch packs to always be engaged. Within the clutch pack, there is one composite clutch plate, comprised of steel on one side and friction material on the other. During assembly, the composite plate is positioned with its fiber side against the hub. The other plates are placed in alternating order with a friction disc followed by a steel one.

After the clutch pack is assembled, check its total width with a template or other suitable gauge (Figure 7-61) and use a feeler gauge to determine the correct thickness of the shim needed to maintain preload. Always refer to your shop manual to determine the proper way to measure the clutch pack preload. Clutch packs should be allowed to soak in gear lube for at least 30 minutes prior to installation and before making this check.

SPECIAL TOOLS

Clutch gauge kit
Feeler gauge
Torque wrench

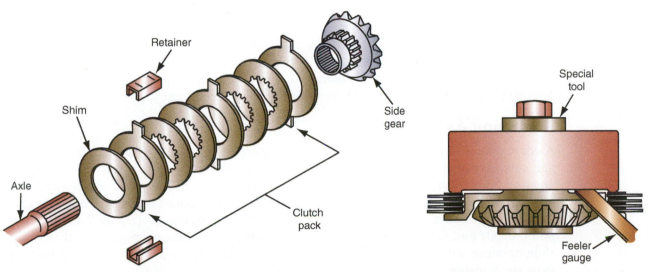

FIGURE 7-60 Friction plates are splined to the side gears that are splined to the drive axles.

FIGURE 7-61 Measuring a clutch pack to determine proper preload shim size.

Before assembling the clutch pack, carefully inspect the clutch plates and side gear retainers for excessive wear and cracks. All damaged parts need to be replaced. After the unit is assembled, the side bearing preload, backlash, and tooth pattern should be checked by using the same methods as for an open differential.

An additional check should be made on the unit to determine if it is working properly. Use a torque wrench and measure the torque required to rotate one side gear while the other is held stationary. Compare this reading of initial breakaway torque to the specifications given in your service manual.

Break-In Procedure

Many manufacturers recommend that a break-in procedure be followed after rebuilding a LSD unit or when a new unit has been installed. Manufacturers give detailed instructions on how to break the differential in. If these are not followed, the LSD may be permanently damaged and may engage and disengage erratically due to irregularities on and damage to the clutch surfaces. Typically, if gear-based units are properly set up, no break-in period is required. Along with these procedures, manufacturers will give a timetable for the breaking-in process.

AXLE SHAFTS AND BEARINGS

Classroom Manual
Chapter 7, page 183

SPECIAL TOOLS
Vee-blocks
Dial indicator
Axle puller tool

Rear drive axles involves unique procedures when compared with servicing FWD axles. RWD axles are not designed to allow the wheels to move independently in response to changes in road surfaces, unless they are part of an IRS system. Most RWD axles are enclosed in a single housing and rotate on bearings. The axles seldom require repair; however, axle bearings and seals are commonly replaced.

RWD Axle Inspection

Axle replacement may be necessary if the shaft is damaged or bent. Placing it in a pair of Vee-blocks and positioning a dial indicator at the center of the shaft can check the trueness of a shaft. Rotating the shaft and observing the indicator will identify any warping or bends. Slight indications on the dial indicator may be caused by casting imperfections and do not indicate the need to replace the shaft.

An axle should be replaced if there is excessive runout of the axle's **flange**. Position a dial indicator against the outer surface of the flange. Apply slight pressure to the center of the axle to remove the end play and zero the indicator. Slowly rotate the axle one complete revolution and observe the readings on the indicator. The total amount of indicator movement is the total amount of axle flange lateral runout. Compare the measured runout with specifications.

Inspect the wheel studs in the axle flanges. If they are broken or bent, they should be replaced. Also check the condition of the threads. If they have minor distortions, run a die over the stud. If the threads are severely damaged, the stud should be replaced. Studs are normally pressed in and out of the flange. Make sure you install the correct size stud.

Removing Axles

Axles in a removable carrier are commonly retained in the housing by a retaining plate. Integral carriers normally have C-locks placed on the inner ends of the axle shafts. If a retainer and four bolts retain the axle (Figure 7-62), it is not necessary to remove the differential cover to remove the axle. Most ball and tapered roller bearing-supported axle shafts are retained this way. To remove the axle, remove the four bolts at the retainer (Figure 7-63) and then pull the axle out. Sometimes the shaft will slide out without using a puller. In other cases a slide hammer with a special attachment is used to remove the axle and bearing assembly from the housing.

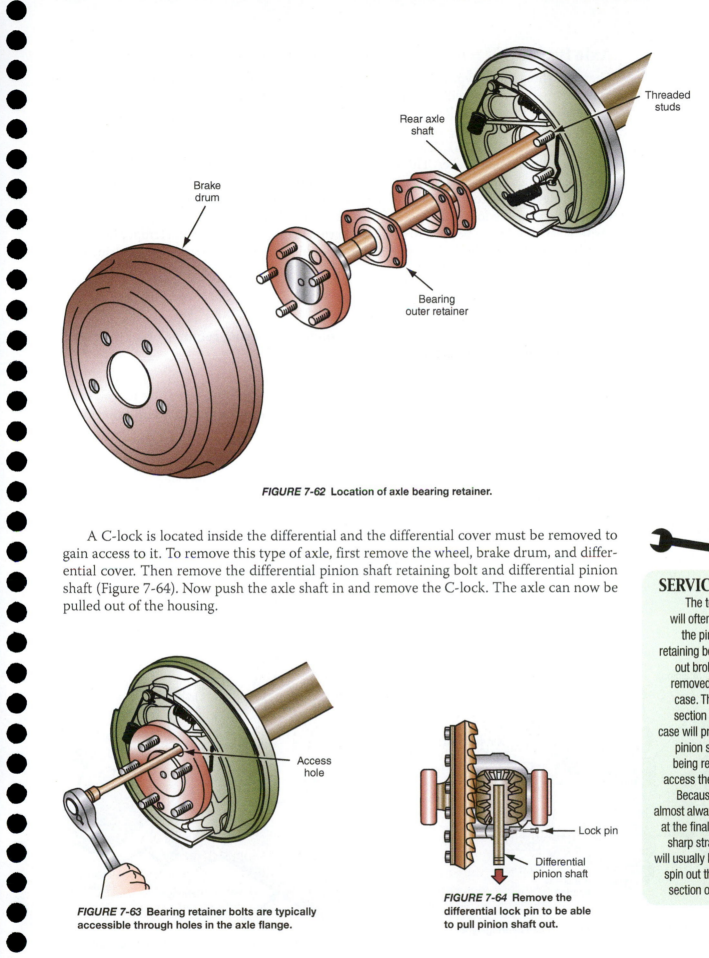

FIGURE 7-62 Location of axle bearing retainer.

A C-lock is located inside the differential and the differential cover must be removed to gain access to it. To remove this type of axle, first remove the wheel, brake drum, and differential cover. Then remove the differential pinion shaft retaining bolt and differential pinion shaft (Figure 7-64). Now push the axle shaft in and remove the C-lock. The axle can now be pulled out of the housing.

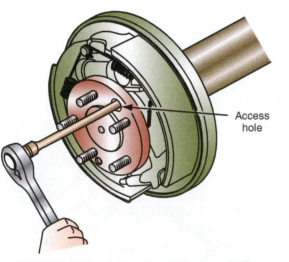

FIGURE 7-63 Bearing retainer bolts are typically accessible through holes in the axle flange.

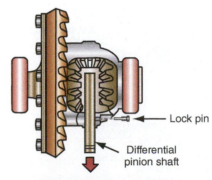

FIGURE 7-64 Remove the differential lock pin to be able to pull pinion shaft out.

SERVICE TIP:
The technician will often find that the pinion shaft retaining bolt comes out broken when removed from the case. The broken section left in the case will prevent the pinion shaft from being removed to access the C-locks. Because the bolt almost always breaks at the final thread, a sharp straight pick will usually be able to spin out the broken section of the bolt.

Axle Bearings

There are three major types of rear axle bearings used in passenger cars today: ball bearings, roller bearings, and tapered roller bearings. Ball bearings are lubricated with grease packed in the bearing at the factory. Rear axle gear oil does not lubricate this type of bearing. An inner seal, designed to keep the gear oil from the bearing, rides on the axle shaft just in front of the retaining ring (Figure 7-65). This type of bearing also has an outer seal to prevent grease from spraying onto the rear brakes. This type of axle bearing is pressed on and off the axle shaft.

The retainer ring is made of soft metal and is pressed onto the shaft against the wheel bearing. Never use a torch to remove the ring: rather, drill into it or notch it in several places with a cold chisel to loosen the press fit (Figure 7-66). The ring can then be slid off the shaft easily. Heat should not be used to remove the ring because it can take the temper out of the shaft and thereby weaken it. Likewise, a torch should never be used to remove a bearing from an axle shaft.

 WARNING: **Never use a chisel to notch the bearing retainer of an axle shaft without wearing safety glasses.**

 WARNING: **Never hammer on a bearing. The races are very hard and therefore, very brittle. They can shatter easily.**

Roller axle bearings are lubricated by the gear oil in the axle housing. Therefore, only a seal to protect the brakes is necessary with these bearings. These bearings are typically pressed into the axle housing and not on the axle. To remove them, the axle must first be removed and then the seal and bearing are pulled out of the housing (Figure 7-67). With the axle out, inspect the area where it rides on the bearing for pits or scores. If pits or score marks are present, replace the axle. Sometimes an "axle saver" bearing is available. This rides on a different spot on the axle and eliminates the need to replace the axle.

Classroom Manual

Chapter 7, page 184

SPECIAL TOOLS

Cold chisel
Axle puller
Slide hammer
Bearing adapters
Pilot tools

⚠ **CAUTION:**

Only apply puller force to the inner race of the axle housing.

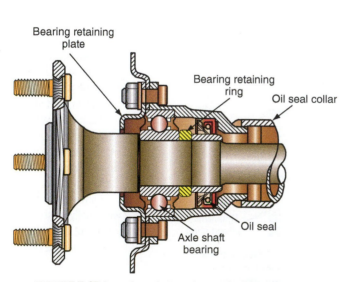

FIGURE 7-65 Location of oil seals on a ball bearing–type axle shaft.

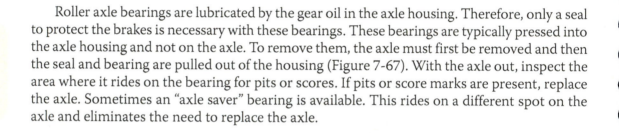

FIGURE 7-66 Freeing the retainer ring from the axle shaft.

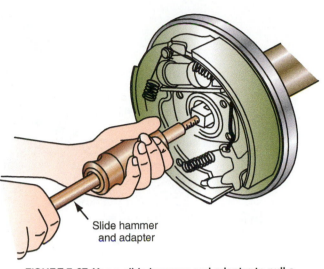

FIGURE 7-67 Use a slide hammer and adapter to pull a bearing out of the housing.

Tapered roller axle bearings (Figure 7-68) are lubricated by gear oil. This type of bearing uses two seals, like the ball bearing type and normally must be pressed on and off the axle shaft using a press.

After packing the bearing, install the axle in the housing and check the shaft's end play (Figure 7-69). Use a dial indicator and adjust the end play to the specifications given in the service manual. If the end play is not within specifications, change the size of the bearing shim (Figure 7-70). On many vehicles, there is an adjustment nut, rather than shims, to adjust preload.

Classroom Manual
Chapter 7, page 185

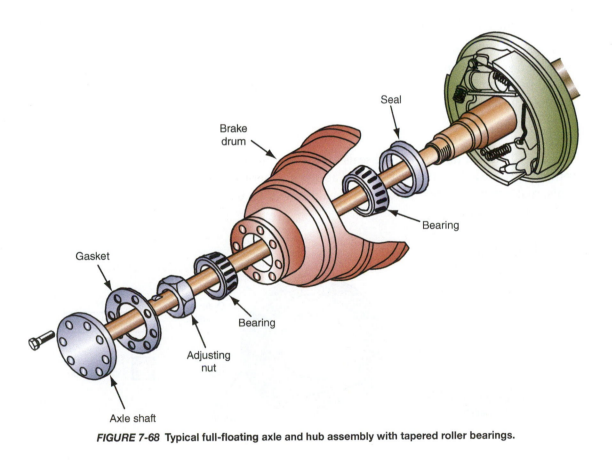

FIGURE 7-68 Typical full-floating axle and hub assembly with tapered roller bearings.

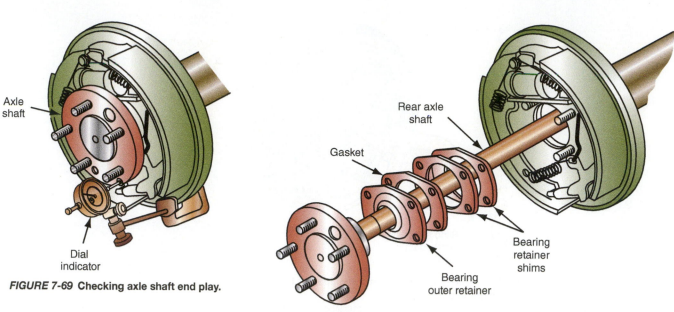

FIGURE 7-69 Checking axle shaft end play.

FIGURE 7-70 Installing gasket and axle bearing retainer shims.

BEARING INSPECTION

Axle bearings should be inspected for wear or other damage and if they show any evidence of damage, they should be replaced. There is no other service or repair recommended for them. Normally the condition of the bearings will indicate the need for servicing other components of the axle assembly, in addition to replacing the bearings (Figure 7-71). Table 7-1 lists the terms used to define the possible conditions of a bearing, a description of each term, and the recommended repair service.

Servicing the Bearing Bore

When an axle is removed from a housing, the bearing's bore in the axle housing should be inspected and cleaned. If the bore is corroded, sand it lightly with fine emery cloth. If the bearing is pressed into the housing, it should not be able to move in its bore. If the bore is worn, the housing should be replaced.

Axle Shaft Seals

Classroom Manual
Chapter 7, page 184

The installation of new axle shaft seals is recommended whenever the axle shafts have been removed. Some axle seals are identified as being either right or left side. When installing new seals, make sure to install the correct seal in each side. Check the seals for markings of *right* or *left* or for color coding.

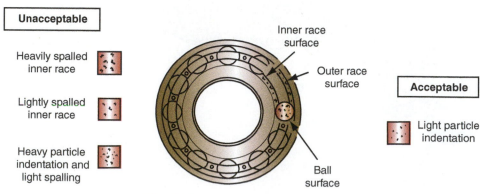

FIGURE 7-71 Rear axle bearing diagnostic chart.

TABLE 7-1   BEARING DIAGNOSTIC CHART

Condition	Description	Recommended Service
FATIGUE SPALLING	Flaking of surface metal resulting from fatigue.	Replace the bearings and clean the related parts.
ETCHING	Bearing surfaces are gray or grayish black in color with etching-away of material, usually at roller spacing.	Replace the bearings and check the seals and check for proper lubrication.
INDENTATIONS	Surface depressions on the race and rollers caused by particles of hard foreign material.	Clean all parts and the housing. Check the seals and replace the bearings if they are rough or noisy.
ABRASIVE STEP WEAR	Pattern on the ends of the rollers caused by fine abrasives.	Clean all parts and the housing. Check the seals and bearings and replace them if they were leaking, rough, or noisy.
MISALIGNMENT	Outer race misalignment due to the presence of something in the bore.	Clean all related parts and replace the bearing. Make certain the races are properly seated and the bore is clean.
HEAT AND DISCOLORATION	Heat discoloration can range from a faint yellow to dark blue color, resulting from overload or the use of an incorrect lubricant.	Replace the bearings if overheating and loss of temper was found. Check the seals and other parts for damage.
STAIN DISCOLORATION	Discoloration can range from light brown to black and is caused by moisture or incorrect lubricant.	Reuse the bearings if the stains can be removed by light polishing or if no evidence of overheating is observed. Check the seals and related parts for damage.
BRINELLING	Surface indentations in the raceway that were caused by the rollers either under impact loading or vibration while the bearing is not rotating.	Replace the bearing and inspect the axle and other parts for damage.
GALLING	Metal smears on the roller ends due to overheating, lubricant failure, or overload.	Replace the bearing and check the seals and check for proper lubrication.
ABRASIVE ROLLER WEAR	Pattern on races and rollers caused by fine abrasives.	Clean all parts and the housing. Check the seals and bearings and replace them if they were leaking, rough, or noisy.
SMEARS	Smearing of metal due to slippage, which was caused by poor fit, lubrication, overheating, or overloads.	Replace the bearings and clean all related parts. Check for proper fit and lubrication. Replace the axle shaft if it was damaged.
CAGE WEAR	Wear around the outside diameter of the cage and roller pockets caused by abrasive material and inefficient lubrication.	Check the seals and replace the bearings.
BENT CAGE	Cage damage due to improper handling or tool usage.	Replace bearing.
CRACKED INNER RACE	Race cracked due to improper fit, cocking, or poor seating of the bearing.	Replace bearing and inspect the housing and axle for damage.
FRETTAGE	Corrosion set up by the axle and bearing experiencing small amounts of movement without proper lubrication.	Replace the bearings and clean the related parts. Check the seals and for proper lubrication.

Coat the outer edge of the new seal with gasket sealer and apply a coat of lubricant to the seal's inner lip (Figure 7-72). Make sure the lip seal is facing the inside of the axle housing, then use the correct-size seal driver to install the seal squarely in the axle tube. Some General Motors products have axle seals on the inside of the bearing and an O-ring on the outside of the bearing. Both of these seals should be lubricated prior to installation. Newer seals may have a red or green sealant bead on the casing. These seals do not require additional sealant.

SERVICE TIP:
Take extra care not to damage the new seal when reinstalling the axle shaft. It is helpful to support the entire length of the axle and keep it level while inserting it into the axle tube.

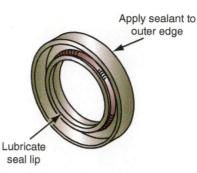

FIGURE 7-72 Apply sealant to the outer edge and lubricate the inner seal lip prior to installing a new oil seal.

Apply sealant to outer edge

Lubricate seal lip

Installing Axles

In most cases, the installation of axle shafts is a simple procedure. Prior to installing the shafts, make sure you installed all of the bearings, seals, and retaining plates on the shaft.

Some tapered roller bearing–equipped axles require an end play adjustment after installation. This adjustment is made by an adjusting nut or by selective shims. The shims are normally positioned between the retainer plate and the axle housing. The adjusting nut threads into the axle retainer plate.

To check the end play, position the dial indicator so it is able to measure the end-to-end movement of the axle. Push the axle into the housing and set the indicator to zero. Then, pull the axle out and note the reading on the indicator. This reading is the amount of end play in the shaft. Compare the reading against specifications and correct the end play as necessary.

End play adjustments are made by adding or subtracting shims or by turning the adjusting nut. This adjustment is done on one side of the housing but sets the end play for both sides.

TERMS TO KNOW

Bearing rumble

Bearing whine

Chatter

Chuckle

Clunk

Coast

Companion flange

Drive

Face

Flange

Flank

Gear noise

Heel

Knocking

Multiviscosity

Noise

Pinion flange

Toe

CASE STUDY

A customer brings her car in the shop complaining of an abnormal noise coming from the rear of the car. The service writer asks the customer the usual questions: When did it start? When does it make the noise? How often is the noise noticeable? The customer answers the questions and notes that the noise seems to have grown louder in recent weeks. She is not sure about when the noise was first noticed. She tells the service writer that the noise seems to be loudest when the car is moving at slow speeds.

The service writer notes the mileage of the car and checks its service record. According to the records, no major work has been performed on the car and it seems to have been maintained well.

The service writer and the customer then take the car for a test drive. Immediately upon pulling the car out of the shop, the noise is evident. As the speed of the car increases, the noise does seem to decrease. Having confirmed the problem, the service writer brings the car back to the shop and assigns a technician to

the problem. While informing the technician about the problem, he describes the noise as gear chuckle.

Knowing that the cause of chuckle is worn differential parts, the technician informs the service writer that the repair will be costly as the differential will need to be disassembled. Proceeding to determine the exact cause of the problem by disassembling the axle assembly, the technician notices the exhaust system hanging freely in the back. Upon inspection he finds that one exhaust hanger is broken. Before proceeding to continue the disassembly of the axle, he fastens the exhaust in place and takes the car for a test drive. No noise is evident.

Upon returning to the shop, he asks the service writer to take the car for a test drive as he cannot hear any chuckle. The service writer does so and explains that the noise is gone. The technician knows he has fixed the problem by reattaching the exhaust. The problem is not gear chuckle and the customer is saved quite a bit of money.

ASE-STYLE REVIEW QUESTIONS

1. When discussing the procedure for removing a differential unit,

   Technician A says that the same procedure should be followed on both a removable and integral carrier housing.

   Technician B says that the axle shafts must be removed before removing the differential case.

   Who is correct?

   A. A only   C. Both A and B
   B. B only   D. Neither A nor B

2. When discussing the retention of axle shafts,

   Technician A says that some axle shafts are retained in the housing by a plate and bolts.

   Technician B says that some axle shafts are retained in the housing by a C-washer or clip.

   Who is correct?

   A. A only   C. Both A and B
   B. B only   D. Neither A nor B

3. When discussing the proper timing of a ring and pinion gear set,

   Technician A says that if the gear set does not have timing marks, it is a nonhunting gear set.

   Technician B says that if that there are timing marks, one tooth of the pinion gear may be grooved or painted and there will be a notch between two ring gear teeth.

   Who is correct?

   A. A only   C. Both A and B
   B. B only   D. Neither A nor B

4. When examining a pinion gear, a marking "+2" is found on the small end of the gear,

   Technician A says that this indicates that the gear has been remanufactured.

   Technician B says that this indicates that 0.002 inches should be added to the measured nominal pinion depth.

   Who is correct?

   A. A only   C. Both A and B
   B. B only   D. Neither A nor B

5. When reviewing the procedure for setting backlash,

   Technician A says that backlash can be adjusted along with side bearing preload by loosening or tightening side bearing adjusting nuts.

   Technician B says that to decrease the amount of backlash, a thinner shim is normally installed on the ring gear side of the case and a thicker shim installed on the other side.

   Who is correct?

   A. A only   C. Both A and B
   B. B only   D. Neither A nor B

6. When checking the pattern of a gear set,

   Technician A says that the markings on the coast side of the tooth are the most important.

   Technician B says that if the marking is toward the outside diameter of the ring gear, it is referred to as being toward the toe of the gear.

   Who is correct?

   A. A only   C. Both A and B
   B. B only   D. Neither A nor B

7. When discussing axle bearing wear,

   Technician A says that axle bearings that show signs of galling have been overheated and should be replaced.

   Technician B says axle bearings that show signs of brinelling have been subjected to impact loading while it was not rotating and should be replaced.

   Who is correct?

   A. A only   C. Both A and B
   B. B only   D. Neither A nor B

8. While discussing the cause of an oil leak in an axle housing,

   Technician A says if the fluid leaks past the threads of the drive pinion nut, the leak can be corrected by installing a copper washer under the nut.

   Technician B says fluid leakage past the stud nuts that hold the carrier assembly to the axle housing can be corrected by applying thread or gasket sealer on the threads and tightening the nuts to specifications.

   Who is correct?

   A. A only   C. Both A and B
   B. B only   D. Neither A nor B

9. While inspecting a planetary gear-type final drive,

   Technician A says blackened pinion shafts indicate severe overloading.

   Technician B says bluish gears indicate overheating.

   Who is correct?

   A. A only   C. Both A and B
   B. B only   D. Neither A nor B

10. While checking the end play of a transfer gear assembly,

Technician A says if the end play of the output transfer shaft is incorrect, a different-size shim should be installed behind the output shaft transfer gear.

Technician B says after end play of the output shaft has been corrected, the turning torque of the shaft should be measured. If the turning torque is too high, a slightly thinner shim should be installed. If the turning torque is too low, a slightly thicker shim should be installed.

Who is correct?

A. A only
B. B only
C. Both A and B
D. Neither A nor B

ASE CHALLENGE QUESTIONS

1. When discussing drive chain and sprocket inspection,

Technician A says a drive chain that is too loose should be shortened by removing a pair of links in the chain.

Technician B says the drive sprockets should be replaced if the gear teeth are grooved or show any other signs of wear.

Who is correct?

A. A only
B. B only
C. Both A and B
D. Neither A nor B

2. A *knocking* noise is coming from a differential only when the vehicle is cornering.

Technician A says that the differential side gears could be damaged.

Technician B says that the noise could be due to damage on the ring gear teeth.

Who is correct?

A. A only
B. B only
C. Both A and B
D. Neither A nor B

3. A differential is being assembled.

Technician A says that carrier bearing preload must be adjusted prior to making the pinion gear depth adjustment.

Technician B says that if the pinion bearing preload is excessive, the pinion nut will need to be backed off until the correct preload is reached.

Who is correct?

A. A only
B. B only
C. Both A and B
D. Neither A nor B

4. The adjustment of the ring and pinion gear set on a shim-style carrier is being discussed.

Technician A says that increasing the size of both shims an equal amount will result in a reduction in carrier bearing preload.

Technician B says that placing a thinner shim on one side of the carrier and a thicker shim on the other side of the carrier will affect the ring and pinion gear contact pattern.

Who is correct?

A. A only
B. B only
C. Both A and B
D. Neither A nor B

5. The interpretation of a ring and pinion gear contact pattern test is being discussed.

Technician A says that excessive toe contact will require that the drive pinion gear be moved closer to the ring gear.

Technician B says that the ideal contact pattern will be centered on the coast side of the ring gear.

Who is correct?

A. A only
B. B only
C. Both A and B
D. Neither A nor B

Name _____ Date _____

DIFFERENTIAL HOUSING SERVICE

Upon completion of this job sheet, you will be able to inspect and flush a differential housing and refill it with the correct lubricant.

NATEF MAST Task Correlation

Drive Axle Diagnosis and Repair

Task #2, Check and adjust differential housing fluid level.

Task #3, Drain and refill differential housing.

Tools and Materials

Basic hand tools

Clean rag

Protective Clothing

Goggles or safety glasses with side shields

Describe the Vehicle Being Worked On

Year _____ Make _____ VIN _____

Model _____

Procedure

                                                                    _____
                                                                    Task Completed

1. On a RWD differential and final drive assembly, check the fluid level with the vehicle on a level surface and the axle at normal operating temperature. The fluid level should be even with the bottom of the fill plug opening in the axle housing unless otherwise specified. State the type of recommended fluid.

   _____

   _____

2. If the fluid level is low, inspect the differential and axle housing for signs of leakage. What did you find?

   _____

   _____

3. State the condition of the fluid and explain what you think may have caused the unusual smell or the contamination.

   _____

   _____

4. On most FWD models with manual transaxles, the transmission and final drive assembly are lubricated with the same fluid and share the same fluid reservoir. State the type of recommended fluid.

    _____

    _____

5. Check the fluid level at the transaxle fill plug or dipstick. The fluid should be even with the bottom of the fill plug opening unless otherwise specified. What did you find?

    _____

    _____

6. State the condition of the fluid and explain what you think may have caused the unusual smell or the contamination.

    _____

    _____

☐ 7. If the fluid was contaminated and the problem was not caused by the destruction of parts in the differential, the housing should be flushed. To do this, drain the fluid from the housing.

☐ 8. Then refill the housing with clean fluid and rotate the wheels and/or drive shaft to circulate the fluid.

☐ 9. Drain this fluid and refill the housing with a fresh supply of fluid.

☐ 10. Rotate the wheels and/or drive shaft several times, and then reexamine the condition of the fluid. If the fluid still has traces of residue, flush the housing again or until the fluid is clean.

11. Is the differential a limited-slip unit? If so, what fluid is recommended and what additive should you mix with the fluid? Before adding friction modifiers to the lubricant of a limited-slip differential, always check its compatibility. Using the wrong additive may cause the axle's lubricant to break down, which would destroy the axle very quickly.

    _____

    _____

Instructor's Response: _____

_____

_____

JOB SHEET

27

Name _____ Date _____

ROAD CHECK DIFFERENTIAL NOISE

Upon completion of this job sheet, you should be able to demonstrate the ability to road-check a vehicle and identify noises in the differential.

NATEF MAST Task Correlation

Drive Axle Diagnosis and Repair

Task #4, Diagnose noise and vibration concerns; determine necessary action.

Tools and Materials

An AWD vehicle students are authorized to road-check

Permission for student to road-test the vehicle

Describe the Vehicle Being Worked On

Year _____ Make _____ VIN _____

Model _____

Procedure

Task Completed

1. Fluid level should be checked. Check vent cap on axle housing to be sure it is open. ☐

2. Road-test the vehicle. It should be driven for enough time to warm it up. Drive about ☐
   55 mph, change the accelerator position, increasing and decreasing speed. Drive at a
   set speed for a time. Do not accelerate at a high rate.

3. During road test, does the differential seem to be making a noise?   ☐ Yes   ☐ No

4. If yes, does it make a noise:
   a. On acceleration?   ☐ Yes   ☐ No
   b. At a set speed (which might be called a float)?   ☐ Yes   ☐ No
   c. On deceleration?   ☐ Yes   ☐ No

5. Could the noise be made by the tires?   ☐ Yes   ☐ No

6. What will you tell the customer?

_____

_____

_____

_____

_____

_____

_____

_____

_____

_____

_____

Instructor's Response: _____

_____

_____

Name _____ **Date** _____

CHECK DIFFERENTIAL NOISES ON A HOIST

Upon completion of this job sheet, you should be able to demonstrate the ability to check for noises in a differential on a hoist or stand.

NATEF MAST Task Correlation

Drive Axle Diagnosis and Repair

Task #4, Diagnose noise and vibration concerns; determine necessary action.

Tools and Materials

Hoist or jacks

Jack stands

Stethoscope

Protective Clothing

Safety glasses with side shields

Describe the Vehicle Being Worked On

Year _____ Make _____ VIN _____

Model _____

Procedure

Task Completed

1. Put a RWD vehicle on a hoist or on jack stands, making sure that it is supported safely. ☐

2. Have an assistant run the drivetrain on the hoist or stands. Use a stethoscope to listen to the rear axle. It will be very noisy, but you must attempt to separate normal noises from noises that are not supposed to be there. ☐

3. Put the stethoscope on the rear axle housing. On either the right or left axle bearing, do you hear any abnormal noise?   ☐ Yes   ☐ No

4. Put the stethoscope on the differential housing near where the carrier bearings are. Do you hear any abnormal noises either from the right or left bearings?   ☐ Yes   ☐ No

5. Put the stethoscope on the pinion area of the housing. Do you hear any abnormal noises?   ☐ Yes   ☐ No

6. Would you say the differential was making a noise?   ☐ Yes   ☐ No

7. Describe the noise you heard.

_____

_____

_____

_____

8. What are you going to tell the customer?

_____

_____

_____

_____

_____

Instructor's Response: _____

_____

_____

Name _____ Date _____

IDENTIFY THE TYPE OF DIFFERENTIAL

Upon completion of this job sheet, you should be able to demonstrate the ability to identify the different types of differentials.

NATEF MAST Task Correlation

Drive Axle Diagnosis and Repair

Task #1, Clean and inspect differential housing; check for leaks; inspect housing vent.

Task #4, Diagnose noise and vibration concerns; determine necessary action.

Tools and Materials

Hoist or jack

Jack stands

Protective Clothing

Safety glasses with side shields

Describe the Vehicle Being Worked On

Year _____ Make _____ VIN _____

Model _____

Procedure

Task Completed

1. Put a RWD vehicle up on a hoist or safety stands. ☐

2. Position yourself underneath the vehicle. ☐

3. Identify the rear axle assembly. Is it:
   a. Integral? ☐ Yes ☐ No
   b. Removable? ☐ Yes ☐ No

4. Does the vehicle have a limited-slip differential? ☐ Yes ☐ No

Instructor's Response: _____

_____

_____

Name _____ Date _____

DRIVE AXLE LEAK DIAGNOSIS

Upon completion of this job sheet, you should be able to identify the cause of drive axle fluid leakage problems.

NATEF MAST Task Correlation

Drive Axle Diagnosis and Repair

Task #1, Clean and inspect differential housing; check for leaks; inspect housing vent.

Task #2, Check and adjust differential housing fluid level.

Tools and Materials

Basic hand tools

Trouble light or flashlight

Lift

Protective Clothing

Goggles or safety glasses with side shields

Describe the Vehicle Being Worked On

Year _____ Make _____ VIN _____

Model _____

Procedure

1. To identify the exact source for a leak, a careful inspection must be completed. Raise the vehicle on a lift or hoist. Look at the drive axle and note the general areas where fluid is present.

   _____

   _____

2. Check the area around the pinion shaft. An improperly installed or damaged drive pinion seal will allow the lubricant to leak past the outer edge of the seal. Any damage to the seal's bore, such as dings, dents, and gouges, will distort the seal casing and allow leakage. Also, the spring that holds the seal lip against the companion flange may be knocked out and allow leakage past the seal's lip. Clean the area around the leak so you can determine the location of the leak and record your findings.

   _____

   _____

3. It is also possible for oil to leak past the threads of the drive pinion nut or the pinion retaining bolts. Removing the nut or bolts, applying thread or gasket sealer on the threads, and torquing the nuts or bolts to specifications can stop these leaks. Clean the area around the leak so that you can determine the location of the leak and record your findings.

_____

_____

4. Leakage past the stud nuts that hold the carrier assembly to the axle housing can be corrected by installing copper washers under the nuts. Always make sure there is a copper washer under the axle ID tag. Clean the area around the leak so you can determine the location of the leak and record your findings.

_____

_____

5. Check the housing cover for signs of leakage. Check for poor gasket installation, loose retaining bolts, or a damaged mating surface. Because most late-model vehicles do not use a gasket on the housing cover, silicone sealer is used and the old sealant should be cleaned off before applying a new coat to the surface. If a housing cover is leaking, inspect the surface for imperfections, such as cracks or nicks. File the surface true and install a new gasket. If the surface cannot be trued, apply some gasket sealer to the surface before installing the new gasket. Always follow the recommended tightening sequence and torque specifications when tightening an axle housing cover. Describe your findings.

_____

_____

6. Check for lubricant leaks on the housing itself. At times, lubricant will leak through the pores of the housing. To correct this problem if the porous area is small, force some metallic body filler into the area. After the filler has set, seal the area with an epoxy-type sealer. If the leaking pore is rather large, drill a hole through the area and tap an appropriately sized setscrew into the hole, then cover the area with an epoxy sealer. Describe your findings.

_____

_____

7. Check the level of the fluid in the assembly. Describe your findings and state what type of fluid is recommended for this vehicle.

_____

_____

_____

8. Check the vent for signs of leakage. If the wrong vent was installed in the axle housing or if there is an excessive amount of lubricant or oil turbulence in the axle, lubricant may leak through the axle vent hose. Also check for a crimped or broken vent hose. If the cause of the leakage is an overfilled axle assembly, drain the housing and refill the unit with the specified amount and type of lubricant. Describe your findings.

_____

_____

_____

9. Check the brake backing plates and drums for signs of fluid. This is usually caused by a worn or damaged axle seal. However, if the seal's bore in the axle tube is damaged, the seal will be unable to seal properly. Whenever fluid is leaking past the axle seal, the bore should be checked and the seal replaced. Describe your findings.

_____

_____

10. Some drive axle units are fitted with an ABS sensor. Lubricant can leak from around the sensor's O-ring if it is damaged. Check here and describe your findings.

_____

_____

_____

Instructor's Response: _____

_____

_____

Name _____ Date _____

MEASURE RING GEAR RUNOUT AND BACKLASH

Upon completion of this job sheet, you should be able to demonstrate the ability to measure ring gear runout and backlash.

NATEF MAST Task Correlation

Drive Axle Diagnosis and Repair

Task #6, Inspect ring gear and measure runout; determine necessary action.

Task #10, Measure and adjust side bearing preload and ring and pinion gear total backlash and backlash variation on a differential carrier assembly (threaded cup or shim types).

Tools and Materials

Appropriate hand tools

Lift or jack

Jack stands

Dial indicator

Service manual

Protective Clothing

Safety glasses with side shields

Describe the Vehicle Being Worked On

Year _____ Make _____ VIN _____

Model _____

Procedure

Using the service manual for the vehicle you are working on, do the following:

Task Completed

Ring Gear Runout

1. Mount dial indicator on carrier housing. ☐

2. Set dial indicator on back side of ring gear. ☐

3. Rotate ring gear one complete revolution. ☐

4. Record the highest and lowest reading. ☐

5. Subtract lowest reading from highest: The result is runout. ☐

6. Compare runout to specifications. ☐

7. If not within specs, check case runout. ☐

Ring and Pinion Backlash

☐   1. Mount dial indicator on carrier housing.

☐   2. Set dial indicator on a ring gear tooth.

☐   3. Measure backlash at four positions.

☐   4. Average the readings.

☐   5. Compare to specifications.

Instructor's Response: _____

_____

_____

Name _____ Date _____

LIMITED-SLIP DIFFERENTIAL DIAGNOSTICS

Upon completion of this job sheet, you will be able to diagnose noise, slippage, and chatter concerns in a limited-slip differential assembly. You will also be able to measure the rotating torque of the assembly.

NATEF MAST Task Correlation

Drive Axle Diagnosis and Repair

Task #1, Diagnose noise, slippage, and chatter concerns; determine necessary action.

Task #2, Measure rotating torque; determine necessary action.

Tools and Materials

Axle shaft puller

Torque wrench

Service information

Protective Clothing

Goggles or safety glasses with side shields

Describe the Vehicle Being Worked On

Year _____ Make _____ VIN _____

Model _____

Procedure

1. Take the vehicle for a road test. Pay attention to the noise level and action of the differential unit as you make a right and left turn. Describe your findings.

   _____

   _____

2. Now drive the vehicle on a slick surface and accelerate hard enough to lose traction. Pay attention to the noise level and action of the differential unit. Describe your findings.

   _____

   _____

3. Based on the above tests, is the unusual noise or action caused by the limited-slip function of the differential or is it caused by something else? Explain your answer.

   _____

   _____

4. A limited-slip differential can be checked for proper operation without removing the differential from the axle housing. Be sure the transmission is in neutral, one rear wheel is on the floor, and the other rear wheel is raised off the floor. Specifications should give the breakaway torque reading required to start the rotation of the wheel that is raised off the floor. What are the specifications?

_____

_____

5. Using a torque wrench, measure the torque required to turn the wheel. What was your measurement and how does it compare with the specification?

_____

_____

Instructor's Response: _____

_____

_____

Chapter 8

SERVICING FOUR-WHEEL-DRIVE SYSTEMS

UPON COMPLETION AND REVIEW OF THIS CHAPTER, YOU SHOULD BE ABLE TO:

- Diagnose four-wheel-drive assembly noise, vibration, hard shifting, unusual steering problems, and determine needed repairs.

- Inspect, adjust, and repair transfer case shifting mechanisms, bushings, mounts, levers, and brackets.

- Inspect and service transfer case and components and check lube level.

- Inspect, service, and replace front-drive shafts and Universal joints.

- Inspect, service, and replace front-drive axle knuckles and driving shafts.

- Inspect, service, and replace front wheel bearings and locking hubs.

- Check four-wheel-drive unit seals.

INTRODUCTION

This chapter covers the basic diagnostic, service, and repair procedures for four-wheel-drive (4WD) systems. By no means are the procedures given intended to be used on all systems. Always refer to the shop manual to determine the specific procedures for the system being worked on. Service procedures for those components that would be part of a 2WD system should be followed as if the vehicle were not 4WD. This includes the transmission, drive shafts, differentials, and drive axles. A 4WD vehicle has two drivelines instead of one, so there is twice the potential driveline problems.

MAINTENANCE

The lubricants in 4WD drivelines and the transfer case (Figure 8-1) must be checked and changed on a regular basis. The fluid is drained and added at the transfer case's drain plug and filled through the fill plug. The interval for changing the fluid will vary with the manufacturer and how the vehicle is operated. Also, the Universal joints on the axles and drive shafts need to be lubricated as recommended by the manufacturer.

Common 4WD problems are the result of fluid contamination and low fluid levels. The recommended oil for the transfer case varies with the manufacturers. Most transfer cases require ATF while others require 50W oil manufacturer-specific fluids. Automatic transfer cases with multiplate clutch assemblies typically require fluids with special friction modifiers to ensure correct operation.

Most drive axle assemblies require 80W–90W or similar hypoid-type oil. However, some have ATF or engine oil as the recommended lubricant for the front-drive axles. Always refer to and follow the recommendations of the manufacturer.

Limited-slip units are often used in the differential units. These units can be found in any differential in the vehicle. Always check the specifications and be sure to add the correct additive to the differential fluid if the unit is a limited-slip unit.

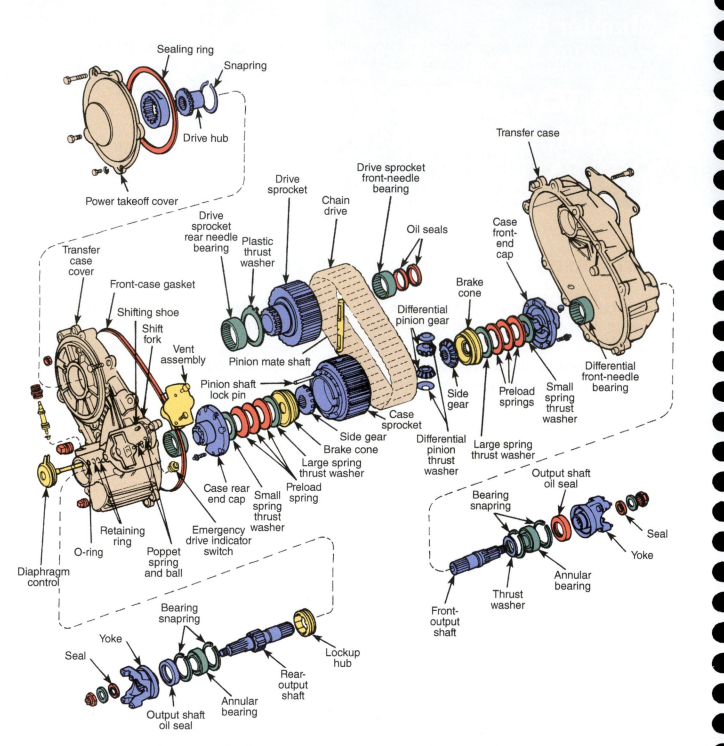

FIGURE 8-1 An exploded view of a transfer case with an integral differential and cone brakes for limited slip.

When a 4WD vehicle is operated in water or mud that gets over the top of the axle housings, the hubs, wheel bearings, and Universal joints should be immediately checked for fluid contamination.

Classroom Manual

Chapter 8, page 192

DIAGNOSIS

Diagnosis of 4WD systems can be quite tricky because of the increased load on the vehicle's suspension and driveline, as well as the operating conditions a 4WD vehicle usually faces. Often complaints will not be caused by the 4WD system; rather, they will be caused by related systems or parts. Begin your diagnosis of a problem by thoroughly questioning the customer about the problem. If at all possible, road-test the vehicle with the customer to check the function of the unit and to make sure you know what the complaint is. Always check for relevant service bulletins and look at the service history of the vehicle. Most drivetrain and suspension problems show up initially as noises or vibrations.

Like 2WD vehicles, certain handling symptoms on a 4WD vehicle are most often caused by suspension problems. Here are some common problems and their most probable causes:

- *Bottoming-out over rough terrain*—This may be caused by an overloaded vehicle, weak springs, or worn shock absorbers.
- *Swaying*—Probable causes are worn or improperly inflated tires, a bent or broken stabilizer bar, or a worn shock absorber.
- *Front-wheel shimmy*—Typically caused by worn or improperly inflated tires, out-of-balance wheel and tire assemblies, worn shock absorbers, incorrect wheel alignment, worn ball joints, worn hub bearings, or loose or worn steering linkage parts.
- *Abnormal tire wear*—This problem is normally caused by improperly inflated tires, incorrect wheel alignment, worn shock absorbers, or worn suspension parts.

Certain axle and drive shaft symptoms are related to the engine, transmission, tires, and other parts of the vehicle. During diagnosis it is important that the condition be isolated into its specific area as soon as possible. A road test followed by a detailed inspection may readily pinpoint the source of the problem. It is important to check all external parts of the system before suspecting something in the transfer case or electronic control module.

Before attempting to diagnose any 4WD system problem, make sure the tire pressures are equal on all four wheels. Measure the tire diameters of all four wheels to make sure all tires are exactly the same diameter. Tire diameter can be affected by wear, different tread design, or manufacturer even on the same-size tires. There are various methods to measure tire diameter. One common method is to mark the shop floor and the tire, then roll the vehicle forward until the tire mark again contacts the floor. Measure the distance between the marks. Another method uses a special thin tire stagger tape stretched around the tire. Some manufacturers recommend no more than ⅛-inch diameter difference between all four tires. A quick and fairly accurate estimate of tire diameter can be made with a sliding **stagger gauge** (Figure 8-2). On many late-model transfer cases, uneven tire pressure, tire

A 4WD vehicle has two drivelines instead of one, so there are twice the number of potential driveline problems.

FIGURE 8-2 A sliding stagger gauge with a graduated scale.

wear or unmatched tire diameter will cause problems in the transfer cases and will result in poor handling or damage to the transfer case. Proper tire rotation is especially important on AWD and 4WD vehicles.

Diagnosis of 4WD systems is similar to those of 2WD systems. Transmission, drive shaft, and axle problems are diagnosed as if they were 2WD problems and all service procedures for these components are the same.

Find out the speed or condition at which the noise or vibration occurs. Also find out if the problem is more noticeable when the vehicle is making a turn or going over bumps. This type of information will be helpful during your diagnosis of the problem.

Noise Diagnosis

Noise problems are more difficult to diagnose and service than mechanical failures. A slight axle noise heard only at certain speeds or during specific operating conditions may be normal. A normal axle will tend to be louder at specific speeds and the noise may not be an indication of a problem in the axle. There are several different noises that can be produced by a drive axle or differential. These include gear and bearing noise, chuckle, knock, and clunk.

If the noise or vibration is more noticeable at higher speeds, suspect a problem in the transfer case, a bent drive shaft, or bad wheel bearings. A noise or vibration at low speeds is usually often caused by bad U-joints. If the problem is more noticeable when the vehicle is traveling on a rough road or over bumps, worn shocks or sway bars are the likely cause. Worn outboard axle joints will cause a noise or vibration when the vehicle is making a turn.

Classroom Manual
Chapter 8, page 194

If the problem seems to get worse with an increase in throttle position, worn engine mounts or U-joints are likely causes. Worn U-joints can also cause a noise that only appears when the vehicle is pulling a load or climbing a hill. Often, worn splines on the drive shaft's slip joint can rumble and grind and sound like a worn U-joint. Vibrations under a load may also be caused by a worn slip joint, which can cause a drive shaft to be out-of-balance as it shifts to one side.

Front hubs may make a ratcheting sound when water or dirt have entered the hub and contaminated the lubricant. This prevents free movement of the components in the hub. A ratcheting sound from an automatic locking hub may indicate that the hub on the opposite side of the axle is not disengaging.

Classroom Manual
Chapter 8, page 211

Most transfer cases use a planetary gear set with straight-cut pinion gears that are inherently noisy. If the customer complains about a whirring noise, it may be the natural sound of the pinion gears in the gear set. This noise is most evident when the transfer case is in 4WD-Low.

Abnormal noises coming from the transfer case are typically caused by incorrect tire inflation or tire size, excessive tire wear, worn transfer case bearings, and/or gear damage. A loud ratcheting sound under load in 4WD may be caused by a stretched transfer case chain slipping on the sprockets.

Shifting Problems

Classroom Manual
Chapter 8, page 207

The locking of the splines is typically caused by driveline windup.

A common complaint on all transfer cases is the inability to go from 4WD to 2WD. Before condemning the transfer case, back it up about 60 feet and then try to take it from 4WD to 2WD. It is common for these units to get spline-locked, which results in the torque load on the shafts not relaxing to let the splines line up for a shift. By changing directions and reversing the torque load, the unit should shift freely.

If the transfer case jumps out of gear, suspect an improperly adjusted shift linkage, loose mounting bolts or brackets, worn front- or rear-drive shaft slip yokes, a damaged sliding clutch hub (Figure 8-3) or gear teeth, or a worn range or mode fork.

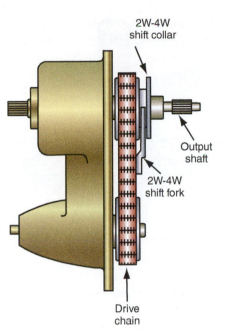

FIGURE 8-3 A worn or damaged shift collar, shift fork, or clutch hub can cause the transfer case to jump out of gear.

INSPECTION

Fluid Leaks

Fluid leaks are a major concern on 4WD vehicles. Check under the vehicle for signs of leaks. Fluid leaks can cause transfer cases and other vital 4WD components to overheat. Most transfer cases are filled with ATF or manufacturer-specific fluid. First, carefully inspect the transfer case and the area around it (Figure 8-4). Check for red spots of ATF on the transfer case, especially where it connects to the front- and rear-drive shafts, which would indicate that its external seals are leaking. Then check the connections to the front and rear differentials. Black or brown spots indicate leaking seals at those points. Also check for leaking rear-axle seals by removing the rear wheels and brake drums or discs and inspecting the brake backing plates. Brown spots indicate leaks.

⚠ **WARNING:** Whenever working under a vehicle, wear safety glasses or goggles.

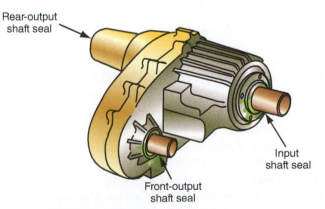

FIGURE 8-4 Carefully inspect the transfer case for signs of leakage.

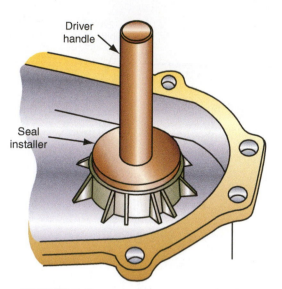

Driver handle

Seal installer

FIGURE 8-5 Always use the correct tools when installing a seal.

Fluid leaks are a common problem that can be prevented by following proper service procedures. Using only the gaskets or sealing materials recommended by the manufacturer can also prevent seal leakage. Never use a sealant in place of a gasket unless the manufacturer says to. Always use the proper tools when installing a seal (Figure 8-5). A cocked seal or one that is installed in a damaged housing will leak. When installing a seal, make sure the surface the seal lip rides on is smooth. The slightest burr can tear the seal and cause it to leak.

Be sure to inspect the air vent for the transfer case. If the vent is plugged with dirt or undercoating, pressure will build up inside the case and cause the seals to leak. Some transfer cases are equipped with a remote vent. These vents are placed above the transfer case to reduce the chance of water entering the transfer case while driving off the road or through water. Make sure the vent is not plugged and the tubing to the vent is not bent or restricted.

Transfer Case

Classroom Manual
Chapter 8, page 207

Internal failures in the transfer case are sometimes caused by faulty shift controls. If the transfer case is shifted by a vacuum control, make sure the engine has at least 15 inches Hg. This is done by connecting a vacuum gauge to the intake manifold and measuring the vacuum with the engine at idle. Also make sure the shift linkage is tight and well lubricated. Internal failures in the transfer case may also be related to improper tire size, unevenly worn tires, or incorrect tire inflation. The driveline windup created by tire problems puts stress on transfer case components.

Drive Shafts

SERVICE TIP:
Fitting a truck with a raised suspension (Figure 8-7) changes the drive shaft yoke and steering linkage angles and may be the cause of the customer's complaints.

Support the vehicle on jack stands and hose off any mud, road salt, or other debris from under the vehicle. Then, with the transmission in Neutral, check each U-joint (Figure 8-6) by twisting and shaking both ends back and forth. Any looseness is an indication of a worn joint, as is a steady squeak during acceleration from a stop, or clunking when driving.

Move the slip joints up and down. Because these are exposed to much abrasion and vibration and are not always lubricated as often as they should be, wear in the slip joints is common. If there is any movement at the joints, the joint and drive shaft should be replaced. If the slip joint is seized, penetrating oil or heat may be needed to free it up.

Rotate the shaft and look for any signs of uneven wear on the splines of the slip joint. Also check the drive shafts for evidence of damage caused by an impact, especially if the vehicle is used off-road. On two-piece drive shafts, the center bearing assembly and joints at the bearing should also be checked. All of the U-joints should be inspected and checked.

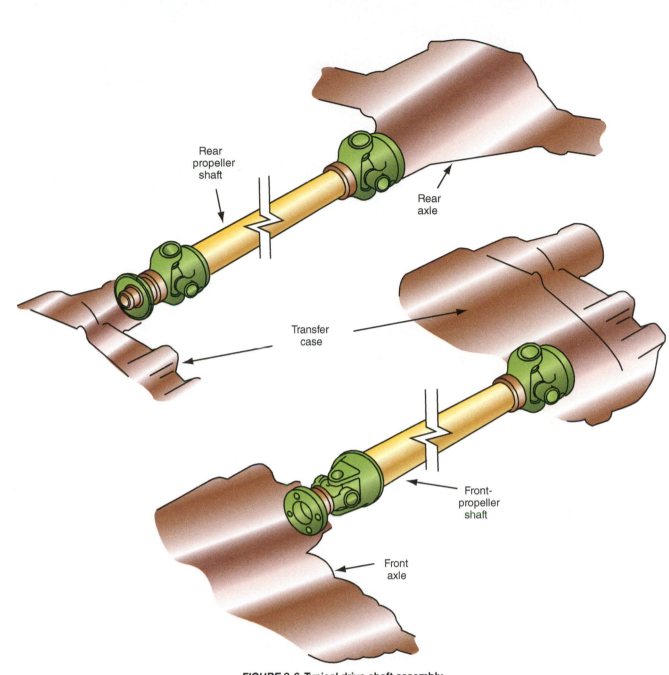

Rear
propeller
shaft

Rear
axle

Transfer
case

Front-
propeller
shaft

Front
axle

FIGURE 8-6 Typical drive shaft assembly.

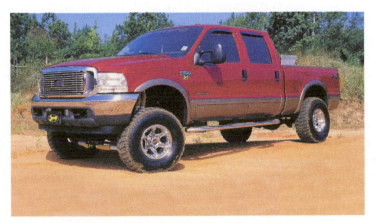

FIGURE 8-7 A pickup truck with a raised suspension.

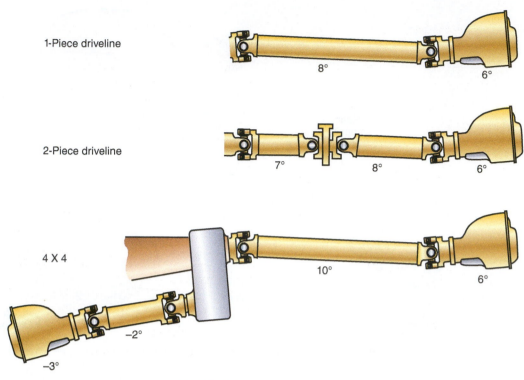

1-Piece driveline
8° 6°

2-Piece driveline
7° 8° 6°

4 X 4
10° 6°
−2°
−3°

FIGURE 8-8 Typical driveline angles.

U-joint operating angle plays an important part in the driving performance of a 4WD vehicle (Figure 8-8). Because When U-joints are used instead of CV joints, their operating angle will directly affect the ride of a 4WD vehicle. The angle of the driveline should be checked whenever the angle is suspect. Pay special attention to operating angles if a lift kit has been installed on the vehicle.

Suspension

Often 4WD and AWD vehicle problems are caused by, or are associated with, suspension system problems. Begin your inspection of the suspension system by measuring the vehicle's standing suspension height and comparing the measurements to the specifications for the vehicle. If the standing height is not within specifications, look for cracked springs and/or shackles (Figure 8-9) and for damaged **spring pads**. The spring pads can be seen by prying the leaves apart when the spring is unloaded.

Inspect the condition of the suspension and make sure it is properly lubricated. Look for evidence that heavy-duty suspension components have been installed on the vehicle. Competition or heavy-duty suspension components have hard ride characteristics and if these components were installed through error, they should be replaced with standard suspension parts.

To check the **ball joints** or **king pins** (Figure 8-10), grasp each tire at the bottom and top and shake the wheel while watching the movement of the disc brake support or the front spindle assembly. If the front spindle assembly moves more than $\frac{1}{32}$ inch at the upper or lower arms relative to the axle or the king pin, then rebush the king pin spindles or replace both ball joints on that side.

Shock absorbers used on current 4WD vehicles are the twin-tube, low-pressure nitrogen gas type (Figure 8-11). This type of shock absorber is nonadjustable and nonrefillable and cannot be serviced, except for replacement. Early-model 4WD vehicles used the conventional hydraulic direct-acting shock absorber. Check the shock absorber to be sure it is securely and properly installed. Also check the shock absorber insulators for damage and wear. Replace any worn or damaged shock absorbers or insulators and tighten the attaching points to the torque specified in the service manual.

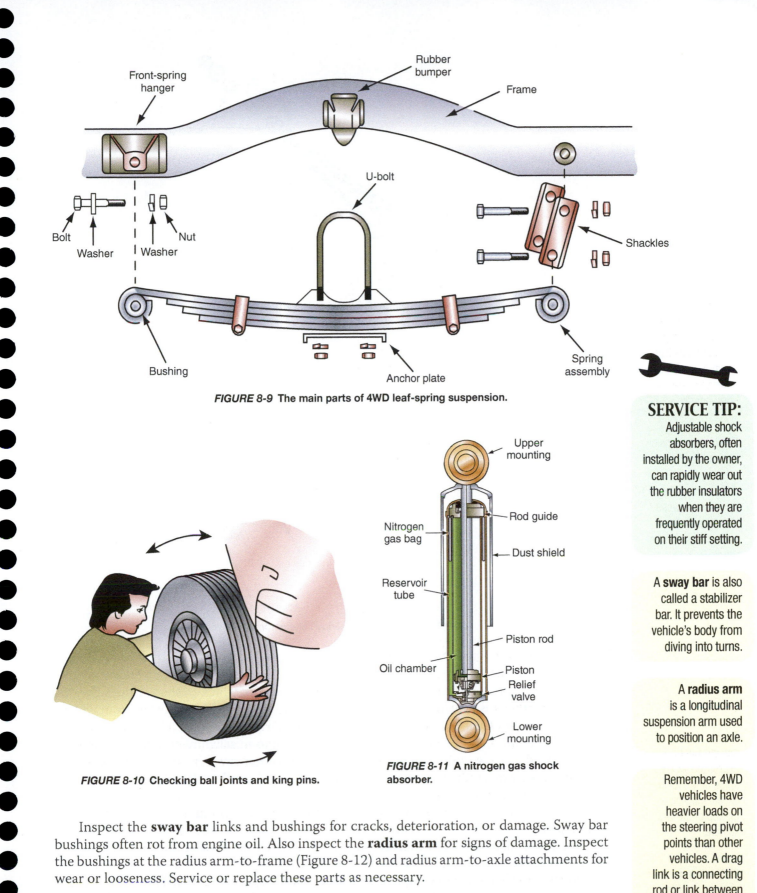

FIGURE 8-9 The main parts of 4WD leaf-spring suspension.

FIGURE 8-10 Checking ball joints and king pins.

FIGURE 8-11 A nitrogen gas shock absorber.

Inspect the **sway bar** links and bushings for cracks, deterioration, or damage. Sway bar bushings often rot from engine oil. Also inspect the **radius arm** for signs of damage. Inspect the bushings at the radius arm-to-frame (Figure 8-12) and radius arm-to-axle attachments for wear or looseness. Service or replace these parts as necessary.

Steering System

Check the entire steering linkage for wear or misadjustment. Check the steering gear mounting bolts and the front-spring U-bolts and tie bolts and tighten all bolts as needed.

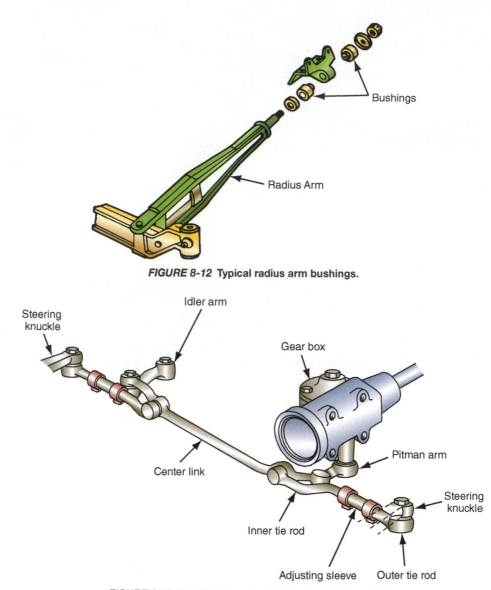

FIGURE 8-12 Typical radius arm bushings.

Bushings

Radius Arm

Steering knuckle

Idler arm

Gear box

Pitman arm

Steering knuckle

Center link

Inner tie rod

Adjusting sleeve   Outer tie rod

FIGURE 8-13 A typical steering linkage for a 4WD vehicle.

A **Pitman arm** is a short arm in the steering linkage that connects the steering gear to other steering components.

Toe is a suspension dimension that reflects the difference in the distance between the extreme front and extreme rear of the tires.

Camber is the amount that the centerline of the wheel is tilted inward or outward from the true vertical.

Caster is a measurement expressed as an angle of the forward or rearward tilt of the top of the wheel spindle.

A **steering damper** helps to stabilize the movement of the wheels as a vehicle is making turns.

Next, check the steering linkage (Figure 8-13) for play. Most manufacturers allow ⅛-inch play. Most full-size 4WD trucks use a drag link for the steering and are set up differently than cars. In this setup, the steering gear turns the **Pitman arm**, which pushes or pulls the drag link. The drag link is usually attached directly to the left-side steering knuckle, from which a transverse tie-rod transfers the steering forces onto the right wheel.

Many handling problems on 4WD vehicles are due to misalignment. Some wheel alignment procedures are quite straightforward, whereas others are not. In fact, on some, only toe can be adjusted. Others allow only changes to **toe** and **camber**. Still others allow for toe, camber, and **caster** adjustments.

Although some vehicles allow for caster adjustments, a few require the suspension and steering systems be modified to correct caster. Improper side-to-side caster can contribute to vehicle drift or pull in one direction on the side with less caster. Misalignment is often evident by tire wear patterns (Figure 8-14).

Most 4WD vehicles are equipped with **steering dampers** (Figure 8-15). These are typically nonadjustable, nonrefillable, and nonrepairable. Their mountings should be inspected and tightened, as should their attachment points. If the rubber bushings are worn, the damper should be replaced. Inspect the damper for leaks. A light film of fluid is permissible on the body of the damper near the shaft seal; a dripping damper should be replaced.

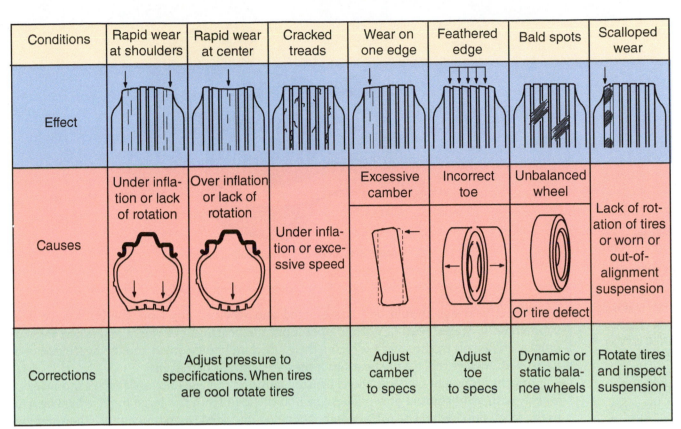

Conditions	Rapid wear at shoulders	Rapid wear at center	Cracked treads	Wear on one edge	Feathered edge	Bald spots	Scalloped wear
Effect							
Causes	Under inflation or lack of rotation	Over inflation or lack of rotation	Under inflation or excessive speed	Excessive camber	Incorrect toe	Unbalanced wheel / Or tire defect	Lack of rotation of tires or worn or out-of-alignment suspension
Corrections	Adjust pressure to specifications. When tires are cool rotate tires			Adjust camber to specs	Adjust toe to specs	Dynamic or static balance wheels	Rotate tires and inspect suspension

FIGURE 8-14 Tire-wear patterns are a good indicator of wheel alignment problems.

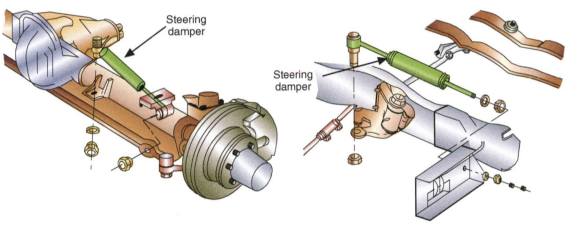

FIGURE 8-15 Location of steering damper.

To check the operation of a steering damper, disconnect the frame or axle end of the damper. Then extend and compress the damper using as much travel as possible. The damper action should be smooth throughout each stroke and there should be no evidence of fluid leakage. The damper should be replaced if it leaks any fluid during the test, if it seizes or binds while moving, or if there is a skip or lag near the downstroke of the damper.

Wheel Bearings

To check the adjustment of the wheel bearings (Figure 8-16), raise the front of the vehicle. Grasp each front tire at the front and rear and push the wheel inward and outward. If any free play is noticed between the hub, rotor, and front spindle, adjust the wheel bearings. Any play noticed on a unit-type wheel bearing will require bearing replacement.

FIGURE 8-16 Checking wheel bearing adjustment.

SERVICE TIP:
If a customer complains of steering problems after the installation of larger tires and wheels, suggest that a steering damper be added. This device helps to stabilize the movement of the wheels as a vehicle is making turns.

Wheel offset is basically the amount of the wheel assembly that is to the side of the wheel's mounting hub.

The circumference of a wheel is actually the distance it travels when it completes one full revolution.

SPECIAL TOOLS
Tire pressure gauge
Tire balancing tools
Dial indicator

Wheel bearings should be disassembled and serviced any time the hubs have been submerged in water. During normal operation, the bearings get warm and when they are quickly cooled off by the splash of water, their lubricant breaks down and the bearings can be destroyed. Always replace the bearings if they are worn or damaged.

Another frequent cause of wheel bearing failure is the use of oversize tires mounted on wheels with a substantial offset. These switch the load from the large inner wheel bearing to the small outer wheel bearing, which was never intended to do much more than stabilize the wheel.

Tires and Wheels

The effective axle ratio is the total gear ratio measured at the outer circumference of the driving wheels. This takes into account differences in tire size, inflation pressures, and the number of teeth on the ring and pinion gears. All of the tires should be the same size and at the proper inflation. Different tire inflation pressures can cause a 0.25 difference in the effective gear ratio between the front and rear wheels, which can cause severe driveline windup problems.

Check tire radial runout (Figure 8-17). This check is performed with a dial indicator placed at the center and outside ribs of the tread face. Move the dial indicator onto the rim of the wheel and measure radial and lateral runout of the wheel (Figure 8-18), which

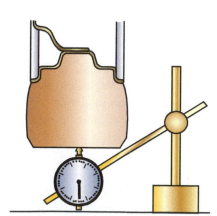

FIGURE 8-17 Checking tire runout.

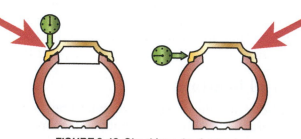

FIGURE 8-18 Checking wheel runout.

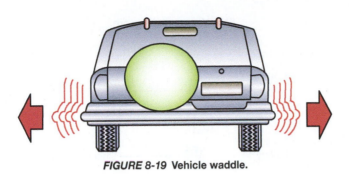

FIGURE 8-19 Vehicle waddle.

SERVICE TIP:
Most transfer cases that shift into and out of 4WD with spring pressure on the shift collar will not shift out of 4WD until driveline windup is relaxed. This means the vehicle will stay in 4WD until the collar is free to move. In a situation like this, there is nothing wrong with the transfer case, it is just the nature of a 4WD vehicle.

should not exceed specifications. If the runout is excessive, check the runout of the axle or hub before replacing the tire or wheel. Check and balance each front wheel as required. Large tires should be balanced frequently because it is characteristic of tires with a thick tread to wear unevenly, which changes the balance of the tires and causes further uneven tire wear.

4WD vehicles have a minimal amount of **toe-out on turns**. This is the main cause of accelerated tire wear, especially of the right front tire on full-time 4WD vehicles. Another cause of excessive tire wear is the additional **unsprung weight** of the differential. This tire wear is normal and cannot be corrected; however, you can minimize the wear by frequent rotation of the tires.

A 4WD vehicle may display a side-to-side movement at the front or rear of the vehicle. This is called *waddle*. This problem can also occur on passenger cars but is more evident on taller and heavier 4WD vehicles. Waddle (Figure 8-19) can be caused by a defective tire or excessive tire or wheel lateral runout. It is most noticeable at low speeds but may cause a rough ride at highway speeds.

Defective tires also can cause the vehicle to pull or lead to one side. Of course, this problem can also be caused by poor wheel alignment or uneven brake adjustments. Tire pull will only occur at the front-drive axle.

Toe-out on turns allows the front wheels to be at different angles when the vehicle is turning a corner.

Unsprung weight is the weight of the tires, wheels, axles, control arms, and springs.

Axle Hub Diagnosis

Front hubs may make a ratcheting sound when water or dirt has entered the hub and contaminated the lubricant. This prevents free movement of the components in the hub. A ratcheting sound from an automatic locking hub may indicate that the hub on the opposite side of the axle is not disengaging.

Manual locking hubs can be quickly checked by rotating the brake drum or rotor slightly and turning the hub selector into the lock position. A click should be heard when the hub engages the axle, and the axle should now turn with the hub. Next, with the hub still turning, turn the selector to the free position. The axle should now be free of the hub. If both of these events do not happen, the hub assembly needs to be repaired or replaced.

Centrifugal automatic locking hubs can be difficult to diagnose. Correct operation requires that the close tolerances of hub components be maintained. A visual inspection may not identify wear, but if the components are cleaned and properly lubed and the hub still does not lock, it must be replaced. Vacuum-operated hub problems can be caused by vacuum supply problems, leaky vacuum hoses, faulty vacuum solenoids, and vacuum leaks inside of the hub assembly (Figure 8-20).

Automatic locking hub assemblies can be very expensive to replace, or may no longer be available from the manufacturer. Aftermarket suppliers have manual locking hub replacement kits for many vehicles.

IWE Vacuum diagram

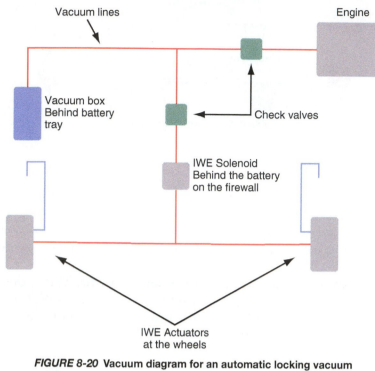

FIGURE 8-20 Vacuum diagram for an automatic locking vacuum hub system.

Vehicles equipped with a **front-axle disconnect** do not require locking hubs. If the front drive shaft spins when the vehicle is placed into 4WD, but the front wheels do not turn, suspect an inoperative actuator. The actuator may be controlled by vacuum, or may be a motor. The vacuum or electrical control system should be diagnosed before removing the actuator. In some cases, the shift fork that moves the sliding collar to connect the two axle sections may be worn or broken.

TRANSFER CASES

Various designs of transfer cases (Figure 8-21) have been and are used. Each has its own features, characteristics, and service procedures. Proper diagnostics and service begin with the proper identification of the transfer case.

Typically, there is an identification tag riveted to the back half of the case that gives the model number and the gear ratio of the transfer case. If the tag is not there, it is possible to identify the unit by the vehicle's VIN or by matching the appearance of the unit to those shown in a service manual.

The many basic designs of transfer cases vary in construction, operation, and appearance. Many part-time transfer cases are aluminum two- or three-piece cases with a chain drive and planetary gear reduction. Many of these units have an integral oil pump (Figure 8-22) driven off the rear output shaft that allows the vehicle to be towed for distances without removing the drive shaft. They are four-mode units with 4WD High and Low, 2WD, and Neutral. The shifter may be floor-mounted or electric with buttons on an overhead console or on the dash. Electric models have an electric shift motor on the transfer

Classroom
Manual
Chapter 8, page 202

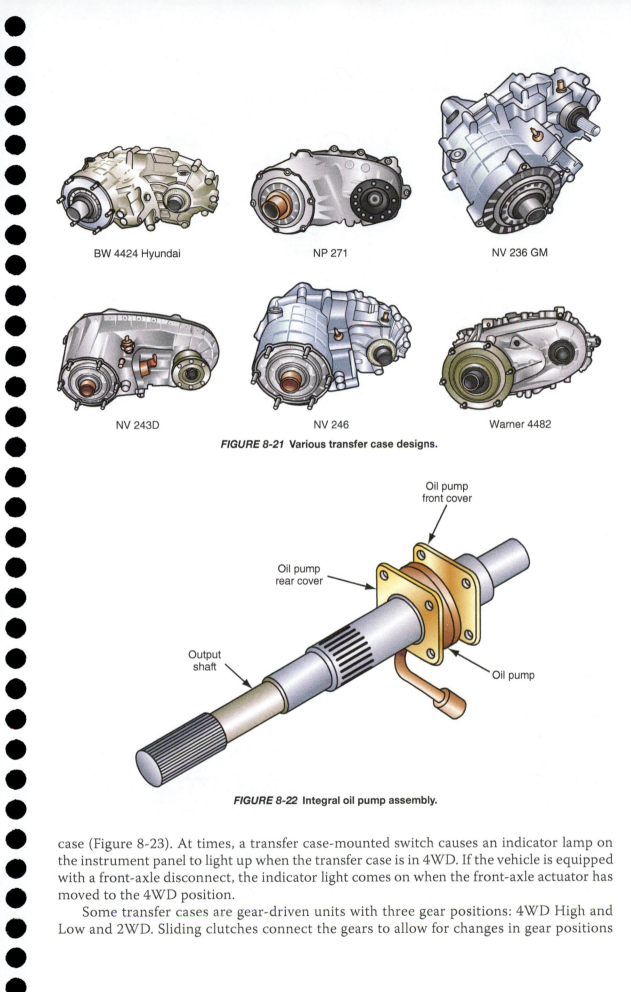

BW 4424 Hyundai

NP 271

NV 236 GM

NV 243D

NV 246

Warner 4482

FIGURE 8-21 Various transfer case designs.

Oil pump front cover

Oil pump rear cover

Output shaft

Oil pump

FIGURE 8-22 Integral oil pump assembly.

case (Figure 8-23). At times, a transfer case-mounted switch causes an indicator lamp on the instrument panel to light up when the transfer case is in 4WD. If the vehicle is equipped with a front-axle disconnect, the indicator light comes on when the front-axle actuator has moved to the 4WD position.

Some transfer cases are gear-driven units with three gear positions: 4WD High and Low and 2WD. Sliding clutches connect the gears to allow for changes in gear positions

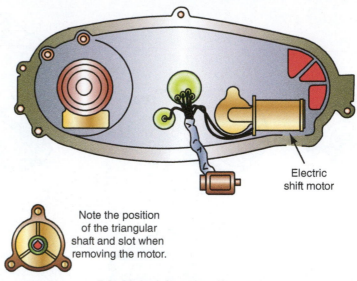

Note the position
of the triangular
shaft and slot when
removing the motor.

FIGURE 8-23 Electric shift motor.

**Classroom
Manual**
Chapter 8, page 202

(Figure 8-24). Most units are chain driven (Figure 8-25) but also have only three modes of operation. These are often equipped with a vacuum motor-operated gear-selection system.

A few units are actually transmissions equipped with transfer cases. The basic design of the powertrain section is the same as that of the five-speed manual transmission, which is used on conventional RWD models. It is a part-time system with a two-speed counter gear system. A chain is used to drive the front wheels, whereas the rear wheels are driven through direct engagement with the transmission.

Determining the exact cause of a problem may require disassembly of the transfer case (Figure 8-26). All parts of the case should be carefully inspected and all damaged parts should be replaced. However, before disassembling the case, make certain that the problem is in the case and not in the transmission, drive shafts, drive axles, or electric control systems. Also, always refer to your service manual prior to taking a transfer case apart.

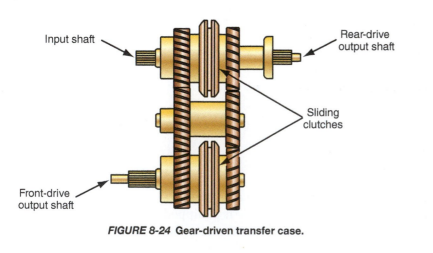

FIGURE 8-24 Gear-driven transfer case.

Input shaft

Rear-drive output shaft

Sliding clutches

Front-drive output shaft

Rear-drive output shaft

Chain

Input shaft

Front-drive output shaft

FIGURE 8-25 Chain-driven transfer case.

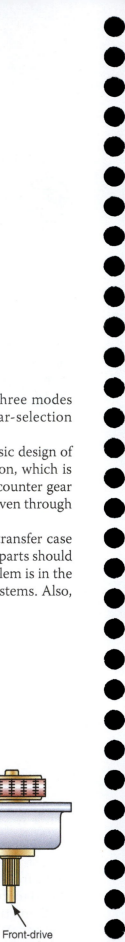

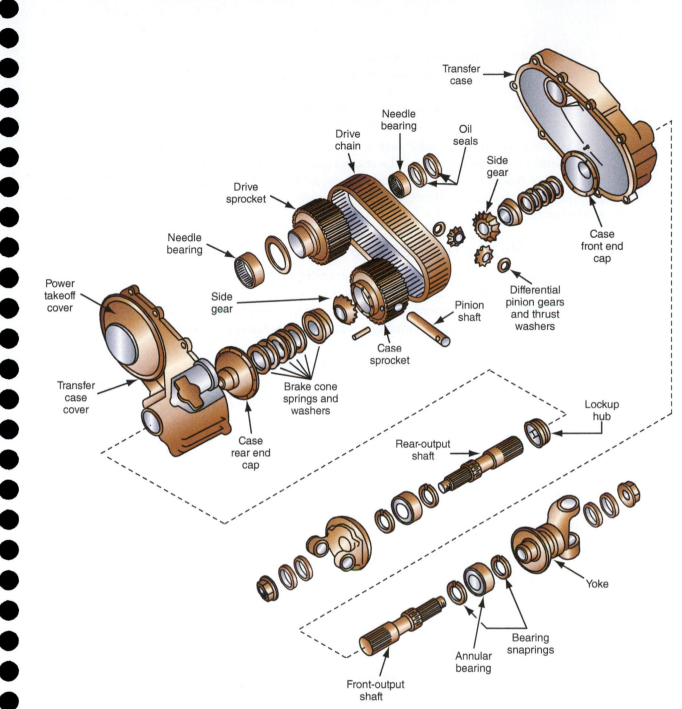

FIGURE 8-26 An example of the parts that must be inspected after transfer case disassembly.

REMOVAL OF A TRANSFER CASE

The procedure for removing a transfer case varies with each application. The specific procedures are outlined in the service manual for the particular model vehicle being worked on. However, certain steps apply to all models. Some general guidelines for removing a typical transfer case follow:

1. Raise the vehicle.
2. Place a drain pan under the transfer case and remove the drain plug.
3. Disconnect and label all the wires leading to electrical switches in and the speed sensor on the transfer case.
4. Disconnect the front-drive shaft from the transfer case.
5. Disconnect and remove the rear-drive shaft from the transfer case.

SPECIAL TOOLS
Drain pan
Clean rags
Transmission jack

6. Disconnect the speedometer cable.
7. Disconnect the vent hose from the case.
8. Support the transfer case on a transmission jack or equivalent.
9. Remove the bolts that attach the transfer case to the transmission or support.
10. Slide the transfer case rearward and remove it from the vehicle.

Reinstallation of the transfer case is the reverse procedure to removing it. When installing a transfer case, always replace the sealing gaskets. Normally, transfer cases are lubricated with ATF or manufacturer-specific oil and should be filled to the bottom of the fill opening. Always refer to the service manual to determine what type of fluid is required and how much to put into the case.

SPECIAL TOOLS
Seal remover
Pry bar
Bearing driver
Snapring pliers

Classroom Manual
Chapter 8, page 208

A **thrust bearing race** is the surface the bearings ride against.

A **drive sprocket** is a toothed pulley that drives a shaft via a toothed drive belt or chain.

A **sprocket carrier** is a device used on some shafts to connect the shaft to the drive sprocket.

Disassembly of a Transfer Case

The exact procedures for disassembling a transfer case will vary with the different models. Photo Sequence 23 shows the procedure for disassembling a Warner 13-50 transfer case. This design transfer case is commonly used in domestic 4WD vehicles. Always refer to your service manual and follow the specific procedures given for the transfer case being serviced.

Guidelines for Transfer Case Disassembly

Thoroughly clean and inspect all parts as the transfer case is being disassembled. Make sure to clean off all old gasket material and sealant that may be present on mating parts, such as the bearing retainer and its cover and the front case and its rear cover.

If the transfer case is chain driven, pay attention to the thick and thin **thrust bearing races** on both sides of the sprocket (Figure 8-27) as you pull the drive chain and driven sprocket out of the case. The thin race is normally located with the sprocket while the thick one is normally attached to the case. Bearing failure will result if the thrust races are installed in the wrong position.

When removing the main shaft assembly, remove the entire assembly as one piece with the side gear, clutch gear, **drive sprocket**, and spline gear, to avoid the loss of the needle bearings from the side gear and **sprocket carrier**. Also index the carrier to the sprocket before separating the two parts. It is easy to install the carrier backward during reassembly. The carrier will fit either way but will prevent engagement of 4WD if it is installed incorrectly.

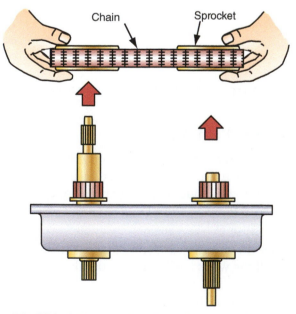

FIGURE 8-27 Remove chain assembly as a single unit.

TYPICAL PROCEDURE FOR DISASSEMBLING A WARNER 13–50 TRANSFER CASE

P23-1 If not previously drained, remove the drain plug and allow the oil to drain, then reinstall the plug. Loosen the flange nuts.

P23-2 Remove two output-shaft yoke nuts, washers, rubber seals, and output yokes from the case.

P23-3 Remove the 4WD indicator switch from the cover.

P23-4 Remove the wires from the electronic shift-harness connector.

P23-5 Remove the speed-sensor retaining-bracket screw, bracket, and sensor.

P23-6 Remove the bolts securing the electric shift motor and remove the motor.

P23-7 Note the location of the triangular shaft in the case and the triangular slot in the electric motor.

P23-8 Loosen and remove the front-case to rear-case retaining bolts.

P23-9 Separate the two halves by prying between the pry bosses on the case.

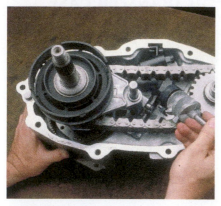

P23-10 Remove the shift rail for the electric motor.

P23-11 Pull the clutch coil off the main shaft.

P23-12 Pull the 2WD/4WD shift fork and lockup assembly off the main shaft.

P23-13 Remove the chain, driven sprocket, and drive sprocket as a unit.

P23-14 Remove the main shaft with the oil pump assembly.

P23-15 Slip the High-Low range shift fork out of the inside track of the shift cam.

P23-16 Remove the High-Low shift collar from the shift fork.

P23-17 Unbolt and remove the planetary gear mounting plate from the case.

P23-18 Pull the planetary gear set out of the mounting plate.

If the transfer case is fitted with a viscous coupling and the coupling was overheated, expect the shift mode sleeve's teeth and corresponding gear splines to be broken or worn. The teeth of the shift mode sleeve engage with the splined gear and sprocket carrier. A worn engine, out-of-adjustment linkage, missing fork pads, or a seized vacuum motor can prevent precise engagement by the sleeve. More sleeve damage can also occur if the driver does not lift off the throttle when shifting into 4WD.

A locking plate is used to hold the planetary gear set in certain shift modes. If the transfer case is equipped with a locking plate, make sure the plate's teeth are pointed. The area behind the plate should be cleaned and the plate's bolts should be resealed with silicone to prevent leaks out of the case.

Check the input shaft for damage (Figure 8-28). Check the area on which the seal rides. This area should not be worn into the metal. A mark on the surface is okay but a groove is not. Also inspect the teeth on the input shaft. Inspect the range and mode sectors—they should have sharp edges. Pay careful attention to the condition of the range and mode forks. Excessive wear often indicates another problem in the transfer case or linkage that caused the sleeve to be forced by the fork (Figure 8-29).

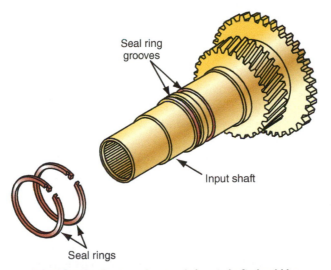

Seal ring grooves

Input shaft

Seal rings

FIGURE 8-28 The transfer case's input shaft should be carefully inspected.

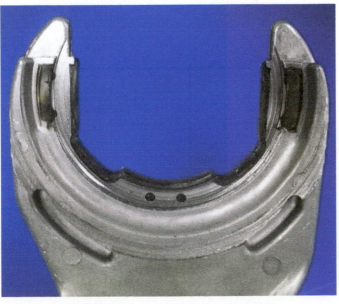

FIGURE 8-29 A worn transfer case shift fork.

ASSEMBLY OF A TRANSFER CASE

Prior to reassembling the transfer case, all parts should be thoroughly inspected, cleaned, and covered with a fresh coat of automatic transmission fluid. Replace all gaskets and seals. Also, to help prevent fluid leaks, coat the bolt threads with a thread sealer.

The exact procedures for assembling a transfer case will vary with the different models. Photo Sequence 24 shows the procedure for assembling a Warner 13-50 transfer case. Always refer to your service manual and follow the specific procedures given for the transfer case being serviced.

Guidelines for Transfer Case Assembly

Never reuse snaprings (Figure 8-30). Always replace them during the assembly of the transfer case. After they have been spread during removal, they lose some of their tension and may be unable to hold the intended components together.

When placing the speedometer gear over the shaft, be sure the locating ball is in its bore in the shaft, then pull the gear over the locating ball.

During reassembly, it is wise to coat the main shaft's thrust bearing with petroleum jelly. Make sure the bearing is centered over the shaft's bore before you lower the main shaft through the sleeves. Also, use petroleum jelly while installing the needle bearings for the output shaft onto the main shaft. The jelly will hold the needle bearings in place while you slide the output shaft over the end of the main shaft.

While reassembling the transfer case, pay attention to the transmission mounting studs. Make sure they are not cross threaded and are not able to contact the ring gear. Often the studs are overtightened or have been tightened too far into the case, allowing the bottoms of the studs to hit the ring gear.

Whenever you are pressing a new bearing into the case, pay close attention to any oil feed holes in the bearing bores. Pressing the bearing too far into the case will block off oil flow and cause the bearings to overheat.

Pay close attention to the recommendations of the manufacturer. Some transfer case housings are made of magnesium and special considerations are taken to prevent **galvanic corrosion**. Bolts are fitted with aluminum washers beneath their heads to prevent contact between the magnesium case and the steel bolts. Special seals with rubber outside diameters are used to prevent contact between the steel housing of the seal and the magnesium case.

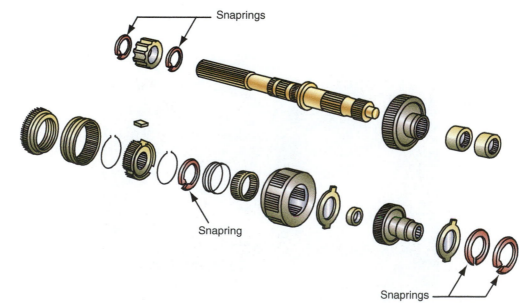

FIGURE 8-30 Snaprings must be expanded to be removed and should be replaced during assembly.

A seized vacuum motor is often caused by a buildup of ice in the motor. If moisture is present in the vacuum system, it will freeze when it is cold.

SPECIAL TOOLS

Snapring pliers
Seal installer
Bearing driver
Silicone sealer
Torque wrench

Petroleum jelly is often recommended because it is sticky and it dissolves quickly once the components get hot.

Galvanic corrosion is a type of corrosion that occurs when two dissimilar metals, such as magnesium and steel, are in contact with each other.

TYPICAL PROCEDURE FOR ASSEMBLING A WARNER 13-50 TRANSFER CASE

P24-1 Install the input shaft and front-output shaft bearings into the case.

P24-2 Apply a thin bead of sealer around the ring-gear housing.

P24-3 Install the input shaft with the planetary gear set and tighten the retaining bolts to specifications.

P24-4 Install the High-Low shift collar into the shift fork.

P24-5 Install the High-Low shift assembly into the case.

P24-6 Install the main shaft with the oil pump assembly into the case.

P24-7 Install the drive and driven sprockets and chain into position in the case.

P24-8 Install the shift rails.

P24-9 Install the 2WD/4WD shift fork and lock-up assembly onto the main shaft.

P24-10 Install the clutch coil onto the main shaft.

P24-11 Clean the mating surface of the case.

P24-12 Position the shafts and tighten the case halves together. Tighten the attaching bolts to specifications.

P24-13 Apply a thin bead of sealer to the mating surface of the electric shift motor.

P24-14 Align the triangular shaft with the motor's triangular slot.

P24-15 Install the motor over the shaft. Wiggle the motor to ensure that it is fully seated on the shaft.

P24-16 Tighten the motor's retaining bolts to specifications.

P24-17 Reinstall the wires into the connector and connect all the electric sensors.

P24-18 Install the companion flanges' seal, washer, and nut. Then tighten the nut to specifications.

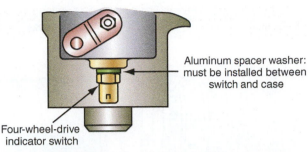

FIGURE 8-31 Location of aluminum washer beneath indicator light switch.

An aluminum washer also may be located beneath the indicator light switch (Figure 8-31). This washer serves as a sealing device. It is a *squash washer* that sets the distance and deforms enough under torque to form a seal.

AXLE HOUSINGS AND DIFFERENTIALS

Many complaints of noise and vibration are caused by the drive axles. These problems are serviced in the same way as with 2WD drive axles. Most axle housings used at the front and rear of a vehicle are integral carrier types. The carrier housing may be made of aluminum or magnesium, which requires special service procedures to prevent damage to the housing.

Normal adjustments of the gears and bearings in the front, rear, or center differentials (Figure 8-32) are the same as those on a 2WD vehicle. Typically, adjustment of the side carrier bearings is accomplished by means of shims placed between the bearings and the differential case. The side bearings are preloaded by the clamping action of the housing. A spreader must be used on the housing to remove or install the differential assembly. Pinion depth and preload are controlled by shims.

Fluid leaks are common problems that can be prevented by using only the gaskets and/or sealing materials recommended by the manufacturer. Never use a sealant in place of a gasket, unless the manufacturer says to. Always use the proper tools when installing a seal. A cocked seal or one that is installed in a damaged housing will leak. When installing a seal, make sure the surface that the seal lip rides on is smooth. The slightest burr can tear the seal and cause it to leak. Always check for a plugged vent when axle leaks are found. 4WD vehicles are often driven in an environment causing mud to block an axle vent.

Classroom Manual
Chapter 8, Page 199

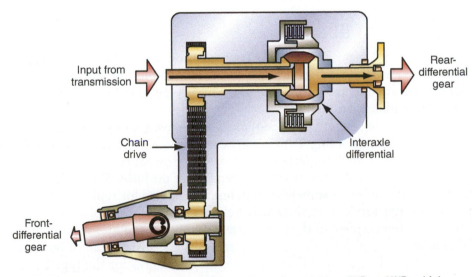

FIGURE 8-32 A common location for the center differential in a 4WD or AWD vehicle.

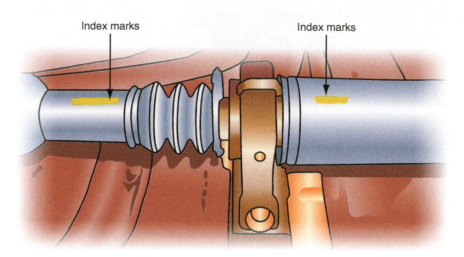

FIGURE 8-33 Mark the drive shaft before disassembling it to ensure proper phasing of the U-joints.

Drive Shafts

The drive shafts on a 4WD vehicle are serviced in the same way as other drive shafts. They rely on U- and slip joints to transfer engine torque to a set of drive wheels and allow for changing operating length and angles. On some 4WD vehicles, you may find a slip joint in the middle of the shaft assembly that will be covered by a rubber boot that keeps debris out when allowing the shaft to change length. The boot is similar to those used on CV joints and should be replaced when it is damaged.

To replace the boot, the shaft is removed from the vehicle. The shaft on both sides of the boot should be marked to ensure proper U-joint phasing during reassembly (Figure 8-33). Using a dull screwdriver, pry the loose ends of the boot clamps away from the boot and remove them. Pull the two sections of the shaft apart, then remove the boot. Carefully check the splines of the slip joint before installing the new boot and clamps. Also make sure the shaft is properly phased before assembly.

CV boots on 4WD front-axle shafts are often made of a hard, tough material called Hytrel. The clamps that hold these boots on the shaft must be crimped with special pliers, a breaker bar, and a torque wrench capable of 100 ft/lbs.

FRONT AXLES AND HUBS

Service procedures for many components (transmission, drive shafts, and differentials) of a 4WD powertrain are the same as the identical units used in 2WD systems. However, the removal of these components is often different. This is especially true regarding the removal of the front differential and front axles.

The axles and locking hubs (if equipped) must be removed before the front differential can be removed from the axle housing. Removing these components requires special procedures, and each type of 4WD system requires its own specific procedure.

Removal of the axle shafts begins with the removal of the hubs. Normally, the procedures for replacing manual hubs are somewhat different than those for replacing automatic hubs. Locking hubs are not serviceable; therefore, hubs are serviced by replacing them. Also, the procedure for the replacement of the hubs varies with the type of hub and the manufacturer. The following are general procedures for the removal and installation of both manual and automatic locking hubs. Always follow the specific procedures for the type of hub you are servicing.

SPECIAL TOOLS

Snapring pliers
Slide hammer
Clean multipurpose grease
Torque wrench

Classroom Manual
Chapter 8, Page 211

To remove a manual locking hub:

1. Set the control knob to the FREE or unlocked position (Figure 8-34).
2. Remove the cover bolts (Figure 8-35) and the outer snapring.
3. If hubs are equipped with shims, remove them.
4. Remove the drive flange or gear from the axle.
5. Remove the bolts that attach the hub body to the hub.
6. Remove the hub assembly from the axle shaft.

To install a manual locking hub (Figure 8-36):

1. Separate the base and handle units of the hub lock assembly.
2. Apply a light coat of grease to the axle shaft splines and the hub lock base. Also apply a light coating of lubricant to the O-ring.
3. Make sure the gasket surface is smooth and clean, then place a new hub gasket onto the hub.
4. Set the control handle to the unlocked position and install the hub assembly.

SERVICE TIP:
After the hub assembly has been removed, inspect the splines of the axle shaft for nicks and burrs.

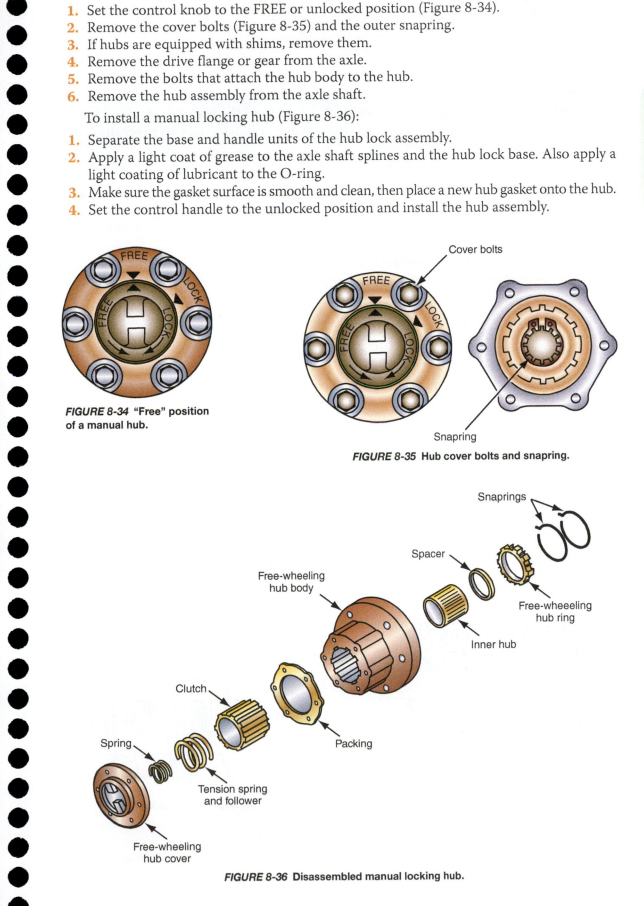

FIGURE 8-34 "Free" position of a manual hub.

FIGURE 8-35 Hub cover bolts and snapring.

FIGURE 8-36 Disassembled manual locking hub.

5. Install the snapring onto the axle shaft.
6. Tighten the attaching bolts to the specified torque.
7. Install the hub cover with a new gasket onto the hub.
8. Tighten the remaining attaching bolts to specifications.
9. Check the control handle for ease of operation.

> **CUSTOMER CARE:** Inform the customer that to extend the life of the locking hubs, they should be disassembled, cleaned, inspected, and lubricated whenever the hubs have been exposed to high water.

Classroom Manual
Chapter 8, page 211

There are basically two designs of automatic locking hubs: internally and externally retained types (Figure 8-37). The procedure for servicing automatic hubs depends on how they are retained. The following procedure is typical.

To remove an automatic locking hub:

1. Remove the bolts from the outer cap assembly and pull the cap from the body assembly.
2. Remove the bolt, lock washer, and axle shaft stop from one end of the axle shaft.
3. Remove the body assembly lock ring from the groove in the wheel hub.
4. Slide the body assembly from the wheel hub.
5. Loosen the set screws (Figure 8-38) in the spindle locknut until their heads are flush with the face of the locknut.
6. Remove the spindle locknut.

To reinstall an automatic locking hub:

1. Adjust the wheel bearings, then install the spindle locknut and torque to specifications.
2. Tighten the locking set screws firmly.
3. Grease the hub inner splines with multipurpose grease. The hubs should not be packed with grease.
4. Slide the body assembly into the wheel hub. Push firmly on the body until it seats in the hub.
5. Install the lock ring in the groove of the wheel hub.
6. Place the lockwasher and axle shaft stop onto the axle bolt. Install the bolt and torque it to specifications.

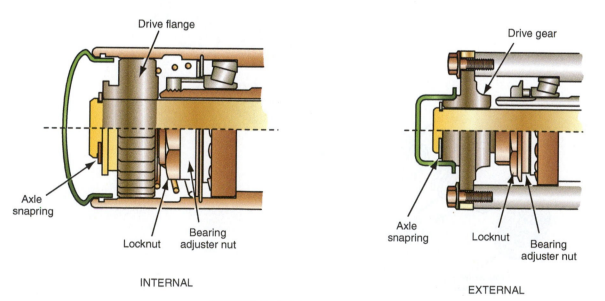

INTERNAL   EXTERNAL

FIGURE 8-37 Different hub designs.

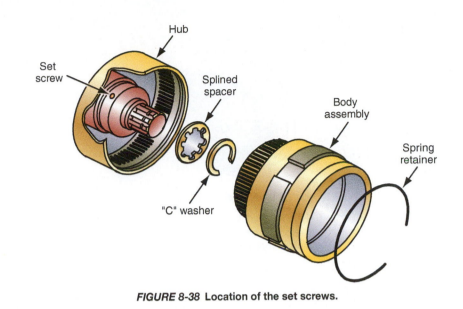

FIGURE 8-38 Location of the set screws.

7. Apply a small amount of lube to the seal on the cap assembly. The cap assembly should not be greased.
8. Install the cap assembly over the body assembly and into the wheel hub.
9. Install the attaching bolts and torque them to specifications.
10. Firmly turn the control from stop to stop to check the assembly. Then, set both hub controls in the same position, AUTO or LOCK.

Drive Axles

The following procedure is given as a guideline for the removal of the axle shafts from a front-drive axle housing (Figure 8-39).

1. Raise the vehicle and remove the wheels.
2. Remove the manual or locking hub assemblies.
3. Remove the brake caliper with the brake line attached and secure the assembly to the frame with wire.
4. Remove hub and rotor.
5. Unbolt the spindle from the steering knuckle.

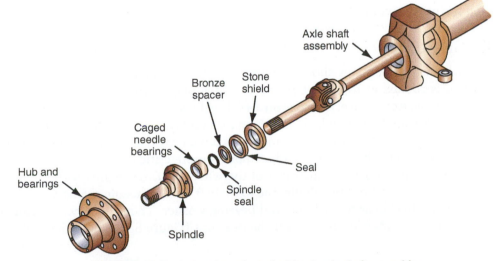

FIGURE 8-39 Exploded view of a typical front-axle shaft assembly.

> ⚠️ **CAUTION:**
> Never rotate the axle shaft when it remains engaged to the differential side gears. This could cause the pinion gears to turn to the opening in the case and drop out.

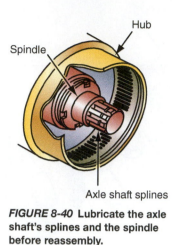

FIGURE 8-40 Lubricate the axle shaft's splines and the spindle before reassembly.

6. Remove the splash pan.
7. On the left side, pull the axle shaft and joint assembly out through the steering knuckle.
8. If the axle is an IFS type, remove the clamps that hold the right-side rubber boot in place and slide the boot onto the stub shaft. If the axle is not an IFS type, remove the boot from the left side. Pull the right axle shaft and joint assembly from the splined stub shaft.
9. Remove the seal and needle bearing from the spindle end of the axle shaft.

To reinstall a drive axle shaft into a front-axle housing:

1. Thoroughly clean the spindle bearing bores.
2. Drive the spindle bearing into the spindle.
3. Drive a new grease seal into the bore.
4. Coat the lip of the seal and the axle shaft's splines (Figure 8-40) with multipurpose grease.
5. Slide the left axle shaft and joint assembly through the steering knuckle. Make sure the shaft splines engage into the differential carrier.
6. Install a new axle shaft seal on the right side and install the rubber boot with new clamps to the stub shaft.
7. Install the right axle and joint assembly.
8. Align axle splines with the splines of the stub shaft and slide the rubber boot over this junction. Then tighten the clamps.
9. Install the splash shield and fasten the spindle to the steering knuckle.
10. Install the hub and adjust the hub bearings.
11. Install the brake rotors, calipers, and wheels.

WHEEL BEARINGS

SERVICE TIP:
Most wheel bearing locknuts on 4WD vehicles require the use of a four- or six-pronged spanner wrench. Typically, heavy-duty vehicles require the six-pronged wrench.

Unit-style front wheel bearings should be inspected for any play. Movement of the wheel when the tire is rocked at top and bottom indicates play in the bearing and it must be replaced. Adjustable front-wheel bearings on a 4WD vehicle should be disassembled, cleaned, inspected, and lubricated on a regular basis. Upon reassembly, it is important that the bearings be properly adjusted. The procedure for doing this is similar to other wheel bearings. However, because of the load on the bearings, the adjustment is more critical.

A typical procedure for the adjustment of the wheel bearings begins with the removal of the snapring at the end of the spindle, the axle shaft spacer, needle thrust washer, bearing spacer, outer wheel bearing locknut, and bearing washer. Then the inner bearing locknut is loosened and tightened fully to seat the bearings (Figure 8-41). Spin the brake rotor

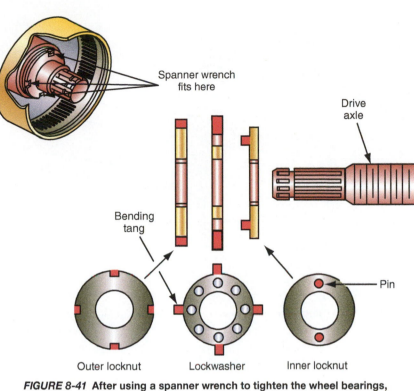

Spanner wrench fits here

Drive axle

Bending tang

Pin

Outer locknut   Lockwasher   Inner locknut

FIGURE 8-41 After using a spanner wrench to tighten the wheel bearings, secure the adjustment by securing the locknut or lockwasher.

on the spindle, then loosen the inner bearing locknut approximately one-quarter turn. Install the outer bearing locknut and tighten it to the specified amount, usually 70–90 ft.-lb. Assemble the remaining parts of the hub assembly, making sure the snapring is fully seated in its groove on the axle.

TYPICAL 4WD MODIFICATIONS

The original equipment manufacturers and the aftermarket industry provide an enormous variety of performance components that can enhance the off-road capabilities of many 4WD vehicles. The proper installation of these components is the responsibility of the technician. Performance enhancements can be roughly categorized into three categories: tires and wheels, suspension modifications, and drivetrain components. Customers often replace the standard-sized wheels and tires for improved traction. Also, they will often raise the body to gain ground clearance. As a result of these modifications, the vehicle's handling may be difficult and the ride may be rough.

Installing oversized tires or **lift kits** alters the vehicle's relationship of sprung and unsprung weight. The addition of unsprung weight changes the way the suspension rides and responds. Oversized tires are very heavy and four of them will add a considerable amount of unsprung weight. This added weight can slow down the suspension's response to bumps while increasing the harshness felt in the chassis and steering wheel. To compensate for this, stiffer shock absorbers and springs should be installed. However, these will make the ride rough. Larger tires also will change the overall final drive ratio. As the diameter of a tire increases, the numerical overall drive ratio becomes lower and the torque multiplication caused by the final drive gears decreases. This can affect driveability and speedometer readings. Large tires can have an adverse effect on steering, especially steering effort. Also stability can be a problem if the tires have an aggressive tread pattern. These tires tend to wander and vibrate on dry

Lift kits can either raise the axle housings or the body of the vehicle. If a kit is to be installed, a body lift is preferred because it does not alter the angle of the driveline.

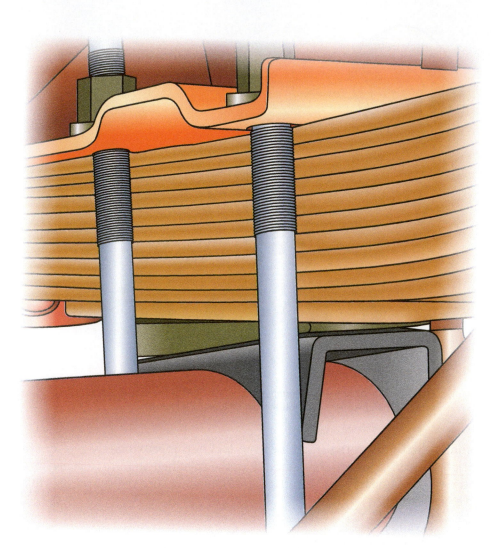

FIGURE 8-42 Installation of a tapered spring shim to change pinion angle.

Classroom Manual
Chapter 8, Page 216

surfaces. Remember that although large tires can cause a number of issues, tire size has no effect on drive shaft angles.

Some lift kit modifications require adjustments to compensate for increased drive shaft working angles. **Tapered shims** at the rear spring pads can be used to change the pinion nose angle (Figure 8-42). The transmission cross member can be dropped using spacers to help reduce severe drive shaft angles (Figure 8-43). Other modifications include the use of a slip yoke eliminator kit that requires a different transfer case output shaft and a bolt-on yoke (Figure 8-44).

Extreme drive shaft angles may require replacing a U-joint drive shaft with one that is fitted with CV joints or double Cardan joints (Figure 8-45). The customer should be aware that U-joints can only operate within designed angles or shaft speed fluctuations can cause vibration and U-joint damage.

Larger tires also will change the overall final drive ratio. As the diameter of a tire increases, the numerical overall drive ratio becomes lower and the torque multiplication caused by the final drive gears decreases. This can affect driveability and speedometer readings.

Large tires can have an adverse effect on steering, especially steering effort. Also stability can be a problem if the tires have an aggressive tread pattern. These tires tend to wander and vibrate on dry surfaces.

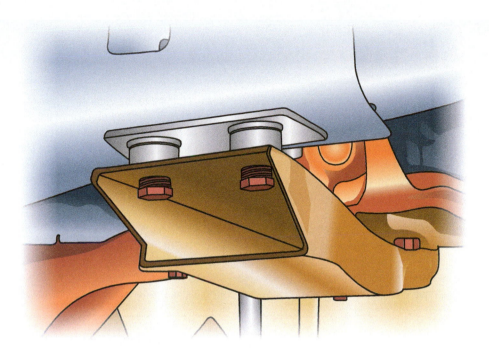

FIGURE 8-43 Spacers can be used to lower the cross member.

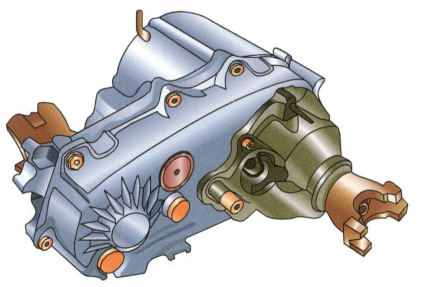

FIGURE 8-44 A transfer case with a slip yoke eliminator kit.

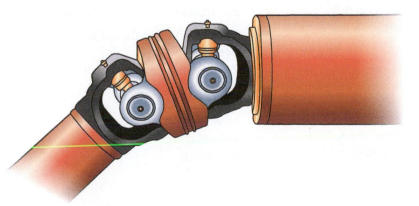

FIGURE 8-45 A replacement drive shaft with a double Cardan joint.

CASE STUDY

A 4WD pickup is brought into the shop. There is a severe vibration through the truck when it is traveling at road speeds. The new wheels and tires that were recently installed on the truck are drastically oversized. Lift kits were installed by the owner to provide the necessary tire clearance. The owner would like the alignment checked and the wheels balanced.

The technician carefully balances and aligns the wheels. She takes the truck for a road test to verify the repair. The truck still has a harsh vibration that seems to get worse as speed increases. A bent wheel or bad wheel bearing or hub is suspected.

After returning to the shop, she measures the runout and checks the play of each wheel. Finding nothing wrong, she carefully checks the roundness of the tires. When doing this on the rear tires, she discovers the cause of the problem. When the lift kit was installed, the operating angle of the rear U-joint was affected. As the drive shaft turns, it hesitates slightly. A binding of the U-joint causes this hesitation.

The technician informs the owner of the cause and suggests that the lift kits be removed or the drive shaft operating angle be corrected. The owner elects to have the drive axles repositioned to correct the operating angle. This took care of the problem.

ASE-STYLE REVIEW QUESTIONS

1. The front wheels will not drive when a vehicle is placed in to 4WD.

   Technician A says that the position switch on the front axle disconnect could be faulty.

   Technician B says that the problem could be a vacuum leak at the front axle disconnect actuator.

   Who is correct?

   A. A only   C. Both A and B

   B. B only   D. Neither A nor B

2. When discussing 4WD front hubs,

   Technician A says that manual locking hubs may make a ratcheting sound when water or dirt have entered the hub.

   Technician B says that a ratcheting sound from an automatic locking hub may indicate that the hub on the opposite side of the axle is not disengaging.

   Who is correct?

   A. A only   C. Both A and B

   B. B only   D. Neither A nor B

3. When discussing noises coming from the transfer case, *Technician A* says that whining in low range can be expected in many transfer cases.

   Technician B says that incorrect tire inflation is a typical cause of abnormal noises.

   Who is correct?

   A. A only   C. Both A and B

   B. B only   D. Neither A nor B

4. When discussing the cause of a damaged shift mode sleeve,

   Technician A says that high-speed shifts into 4WD can cause the problem.

   Technician B says that a misadjusted shift linkage can cause this problem.

   Who is correct?

   A. A only   C. Both A and B

   B. B only   D. Neither A nor B

5. When discussing service on 4WD vehicles,

   Technician A says when a 4WD vehicle is operated in water or mud that gets over the top of the axle housings, the hubs, wheel bearings, and Universal joints should be immediately checked for fluid contamination.

   Technician B says the front-wheel hubs must be locked before a 4WD vehicle is towed on its wheels.

   Who is correct?

   A. A only

   B. B only

   C. Both A and B

   D. Neither A nor B

6. When diagnosing the cause of noise from a transfer case,

   Technician A says that this may be caused by incorrect tire size.

   Technician B says that a faulty bearing or gear may be the cause.

   Who is correct?

   A. A only

   B. B only

   C. Both A and B

   D. Neither A nor B

7. When discussing 4WD front axle repair,

   Technician A says that to remove the axles from a front-drive axle equipped with locking hubs, the hubs must be removed first.

   Technician B says that the hubs must be disassembled in order to free the axles from the housing.

   Who is correct?

   A. A only

   B. B only

   C. Both A and B

   D. Neither A nor B

8. When discussing transfer case diagnosis,

   Technician A says that if a transfer case jumps out of gear, suspect an improperly adjusted shift linkage.

   Technician B says that when a transfer case will not shift from 4WD to 2WD, the transfer case may be spline-locked.

   Who is correct?

   A. A only

   B. B only

   C. Both A and B

   D. Neither A nor B

9. When diagnosing the cause of the transfer case jumping out of gear,

   Technician A says that a possible cause is a worn shift fork.

   Technician B says that a possible cause is an improperly adjusted shift linkage.

   Who is correct?

   A. A only

   B. B only

   C. Both A and B

   D. Neither A nor B

10. When discussing why a recently rebuilt transfer case will not shift into 4WD,

   Technician A says that the drive chain could be stretched.

   Technician B says that the transfer case shift motor might be seized. Who is correct?

   A. A only

   B. B only

   C. Both A and B

   D. Neither A nor B

ASE CHALLENGE QUESTIONS

1. When discussing a 4WD vehicle with a vibration problem that is more noticeable when cornering,

   Technician A says that the rear drive shaft U-joints may be worn.

   Technician B says that the outboard front-axle joints may be worn.

   Who is correct?

   A. A only

   B. B only

   C. Both A and B

   D. Neither A nor B

2. All of the following statements regarding 4WD front-drive axles and joints are true *except*:

   A. The inner tripod joint is held on the axle shaft with a snapring.

   B. A special swaging tool may be required to tighten the outer Hytrel CV boot clamps.

   C. The new joint is coated with the amount and type of grease required for the joint.

   D. A worn CV joint may cause a clicking noise when cornering.

3. When discussing transfer case removal,

   Technician A says that on some vehicles the transmission and transfer case are removed as one assembly.

   Technician B says that transfer cases may be removed as separate assemblies to provide for easier handling of these heavy components. Who is correct?

   A. A only

   B. B only

   C. Both A and B

   D. Neither A nor B

4. When discussing transfer case vents,

   Technician A says that a plugged transfer case vent may cause seal leakage.

   Technician B says that a remote transfer case vent helps prevent moisture from entering the transfer case when driving through water.

   Who is correct?

   A. A only

   B. B only

   C. Both A and B

   D. Neither A nor B

5. On 4WD vehicles, all of the following may cause a vibration when the throttle position is changed *except*:

   A. Worn U-joints

   B. Worn front-axle drive joints

   C. Incorrect drive shaft angles

   D. Worn drive shaft slip joint splines

CHECK FLUID IN A TRANSFER CASE

Name _____ Date _____

Upon completion of this job sheet, you will be able to inspect a transfer case for leaks and properly check its fluid level.

ASE NATEF MAST Task Correlation

Four Wheel Drive/All-Wheel Drive Component Diagnosis and Repair

Task #3, Check for leaks at drive assembly seals; check vents; check lube level.

Tools and Materials

Service information

Protective Clothing

Goggles or safety glasses with side shields

Describe the Vehicle Being Worked On

Year _____ Make _____ VIN _____

Model _____

Describe the type of system and the model of the transfer case:

_____

Procedure

1. Raise and support the vehicle. Then carefully inspect the area around the transfer case for signs of oil leakage. Summarize your findings.

   _____

   _____

2. Check the connections for the drive shafts to the transfer case. The presence of oil may indicate bad seals. Summarize what you found.

   _____

   _____

3. Check the mounting point for the transfer case at the transmission, looking for signs of fluid that may indicate a gasket problem at this connection. Summarize what you found.

   _____

   _____

4. Locate the fill plug on the transfer case. (Refer to the service information for the location of the plug.) Then, remove the plug. Where is the plug?

   _____

   _____

5. Using your little finger, feel in the hole to determine if you can touch the fluid. Summarize your findings.

_____

_____

6. If you cannot touch the fluid, refer to the service manual for fluid type. Fill the transfer case. The recommended fluid is _____

7. If the transfer case is low on fluid, visually inspect it to locate the leaks. What did you find?

_____

_____

8. If you can reach the fluid with your finger, check the smell, color, and texture of the fluid. Summarize your findings.

_____

_____

9. If the fluid is contaminated, determine what is contaminating it. Then correct that problem and drain the fluid from the transfer case and refill it with clean fluid. Summarize your findings.

_____

_____

Instructor's Response: _____

_____

_____

Name _____ **Date** _____

ROAD-CHECK TRANSFER CASE

Upon completion of this job sheet, you should be able to demonstrate the ability to road-test a transfer case for correct operation.

ASE NATEF MAST Task Correlation

Four Wheel Drive/All-Wheel Drive Component Diagnosis and Repair

Task #5, Diagnose noise, vibration, and unusual steering concerns; determine necessary action.

Equipment Needed

A vehicle with a transfer case that the owner has authorized to be road-tested

Describe the Vehicle Being Worked On:

Year _____ Make _____ VIN _____

Model _____

Procedure

Task Completed

_____

Road-test the vehicle; attempt to drive the vehicle in all the different drives.

1. Does the transfer case make a noise in:
   a. Low drive in rear-wheel drive?   ☐ Yes   ☐ No
   b. High drive in rear-wheel drive?   ☐ Yes   ☐ No
   c. Low drive in four-wheel drive?   ☐ Yes   ☐ No
   d. High drive in four-wheel drive?   ☐ Yes   ☐ No

2. Does it make a noise as your turn corners with it in four-wheel drive?   ☐ Yes   ☐ No

3. Does it make a noise as your drive straight down the road in high gear?   ☐ Yes   ☐ No

4. Does it make a noise when the transfer case is in neutral?   ☐ Yes   ☐ No

Instructor's Response: _____

_____

_____

Name _____ Date _____

CHECK THE TRANSFER CASE FOR OIL LEAKS

Upon completion of this job sheet, you should be able to demonstrate the ability to check a transfer case for oil leaks.

ASE NATEF MAST Task Correlation

Four Wheel Drive/All-Wheel Drive Component Diagnosis and Repair

Task #3, Check for leaks at drive assembly seals: check vents; check lube level.

Tools and Materials

Lift or jack

Jack stands

A vehicle with a transfer case that can be put up on a hoist or on jack stands to be inspected for oil leaks

Describe the Vehicle Being Worked On

Year _____ Make _____ VIN _____

Model _____

Procedure

_____
Task Completed

1. Raise the vehicle up on a hoist or jack it up and put safety stands under it.

2. Check the transfer case for oil leaks.

3. Is it leaking at the forward drive shaft?   ☐ Yes   ☐ No

4. Is it leaking at the rear drive shaft?   ☐ Yes   ☐ No

5. Is it leaking where it fastens to the transmission?   ☐ Yes   ☐ No

6. Does it look as if it is leaking in any other place, such as a breather or fill hole?
   ☐ Yes   ☐ No

Instructor's Response: _____

_____

_____

Name _____ **Date** _____

IDENTIFYING THE TYPE OF TRANSFER CASE

Upon completion of this job sheet, you should be able to demonstrate the ability to identify the type of transfer case.

ASE NATEF MAST Task Correlation

Four-wheel Drive/All-Wheel Drive Component Diagnosis and Repair

Task # 1 Inspect, adjust, and repair shifting controls (mechanical, electrical, and vacuum), bushings, mounts, levers, and brackets.

Tools and Materials

Lift or jack

Jack stands

A vehicle with a transfer case that can be put on a hoist or on jack stands

Describe the Vehicle Being Worked On

Year _____ Make _____ VIN _____

Model _____

Procedure

_____
Task Completed

1. Hoist the vehicle, or jack it up and put safety stands under it.

2. Inspect the transfer case.

3. If it has a tag on it, write down the tag number.

4. If it has a stamp on the case, record what it says.

5. Record any other identifying marks.

6. Use the numbers and marks to identify it in a book that gives the information on transfer cases.

7. Determine what kind it is—gear type or chain type.

Instructor's Response: _____

_____

_____

REMOVING AND INSTALLING A TRANSFER CASE

Upon completion of this job sheet, you should be able to remove and reinstall a transfer case.

ASE NATEF MAST Task Correlation

Four Wheel Drive/All-Wheel Drive Component Diagnosis and Repair

Task #7, Disassemble, service, and reassemble transfer case and components.

Tools and Materials

Drain pan

Clean rags

Transmission jack

Hand tools

Service manual

Protective Clothing

Goggles or safety glasses with side shields

Describe the Vehicle Being Worked On

Year _____ Make _____ VIN _____

Model _____

Procedure

Task Completed
_____

NOTE: These are intended to be basic guidelines. Refer to the service manual and modify the following sequence according to the procedures outlined by the manufacturer.

1. Raise the vehicle on a lift.

2. Place a drain pan under the transfer case and remove the drain plug.

3. Disconnect and label all wires connected to the transfer case.

4. Disconnect and remove the front drive shaft from the transfer case.

5. Disconnect and remove the rear drive shaft from the transfer case.

6. Disconnect the speedometer cable if it is attached to the transfer case.

7. Disconnect the vent hose from the case.

8. Reinsert the drain plug and move the drain pan away from the vehicle.

9. Support the transfer case with the transmission jack.

10. Remove the bolts that attach the transfer case to the transmission or to a support.

11. Slide the transfer case rearward and remove it from the vehicle.

12. Before reinstalling the transfer case, make sure the mounting surface on the transmission is clean and all old gasket material is removed.

13. Look up all installation torque specs and record them here.

_____

_____

14. With the transmission jack, raise the transfer case into position.

15. Install a new gasket at the mounting surface and slide the transfer case forward until it is fully seated on the transmission or support.

16. Install and tighten the bolts that attach the transfer case to the transmission or to the support.

17. Move the transmission jack away from the vehicle.

18. Connect the vent hose to the case.

19. Connect the speedometer cable to the transfer case.

20. Install and connect the rear drive shaft to the transfer case.

21. Install and connect the front drive shaft to the transfer case.

22. Connect all electrical terminals and wires according to the labels put on during removal of the transfer case.

23. Lower the vehicle.

24. Refill the transfer case with the correct fluid and amount. What type of fluid does this transfer case use?

_____

_____

Instructor's Response: _____

_____

_____

Name _____ **Date** _____

INSPECTING FRONT-WHEEL BEARINGS AND LOCKING HUBS

Upon completion of this job sheet, you should be able to inspect front-wheel bearings and locking hubs on a 4WD vehicle.

ASE NATEF MAST Task Correlation

Four Wheel Drive/All-Wheel Drive Component Diagnosis and Repair

Task #2, Inspect front wheel bearings and locking hubs; perform necessary action(s).

Tools and Materials

Lift

Service manual

Protective clothing

Goggles or safety glasses with side shields

Describe the Vehicle Being Worked On

Year _____ Make _____ VIN _____

Model _____

Procedure

Task Completed _____

1. Raise the vehicle to a height that allows you to work comfortably around the vehicle's wheels.

2. If the vehicle has locking hubs, rotate a front wheel slightly and move the hub selector to the "lock" position. Did the hub engage with a click?   ☐ Yes   ☐ No

3. Does it feel like the axle is now rotating with the wheel?   ☐ Yes   ☐ No

4. Do the same thing to the wheel and hub on the other side of the vehicle.

5. Based on these checks, what are your conclusions about the condition of the hubs?

   _____

   _____

6. Grasp the front and rear of a front tire and push it in and out. Do you feel any end play?

   _____

   _____

7. Do the same to other front tire. Did you feel any end play?

   _____

   _____

8. Based on these two checks, do you think the wheel bearings need to be adjusted? Why or why not?

_____

_____

9. If adjustment is needed, remove the tire and wheel assembly.

10. Remove the snapring at the end of the axle's spindle.

11. Remove the axle shaft spacers, thrust washers, and outer wheel bearing lock-nut. Keep the spacers and washers in the order they were so you can assemble the unit correctly.

12. Loosen the inner bearing locknut and then fully tighten it to seat the bearings.

13. Spin the brake rotor on its spindle.

14. Loosen the inner bearing locknut approximately one-quarter turn or the amount specified in the service manual.

15. Install the axle shaft spacers and washers, then install the outer bearing locknut.

16. Tighten the outer locknut to the specified torque. The torque spec is: _____

17. Install a new snapring into the end of the axle's spindle.

18. Repeat the wheel bearing adjustment procedure on the other front wheel.

Instructor's Response: _____

_____

_____

Name _____ Date _____

TIRE SIZE CHANGES

Upon completion of this job sheet, you will be able to understand the effects a change of tire and/or wheel size has on the operation of a vehicle. You will also be able to calculate the difference that a change in size may cause.

ASE NATEF MAST Task Correlation

Four Wheel Drive/All-Wheel Drive Component Diagnosis and Repair

Task #4, Identify concerns related to tire circumference and/or final drive ratios.

Tools and Materials

Service information

Protective Clothing

Goggles or safety glasses with side shields

Describe the Vehicle Being Worked On

Year _____ Make _____ VIN _____

Model _____

Procedure

1. Carefully check the size of the tires on the vehicle. Are they the same size? If not, what and where is the difference?

   _____

   _____

2. When is the only time that it is recommended that just one tire be replaced on a vehicle? Where is that tire normally placed on the vehicle?

   _____

   _____

3. When deciding if to replace just two tires, what factors need to be considered and where should those tires be placed on the vehicle?

   _____

   _____

4. When changing the size of the tires to improve handling or fuel economy, what are the main things to consider when determining an acceptable tire size?

   _____

   _____

5. Most tire-width changes affect the overall diameter of the tire. On passenger vehicles, a 3 percent or less change in tire diameter is acceptable. What can be affected by a greater change than that?

_____

_____

6. Calculate the section height of the following tire sizes by multiplying the tire's aspect ratio by its sectional width. Then calculate the overall diameter of the tires by multiplying the sectional height by two. (This is called the combined sectional height because there are two.) Then add the diameter of the wheel.

TIRE SIZE	SECTION WIDTH	RIM WIDTH	SECTION HEIGHT	OVERALL DIAMETER
P215/70R14	8.7 inches	6.5 inches		
P215/70R15	8.7 inches	6.5 inches		
P265/50R14	10.9 inches	8.5 inches		
P275/50R15	11.2 inches	8.5 inches		
P265/50R16	10.9 inches	8.5 inches		

7. The circumference of a tire is the actual length around the tire. In other words, this is the distance the tire will travel as it completes one full revolution. Calculate the circumference of these tires by using the formula: pi (π) times the diameter, use a value of 3.14 for.

TIRE SIZE	OVERALL DIAMETER	CIRCUMFERENCE
P215/70R14		
P215/70R15		
P265/50R14		
P275/50R15		
P265/50R16		

8. Tread wear has an effect on the circumference of a tire. If a P275/50R15 tire has a tread depth of 11/32 inch when it is new, how much will the circumference decrease when the tire's tread wears down to 3/32 inch?

_____

_____

9. Since the circumference of a tire determines the distance it will travel in one revolution, the number of times the tire will rotate in 1 mile can be calculated. How will you calculate that?

_____

_____

10. Calculate the revolutions per mile for the following tires:

Approximate revolutions per mile for a P215/70R15 tire =

_____

Approximate revolutions per mile for a P275/50R15 tire =

_____

11. Based on the above calculations, what would happen to the readings at wheel speed sensors if the P215/70R15 tires were replaced with P275/50R15 tires?

_____

_____

12. What effect would the difference in revolutions per mile have on the vehicle's overall final drive ratio?

_____

_____

13. If a new P275/50R15 tire is placed on the same axle as one that has its tread worn down to 3/32 inch, what possible effects would this have on the handling of the vehicle?

_____

_____

Instructor's Response: _____

_____

_____

Chapter 9

DIAGNOSING 4WD AND AWD CONTROL SYSTEMS

BASIC TOOLS
Basic mechanic's tool set
Torque wrench
Lift

UPON COMPLETION AND REVIEW OF THIS CHAPTER, YOU SHOULD BE ABLE TO:

- Explain the basic procedures for diagnosing shift problems on 4WD systems.
- Test, service, and replace center differential assemblies.
- Test and replace viscous coupling units.

- Diagnose vacuum-operated "shift-on-the-fly" systems.
- Diagnose on-demand 4WD systems.
- Diagnose AWD systems.
- Diagnose differential lock systems.

BASIC DIAGNOSTICS

Begin the diagnosis of a 4WD or an AWD control system by thoroughly questioning the customer about the problem. If at all possible, road-test the vehicle with the customer to check the operation of the system and to clearly define the problem. Find out the speed or condition when the problem occurs. Also find out if the problem is more noticeable when the vehicle is making a turn or is going over bumps. This type of information will help during your diagnosis of the problem. Always check for relevant technical service bulletins (TSBs) and review the service history of the vehicle. Make sure to refer to the service information to determine how the system is controlled and what inputs are used to govern system operation (Figure 9-1).

Certain axle and drive shaft symptoms are related to the engine, transmission, tires, and other parts of the vehicle. During diagnosis isolate the problem to a specific area, as soon as possible. A road test followed by a detailed inspection may readily pinpoint the source of the problem. It is important to check all external parts of the system before suspecting something in the transfer case or electronic control module. Before attempting to diagnose a transfer case problem, make sure all tire pressures are equal. Tire wear or uneven pressures can cause problems in the transfer case and result in poor handling.

A common complaint is the inability to go from 4WD to 2WD. Back the vehicle up about 60 feet and then try to take it from 4WD to 2WD. It is common for these units to get splines locked; the torque load on the shafts will not relax to allow the splines to line up for a shift. By changing directions and reversing the torque load, the unit should shift freely.

Diagnosing 4WD and AWD Systems

Current 4WD and AWD systems are controlled by the powertrain control module (PCM) or a separate control module. These systems are diagnosed like other computer-controlled systems; faults can be identified by **diagnostic trouble codes (DTCs)**. Make sure the scan tool can communicate with the vehicle by completing a communication test with the scan tool. If the system has a separate diagnostic routine for the 4WD system, run that and retrieve and record all DTCs. Always use the information from the manufacturer when interpreting data.

> **Diagnostic trouble codes (DTCs)** are codes displayed by the vehicle's computer that indicate problems it recognized during its diagnostic routine.

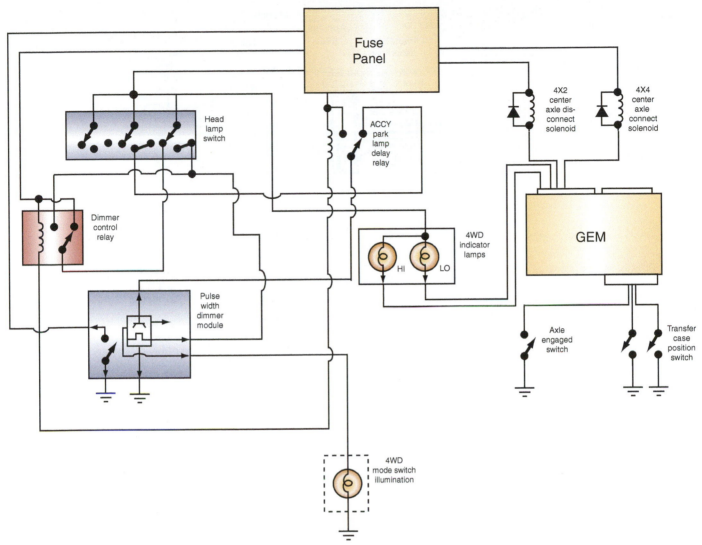

FIGURE 9-1 The basic wiring for a locking center differential.

It is also valuable to understand the power-flow within the transfer case (Figure 9-2). This can help identify what could be going wrong internally. To diagnose late-model 4WD systems:

1. Verify the customer's complaint.
2. Look for obvious signs of mechanical or electrical damage. Check the condition of the following (some systems will not have all of these):

- Half shafts
- Locking hubs
- Drive shaft and U-joints
- Shift lever and/or linkage
- Mode switch
- Shift motor
- Electromagnetic clutch
- Vacuum and fluid lines
- Tires
- Transfer case
- System fuses
- Wiring harnesses and connectors

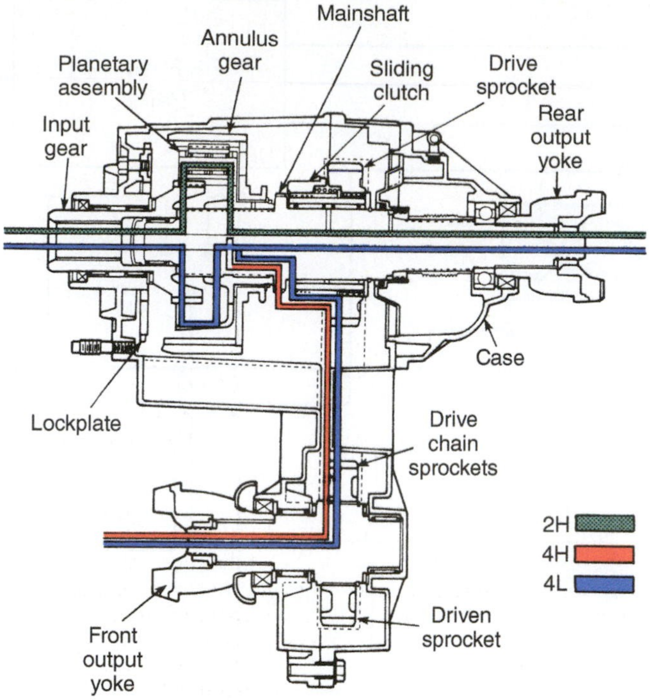

FIGURE 9-2 The power-flow through transfer case in 2H, 4H, and 4L modes.

3. If the cause was not visually evident, connect the scan tool to the data link connector (DLC).
4. Note the DTCs retrieved.
5. Conduct the self-test for the 4WD control module.
6. If the retrieved DTCs are related to the concern, interpret the codes and follow the specific pinpoint tests for those codes.
7. If the DTCs are not related to the concern or there are no DTCs, go to the manufacturer's symptom chart (Figure 9-3).
8. Identify and repair the cause of the problem.
9. Verify the repair.

GENERAL SHIFT CONTROLS

Inspection

Many mechanical failures are caused by faulty transfer case shift mechanisms. Mechanical linkages (Figure 9-4) should be checked to make sure they do not bind or are not too loose. Lubricate the linkage with a good multipurpose grease. A stuck linkage can prevent the transfer case from shifting into 4WD or out of it. Frequent lubrication will keep the transfer case and transmission shift linkages operating smoothly.

Diagnosis

Before beginning to test any electric or electronic circuit or component, refer to the appropriate service manual. Some components should not be exposed to full battery voltage and others may require specific testing procedures. Never use a low-impedance meter or test light on electronic systems.

Electric switches and their circuits (Figure 9-5) should be checked for causes of high resistance if they do not allow for a complete shift of gears. Also, the solenoids' action should be checked and adjusted if possible.

Classroom Manual
Chapter 9, page 223

SERVICE TIP:
Solenoid action can be checked by listening for a click when they are turned on. If they click, they are working. If no click is heard, and power is reaching the solenoid, the solenoid may be mechanically stuck.

Condition	Possible Cause
No communication with the control module (PCM)	Scan tool Data link connector (DLC) Control module (PCM) Circuitry
The 4WD indicators do not operate correctly or do not operate	Indicator lamp Circuitry Control module (PCM) Ignition switch
The vehicle does not shift between modes correctly	Mode select switch Transfer case Transfer case clutch Control module (PCM) Circuitry Locking hubs Ignition switch Transfer case shift lever Mode indication switch Ignition switch
4WD does not engage correctly at speed	Transfer case clutch coil Control module (PCM) Locking hubs Ignition switch
The front axle does not engage or disengage correctly or makes noise in 2WD under heavy throttle	Mode select switch Locking hubs Vacuum leaks Control module (PCM) Front halfshaft Ignition switch
The transfer case jumps out of gear	Transfer case Vacuum solenoid vent Mode select switch Transfer case Shift lever
Driveline wind-up while moving straight	Tire inflation pressure Tire and wheel size Tire wear Axle ratio
Grinding noise during 4WD engagement, especially at high speeds	The front halfshafts are not turning at the same speed
Flashing 4WD indicators	Loss of CAN communication between 4WD control module (PCM) and instrument cluster Loss of high-speed (HS-CAN) communication between 4WD control module (PCM) Ignition switch
The transfer case makes noise	Tire inflation pressure Unmatched tire and wheel size Tire tread wear Internal components Fluid level
The vehicle binds in turns, resists turning, or pulsates or shudders while moving straight	Unmatched tire sizes Unequal amounts of tire wear Unequal tire inflation pressures Unmatched front and rear axle ratios

FIGURE 9-3 **A typical symptom chart.**

If the additional drive axle is engaged or disengaged by a vacuum motor (Figure 9-6), check the engine's vacuum before proceeding. When the vacuum signal pulls the mode fork into the 4WD position and the transfer case is fully engaged, vacuum is diverted to the front axle to lock it into 4WD. If the engine has low vacuum, if there is a vacuum leak in the circuit, or if the shift linkage is out of adjustment, the transfer case cannot shift properly. Make sure all vacuum motors, canisters, and lines are not leaking and check the adjustment of the linkage.

The operation of an AWD transfer case is often controlled by a vacuum and electric shift mechanism. This system normally consists of a selector switch, check valve, vacuum reservoir, control relay, vacuum solenoids, and a vacuum servo mounted on the transfer case. A lamp

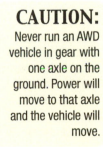

CAUTION:
Never run an AWD vehicle in gear with one axle on the ground. Power will move to that axle and the vehicle will move.

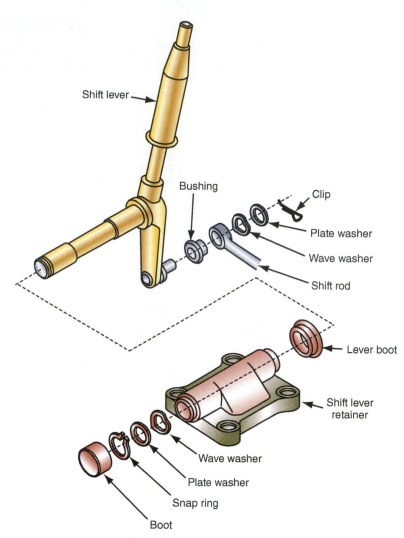

FIGURE 9-4 Typical transfer case linkage.

on the selector switch will light whenever the system is engaged. If shifting problems occur in this type of system, both the vacuum and electrical circuits need to be checked.

Electric checks include power checks at the switch and solenoids. If the circuit's fuse is blown, check for a short in the wiring or solenoids. If the circuit does not deliver power to the solenoids, check for an open circuit in the wires. If there is power to the solenoids, but they do not work, check for an open winding in the solenoid. High resistance, such as corroded or loose connectors, may be the reason the solenoids are unable to be fully energized. Check all connections and wires and repair as necessary.

If the electrical circuits are fine, check the vacuum source at the engine. Then check all lines, the servo, and the reservoir for leaks. All vacuum leaks should be repaired. Check the action of the check valve and the electric solenoids to make sure vacuum is being directed to the proper place. Any defective part should be replaced.

If the transfer case is fitted with a viscous coupling and the coupling was overheated, expect the **shift mode sleeve**'s teeth and corresponding gear splines to be broken or worn. The teeth of the shift mode sleeve engage with the splined gear and sprocket carrier. A worn engine, out-of-adjustment linkage, missing fork pads, or a seized vacuum motor can prevent precise engagement by the sleeve. Mode sleeve damage can also occur on some models if the driver does not lift off the throttle when shifting into 4WD.

If the linkage and vacuum source and circuit are fine, any shifting problems are caused by something internal in the transfer case.

A **shift mode sleeve** is the device that engages or disengages a gear in the transfer case.

A seized vacuum motor is often caused by a buildup of ice in the motor. If moisture is present in the vacuum system, it will freeze when it is cold.

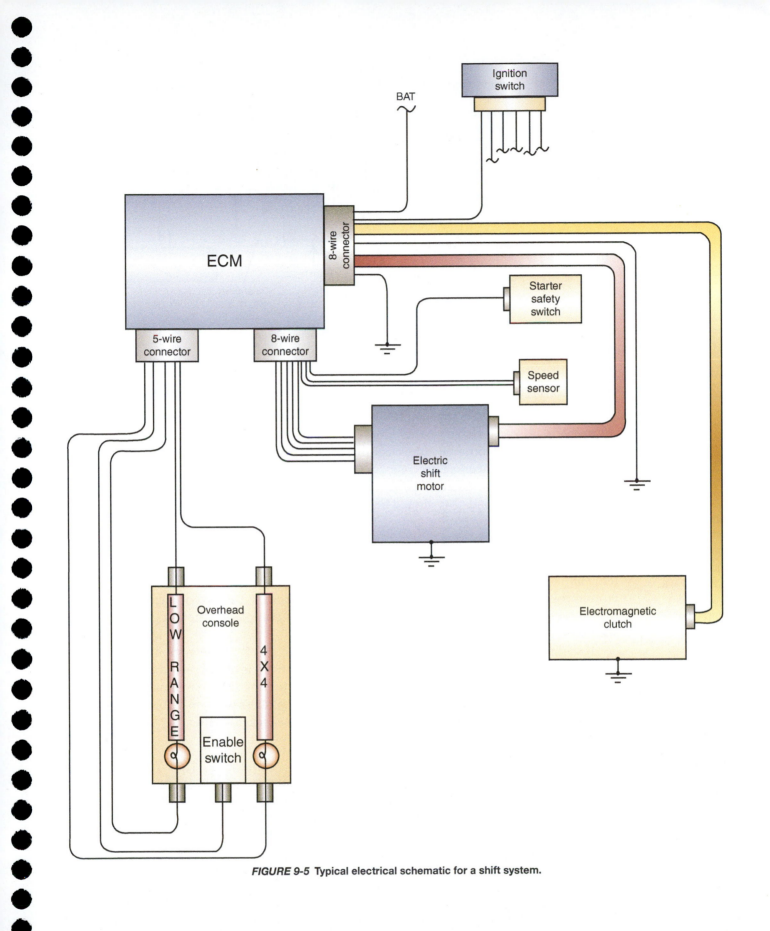

FIGURE 9-5 Typical electrical schematic for a shift system.

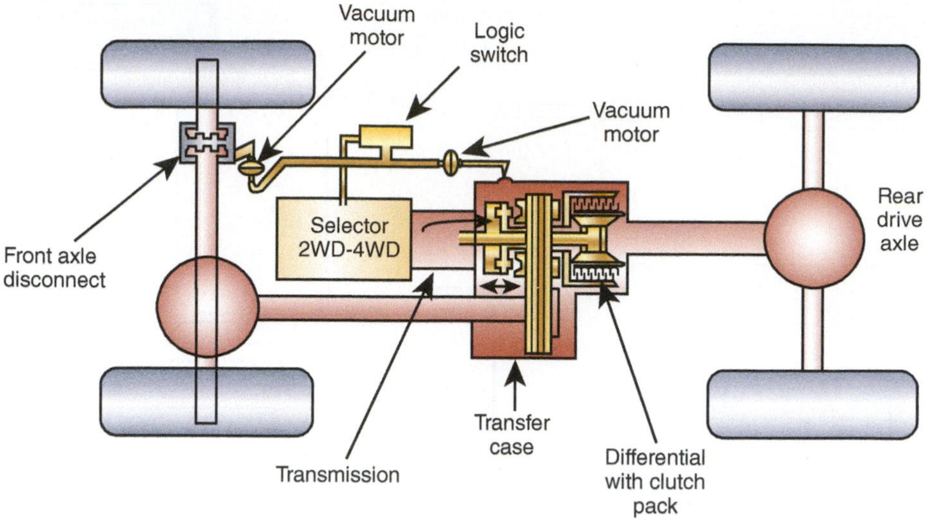

FIGURE 9-6 Schematic for vacuum shift controls.

ELECTRONIC SYSTEM DIAGNOSIS

The conditions present when the problem occurs should be duplicated to verify and define the problem. Also the system should be operated in all conditions and modes. Doing this will allow you to identify something the customer may have missed and all of the conditions during which the system works fine.

All mechanical and electrical parts of the system should be carefully looked at for signs of damage. If any are found, they should be repaired or replaced.

Scan Tool

If no problems were found during the inspection or if those faults were corrected, the system should be monitored with a scan tool. Connect the correct scan tool to the **data link connector (DLC)** and select the vehicle to be tested from the scan tool menu. Then check that the scan tool can communicate with the vehicle by completing the test provided by the scan tool.

If communications are okay, retrieve and record all DTCs. If the system has a separate self-diagnostic routine for the 4WD system, run that and retrieve and record all DTCs. If the DTCs are related to the original complaint or concern, use the service manual and complete the detailed testing of the system as defined by the DTCs. If there were no DTCs or if the DTCs were not related to the concern, the system needs to be checked further.

Component Testing

Inputs should be tested according to the recommended procedures. The speed sensors are typically Hall-effect or magnetic pulse units. These are best checked with a lab scope. Often, the manufacturer will recommend testing the sensors with a voltmeter or ohmmeter. Normally these tests involve taking readings at specific terminals in a multiple-pin connector.

The selector switch can be checked by connecting an ohmmeter across the switch's terminals. As the switch is moved to various positions, the circuits connected to the switch should be open or closed. By referring to the wiring diagram, you can identify what should happen in the different positions. If the switch does not function as it should, it must be replaced.

The **data link connector (DLC)** is the connector that allows access to the data stored in a vehicle's on board computer.

Chapter 11 has a detailed discussion on the use of a scan tool, including what to do if the tool cannot communicate with the vehicle.

A detailed discussion of testing transmission and other driveline electrical and electronic components is included in Chapters 10 and 11.

When the transfer case shift motor is suspected as being faulty, begin diagnosis with a careful inspection. Then follow the manufacturer's procedures for testing it. Often testing involves checking the motor for opens, shorts, and high resistance. In some cases, tapping the motor lightly will cause it to operate. This procedure is for diagnosis only; even if the motor frees up when tapped, it is on its way out. If the motor is found to be defective, it must be replaced. Where replacing the motor, make sure it is mounted correctly and that its retaining bolts are tightened to specifications.

If the system uses a magnetic clutch, it can be checked with an ohmmeter through the connector terminals at the transfer case. Noise from the clutch can be caused by a faulty clutch or a clutch that needs adjustment.

CENTER DIFFERENTIALS

Classroom Manual
Chapter 9, page 229

The center or interaxle differential takes the place of a transfer case and can be constructed as a single speed that transfers power to the front and rear axles in accordance to the traction of each axle. Most often, the center differential is fitted with some sort of traction control device that sends the majority of the power to the axle that has the most resistance or traction. These traction devices are similar to those used in a front or rear axle assembly.

An Audi center differential provides a good example of the components that a technician would inspect during repair (Figure 9-7). The unit would be removed from the car and disassembled on the bench. The technician would pay special attention to the transfer clutches, housing, and hub for wear or damage (Figure 9-8). All other components would be cleaned and inspected before reassembly.

Often the center or interaxle differential is merely a ring and pinion gear assembly. The ring and pinion serve one major purpose: to change the angle of the torque before it is sent to the additional drive axle.

The center differential also can be fitted with a gear set and shifting mechanisms to provide for a low and high operating range (Figure 9-9). These ranges affect the overall gear ratio of the driveline. The shifting can be manually or electrically controlled.

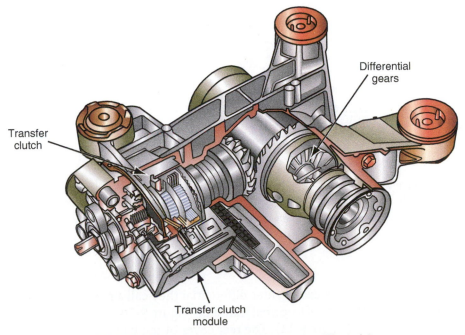

Differential gears

Transfer clutch

Transfer clutch module

FIGURE 9-7 One type of Audi Quattro center differential.

FIGURE 9-8 Inspecting the Quattro transfer clutch, housing, and hub.

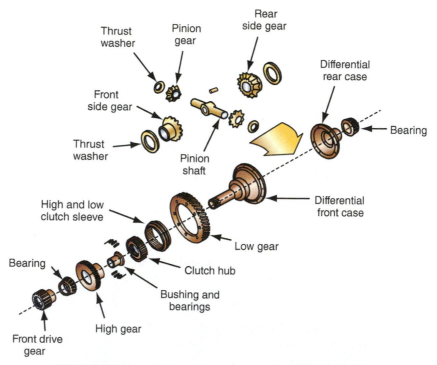

FIGURE 9-9 A two-speed center differential (transfer) assembly.

Chrysler's AWD minivans use a center differential they call a power transfer unit (PTU). This unit is mounted to the side of a normal transaxle (Figure 9-10). The output from the transaxle powers the ring gear (Figure 9-11). The rear cover of the PTU positions the pinion gear.

Service to center differentials is much like service to any other differential. Setup of the gears may require special procedures. The manufacturer's recommendations should always

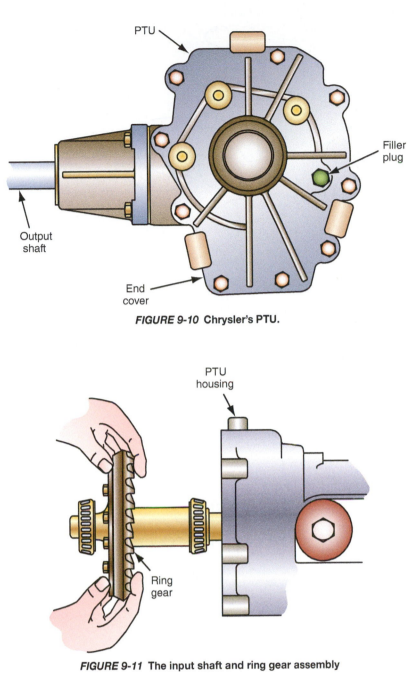

FIGURE 9-10 Chrysler's PTU.

PTU

Filler plug

Output shaft

End cover

PTU housing

Ring gear

FIGURE 9-11 The input shaft and ring gear assembly from a center differential.

be followed. Also because some of these units do have side gears, some normal procedures cannot be done to these units.

All differentials should be carefully inspected, especially for leaks. A lack of lubricant will destroy the unit. The procedures for installing new seals and gaskets may also be unique when compared to other differential assemblies. For example, because the pinion gear assembly is also the output shaft assembly in Chrysler's PTU, removal and installation of the pinion seal is unlike that for most differentials (Figure 9-12). To provide additional sealing, the rear cover or housing for the pinion shaft may have O-rings (Figure 9-13) to seal the cover to the main differential housing. These seals must be carefully installed to ensure a good seal and to provide the correct tooth contact between the pinion and ring gears. After the leaks have been corrected, refill the unit with the lubricant specified by the manufacturer.

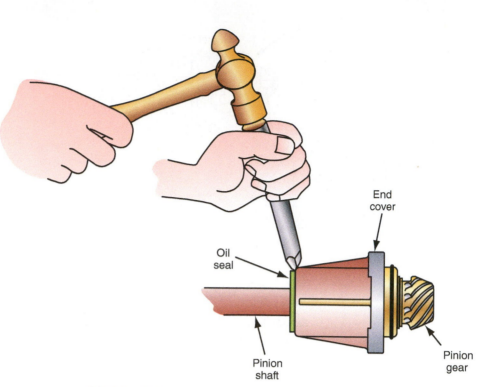

FIGURE 9-12 Removing the pinion seal on a Chrysler's PTU.

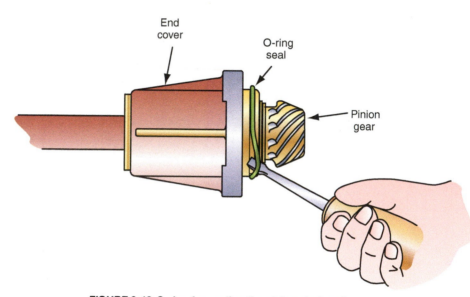

FIGURE 9-13 O-ring for sealing the pinion shaft and rear cover.

VISCOUS COUPLING

Classroom Manual

Chapter 9, page 228

A viscous coupling (Figure 9-14) is used on many AWD systems and is also found in some transfer cases. Under normal driving the vehicle maintains either FWD or RWD (depending on the setup), with most of the engine's torque sent to the front or rear drive axles. A viscous coupling allows more torque to the opposite drive axles when the primary drive axle experiences slippage. A viscous coupling provides for variable torque distribution automatically without intervention by the driver.

A viscous coupling cannot be rebuilt because fluid is added to the unit by weight, not volume. This is a complicated procedure and shops will not have the necessary instruments,

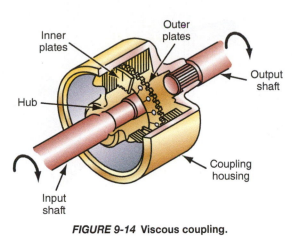

FIGURE 9-14 Viscous coupling.

nor will they have the fluid. At the bottom of the coupling is a factory fill plug. Never remove it to drain the fluid. When a viscous coupling is bad, it must be replaced.

The greatest cause of viscous coupling failure is incomplete engagement of 4WD. If the vacuum system or linkage adjustments are not correct, the coupling will slip and not fully engage, causing the unit to overheat and usually blow out a seal. The viscous coupling will overheat as it tries to compensate for a free-running axle, which is caused by partial engagement into the 4WD mode. Another common cause of viscous clutch failure is incorrect tire size or inflation. Constant shear forces inside the coupling will cause overheating and failure. Front or rear lift towing of an AWD vehicle with a viscous coupling will destroy the coupling in a very short period of time.

Viscous couplings cannot be accurately tested. The best way to diagnose a faulty unit is to observe the behavior of the vehicle. Since the unit acts automatically, there are no controls involved. If the torque split is not normal, the viscous unit is undoubtedly bad. A road test on dry and slippery surfaces is the best way to determine if the unit is functioning normally.

Removal

The viscous coupling is typically mounted to the drive shaft for the secondary drive axle or inside the transfer case. To remove the unit, it must be disconnected from the drive shaft. To do this, a special tool must be used to prevent the shaft from turning while loosening the retaining bolts for the viscous coupling (Figure 9-15). This tool is also used to install the coupling. Without the tool it is impossible to properly tighten the retaining bolts.

AUTHOR'S NOTE: If the coupling is inside the transfer case, it can be removed when the transfer case is disassembled.

Once the viscous coupling is loosened from the drive shaft, the drive shaft is moved away to gain access to the retaining nut for the coupling (Figure 9-16). Once again, to loosen this nut the drive shaft must be held stationary. Once the nut is removed, the coupling can be slid off the drive shaft. Installation of the unit is completed by reversing the removal procedure.

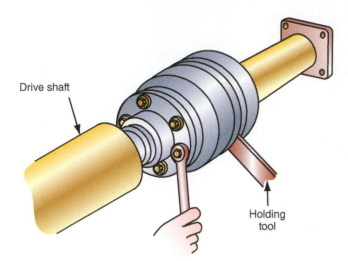

FIGURE 9-15 Disconnecting the drive shaft from the viscous-coupling unit.

FIGURE 9-16 Removing the retaining nut for a viscous coupling.

SHIFT-ON-THE-FLY SYSTEMS

CUSTOMER CARE: If a customer complains of improper operation of a shift-on-the-fly system, review the operating instructions with the customer to determine if there is indeed a problem and to better identify the possible causes.

SPECIAL TOOLS
Volt-ohmmeter
Vacuum gauge
Vacuum pump

Classroom Manual
Chapter 9, page 224

A shift-on-the-fly system switches between 2WD and 4WD electrically via a driver-activated switch on the instrument panel. These systems engage and disengage the additional drive axle and connect and disconnect the wheels to the additional drive axle. Some of these systems use electrically controlled vacuum motors to connect or disconnect the wheels from the axle and an electrical motor or electromagnet at the transfer case to transfer power to the additional axle.

In the system shown in Figure 9-17, the vehicle operates as a RWD vehicle until 4WD is activated. At that time, power is transmitted to the front axle and that axle is connected to the wheels. When the 4WD selector switch is activated, the motor at the rear of the transfer case is energized and causes the transfer case to split the engine's power between the front and rear drive axles. Once this is accomplished, the vacuum solenoids at the front axle connect the axle to the wheels (Figure 9-18).

The 4WD indicator light will illuminate when 4WD is engaged. If there is a problem with the system or if 4WD cannot be engaged, the indicator lamp will blink to notify the driver of a problem. Diagnosis begins with watching the frequency of the blinking lamp.

- If the indicator lamp does not illuminate or blink when 4WD is selected, verify its operation by turning on the ignition. The lamp, as well as the other indicator lamps, should illuminate for about 2 seconds then turn off. If the bulb does not light during this check, check the circuitry to the bulb and the bulb itself. If the bulb does illuminate with the ignition, check the circuit for the 4WD switch.
- If the bulb does not turn off when 2WD is selected after 4WD operation, check the circuit for the switch. Because the bulb remains on, the most likely problem is an electrical short.
- Sometimes 4WD does not engage simply because of extreme cold or ice buildup on the actuators. If the driver orders 4WD during extremely cold temperatures or when cornering, the indicator may blink rapidly during the attempted engagement. If engagement cannot be completed, the lamp will blink but at a slower rate. To identify this as the cause of the blinking lamp, stop the vehicle and put the gear selector in the High position, move

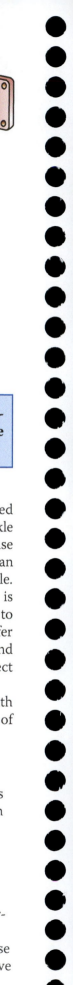

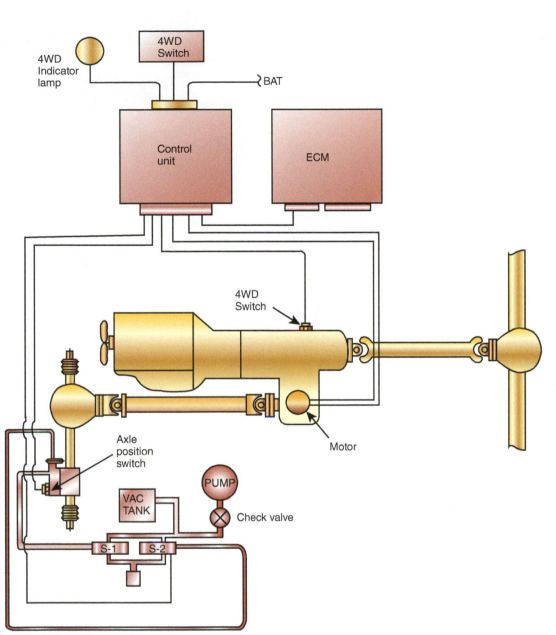

FIGURE 9-17 System layout for an electrical- and vacuum-operated shift-on-the-fly system.

the vehicle forward and backward a short distance. If the cause of the blinking lamp is related to cold, cornering, or low range selection, the lamp should stop blinking and 4WD should engage after the stop.

- When there is a rotational difference in the wheels on the front axle or if there is a phase difference between the front wheels and the axles, it is difficult for the unit to connect the axles to the wheels. In this case, the lamp will blink until the condition changes or 4WD is engaged. This can be caused by uneven tire pressures or uneven wear of the front tires.
- The lamp will also blink if the transfer case gear selector is in 4WD-Low and the system will not allow a shift from 4WD to 2WD when the selector is in low.
- If blinking of the bulb is rather rapid, the axle connect and disconnect system may be to blame. Check the condition of the tires before continuing your diagnosis. If the tires have the same air pressure and wear, check the operation of the gear selector. Make sure it moves freely from high to low and back. If the gear selector does not move freely, the problem is inside the transfer case or in its motor circuit. If the selector moves freely, check the vacuum actuator.

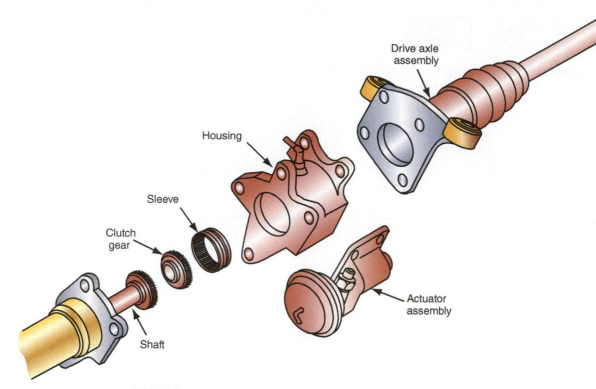

FIGURE 9-18 Axle engagement components on a shift-on-the-fly system.

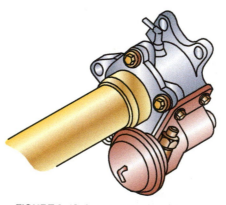

FIGURE 9-19 A vacuum solenoid assembly for a shift-on-the-fly system.

If the vehicle still will not engage into 4WD, there may be a faulty transfer case actuator motor or faulty vacuum solenoid assembly (Figure 9-19). If the lamp's frequency is slow, the problem could be the motor at the transfer case. Remove it and check its operation. If the problem was found and repaired, try the operation of the system again. If no problem was found, check the transfer case and repair or replace any faulty components. If no problem is found in the transfer case, it is likely that the electronic control unit is faulty.

Vacuum Systems

Begin your check of the vacuum actuator by checking for engine vacuum at the hoses to the vacuum solenoid. If no vacuum is present when the engine is running, check the transfer position switch. A voltage signal should be sent to the vacuum solenoid assembly when 4WD is selected. If no signal is being sent, the switch or switch circuit is faulty. If the solenoid is receiving a voltage signal, suspect a faulty solenoid assembly. If there are no electrical problems, check the vacuum circuit for leaks (Figure 9-20).

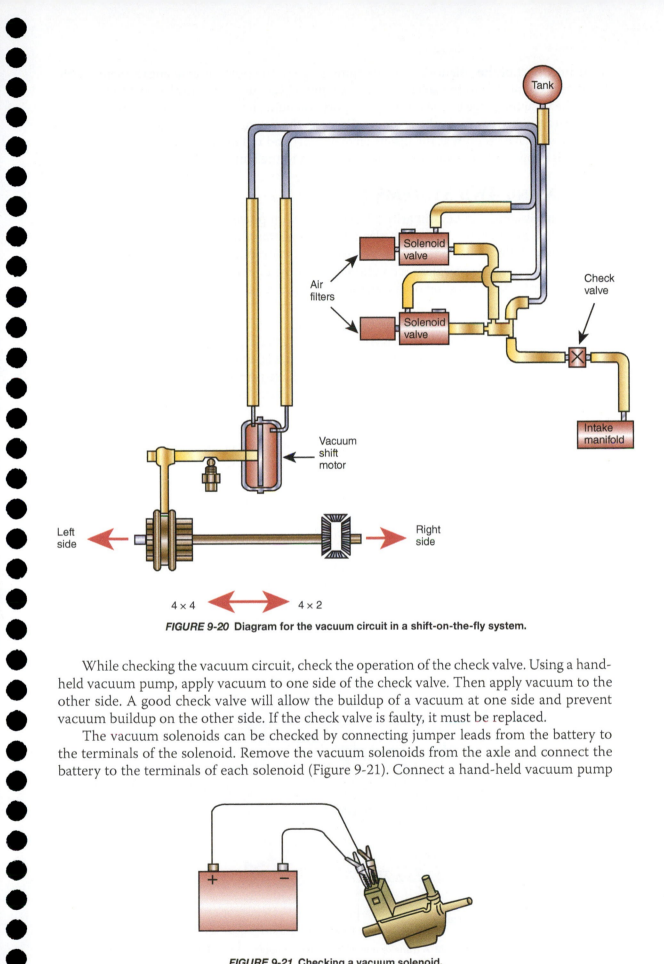

FIGURE 9-20 Diagram for the vacuum circuit in a shift-on-the-fly system.

While checking the vacuum circuit, check the operation of the check valve. Using a hand-held vacuum pump, apply vacuum to one side of the check valve. Then apply vacuum to the other side. A good check valve will allow the buildup of a vacuum at one side and prevent vacuum buildup on the other side. If the check valve is faulty, it must be replaced.

The vacuum solenoids can be checked by connecting jumper leads from the battery to the terminals of the solenoid. Remove the vacuum solenoids from the axle and connect the battery to the terminals of each solenoid (Figure 9-21). Connect a hand-held vacuum pump

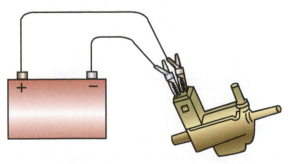

FIGURE 9-21 Checking a vacuum solenoid.

to the inlet port of the solenoid and a vacuum gauge to the port for axle engagement. With the battery connected to the solenoid, there should be vacuum at the outlet port to the axle. Move the vacuum gauge to the other outlet port. Vacuum should not be present there. Now disconnect the battery from the solenoid. Recheck the vacuum at the outlet ports. A good solenoid will have vacuum at the disengagement port and not at the engagement port when vacuum is applied and the solenoid is not connected to the battery.

On-Demand 4WD Systems

Classroom Manual
Chapter 9, page 227

On-demand systems can be primarily 2WD systems with power delivered to the other drive wheels as the need occurs. They can also be 4WD systems whose torque split between the axles varies with the conditions. The standard 4WD system splits the engine's power at a fixed and consistent amount regardless of conditions. Some on-demand systems allow the driver to select standard 4WD operation or automatic 4WD operation. In the latter, the split of power varies with road conditions.

Power is split by controlling the output clutch (Figure 9-22) at the transfer case or center differential. The activity of the clutch is controlled by regulating its duty cycle. During normal operation, the duty cycle is low. This allows for a slight speed difference between the front and rear drive shafts, which normally occurs when the vehicle is moving through a curve. When slip is detected at the rear wheels, the duty cycle to the clutch is increased until the difference between the front and rear drive axles is reduced.

SPECIAL TOOLS

Scan tool

Proper module for the specific vehicle (if necessary)

Various adapters for DLC

To vary the torque split, a computer monitors many things, especially the rotational speeds of the front and rear drive axles. Some systems rely on wheel speed sensors for this, while others have additional speed sensors on the front and rear output shafts from the transfer case. Many of the inputs to the electronic module for 4WD are shared and work with other systems such as the vehicle speed sensor, which may be used by the powertrain control unit, cruise control system, power-steering controls, instrument panel displays, and the 4WD control module. Diagnosis is best done with a scan tool. The typical procedure for initial checks with a scan tool is shown in Photo Sequence 25.

The system also uses a variety of other switches or sensors to monitor and control its operation. The name, location, and purpose of these switches and sensors vary by the manufacturer. The best place to start when diagnosing one of these systems is the service manual. Go to the section on the system and identify the components involved in the various modes of operation.

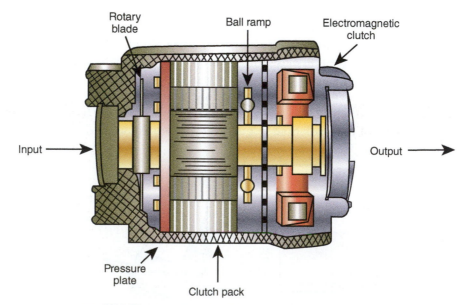

FIGURE 9-22 A transfer unit with an electromagnetic clutch that is controlled by a computer.

DIAGNOSIS WITH A SCAN TOOL

P25-1 Review the repair order information and service history of the vehicle.

P25-2 Be sure the engine is at normal operating temperature and the ignition switch is off.

P25-3 Select the appropriate scan tester data cable for the vehicle.

P25-4 Connect the scan tool to the DLC.

P25-5 Select the vehicle from the scan tool menu.

P25-6 Follow the menu on the scan tool and retrieve the DTCs.

P25-7 Start the engine and obtain the input sensor and output actuator data.

P25-8 The oscilloscope function can be useful for diagnosis.

P25-9 Compare the sensor and actuator data to the specifications given in the service manual. Determine what data is abnormal and continue diagnostics using the guidelines for that condition.

AWD Systems

Many AWD systems function around a viscous coupling. These have been covered previously. Problems with AWD systems normally occur with the controls that provide for braking efficiency and AWD in reverse gear. Chrysler's minivan AWD system and others use a vacuum solenoid setup to provide AWD in reverse. The solenoids are used to bypass the overrunning clutch. If AWD is not available in reverse gear, suspect the vacuum solenoids, their controls, or the dog-clutch assembly.

An overrunning clutch is used to prevent any feedback of front wheel braking torque to the rear wheels. This allows the brake system to control braking as if it were a 2WD vehicle. The controls for this feature vary with the manufacturer and you should always follow the troubleshooting charts provided by the manufacturer of the vehicle (Figure 9-23).

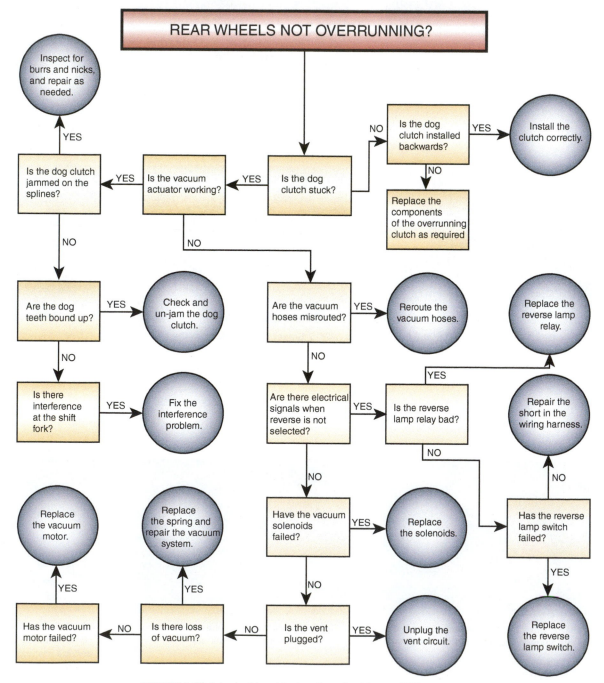

FIGURE 9-23 A typical troubleshooting chart for an AWD system.

DIFFERENTIAL LOCK SYSTEMS

Some vehicles allow the driver to manually lock the rear or center differentials. Doing so stops the unit from having differential action and the power is equally distributed between the two drive axles (right and left or front and rear). The basic locking mechanism is the same for either differential; however, the controls and conditions for operation vary with the design and manufacturer. It is wise to refer to the vehicle's owner manual if you are unsure of the correct way to operate the system. Doing this will help in your diagnosis as well as prevent damage to the system.

Basic diagnosis for all locking differential begins with observing the operation of the warning lamp. When the ignition switch is first turned on, the indicator lamp should light up for approximately 1 second and then it should turn off. To conduct this check on systems with a locking center differential, the transfer case must be in the "H" position. To continue testing, start the engine and press the differential lock button. When the differential lock switch is engaged, the lamp should light (Figure 9-24). On some vehicles, the center differential can only be locked when the transfer case is in the "L" position and on some models, this will happen automatically (Figure 9-25). When the differential lock is turned off or disengaged, the lamp should turn off. To do this on systems with a locking center differential, move the shift lever to the "N" or "H" position.

If the indicator lamp stays on or blinks when the lock is not in the ON position, the system needs to be checked. If the lamp never turns on, follow the manufacturer's recommended procedure for checking the bulb, wiring harness, instrument panel, and the 4WD position or rear differential lock detection switch.

Rear Differential

The rear differential lock system (Figure 9-26) should be used only when wheel spinning occurs in a ditch or on a slippery or ragged surface and one of the rear wheels is spinning. The antilock brake system does not operate when the rear differential is locked. It is normal for the ABS warning light to be on at this time.

If the differential lock indicator did not light up when lock was selected, the possible problems are a faulty differential lock fuse, lock control switch, or electronic control unit, or a bad ground. Begin your diagnosis by checking for battery voltage; it should be about 14 volts. Then check the fuse for the differential lock. Inspect the system circuit. Repair or replace as necessary. If no problems were found, disconnect the differential lock connector at the electronic control unit. Inspect the connector; clean or repair as needed. Using a digital volt ohmmeter,

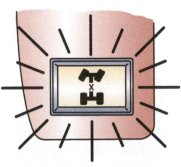

FIGURE 9-24 Diagnosis begins with observing the differential lock indicator lamp.

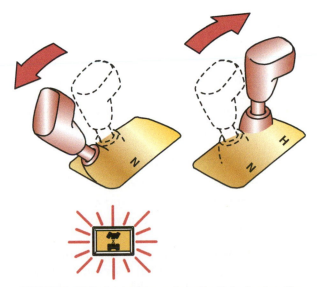

FIGURE 9-25 Some vehicles automatically lock when the transfer shifter is moved into the low position.

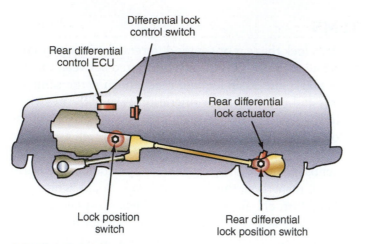

FIGURE 9-26 A 4WD system with a locking rear differential system.

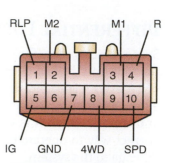

FIGURE 9-27 Terminal identification of the lock differential–related connector to the electronic control unit.

Symbol (terminal no.)	Trouble part	Condition	Specified value
M1 - M2	RR diff. lock actuator		Less than 100Ω
Gnd -Body ground	Body ground		Continuity
Spd - Body ground	Speed sensor	Vehicle moves slowly	1 pulse each 40 cm (15.75 in.)
Ig - Body ground	Diff. fuse	Ignition switch ON	10 -14V
Rlp - Body ground	Rear diff, lock position switch	Ign. swich ON with ind, light ON	About 14V
		Ign. switch ON with ind, light OFF	10 - 14V
4WD - Body ground	L4 position switch	Ign. switch ON with T/F lever except L4	About 0V
		Ign. switch ON with T/F lever L4	10 - 14V
R - Body ground	Diff. lock control switch	Ign. switch ON with diff. lock switch ON	10 - 14V
		Ign. switch ON with diff. lock control switch OFF	About 0V

FIGURE 9-28 Specified results from checking across the various terminals of the connector shown in Figure 9-27.

test the control unit across the terminals specified in the service manual (Figure 9-27), If the readings are not as specified, replace the parts indicated by the procedure (Figure 9-28).

If no problems have been found, use a voltmeter and measure voltage when the differential lock control switch is in the position dictated by the service manual (Figure 9-29). If the readings are not what are specified, replace the control unit (Figure 9-30).

If the differential lock does not engage, the probable causes are a faulty switch, lock actuator, control unit, differential carrier, or wiring or ground. If the lock does engage but does not disengage, suspect a faulty speed sensor, control unit, and wiring or ground.

Center Differential

The four-wheel-drive control lever or button and center differential lock button are used to select the various transfer and center differential modes (Figure 9-31). The high and low position of the four-wheel-drive control provides either lock or unlock mode of the center differential depending on the center differential lock button position. The center differential lock is used when wheels are stuck in a ditch, or when driving on a slippery or bumpy surface. When the center differential is locked, the vehicle's skid control (VSC) system is automatically turned off and the center differential lock and "VSC OFF" indicator lights come on to warn that engine performance will be changed.

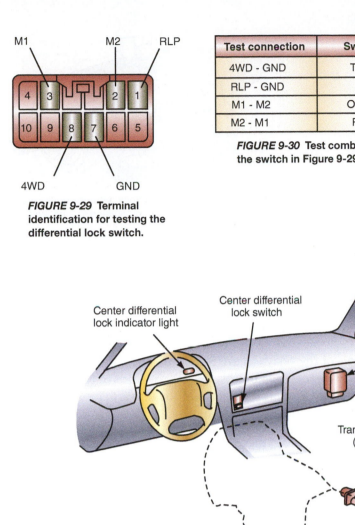

FIGURE 9-29 Terminal identification for testing the differential lock switch.

Test connection	Switch position	Specified value
4WD - GND	T/F lever L4	10 -14V
RLP - GND	ON	0.5V or less → 10 - 14V
M1 - M2	OFF → RR	(Approx. 1 sec.) → 0.5V
M2 - M1	RR → OFF	or less

FIGURE 9-30 Test combinations and expected results from checking the switch in Figure 9-29.

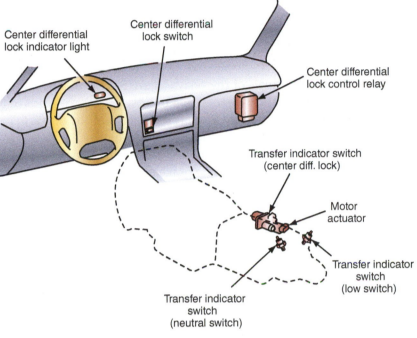

FIGURE 9-31 A 4WD system with a locking center differential.

⚠️ **WARNING:** For normal driving on dry and hard surface roads, the center differential should be unlocked. To prevent damage to the center differential lock system, never push the center differential lock button when the vehicle is cornering or when its wheels are spinning freely off the ground. If the indicator light does not go off when unlocking the center differential, the vehicle should be driven straight ahead while accelerating or decelerating, or driven in reverse.

Like a rear differential locking system, diagnosis of a locking center differential begins with observing the indicator lamp. The lamp should light when the system is engaged and go off when it is turned off. The lamp also should come on when the transfer shifter is moved to the "L" position on many vehicles. If the lamp never comes on, check the differential lock fuse, lock control switch, electronic control unit, and wiring.

If the activity of the lamp is correct but the differential does not operate correctly, continue your diagnosis by checking the relay. This check is done with an ohmmeter connected

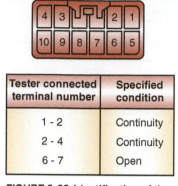

Tester connected terminal number	Specified condition
1 - 2	Continuity
2 - 4	Continuity
6 - 7	Open

FIGURE 9-32 Identification of the terminals of the differential lock relay and the expected results. If the readings do not match the specified values, the relay should be replaced.

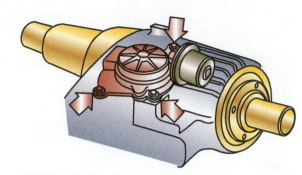

FIGURE 9-33 The differential lock actuator motor must be unbolted and removed to be accurately checked.

across the various specified terminals of the relay (Figure 9-32). At times, there is a diode in the circuit, so if there is no continuity between two points, reverse the leads of the ohmmeter and check again before coming to a conclusion. If the results of the test are not as specified, replace the relay. If continuity is not as specified, replace the relay.

To check the operation of the locking actuator, jack up the front wheels, block the rear wheels so they can move, and apply the parking brake. Disconnect the connector of the actuator and connect it to the relay using jumper wires. Operate the switch when attempting to turn the drive shaft. If the center differential is locked, you should not be able to turn the drive shaft. If the differential is not locked, the front drive shaft should turn freely. If the system worked as it should, check the wiring and connectors.

If the system did not work properly, check the actuator. The actuator must be removed (Figure 9-33) to accurately test it. Once it is removed, check the continuity through the various terminals as specified. If any of the readings do not meet specifications, replace the actuator.

SUBARU SYMETRICAL AWD

The Subaru Symmetrical AWD system is one of the most popular AWD systems on the market, and variations of this system have been used for years. A close look at a particular issue and the components involved provides a good example of the type of diagnosis a technician may go through in the field. This system uses a multiplate transfer clutch in the automatic transmission that automatically adjusts power to the rear axle as needed. The transfer clutch is applied hydraulically by a piston, and runs in the same ATF as the rest of the transmission. The transfer clutch pressure is adjusted by a duty cycle solenoid, which is controlled by the TCU (Figure 9-34). The TCU receives input from the Vehicle Dynamics Control System unit. Wheel speed sensors, steering angle sensor, lateral g-force sensor, and a yaw rate sensor all provide input to the system (Figure 9-35). Let's say the car comes in with a torque-bind condition, that is, the vehicle "jumps" when going around a sharp turn. Clearly the front and rear axles are not free to turn at different speeds. The technician must consider a number of factors. First, are tire size and inflation correct? What is the condition of the transmission fluid? If the fluid is deteriorated, the transfer clutches might bind. What are the input readings for all the sensors that affect the system? Even if all the above variables prove to be fine, is there a possibility that the transfer clutch solenoid is faulty

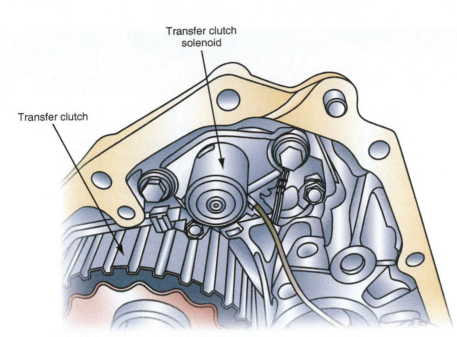

FIGURE 9-34 A Subaru Symmetrical AWD transfer clutch and transfer clutch duty solenoid.

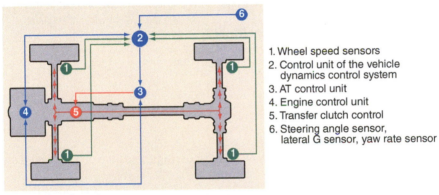

1. Wheel speed sensors
2. Control unit of the vehicle dynamics control system
3. AT control unit
4. Engine control unit
5. Transfer clutch control
6. Steering angle sensor, lateral G sensor, yaw rate sensor

FIGURE 9-35 The electronic components of the Subaru Symmetrical AWD system.

and is improperly directing fluid to the clutch piston? This single example illustrates the complexity of diagnosing any 4WD/AWD complaint.

ACTIVE HEIGHT CONTROL SUSPENSION

Some vehicles have an active height control (AHC) suspension that controls the vehicle's ground clearance according to driving conditions. These systems often are linked to the 4WD system. The driver can select vehicle height settings of high, normal, or low. However, on some vehicles, the systems will automatically raise the vehicle when the vehicle is shifted into 4WD and Low gear at the transfer case. Some vehicles will increase the ground clearance even more than its high position when 4WD-Low is selected and operating conditions, such as being stuck, dictate the need. The extra-high mode is automatically disengaged when the vehicle is no longer stuck or when the four-wheel-drive control lever is put in a position other than "Low."

Problems with the 4WD system can cause problems in the active height control suspension system. This is especially true of the various switches used to monitor the activity and position of the transfer case gears and differentials.

CASE STUDY

A technician had just rebuilt the center differential on an AWD car. On test-driving the vehicle to verify the repair, he noticed that the rear wheels did not have much torque applied to them. Also, the indicator lamp did not light during the test drive. He pulled the vehicle back into the shop and began to check all electrical connections, including power checks at the switch and solenoids. The circuit's fuse was not blown; therefore, there was not a short in the wiring or solenoids. There was power to the solenoids, which means there was no open in the circuit. The solenoids clicked; therefore they were probably good and operating correctly. There was no evidence of high resistance, such as corroded or loose connectors, that might be the reason the solenoids were unable to be fully energized. The technician was confused. He knew he had put the differential together correctly and all of the electrical connections and wires were good. Why was no power being delivered to the rear wheels?

The operation of an AWD transfer case is often controlled by a vacuum and electric shift mechanism. This system normally consists of a selector switch, check valve, vacuum reservoir, control relay, vacuum solenoids, and a vacuum servo mounted on the transfer case. The lamp on the selector switch lights whenever the system is engaged; therefore, this system was not engaging. If shifting problems occur in this type of system, both the vacuum and electrical circuits need to be checked.

Because the electrical circuits are fine, the engine's vacuum should be checked. Then all lines, the servo, and the reservoir should be checked for leaks. The action of the check valve and the electric solenoids was checked to make sure vacuum was being directed to the proper place. At this point, the technician discovered the problem. He had reversed the vacuum lines to the solenoids. One solenoid was venting when it should be directing the vacuum. The other solenoid also was working in the opposite manner to what it should. That was why the system was not engaging.

TERMS TO KNOW

Data link connector (DLC)

Diagnostic trouble codes (DTC)

Shift mode sleeve

ASE-STYLE REVIEW QUESTIONS

1. When discussing computer-controlled AWD systems,

   Technician A says that diagnostics should begin with connecting a scan tool and retrieving DTCs.

   Technician B says that after the problem or concern has been verified, the next step in diagnostics is a thorough inspection. TSBs and the vehicle service history should be checked.

   Who is correct?

   A. A only
   B. B only
   C. Both A and B
   D. Neither A nor B

2. When discussing how to diagnose viscous couplings,

   Technician A says that the unit should be disassembled and its parts inspected for signs of overheating if the coupling is suspect.

   Technician B says that a road test is the best way to check the operation of a viscous coupling.

   Who is correct?

   A. A only
   B. B only
   C. Both A and B
   D. Neither A nor B

3. When discussing the cause of a damaged shift mode sleeve,

   Technician A says that vacuum supply issues can cause the problem.

   Technician B says that a misadjusted shift linkage can cause this problem.

   Who is correct?

   A. A only
   B. B only
   C. Both A and B
   D. Neither A nor B

4. When diagnosing an electrical switch,

   Technician A says that the switch can be tested with an ohmmeter.

   Technician B says that switches can be checked with a voltmeter.

   Who is correct?

   A. A only
   B. B only
   C. Both A and B
   D. Neither A nor B

5. When servicing an AWD vehicle,

    Technician A checks the service manual to identify the correct lubricant for the center differential.

    Technician B says that the same lubricant is used for the viscous coupling, transfer case, and differentials on most vehicles.

    Who is correct?

    A. A only
    B. B only
    C. Both A and B
    D. Neither A nor B

6. When discussing 4WD shift systems,

    Technician A says that a seized shift linkage may cause the transfer case to operate only in 4WD.

    Technician B says that a worn and loose linkage may cause a transfer case to operate only in 2WD.

    Who is correct?

    A. A only
    B. B only
    C. Both A and B
    D. Neither A nor B

7. When discussing electrical checks on 4WD systems,

    Technician A says that if the circuit's fuse is blown, check for a short in the wiring or solenoids.

    Technician B says that if there is power to the solenoids but they do not work, check for an open winding or high resistance in the solenoid.

    Who is correct?

    A. A only
    B. B only
    C. Both A and B
    D. Neither A nor B

8. While discussing the possible causes of the 4WD indicator lamp not working properly,

    Technician A says this can be caused by a faulty PCM.

    Technician B says this can be caused by a faulty ignition switch.

    Who is correct?

    A. A only
    B. B only
    C. Both A and B
    D. Neither A nor B

9. While discussing why an on-demand 4WD system does not respond to changes in traction at the wheels,

    Technician A says there may be faulty wheel speed sensors.

    Technician B says the electromagnetic clutch in the transfer case may be faulty.

    Who is correct?

    A. A only
    B. B only
    C. Both A and B
    D. Neither A nor B

10. While discussing why the 4WD indicators are flashing in a vehicle brought in for service,

    Technician A says the system has determined the tires are either underinflated or are a different size.

    Technician B says there may be a loss of CAN communications.

    Who is correct?

    A. A only
    B. B only
    C. Both A and B
    D. Neither A nor B

ASE CHALLENGE QUESTIONS

1. A vacuum-shifted 4WD system does not shift into four-wheel drive.

    Technician A says that the engine vacuum may be low.

    Technician B says that the vacuum motor at the front axle may be defective.

    Who is correct?

    A. A only
    B. B only
    C. Both A and B
    D. Neither A nor B

2. All of the following are true statements about transfer case inspection and assembly, *except*:

    A. The transfer case chain should be inspected for stretching and looseness.
    B. Thrust washer thickness should be measured to check for wear and select fit.
    C. Specification measurements should be taken during disassembly.
    D. All parts should be cleaned before installation and should be assembled dry.

3. A manual-shift transfer case will not shift into 4WD. The cause of this problem may be:

    A. The fluid is low in the transfer case.
    B. The front drive shaft U-joints are bound up.
    C. The shift linkage needs lubrication.
    D. The electric shift motor is defective.

4. An electronic transfer case does not engage. Which of the following would be the *least* likely cause?

    A. A bad electronic shift motor.
    B. A blown fuse.
    C. A defective 4WD engage switch.
    D. A rusted linkage.

5. When discussing electronic transfer case shift systems,

    Technician A says that all linkages on an electronic-shift transfer case are internal.

    Technician B says that an open circuit in the shift switch wiring will prevent the transfer case from engaging.

    Who is correct?

    A. A only
    B. B only
    C. Both A and B
    D. Neither A nor B

Name _____ **Date** _____

CHECK ENGINE MANIFOLD VACUUM

Upon completion of this job sheet, you should be able to measure engine manifold vacuum and determine whether vacuum-controlled devices may be malfunctioning because of an engine problem.

NATEF MAST Task Correlation

Four-Wheel Drive/All-Wheel Drive Component Diagnosis and Repair

Task #1, Inspect, adjust, and repair shifting controls (mechanical, electrical, and vacuum), bushings, mounts, levers, and brackets.

Tools and Materials

Vacuum gauge

Clean rag

Describe the Vehicle Being Worked On

Year _____ Make _____ VIN _____

Model _____

Engine size _____ # of cycles _____ compression ratio _____ : _____

Expected manifold vacuum readings at idle: _____

Source of information: _____

Procedure

1. Describe the general running condition of the engine:

   _____

   _____

2. What does manifold vacuum represent?

   _____

   _____

3. What is the difference between manifold and ported vacuum?

   _____

   _____

4. Connect a vacuum gauge to a manifold vacuum source. Where is the source you used?

   _____

5. What is the vacuum reading with the engine at idle? _____

6. What is the vacuum reading with the engine at 1500 rpm? _____

7. In what units is vacuum measured? _____

8. Summarize the results and state the general condition of the engine based on the vacuum tests. _____

_____

_____

9. Would the condition of engine hinder the operation of a vacuum-controlled device or system?   ☐ Yes   ☐ No

Instructor's Response: _____

_____

_____

Name _____ Date _____

Servicing Transfer Case Shift Controls

Upon completion of this job sheet, you should be able to inspect, adjust, and repair the shifting controls (including mechanical, electrical, and vacuum), as well as the bushings, mounts, levers, and brackets for a transfer case.

ASE NATEF MAST Task Correlation

Four-Wheel Drive/All-Wheel Drive Component Diagnosis and Repair

Task #1, Inspect, adjust, and repair shifting controls (mechanical, electrical, and vacuum), bushings, mounts, levers, and brackets.

Tools and Materials

DMM

Wiring diagram

Vacuum gauge

Hand-operated vacuum pump

Service manual

Protective Clothing

Goggles or safety glasses with side shields

Describe the Vehicle Being Worked On

Year _____ Make _____ VIN _____

Model _____

_____

Task Completed

Procedure

1. Raise the vehicle on a lift.

2. Locate and check the mechanical linkages at the transfer case for looseness and damage. Record your findings here.

   _____

   _____

3. Carefully check the linkage bushings and brackets. Record your findings here.

   _____

   _____

4. Check the mounts for the transfer case and record your findings here.

   _____

   _____

5. If the mechanical linkage is fine or if there is no direct linkage, proceed to check the electrical controls by lowering the vehicle.

_____

_____

6. Locate the shift controls and circuit on the wiring diagram. What main components are part of the circuit?

_____

_____

7. Check the control switch for high resistance; this could be the cause of poor shifting. How did you test the switch and what did you find?

_____

_____

8. With the ignition on, use the switch to move from 2WD to 4WD. You should have heard the click of the solenoid or motor. If the clicking was heard, what is indicated?

_____

_____

9. If there was no click, check the circuit fuse. If the fuse is bad, there must be a short in the circuit. What was the condition of the fuse?

_____

_____

10. Raise the vehicle and check for power at the solenoids and record your findings here.

_____

_____

11. What would be indicated by having applied voltage to the solenoids but they are not working?

_____

_____

12. Visually inspect all electrical connections and terminals. Record your results.

_____

_____

13. If the electrical circuit is fine, check engine manifold vacuum. Record your results and summarize what the results indicate.

_____

_____

14. Visually inspect the vacuum lines for tears, looseness, and other damage. Record your findings here.

_____

_____

15. With the vacuum pump, create a vacuum at the servo and observe its activity and ability to hold a vacuum. Record your results here.

_____

_____

16. If the system has a vacuum motor, apply a vacuum to it and observe its activity and ability to hold a vacuum. Record your results here.

_____

_____

17. Based on all of these checks, what are your service recommendations for this vehicle?

_____

_____

Instructor's Response: _____

_____

_____

Name _____ Date _____

Servicing 4WD Electrical Systems

Upon completion of this job sheet, you should be able to diagnose, test, adjust, and replace the electrical/electronic components of four-wheel-drive systems.

ASE NATEF MAST Task Correlation

Four-Wheel Drive/All-Wheel Drive Component Diagnosis and Repair

Task #6, Diagnose, test, adjust, and replace electrical/electronic components of four-wheel-drive systems.

Tools and Materials

Component locator

Small mirror

Flashlight

Scan tool

DMM

Wiring diagram

Air nozzle

Service manual

Protective Clothing

Goggles or safety glasses with side shields

Describe the Vehicle Being Worked On

Year _____ Make _____ VIN _____

Model _____

_____

Task Completed

Procedure

1. Check all electrical connections to the transfer case. Make sure they are tight and not damaged. Record your findings.

   _____

   _____

2. Now release the locking tabs of the connectors and disconnect them, one at a time, from the transfer case. Carefully examine them for signs of corrosion, distortion, moisture, and transmission fluid. A connector or wiring harness may deteriorate if ATF reaches it. Using a small mirror and flashlight may help you get a good look at the inside of the connectors. Record your findings.

   _____

   _____

3. Inspect the entire wiring harness for tears and other damage. Record your findings.

_____

_____

4. Because the operation of the engine, transmission, and transfer case are integrated through the control computer, a faulty engine sensor or connector may affect the operation of all of these. With a scan tool, retrieve any DTCs saved in the computer's memory. Were there any present? If so, what is indicated by them?

_____

_____

5. The engine control sensors that are the most likely to cause shifting problems are the throttle-position sensor, MAP sensor, and vehicle speed sensor. Locate these sensors and describe their location

_____

_____

_____

6. Remove the electrical connector from the TP sensor and inspect both ends for signs of corrosion and damage. Record your findings.

_____

_____

7. Inspect the wiring harness to the TP sensor for evidence of damage. Record your findings.

_____

_____

8. Check both ends of the three-pronged connector and wiring at the MAP sensor for corrosion and damage. Record your findings.

_____

_____

9. Check the condition of the vacuum hose. Record your findings.

_____

_____

10. Check the speed sensor's connections and wiring for signs of damage and corrosion. Record your findings.

_____

_____

11. Record your conclusions from the visual inspection.

_____

_____

_____

_____

12. Remove the selector switch and check its action with an ohmmeter. Move the switch to all of its possible positions and observe the ohmmeter. Compare the action of the switch to the circuit shown in the wiring diagram. Summarize the condition of the switch.

_____

_____

_____

13. Locate the transfer case shift motor or solenoid. ☐

14. Carefully inspect this for damage and poor mounting. Record your findings here.

_____

_____

15. Check the solenoid or motor following the procedures given by the manufacturer. Summarize this check and your results.

_____

_____

16. If the solenoid or motor must be removed for testing, make sure it is mounted correctly and the mounting bolts tightened to specs when you reinstall it.

17. Check the service manual and identify all switches located on the transfer case. List these here.

_____

_____

_____

18. Determine the function of each switch. Is it a grounding switch or does it complete or open a power circuit?

_____

_____

_____

19. An ohmmeter can be used to identify the type of switch being used and can be used to test the operation of the switch. There should be continuity when the switch is closed and no continuity when the switch is open. Record your findings.

_____

_____

20. Apply air pressure to the part of the switch that would normally be exposed to oil pressure and check for leaks. Record your findings.

_____

_____

21. The transfer case may be equipped with a speed sensor. Typically, the speed sensor is a permanent magnetic (PM) generator. Locate the speed sensor and describe its location.

_____

_____

☐   **22.** With the vehicle raised on a lift, allow the wheels to be suspended and free to rotate.

☐   **23.** Set your DMM to measure AC voltage.

☐   **24.** Connect the meter to the speed sensor.

25. Start the engine and put the transmission in gear. Slowly increase the engine's speed until the vehicle is at approximately 20 mph, and then measure the voltage at the speed sensor. Record your findings.

_____

_____

26. Slowly increase the engine's speed and observe the voltmeter. The voltage should increase smoothly and precisely with an increase in speed. Record your findings.

_____

_____

27. A speed sensor also can be tested with it out of the vehicle. Connect an ohmmeter across the sensor's terminals. What was the measured resistance?

_____

_____

28. Locate the specifications for the sensor and compare your readings with specifications.

_____

_____

Instructor's Response: _____

_____

_____

Chapter 10

Servicing Drive-Train Electrical Systems

BASIC TOOLS
Basic mechanic's tool set
DMM
Jumper wires
Appropriate wiring diagrams
Circuit testers

Upon completion and review of this chapter you should be able to:

- Diagnose electrical problems by logic and symptom description.
- Perform troubleshooting procedures using meters, test lights, and jumper wires.
- Repair electrical wiring.
- Replace electrical connectors.

- Locate, test, adjust, and replace electrical switches on a transmission.
- Locate, test, and replace transmission-related electrical solenoids.
- Test and replace electromagnetic clutches.
- Diagnose transmission-related electronic control circuits.

Basic Electrical Troubleshooting

Certain diagnostic basics apply to all electrical systems. Many who have a difficult time diagnosing electrical problems do not understand what electricity is, what electrical meters can and do tell you, and how the different types of electrical problems affect a circuit.

Electrical **current** is a term used to describe the movement or flow of electricity. The greater the number of electrons flowing past a given point in a given amount of time, the more current the circuit has. This current, like the flow of water or any other substance, can be measured. The unit for measuring electrical current is the **ampere**. The instrument used to measure electrical current flow in a circuit is called an **ammeter**.

When any substance flows, it meets **resistance**. The resistance to electrical flow can be measured. The resistance to current flow produces heat, which can be measured to determine the amount of resistance. A unit of measured resistance is called an **ohm**. Resistance can be measured by an instrument called an ohmmeter. **Voltage** is electrical pressure—the force developed by the attraction of electrons to protons. The more positive one side of the circuit is, the more voltage is present in the circuit. Voltage does not flow; rather, it is the pressure that causes current flow. To have electricity, some force is needed to move the electrons between atoms. This force is the pressure that exists between a positive and negative point within an electrical circuit. This force, also called **electromotive force (EMF)**, is measured in units called **volts**. One volt is the amount of pressure (force) required to move 1 ampere of current through a resistance of 1 ohm. Voltage is measured by an instrument called a **voltmeter**.

The amount of current that flows in a circuit is determined by the resistance in that circuit. As resistance goes up, the current goes down. The energy used by a **load** is measured in volts. Amperage stays constant in a circuit, but the voltage is dropped as it powers a load. Measuring voltage drop determines the amount of energy consumed by the load.

> An **ampere**, the unit of measure for electrical current, is usually called an amp.

> **Load** is a term normally used to describe an electrical device that is operating on a circuit.

BASIC ELECTRICAL CIRCUITS

When electrons are able to flow along a path between two points, an electrical circuit is formed. An electrical circuit is considered complete when there is a path that connects the positive and negative terminals of the power source. Somewhere in the circuit there must be a load or resistance to control current flow in the circuit. The flow of electricity can be controlled and applied to do work. Components that use electrical power put a load on the circuit and change the electrical energy into a different form of energy.

All electrical circuits are connected as a series or parallel circuit. In a series circuit (Figure 10-1), current follows only one path and the amount of current flow through the circuit depends on the circuit's total resistance. To calculate the total resistance in a series circuit, all resistance values are added together. At each resistor, voltage is dropped and the total amount of voltage dropped in a series circuit is equal to the voltage of the source. Current in a series circuit is always constant.

Parallel circuits (Figure 10-2) allow current to flow in more than one path. This allows one power source to power more than one circuit or load. A vehicle's accessories and other electrical devices can be individually controlled through the use of parallel circuits. Within a parallel circuit, there is a common path to and from the power source. Each branch or leg of a parallel circuit behaves as if it were an individual circuit. Current flows only through the individual circuits when they are completed. All of the legs in the circuit do not need to be complete in order for current to flow through one of them.

In parallel circuits, total circuit current is equal to the sum of the amperages in all of the legs of the circuit. Equal amounts of voltage are applied to each leg or branch of the circuit. The total resistance is always less than the resistance of the leg with the smallest amount of resistance. The total resistance of two resistors in parallel can be calculated by dividing their product by their sum.

The legs of a parallel circuit may contain a series circuit. The resistance of that leg is the sum of the resistances in that leg. The resistance in each leg is used to calculate the total resistance of the circuit. Circuit current flows only through the common power and ground paths, therefore a change in a branch's resistance will not only affect the current in that branch but will also affect total circuit current.

Grounds

Most automotive electrical circuits use the chassis as the path to the negative side of the battery. These have a lead connected to the chassis, which are called **chassis grounds**. The use of the chassis as a ground eliminates the need to have a wire from a component back to the battery.

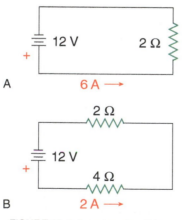

FIGURE 10-1 A series circuit has load (A) or loads (B) connected inline with the flow of current.

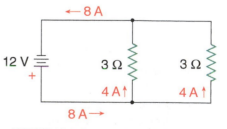

FIGURE 10-2 Current through the loads in a parallel circuit flows through separate paths.

Conductors

Two basic types of wires are used in automobiles: solid and stranded. Solid wires are single-strand conductors and are only used inside components. Stranded wires are made up of a number of small solid wires twisted together to form a single conductor. Stranded wires are the most commonly used type of wire in an automobile. Electronic units, such as computers, use specially shielded, twisted cable for protection from unwanted induced voltages that can interfere with computer functions. In addition, some solid-state components use printed circuits.

The current-carrying capacity and the amount of voltage drop in an electrical wire are determined by its length and gauge (size). The wire sizes are established by the Society of Automotive Engineers (SAE). Wire size is defined by the metric system or the **American wire gauge (AWG)** system. Sizes are identified by a numbering system ranging from number 0 to 20, with number 0 having the largest cross-sectional area and number 20 the smallest. Most automotive wiring ranges from number 10 to 18, with battery cables that are at least number 4 gauge. Battery cables need to be large-gauge wires capable of carrying high current for the starter motor.

Electrical Wiring Diagrams

Wiring diagrams, sometimes called **schematics**, are used to show how circuits are wired and how the components are connected. A typical service manual contains dozens of wiring diagrams vital to the diagnosis and repair of the vehicle.

A wiring diagram does not show the actual position of the parts on the vehicle or their appearance nor does it indicate the length of the wire that runs between components. It usually indicates the color of the wire's insulation and sometimes the wire gauge size. The first letter of the color coding is a combination of letters usually indicating the base color. The second letter usually refers to the strip color (if any). Tracing a circuit through a vehicle is basically a matter of following the colored wires.

Many different symbols are also used to represent components, such as motors, batteries, switches, transistors, and diodes. Common symbols are shown in Figure 10-3.

SERVICE TIP: Keep in mind that electrical symbols are not standardized throughout the automotive industry. Different manufacturers may have different methods of representing certain components, particularly the less common ones. Always refer to the symbol reference charts, wire color-code charts, and abbreviation tables listed in the vehicle's service manual to avoid confusion when reading wiring diagrams.

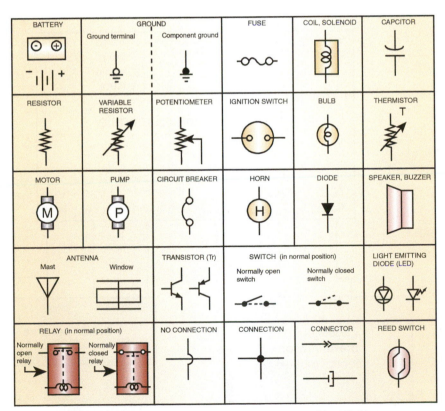

FIGURE 10-3 Common symbols used in electrical wiring diagrams.

Part of a typical wiring diagram is shown in Figure 10-4; notice that the components are also labeled.

Wiring diagrams can become quite complex. To avoid this, the vehicle's electrical system may be divided into many diagrams, each illustrating only one system, such as the back-up light circuit, oil pressure indicator light circuit, or wiper motor circuit. In more complex ignition, electronic fuel injection, and computer-control systems, one diagram may be used to illustrate only part of the entire circuit.

Getting the Right Diagram. Wiring diagrams are only valuable if they are the correct one for the vehicle being diagnosed. Most electronic service information systems will match the wiring diagram to the VIN. To view a wiring diagram for a particular system or component, refer to the index for the wiring diagram. The index will list a letter and number for each major component and the different connection points.

For some vehicles, only total vehicle wiring diagrams are available. These can make it difficult to locate the circuit you need. These diagrams also have an index that will identify the grid where a component can be found. The diagram is marked into equal sections by

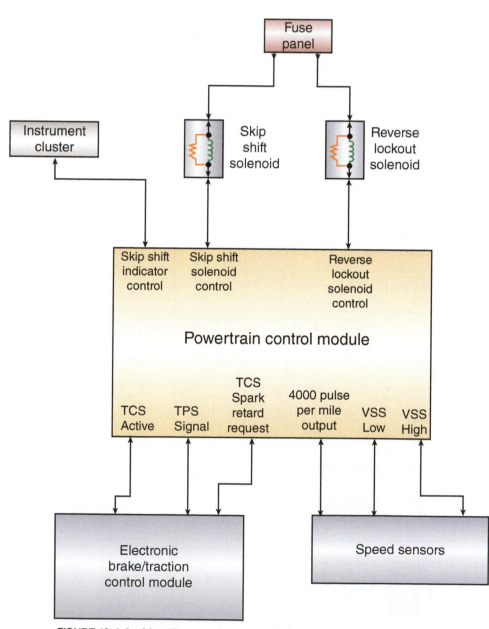

FIGURE 10-4 A wiring diagram of a reverse lockout and shift blocking circuit.

grid letters and numbers on the outside borders like a street map. If the wiring diagram is not indexed, locate the component by relating its general location in the vehicle to a general location on the diagram.

ELECTRICAL PROBLEMS

All electrical problems can be classified as opens, shorts, or high resistance. Knowing the type of problem helps identify the correct tests to conduct.

Open Circuits

An **open** occurs when a circuit is incomplete. Without a completed path, current cannot flow (Figure 10-5) and the load or component will not work. An open circuit can be caused by a disconnected wire or connector, a broken wire, or a switch in the OFF position. When a circuit is off, it is open. When the circuit is on, it is closed. Switches open and close circuits, but at times a fault will cause an open. Although voltage is applied up to the open point, there is no current flow. Without current flow, there are no voltage drops across the various loads.

Shorted Circuits

When a circuit has an unwanted path for current flow, it has a **short**. When an energized wire accidentally contacts the frame or body of the vehicle or another wire, current can travel in unintended directions. This can cause uncontrollable circuits and high-circuit current. Shorts are caused by damaged wire insulation, loose wires or connections, improper wiring, or careless installation of accessories.

A short creates an unwanted parallel leg or path in a circuit. As a result, circuit resistance decreases and current increases. The amount of current increase depends on the resistance of the short. The increased current flow can burn wires or components. Preventing this is the purpose of circuit protection devices. When a circuit protection device opens due to higher-than-normal current, a short is the likely cause. Also, if a connector or wires show signs of burning or insulation melting, high current is the cause, which is most likely caused by a short.

A short can be caused by a number of things and can be evident in a number of ways. It can be an unwanted path to ground. The short is often in parallel to a load and provides a

A **short** to ground is often called a grounded circuit.

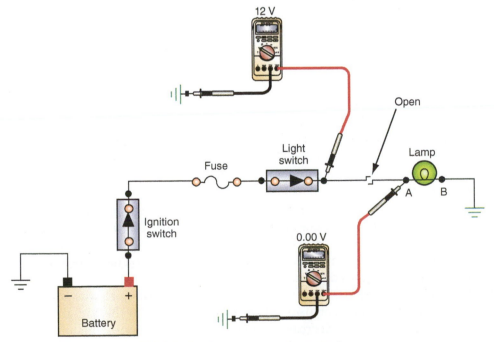

FIGURE 10-5 An open prevents current flow.

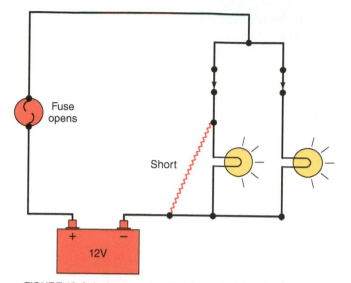

FIGURE 10-6 A short can is often in parallel to a load and provides a low-resistance path to ground. This short is probably caused by bad insulation that is allowing the power feed for the lamp to touch the ground for the same lamp.

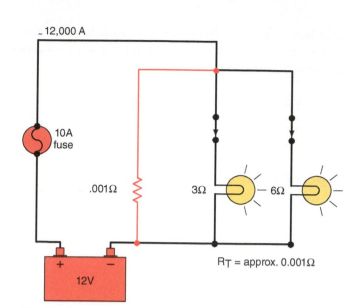

FIGURE 10-7 This is the same circuit as Figure 10-6 but is drawn to show the short as a parallel leg with very low resistance.

low-resistance path to ground. In Figure 10-6, the short is probably caused by bad insulation that is allowing the power feed for the lamp to touch the ground for the lamp. This problem creates a parallel circuit. Figure 10-7 represents Figure 10-6 but is drawn to show the short as a parallel leg with very low resistance. The resistance assigned to the short may be more or less than an actual short, but the value 0.001 ohms is given to illustrate what happens. With the short, the circuit has three loads in parallel: 0.001, 3, and 6 ohms. The total resistance of this parallel circuit is less than the lowest resistance, or 0.001 ohms. Using this value as the total resistance, circuit current is calculated to be more than 12,000 amperes, which is much more than the fuse can handle. The high current will burn the fuse and the circuit will not work. Some call this problem a "grounded circuit."

Sometimes two separate circuits become shorted together. When this happens, each circuit is controlled by the other. This may result in strange happenings, such as the horn sounding every time the brake pedal is depressed (Figure 10-8), or vice versa. In this case, the brake light circuit is shorted to the horn circuit.

High-Resistance Circuits

High-resistance problems occur when there is unwanted resistance in the circuit. The higher-than-normal resistance causes the current flow to decrease and the circuit is unable to operate properly. A common cause of this problem is corrosion at a connector. The corrosion becomes an additional load in the circuit (Figure 10-9). This load uses some of the circuit's voltage, which prevents full voltage to the normal loads in the circuit.

Many sensors on today's vehicles are fed a 5-volt reference signal. The signal or voltage from the sensor is less than 5 volts, depending on the condition it is measuring. A poor ground in the circuit can cause higher-than-normal readings back to the computer. This seems to be contradictory to other high-resistance problems. Look at Figure 10-10. This is a voltage divider circuit with two resistors in series and the voltage reference tap between them. Since the total resistance in the circuit is 12 ohms, the circuit current is 1 ampere. Therefore, the voltage drop across the 7-ohm resistor is 7 volts, leaving 5 volts at the tap.

Figure 10-11 is the same circuit but a bad ground of 1 ohmere was added. This low resistance could be caused by corrosion at the connection. With the bad ground, the total resistance is now 13 ohms. This decreases our circuit current to approximately 0.92 amperes. With this lower amperage, the voltage drop across the 7-ohm resistor is now about 6.46 volts,

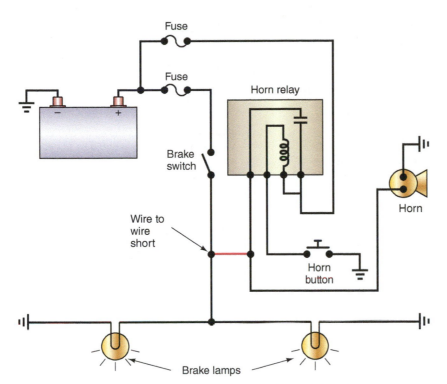

FIGURE 10-8 The above horn circuit is shorted to the brake light circuit.

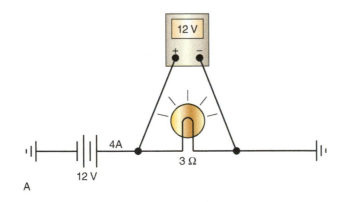

A

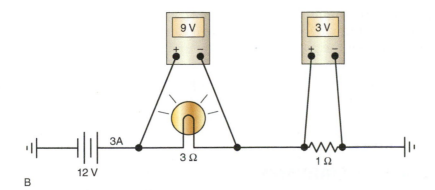

B

FIGURE 10-9 (A) A normal 3-ohm light circuit. (B) The same circuit but with a high-resistance ground. Notice that the voltage drop across the lamp decreased as did the circuit current. The bulb will burn dimmer.

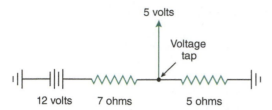

FIGURE 10-10 Two resistors in series with the voltage reference tap between them. Since the total resistance in the circuit is 12 ohms, the circuit current is 1 ampere. Therefore the voltage drop across the 7-ohm resistor is 7 volts, leaving 5 volts at the tap.

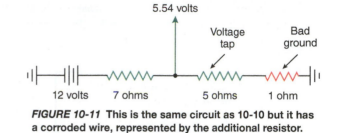

FIGURE 10-11 This is the same circuit as 10-10 but it has a corroded wire, represented by the additional resistor.

leaving 5.54 volts at the voltage tap. This means the reference voltage would be more than one-half volt higher than it should be. As a result, the computer will be receiving a return signal of at least one-half volt higher than it should. Depending on the sensor and the operating conditions of the vehicle, this could be critical.

DIAGNOSIS

Always begin diagnostics by verifying the customer's complaint. Then operate the system and others to get a good understanding of the problem. Often there are problems that are not as evident or bothersome to the customer that will provide helpful information. Refer to the wiring diagrams and study the circuit that is affected. From the diagram, identify appropriate testing points and probable problems. Then test and use logic to identify the cause of the problem.

Making assumptions or jumping to conclusions can be very expensive and a total waste of time. The basic steps for logical diagnosis should be followed:

- **Gather information about the problem.** From the customer find out when and where the problem happens and what exactly happens.
- **Verify that the problem exists.** Take the vehicle for a road test or try the components of the customer's concerns and try to duplicate the problem, if possible.
- **Thoroughly define what the problem is and when it occurs.** Pay attention to the conditions present when the problem happens. Also pay attention to the entire vehicle; another problem may be evident that was not evident to the customer.
- **Research all available information to determine the possible causes of the problem.** Look at all service bulletins and other information related to the problem. Study the system's wiring diagram.
- **Isolate the problem.** Based on an understanding of the problem and circuit, make a list of probable causes. Narrow down this list of possible causes by checking the obvious or easy-to-check items. This includes a thorough visual inspection.
- **Continue testing to pinpoint the cause of the problem.** Once you know where the problem should be, test until you find it! Begin testing to determine whether or not the most probable cause is the problem. If this is not the cause, move to the next most probable cause. Continue this until the problem is solved.
- **Locate and repair the problem, then verify the repair.** Once you have determined the cause, make the necessary repairs. Never assume that your work solved the problem. Operate the circuit to be sure the original problem has been corrected and that there are no other faults.

ELECTRICAL TEST EQUIPMENT

Several meters are used to test and diagnose electrical systems. These are the voltmeter, ohmmeter, and ammeter. A multimeter, often referred to as a **digital multimeter (DMM)**, combines the capabilities of these meters and is the most commonly used meter. Diagnosis may also require the use of a digital storage oscilloscope, scan tool, circuit tester, jumper wires, and other equipment.

⚠ **WARNING:** **When you are testing an electronic circuit or component, make sure the meter you are using is labeled as a high-impedance meter.**

Multimeters

A basic multimeter measures DC and AC voltage, current, and resistance. Most current multimeters also check engine speed, signal frequency, duty cycle, pulse width, diodes, temperature, and pressures. The desired test is selected by turning a control knob or depressing keys on the meter. These meters can be used to test simple electrical circuits, ignition systems, input sensors, fuel injectors, batteries, and starting and charging systems.

Digital Storage Oscilloscope

Digital storage oscilloscopes (DSOs) are also called lab scopes and are fast-reacting meters that measure and display voltages within a specific time frame. The voltage readings appear as a waveform or trace on the scope's screen. An upward movement of the trace indicates an increase in voltage, and a downward movement shows a decrease in voltage. The waveform displays what is happening at that time, and problems can be observed when they happen.

Scan Tools

A scan tool can retrieve fault codes from a computer's memory and digitally display the codes. A scan tool may also perform other diagnostic functions depending on the year and make of the vehicle. Many scan tools have removable modules that are updated each year. These modules are designed to test the systems on various makes of vehicles.

The scan tool must be programmed for the model year, make of vehicle, and type of engine. Some scan tools have a built-in printer to print test results, while other scan tools may be connected to an external printer.

Some scan tools display diagnostic information based on the trouble code. Service bulletins may also be indexed on the tool after vehicle information is entered into the tester. Most of these tools will display sensor specifications and instructions for further testing. Other scan tools can also function as a multimeter.

Trouble codes are only set when a voltage signal is entirely out of its normal range. The codes help technicians identify the cause of the problem. If a signal is within its normal range but is still not correct, the computer will not display a trouble code. However, a problem will still exist. To identify the cause of these problems, most manufacturers recommend that the signals to and from the computer be carefully looked at. This is done by observing the serial data stream or through the use of a breakout box.

Circuit Testers

Circuit testers (test lights) are used to identify shorted and open circuits. Low-voltage testers are used to troubleshoot 6- to 12-volt circuits. High-voltage circuit testers diagnose higher voltage systems, such as the secondary ignition circuit. High-impedance test lights are available for diagnosing electronic systems.

There are two basic types of test lights: the nonpowered and powered test light. Nonpowered test lights are used to check for available voltage. With the wire lead connected to a good ground and the tester's probe at a test point, the light turns on with the presence of voltage (Figure 10-12). The amount of voltage determines the brightness of the light.

A **test light** is used mainly to check for voltage at various points within a circuit.

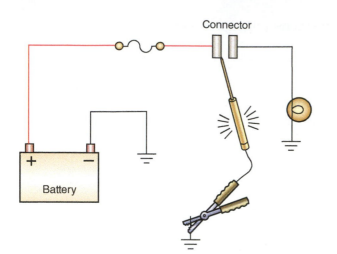

FIGURE 10-12 A test light is mostly used to check for voltage at various points within the circuit.

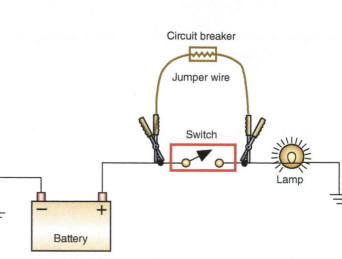

FIGURE 10-13 A jumper wire can be used to bypass part of the circuit or an electrical component.

A self-powered test light is often called a continuity tester.

A **self-powered test light** is used to check for continuity. Hooked across a circuit or component, the light turns on if the circuit is complete. A powered test light should only be used if the power for the circuit or component has been disconnected.

Jumper Wires

Jumper wires are used to bypass individual wires, connectors, or components. Bypassing a component or wire helps to determine if that part is faulty (Figure 10-13). If the symptom is no longer evident after the jumper wire is installed, the bypassed part is faulty. Technicians typically have jumper wires of various lengths; usually they have a fuse or circuit breaker in them to protect the circuits being tested.

Computer Memory Saver

Before disconnecting a vehicle's battery, connect a memory saver to the vehicle. This saver will preserve the memory in the radio, electronic accessories, and the engine, transmission, and body computers.

Two types of memory savers are available or can be made for late-model vehicles. For a power source, use a 12-volt automotive battery or a 12-volt dry-cell lantern battery. If the vehicle's cigar lighter is continuously powered, a cigar lighter adapter plug with suitable wire leads and large alligator clips can be attached to the auxiliary battery. If the cigar lighter is controlled by the ignition switch, a set of jumper wires with alligator clips can be connected to the vehicle's electrical system.

If you connect the memory saver under the hood, do not connect it to the battery cable clamps. Removing and reinstalling the battery will likely dislodge the memory saver's alligator clips and make it useless. Instead, connect the saver's negative (−) lead to a good engine ground and the positive (+) lead to a point that is hot at all times, such as the battery connection at the generator or starter relay. Check a wiring diagram if you are unsure about the connection points.

USING MULTIMETERS

Multimeters are available with either analog or digital displays. Analog meters use a needle to point to a value on a scale. A digital meter shows the measured value in numbers or digits. The most commonly used meter is the DMM, sometimes called a digital volt/ohmmeter (DVOM). Analog meters should not be used to test electronic components. They have low internal resistance (input impedance). The low-input impedance allows too much current to flow through

circuits and can damage delicate electronic devices. Digital meters, on the other hand, have high-input impedance, 1 Meg ohms (million ohms) to 10 Meg ohms. Moreover, metered voltage for resistance tests is well below 5 volts, reducing the risk of damage to sensitive components and delicate computer circuits.

AUTHOR'S NOTE: The DMM used on high-voltage systems in hybrids should be a category III meter. There are basically four categories for electrical meters, each built for specific purposes and to meet certain standards. The categories define how safe a meter is when measuring certain circuits. The standards for the various categories are defined by the American National Standards Institute (ANSI), the International Electrotechnical Commission (IEC), and the Canadian Standards Association (CSA). A CAT III meter is required for testing hybrid vehicles because of the high voltages, three-phase current, and potential for high-transient voltages. Within a particular category, meters have different voltage ratings. These reflect a meter's ability to withstand higher transient voltages. Transient voltages are voltage surges or spikes that occur in AC circuits. To be safe, you should have a CAT III 1000 V meter.

The front of a DMM normally has four distinct sections: the display area, range selectors, mode selector, and jacks for the test leads (Figure 10-14). In the center of the display are large digits that represent the measured value. Normally there will be four to five digits with a decimal point. To the right of the number, the measured units are displayed (V, A, Ω). These units may be further defined by a symbol to denote a multiplier value, such as a "K," which represents 1000.

Setting the range on a DMM is important. If the measurement is beyond the set range, the meter will display a reading of "O.L.," or over limit. The range on some DMMs is manually set, while others have an "auto range" feature, in which the appropriate scale is automatically selected by the meter. Auto-ranging is helpful when you do not know what value to expect. When using a meter with auto range, make sure you note the range being used by the meter. There is a big difference between 10 ohms and 10,000,000 (10 M) ohms.

FIGURE 10-14 A DMM is one of the most versatile electrical test instruments.

> ⚠️ **WARNING:** Many DMMs with auto range display the measurement with a decimal point. Make sure you observe the decimal and the range being used by the meter. A reading of .972 K ohms equals 972 ohms. If you ignore the decimal point, you will interpret the reading as 972,000 ohms.

The mode selector defines what the meter will be measuring. The number of available modes varies with meter design, but nearly all have the following:

- Volts AC
- Volts DC
- Millivolts (mV) DC
- Resistance/continuity (ohms)
- Diode check
- Amps or milliamps AC/DC

Most DMMs have two test leads and four input jacks. The black test lead always plugs into the COM input jack and the red lead plugs into one of the other input jacks, depending on what is being measured. It is good to have multiple sets of test leads, each for specific purposes. For example, a lead set with small tips is ideal for probing in hard-to-reach or tight spaces. Other test leads may be fitted with clips to hold the lead at a point during testing.

The three input jacks are typically:

- "**A**" for measuring up to 10 amperes of current.
- "**A/mA**" for measuring up to 400 mA of current.
- "**V/diode**" for measuring voltage, resistance, conductance, capacitance, and checking diodes.

Measuring Voltage

A DMM can measure source voltage, available voltage, and voltage drops. Voltage is measured by placing the meter in parallel to the component or circuit being tested (Figure 10-15). There are normally several voltage ranges available on a DMM. Most automotive circuits range from 50 mV to 15 volts.

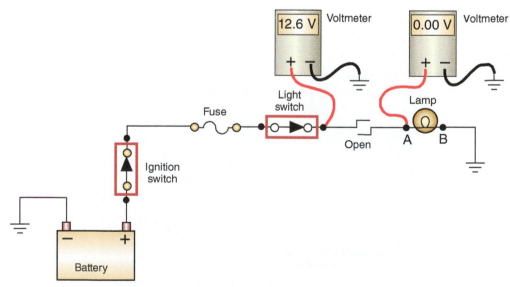

FIGURE 10-15 A voltmeter is used to measure voltage anywhere in a circuit. Notice voltage is present before the open and not after the open.

⚠ **WARNING:** Hybrid vehicles have much higher voltage; always follow all safety precautions and service procedures when working with high-voltage circuits.

A DMM can be used to check for proper circuit grounding. For example, if the voltmeter shows battery voltage at a lamp, but no lighting is seen, the bulb or socket could be bad or the ground connection is faulty.

If the bulb is not defective, the problem is the light socket or ground wires. Connect the voltmeter to the ground wire and a good ground. If zero volts is measured, move the positive meter lead to the power-feed side of the bulb. If zero volts are measured there, the light socket or feed wire is defective. In a normal light circuit, there should be zero volts at the negative side of the bulb. If the socket is not defective and some voltage is measured at the ground, the ground circuit is faulty. The higher the voltage, the greater the problem.

To measure DC voltage:

1. Set the mode selector switch to volts DC.
2. Select the auto-range function or manually set the range to match the anticipated value, normally the range is set to the closest to and higher than 12 volts.
3. Connect the test leads in parallel to the circuit or component. The red lead is connected to the most positive side (side closest to the battery) and the black lead to a good ground.
4. Read the voltage on the meter. If the reading is negative, the leads are probably reversed.

Voltage Drop Test. Measuring voltage drop is a very important test. It can identify circuits with unwanted resistances. This test can be performed between any two points in a circuit and across any component, such as wires, switches, relay contacts and coils, and connectors (Figure 10-16). A voltage drop test can find excessive resistance that may not be detected using an ohmmeter. An ohmmeter works by passing a small amount of current through the component being tested. A voltage drop test is conducted with the circuit operating.

If there is 12 volts available at the battery, and the switch for a lamp is closed, there should also be 12 volts available at the light. If less than 12 volts is measured at the bulb, additional resistance is somewhere else in the circuit. The lamp may light, but not as brightly as it should. This is because the resistance caused a decrease in applied voltage and circuit current.

The procedure for measuring voltage drop is shown in Photo Sequence 26. Remember to always connect the positive test lead to the most positive side of the part or circuit being tested and connect the negative test lead to the ground side. Also, make sure the circuit is turned on.

Current must be flowing in order to have a voltage drop. Voltage drops should not exceed the following:

- 200 mV across a wire or connector
- 300 mV across a switch or relay contacts
- 100 mV at a ground connection
- 0 mV to less than 50 mV across all sensor connections

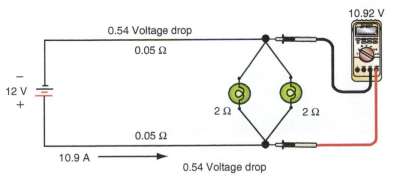

FIGURE 10-16 To measure voltage drop, the voltmeter is connected across the item being tested.

PHOTO SEQUENCE 26

VOLTAGE DROP TEST TO LOCATE HIGH CIRCUIT RESISTANCE

P26-1 Tools required to test for excessive resistance in a starting circuit are fender covers, a DVOM, and a remote starter switch.

P26-2 Connect the positive lead of the meter to the positive battery post. If possible, do not connect the lead to the cable clamp.

P26-3 Connect the negative lead of the meter to the main battery terminal on the starter motor.

P26-4 To conduct a voltage drop test, current must flow through the circuit. In this test, the ignition system is disabled and the engine is cranked using a remote starter switch.

P26-5 With the engine cranking, read the voltmeter. The reading is the amount of voltage drop.

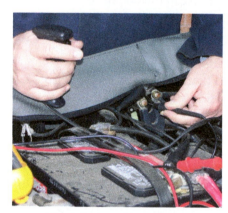

P26-6 If the reading is out of specifications, test at the next connection toward the battery. In this instance, the next test point is the starter side of the solenoid.

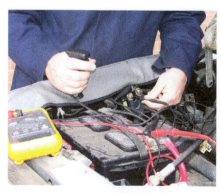

P26-7 Crank the engine and touch the negative test lead to the starter side of the solenoid. Observe the voltmeter while the engine is cranking.

P26-8 Test in the same manner on the battery side of the solenoid. This is the voltage drop across the positive circuit from the battery to the solenoid.

Measuring AC Voltage. There are two ways multimeters display AC voltage: root mean square (RMS) and average responding. When an AC voltage signal is a true sine wave, both methods will display the same reading. Since most automotive sensors do not produce pure sine wave signals, it is important to know how the meter will display AC voltage. RMS meters convert the AC signal to a comparable DC voltage signal. Average responding meters display the average voltage peak. Always identify how the voltage is stated in the specifications.

Measuring Current

Testing the current gives a true picture of what is happening in the circuit because the circuit is under load. Low current means the circuit has higher-than-normal resistance and high current means the circuit has lower-than-normal resistance. Many technicians don't check current because few specifications list what is normal. This is not a problem since current can be calculated if you know the resistance and voltage in the circuit.

Before checking current, make sure the meter is capable of measuring the suspected amount. To check the rating of the meter, look at the rating printed next to the DMM input jacks or check the rating of the meter's fuse (maximum current capacity is typically the same as the fuse rating). If you suspect that a measurement will have a current higher than the meter's maximum rating, use an inductive current probe. An ammeter must be connected in series with the circuit (Figure 10-17); this allows circuit current to flow through the meter. To measure current:

1. Turn off the circuit that will be tested.
2. Connect the test leads to the correct input jacks on the DMM.
3. Set the mode selector to the correct current setting (normally amperes or milliamps).
4. Select the auto-range function or manually select the range for the expected current.
5. Open the circuit at a point where the meter can be inserted.
6. Connect a fused jumper wire to one of the test leads.
7. Connect the red lead to the positive side of the circuit and the black lead to the other side.
8. Turn on the circuit.
9. Read the display on the DMM.
10. Compare the reading to specifications or your calculations.

⚠ WARNING: **Never connect an ammeter across the battery or a load. This puts the meter in parallel with the circuit and will blow the meter's fuse or possibly destroy the meter.**

Inductive Current Probes. Many DMMs have current probes (current clamps) that eliminate the need to insert the ammeter into the circuit. These probes read current by sensing the magnetic field formed by current flow (Figure 10-18). To use a current probe, the DMM's mode selector is set to read millivolts. The probe is then connected to the meter and turned

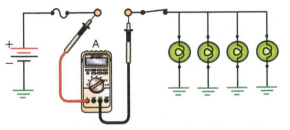

FIGURE 10-17 An ammeter should be connected in series with the circuit.

FIGURE 10-18 A current probe.

on. Some probes must be zeroed before the probe is clamped around a wire. The DMM may have a zero adjust control, which is turned until zero reads on the meter's display. The clamp is placed around a wire in the circuit. Make sure the arrow on the clamp is pointing in the direction of current flow. After the clamp is in place, the circuit is turned on and the voltage read on the display. The voltage reading is then converted to an amperage reading: 1 mV = 1 ampere.

Measuring Resistance

The circuit or component must first be disconnected from power before connecting the ohmmeter (Figure 10-19). Connecting an ohmmeter into a live circuit usually results in damage to the meter.

Checking the resistance checks the condition of a component or circuit. Often specifications list a normal range of resistance values for specific parts. If the resistance is too high,

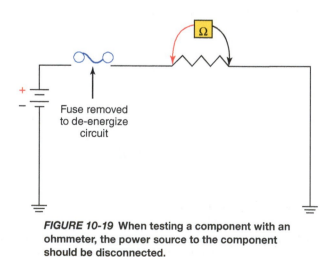

FIGURE 10-19 When testing a component with an ohmmeter, the power source to the component should be disconnected.

check for an open circuit or a faulty component. If the resistance is too low, check for a shorted circuit or faulty component.

Ohmmeters are also used to trace and check wires or cables. Connect one probe of the meter to the known wire at one end of the harness and the other probe to each wire at the other end of the harness. Any evidence of resistance indicates it is the same wire. Using this same method, you can check for a defective wire. If low resistance is shown on the meter, the wire is sound. If no resistance is measured, the wire is open. Medium to high resistance means the wire is shorted to another wire or the harness is defective.

To measure resistance:

1. Make sure the circuit or component to be tested is not connected to any power source.
2. Set the DMM mode selector to measure resistance.
3. Select the auto-range feature or manually select the appropriate range.
4. Calibrate the meter by holding the two test leads together and adjusting the meter to zero. On some DMMs, the calibration should be checked whenever the range is changed.
5. Connect the meter leads across the component or part of the circuit that will be checked.
6. Read the measured value. The DMM will show a zero or close to zero when there is good continuity. If there is no continuity, the meter will display an infinite or over-limit reading.

Continuity Tests. Many DMMs have a continuity test mode that makes a beeping noise when there is continuity through the item being tested. This audible sound will continue as long as there is continuity. This feature can be handy for finding the cause of an intermittent problem. By connecting the DMM across a circuit and wiggling sections of the wiring harness, a problem can be noted when the beeping stops after a particular section or wire has been moved.

MIN/MAX Readings

Some DMMs also feature a MIN/MAX function. This displays the maximum, minimum, and average voltage recorded during the time of the test. This is valuable when checking sensors or when looking for electrical noise. Noise is primarily caused by radio frequency interference (RFI), which may come from the ignition system. RFI is an unwanted voltage signal that rides on a signal. This noise can cause intermittent problems with unpredictable results. The noise causes slight increases and decreases in the voltage. When a computer receives a voltage signal with noise, it will try to react to the minute changes. As a result, the computer responds to the noise rather than the voltage signal.

Other Measurements

Some multimeters may also be able to measure duty cycle, pulse width, and frequency. Duty cycle (Figure 10-20) is the amount of time something is on during one cycle, and it is measured in a percentage. A 60 percent duty cycle means that a device is on 60 percent of the time and off 40 percent of one cycle.

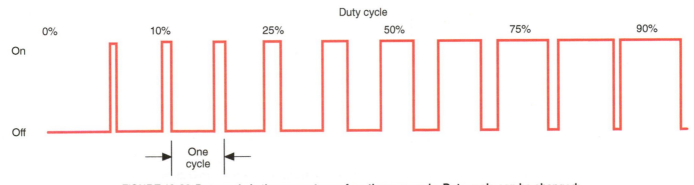

FIGURE 10-20 Duty cycle is the percentage of on-time per cycle. Duty cycle can be changed; however, total cycle time remains constant.

Pulse width is normally measured in milliseconds. When measuring pulse width, you are looking at the amount of time something is on.

To accurately measure duty cycle, pulse width, and frequency, the meter's trigger level must be set. The trigger level tells the meter when to start counting. Trigger levels can be set at certain voltage levels or at a rise or fall in the voltage. Normally, meters have a built-in trigger level that corresponds with the voltage range setting. If the voltage does not reach the trigger level, the meter will not recognize a cycle. On some meters, you can select between a rise and a fall in voltage to trigger the cycle count. Most technicians refer to this as a positive (voltage rise) or negative (voltage fall) slope trigger.

Some DMMs can measure temperature. These meters are equipped with a thermocouple. Temperature readings can be made in Fahrenheit or Celsius. The thermocouple is connected to the DMM and placed on or near the object to be checked. More elaborate DMMs have the ability to store and download data to a PC.

USING LAB SCOPES

When measuring voltage with an analog voltmeter, the meter only displays the average values at the point being probed. Digital voltmeters simply sample the voltage several times each second and update the meter's reading at a particular rate. If the voltage is constant, good measurements can be made with both types of voltmeters. A scope, however, will display any change in voltage as it occurs.

The screen is divided into small divisions of time and voltage (Figure 10-21). These divisions set up a grid pattern on the screen. Time is represented by the horizontal movement of the trace. Voltage is measured by the vertical position of the trace. Since the scope displays voltage over time, the trace moves from the left (the beginning of measured time) to the right (the end of measured time). The value of the divisions can be adjusted to improve the view of the voltage trace. The grid serves as a reference for measurements.

Since a scope displays actual voltage, it will display any electrical noise or disturbances that accompany the voltage signal (Figure 10-22). Electrical disturbances or **glitches** are momentary changes in the signal. These can be caused by intermittent shorts to ground, shorts to power, or opens in the circuit. These problems can occur for only a moment or may last for some time. By observing a voltage signal and wiggling or pulling a wiring harness, any looseness can be detected by a change in the voltage signal.

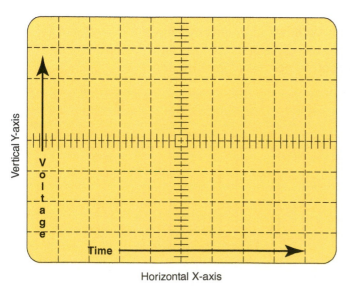

FIGURE 10-21 The screen of a scope is divided into increments of voltage and time.

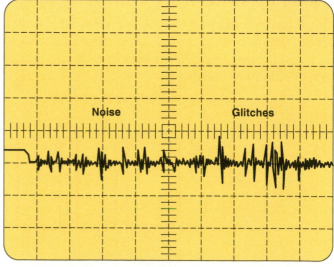

FIGURE 10-22 Noise and glitches caused by EMF or poor connections show in a voltage trace as irregular increases and decreases in voltage.

Analog versus Digital Scopes

Analog scopes show the actual activity of a circuit and are called real-time scopes. This simply means that what is taking place at that time is what you see on the screen. Analog scopes have a fast update rate that allows for the display of activity without delay.

A digital scope, commonly called a digital storage oscilloscope, converts the voltage signal into digital information and stores it into its memory. Some DSOs send the signal directly to a computer or a printer, or save it to a disk. To help in diagnostics, a technician can "freeze" the captured signal for close analysis. DSOs also have the ability to capture low-frequency signals. Low-frequency signals tend to flicker on an analog screen. To have a clean waveform on an analog scope, the signal must be repetitive and occurring in real time. The signal on a DSO is not quite real time. Rather, it displays the signal as it occurred a short time before.

Most DSOs have a sampling rate of 1 million samples per second. This is quick enough to serve as an excellent diagnostic tool, as it allows slight changes in voltage to be observed. Slight and quick voltage changes cannot be observed on an analog scope. Since digital signals are based on binary numbers, the trace appears to be slightly choppy when compared with an analog trace. However, the voltage signal is sampled more often, which results in a more accurate waveform. The waveform is constantly being refreshed as the signal is pulled from the scope's memory.

Both an analog and a digital scope can be dual- or multiple-trace (Figure 10-23) scopes. This means they have the capability of displaying more than one trace at one time. By watching the traces simultaneously, the cause and effect of a sensor is observed, as well as comparing a good or normal waveform to the one being displayed.

Waveforms

Any change in the amplitude of the trace indicates a change in voltage. When the trace is a straight horizontal line, the voltage is constant. A diagonal line up or down represents a gradual increase or decrease in voltage. A sudden rise or fall in the trace indicates a sudden change in voltage.

Often, AC voltage (from RFI) rides on a DC voltage signal. The consistent change of polarity and amplitude of the AC signal causes slight changes in the DC voltage signal. A normal AC signal changes its polarity and amplitude over a period of time. The waveform created by AC voltage is typically called a sine wave. One complete sine wave shows the voltage moving from zero to its positive peak then moving down through zero to its negative peak and returning to zero.

One complete sine wave is a cycle. The number of cycles that occur per second is the frequency of the signal. Checking frequency is one way to look at the operation of some electrical components. Input sensors are the most common components that produce AC voltage.

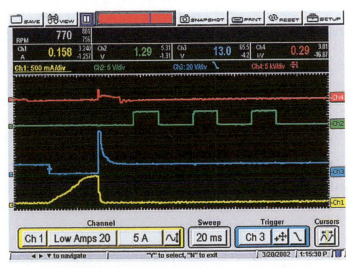

FIGURE 10-23 A scope with multiple channels.

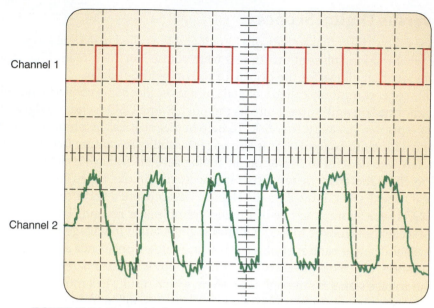

FIGURE 10-24 Channel 1 displays a square wave and channel 2 displays an AC wave.

Permanent magnet voltage generators produce an AC voltage that can be checked on a scope. AC voltage waveforms should also be checked for noise and glitches.

DC voltage waveforms may appear as a straight line or a line showing a gradual change in voltage. Sometimes a DC voltage waveform will appear as square wave that shows voltage making an immediate change (Figure 10-24). Square waves are identified by having straight vertical sides and a flat top. This wave shows voltage being applied (circuit being turned on), voltage being maintained (circuit remaining on), and no voltage applied (circuit is turned off).

Scope Controls

Depending on manufacturer and model of the scope, the function and number of its controls will vary. However, nearly all scopes have intensity, vertical (*y*-axis) adjustments, horizontal (*x*-axis) adjustments, and trigger adjustments. The intensity control is used to adjust the brightness of the trace. This allows for clear viewing regardless of the light around the scope screen.

The vertical adjustment controls the voltage that will be shown per division. If the scope is set at 0.5 (500 milli) volts, a 5-volt signal will need 10 divisions. Likewise, if the scope is set to 1 volt, 5 volts will need only five divisions. While using a scope, it is important to set it so the voltage can be accurately read. Setting the vertical too low may cause the waveform to move off the screen, while setting it too high may cause the trace to be flat and unreadable.

The horizontal position control allows the horizontal position of the trace to be set. The horizontal control is the time control of the trace. If the time per division is set too low, the complete trace may not show across the screen. Also, if the time per division is set too high, the trace may be too crowded for detailed observation. The time per division (TIME/DIV) can be set from very short periods of time (millionths of a second) to full seconds.

Trigger controls tell the scope when to begin a trace. This is important when trying to observe the timing of something. Proper triggering will allow the trace to repeatedly begin and end at the same points on the screen. There are typically numerous trigger controls on a scope. The trigger mode selector has a NORM and AUTO position. In the NORM setting, no trace will appear on the screen until a voltage signal occurs within the set time base. The AUTO setting will display a trace regardless of the time base.

Slope and level controls are used to define the actual trigger voltage. The slope switch determines whether the trace will begin on a rising or falling of the signal. The level control sets when the time base will be triggered according to a certain point on the slope.

A trigger source switch tells the scope which input signal to trigger on. This can be channel 1, channel 2, line voltage, or an external signal. External signal triggering is very useful when observing the trace of a component that may be affected by the operation of another component. An example of this would be observing fuel injector activity when changes in throttle position are made. The external trigger would be voltage changes at the throttle position sensor. The displayed trace would be the cycling of a fuel injector.

Graphing Multimeters

One of the latest trends in diagnostic tools is a **graphing multimeter (GMM)**. A GMM is a DMM that displays voltage, resistance, current, and frequency as a waveform. The display shows the current minimum and maximum readings as a graph. By observing the graph, any undesirable change during the transition from a low reading to a high reading, or vice versa, can be observed. These glitches are some of the more difficult problems to identify without a graphing meter or a lab scope. The waveform on a DSO may miss a change in voltage, resistance, or current that occurs too quickly for the scope to detect and display it. DSOs do not display real-time data. There is always a slight delay between what it measures and what it displays, and depending on the refresh rate of the display it may miss some changes completely. A GMM is perhaps the best tool to use when trying to find the cause of intermittent problems in low-voltage DC circuits.

The capabilities of a GMM depend on the manufacturer and the model. Some GMMs' features include a signal and data recorder, individual component tests, the ability to display measurements along with a graph, glitch capture, and audible alarm. Some also have an electronic library of known-good signals for comparison. Some even have wiring diagrams and a vehicle-specific database of diagnostic and test information.

Transferring Data to a PC. Many DSOs and GMMs allow for the transfer of captured information to a PC through a cable or wireless interface. This feature allows for better viewing of the waveforms and other data. It also allows for the creation of a personal library, which may be helpful in the future.

TROUBLESHOOTING LOGIC

Knowing the type of problem that is causing the customer's concerns will dictate what tests should be conducted. If something does not work, the problem is most likely caused by a short or an open. If the fuse for that circuit is blown, the problem is a short. If the fuse is good, the problem is an open. If a part does not work correctly, such as a dim light bulb, the problem is high resistance.

Quick voltage checks will also help define the problem. Check for voltage at the part that is not working correctly. If source voltage is present, the part is bad or the ground circuit is faulty. If less than source voltage is at the part, there is a fault in the power feed to it. Also, measure the voltage drop across the part; this can indicate a problem with the part. If a faulty part is suspected, it should be checked or replaced.

When making any checks with a meter, follow all safety precautions. Try to take all measurements at a connector. Because the terminals at the connector can be damaged by inserting a meter's test leads into the connector, always use the correct adapter on the ends of the test leads. Adapters are available to match the size of the terminals. Using too large an adapter can deform the terminals. When measurements are taken at the mating side (front) of a disconnected connector, it is called **front probing**. When measurements are taken at the back or wire side of a connected connector, it is called **back probing**. Front probing is the preferred way to take measurements.

The key to identifying the exact cause of the problem is limiting all testing to the components and circuits that could be causing the problem. An understanding of the problem, coupled with an understanding of the system, will lead to the fault. A wiring diagram will serve as the map to the problem. Your understanding and knowledge will tell you where you want to go and the wiring diagram will tell you how to get there.

TESTING FOR COMMON PROBLEMS

It would take thousands of pages to describe all of the possible combinations of electrical problems. But you are in luck: *All* problems can be boiled down to one of three things. Identifying which one you are looking for will define what tests you need to conduct.

Testing for Opens

An open is evident by an inoperative component or circuit. Study the wiring diagram for the component and begin your testing at the most accessible place in the circuit. Check for voltage at the positive side of the load. If there are zero volts, move to the output of the control. If there is at least 10.5 volts, the open is between the control and the load. If the reading is 10.5 volts or more, check the ground side of the load (Figure 10-25). If the voltage is 1 volt or less and the load does not work, the load is bad. If the voltage at ground is greater than 1 volt, there is excessive resistance or an open in the ground circuit. If the voltage at the positive side of the load was less than 10.5 volts, move the positive lead of the voltmeter toward the battery, testing all connections along the way. If 10.5 volts or more are present at any connector, there is an open between that point and the point previously checked. If battery voltage was present at the ground of the load, there is an open in the circuit. Use a jumper wire to verify the location of the open.

Testing for Shorts

Use an ohmmeter to check for an internal short in a component. If the component is good, the meter will read the specified resistance. If it is shorted, it will read lower than normal or zero resistance. Also, if the component has more than two terminals or pins, check for continuity across all combinations of these. Refer to the wiring diagram to see where there should be continuity. Any abnormal readings indicate an internal short.

When a fuse is blown, this is probably due to a wire-to-wire short or a short to ground. To test these circuits, a special jumper wire with a circuit breaker should be used as a substitute

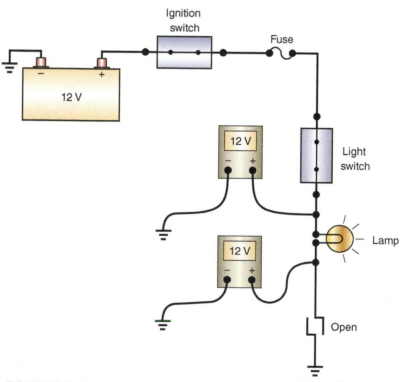

FIGURE 10-25 Since battery voltage is present at both sides of the bulb, there is an open in the circuit after the bulb. Checking this is a good way to identify a bad ground.

for the blown fuse. The jumper wire is fit with a 10- or 20-ampere self-resetting circuit breaker. This will allow for testing the circuit without causing damage to the wires and components. Some tool companies include a buzzer inline with the breaker.

⚠ **WARNING: Never bypass a fuse with a wire. The circuit breaker should be rated at no more than 50 percent higher than the original fuse. This offers circuit protection and provides amperage for testing. After the cause of the problem is corrected, install a fuse with the specified rating.**

Consider a circuit that normally draws 6 amperes and is protected by a 10-ampere fuse. If the circuit constantly blows the fuse, a short exists somewhere in the circuit. Mathematically, each light should draw 1.5 amperes (6 ÷ 4 = 1.5). To find the short, remove all lights from their sockets. Then, close the switch and read the ammeter. With the loads disconnected, the meter should read 0 amperes. If there is any reading, the wire between the fuse and the sockets is shorted to ground.

If 0 ampere was measured, reconnect each light in sequence, the reading should increase 1.5 amperes with each bulb. If, when making any connection, the reading is higher than expected, the problem is in that part of the light circuit.

When a wire-to-wire short is suspected, check the wiring diagram for all of the affected components. Identify all points where the affected circuits share a connector. Check the circuit protection devices for the circuits. Check the wiring for signs of burned insulation and melted conductors. If a visual inspection does not identify the cause of the short, remove one of the fuses for the affected circuits. Install a jumper wire across the fuse holder terminals. Activate that circuit and disconnect the loads that should be activated by the switch. This will create open circuits and there should be no current flow. If there is current flow, there is a short in the circuit. Disconnect all connectors in the circuit, one at a time. If current stops when a connector is disconnected, the short is in that circuit.

If the problem is a short to ground, the circuit's fuse or other protection device will be open. If the circuit is not protected, the wire, connector, or component will be burned or melted. To keep current flowing in the circuit so you can test it, connect the jumper wire across the fuse holder. The circuit breaker will cycle open and closed, allowing you to test for voltage in the circuit. Connect a test light in series with the cycling circuit breaker. Using the wiring diagram, identify the location of the connectors in the circuit. Starting at the ground end of the circuit, disconnect one connector at a time. Check the test light after each connector. The short is in the circuit that was disconnected when the test light went out.

An alternative to this uses a DMM. Remove the bad fuse and disconnect the load. Connect the DMM across the fuse terminals. Refer to the wiring diagram to see if the ignition switch needs to be turned on to energize the circuit. If the meter reads a voltage, there must be a short before the load. Starting with the circuit's wiring closest to the fuse, wiggle the wiring harness. Do this in short steps all along the wire. The point at which the voltage drops to zero is close to where the short is.

AUTHOR'S NOTE: Many manufacturers do not recommend using a circuit breaker as a substitute for the fuse during testing; rather, a sealed beam headlight is connected with jumper wires across the fuse holder. The headlight serves a load and limits the current in the circuit. The headlight will light as long as current is flowing through the circuit.

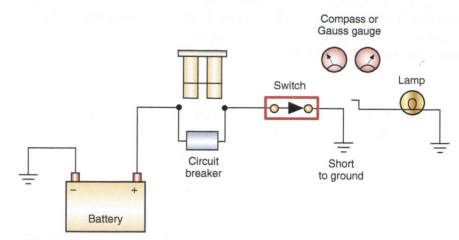

FIGURE 10-26 A Gauss gauge or compass can be used to locate a short in the circuit.

Short Detector. Some technicians use a compass or Gauss gauge to find the location of a short (Figure 10-26). A magnetic field is formed around a current-carrying conductor and a compass reacts to magnetic fields. A shorted circuit will have high current; therefore a large magnetic field will be formed around the shorted circuit. With the wiring diagram, locate the routing of the wires in the affected circuit. Connect a jumper wire across the fuse holder for the blown fuse. Position the compass over or close to the wiring harness. The magnetic field in the wire will cause the needle of the compass to move away from its north position. As the circuit breaker cycles, the needle will fluctuate. As the compass is slowly moved across the wire, it will continue to fluctuate until it passes the point where the short is. Inspect the wire in that area. Look for signs of overheating and broken, cracked, exposed, or punctured wires.

Testing for Unwanted Resistance

High-resistance problems are typically caused by corrosion on terminal ends, loose or poor connections, or frayed and damaged wires. Carefully inspect the affected circuit.

Whenever excessive resistance is suspected, both sides of the circuit should be checked. Begin by checking the voltage drop across the load. It should be close to battery voltage unless the circuit has a resistor before the load. If the voltage is less than desired, check the voltage drop across the circuit from the switch to the load. If the voltage drop is excessive, that part of the circuit contains the unwanted resistance. If the voltage drop was normal, the high resistance is in the switch or in the circuit feeding the switch. Check the voltage drop across the switch. If the voltage drop is excessive, the problem is the switch. If the voltage drop is normal, the high resistance is in the circuit feeding the switch.

If battery voltage is present at the load, the ground circuit for the load should be checked. Connect the red meter lead to the ground side of the load and the black lead to the circuit's ground. If the voltage drop is normal, the problem is the ground. If the voltage drop is excessive, move the black lead toward the red one. Check voltage drop at each step. Eventually you will read high-voltage drop at one connector and then a low-voltage drop at the next. The point of high resistance is between those two test points. If the voltage drop is normal, the high resistance is in the switch or in the circuit feeding the switch.

BASIC ELECTRICAL REPAIRS

Many automotive electrical problems can be traced to faulty wiring. The most common causes are loose or corroded terminals; frayed, broken, or oil-soaked wires; or faulty insulation. Wires, fuses, and connections should be checked carefully during troubleshooting. Keep in mind that the insulation does not always appear to be damaged when the wire inside is broken. Also, a terminal may be tight but still be corroded.

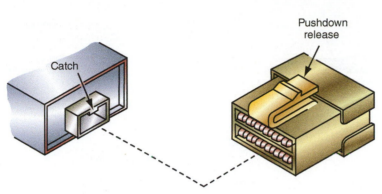

FIGURE 10-27 An example of a pushdown release lock on a connector.

Check all connectors for corrosion, dirt, and looseness. Nearly all connectors have push-down release-type locks (Figure 10-27). Make sure these are not damaged.

> ⚠️ **WARNING:** **Never pull on the wires to separate connectors. This can create an intermittent problem. Always use the tools designed for separating connectors to prevent this problem.**

> **AUTHOR'S NOTE:** **Apply dielectric grease or petroleum jelly to all connections before you assemble them. This will prevent future corrosion problems.**

Replacement Wire Selection

When it is necessary to replace a wire or two, make sure this is done in a way that corrects the problem and does not create a new problem. All replacement wires should be of the same size or larger than the original.

> ⚠️ **WARNING:** **Supplemental Restraint System (SRS) airbag harness insulation and related connectors are color-coded yellow or orange. Do not connect any accessories or test equipment to SRS-related wiring.**

When a section of a wire needs to be replaced, cut the damaged end of the wire from the main wire. Match the dimensions of the new wire to the old one. Measure the required length of the replacement wire; make sure it is slightly longer than the section that was removed. Then connect the two wires together. After the connection is made, the joint should be wrapped with tape or covered with heat-shrink tubing.

Connecting Wires

There are several ways the original wire can be connected to the replacement wire. Some technicians use butt connectors; these provide a good joint between the wires. However, the preferred way to connect wires or to install a connector is by soldering (Figure 10-28).

Soldering. Soldering joins two pieces of metal together by melting a lead and tin alloy and allowing it to flow into the joint. A soldering iron or gun is used to heat the solder. There are different types of solders, but only rosin or resin-type flux core solder should be used for

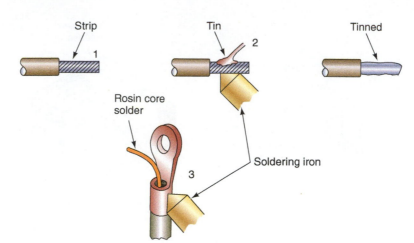

FIGURE 10-28 Proper procedure for soldering a terminal to a wire.

electrical work. Also when making wire repairs on or near an electronic component, use a heat sink to prevent the heat from traveling into the part.

⚠️ **WARNING:** **Never use acid core solder. It creates corrosion and can damage electronic components.**

Make sure the tip of the soldering iron or gun is clean and tinned. The copper tip corrodes through use and this can prevent it from transferring heat as it should. Use a file to remove all residues from the tip. Turn the iron on and allow it to heat. Then dip the tip into soldering rosin flux. Remove the tip from the flux and immediately apply rosin core wire solder to all surfaces. The solder should flow over the tip. The tip is now tinned and the iron is ready to solder.

When joining two wires, twist the wire ends tightly together before soldering the joint. When doing this, it is important to realize the solder does not provide for a mechanical joint. Therefore it is important that the twisting provides a secure joint before soldering. Some manufacturers use aluminum wires. Aluminum cannot be soldered. Follow the manufacturer's guidelines and use the proper repair kits when repairing aluminum wiring.

Splice or Butt Connectors. Make sure the ends of the wires are clean. The correct-size splice clip to use is determined by the outside diameter of the wire, not the insulation. When inserting the wires into the joint, make sure only the conductor enters. Also make sure the ends of the two wires overlap inside the splice. Slip the jaws of the crimping tool over the center of the splice. Squeeze the crimping tool until the contact points of the tool's jaws make contact with the splice. Then check the placement of the wires inside the splice and apply pressure to the crimping tool to form a tight crimp. Figure 10-29 shows the correct way to install a crimp-type wire terminal.

Insulating the Connection. After a connection is made, it must be insulated. This is done with heat shrink tubing or tape. When using tape, place one end on the wire about 1 inch from the joint. Tightly wrap the tape around the wire. While wrapping, cover about one-half of the previous wrap with the tape as it completes one turn around the wire. Once the tape is 1 inch beyond the joint, cut the tape. Firmly press on the tape at that end to form a good seal.

Heat shrink tubing should be slightly larger than the diameter of the splice. Cut a length of tubing that is longer than the splice. Before joining the wires together, slip the tubing over one of the wires. Then make the joint. After the wires are connected, move the shrink tubing over the splice. Use a heat gun to shrink the tubing tightly around the splice. The tubing will

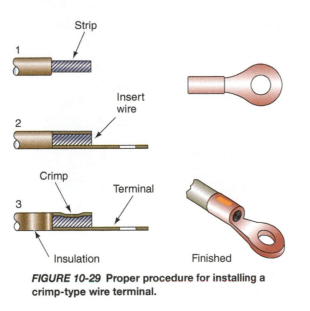

FIGURE 10-29 Proper procedure for installing a crimp-type wire terminal.

only shrink a certain amount; therefore do not continue to heat it after it is in place. Doing this can melt the tubing and/or the insulation of the wire.

Wire Connectors

Many different types of connectors, terminals, and junction blocks are used. Wire end terminals are used as connecting points for wires and come in many shapes and sizes. They are soldered or crimped in place. When installing a terminal, select the appropriate size and type. Be sure it fits the connecting post and has enough current-carrying capacity for the circuit.

⚠️ **WARNING:** **Always follow the manufacturer's wiring and terminal repair procedures. On some components and circuits, manufacturers recommend complete wiring harness replacement rather than making repairs to the wiring.**

When a connector is replaced because it has melted or otherwise damaged, attempt to replace it with the same type and size. Normally the available non-OEM replacement connectors have common shapes with a common number of terminal cavities. Therefore it is best to use a connector that meets the need; this may mean some of the terminal cavities are left empty. For example, if the original connector has six terminals but the available replacement has eight, use the connector and leave two slots open. This means that both halves of the connector will need to be replaced and the wire terminals inside both halves must be arranged so they match as well. Sometimes, the replacement connector will require different terminal ends than the original. This requires the replacement of terminal ends on the wires for the male and female connectors.

Replacing a Terminal

Terminal ends are replaced when damaged or to accommodate the use of a connector. Replacement must provide for good continuity and to prevent future electrical problems.

Use the service information to identify the position of the locking clips and the terminal unlocking procedures for the connector. With a small screwdriver or terminal pick (Figure 10-30), unlock the secondary locking device. Gently push the terminal into the connector and hold it there. With the terminal pick, turn the locking clip on the terminal to the unlock position. Hold it there while carefully pulling the wire to remove the terminal from the connector.

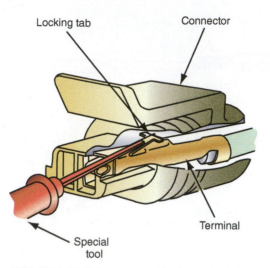

Locking tab   Connector

Terminal

Special
tool

FIGURE 10-30 Using a special tool (terminal pick)
to remove a terminal from a connector.

Measure the diameter of the wire's insulation. Then identify the type and size the replacement terminal should be. Also, select the correct size for the replacement wire. Cut off the old terminal with some of its wire. Use the old wire as a guide and cut the replacement wire slightly longer. Be careful, if the wire is too short, it may pull on the terminal, splice, or connector, which can lead to an open. If the wire is too long, it may get pinched and cause a short.

Strip the insulation from the wire in the harness and both ends of the replacement wire. Normally ⅜ inch of insulation should be removed. Make sure the strands of wire are not damaged while removing the insulation. Place the ends of the wires into the terminal and crimp the terminal. To get a proper crimp, place the open area of the connector facing toward the anvil of the tool. Make sure the wire is compressed under the crimp.

Apply heat shrink tubing over the repair and insert the terminal into the connector. Push the terminal into the connector until a click is heard. Gently pull on the wire. If the terminal is locked in the connector, it will not move.

Connect both sides of the connector and secure the secondary locking device. Tape the new wire to the wiring harness. If the harness is contained in conduit, make sure it is fully enclosed and tape the outside of the conduit.

> **AUTHOR'S NOTE:** Dielectric grease or petroleum jelly should be used to moisture-proof and protect connections from corrosion. If the specifications recommend that a connector be filled with grease, make sure it is. If the old grease is contaminated, replace it.

⚠️
CAUTION:
Fuses and other protection devices normally do not wear out. They go bad because something went wrong. Never replace a fuse or fusible link or reset a circuit breaker without finding out why it went bad.

TESTING CIRCUIT PROTECTION DEVICES

When a circuit does not function at all, there is a good chance that its circuit protection device is faulty. An open protection device prevents voltage from being applied to the circuit. The procedures for checking fuses, fusible links, maxi-fuses, and circuit breakers are somewhat similar.

PROTECTION DEVICES

Fuses

There are three basic types of fuses in automotive use: cartridge, blade, and ceramic. The cartridge fuse is found on most older domestic cars and a few imports. It is composed of a strip of low-melting metal enclosed in a transparent glass or plastic tube. To check this type

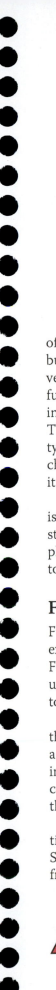

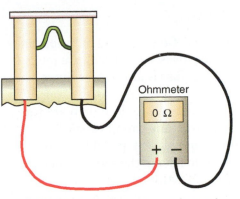

FIGURE 10-31 An ohmmeter can be used to check the condition of a fuse.

of fuse, look for a break in the internal metal strip. Discoloration of the glass cover or glue bubbling around the metal end caps is an indication of overheating. Late-model domestic vehicles and many imports use blade or spade fuses. To check a blade fuse, pull it from the fuse panel and look at the fuse element through the transparent plastic housing. Look for internal breaks and discoloration. The ceramic fuse is used on many European imports. The core is a ceramic insulator with a conductive metal strip along one side. To check this type of fuse, look for a break in the contact strip on the outside of the fuse. All fuses can be checked with an ohmmeter or test light. If the fuse is good, there will be continuity through it (Figure 10-31).

Fuses are rated by the current at which they are designed to blow. A three-letter code is used to indicate the type and size of the fuse. The code and the current rating is usually stamped on the end cap. The current rating for blade fuses is indicated by the color of the plastic case. It may also be marked on the top. The insulator of ceramic fuses is color-coded to indicate different current ratings.

Fusible Links

Fuse or **fusible links** are used in circuits in which limiting the maximum current is not extremely critical. They are often installed in the positive battery lead to the ignition switch. Fusible links are normally found in the engine compartment near the battery. They are also used when it would be awkward to run wiring from the battery to the fuse panel and back to the load.

Because a fuse link is a lighter gauge wire than the main conductor, it melts and opens the circuit before damage can occur in the rest of the circuit. Fuse link wire is covered with a special insulation that bubbles when it overheats, indicating that the link has melted. If the insulation appears good, pull lightly on the wire. If the link stretches, the wire has melted. Of course, when it is hard to determine if the fuse link is burned out, check for continuity through the link with a test light or ohmmeter.

To replace a fuse link, cut the protected wire where it is connected to the fuse link. Then, tightly crimp or solder a new fusible link of the same rating and length as the original link. Since the insulation on the manufacturer's fuse links is flameproof, never fabricate a fuse link from ordinary wire because the insulation may not be flameproof.

⚠️
CAUTION:
Do not mistake a resistor wire for a fuse link. A resistor wire is generally longer and is clearly marked "Resistor— do not cut or splice."

⚠️ **WARNING:** **Always disconnect the battery ground cable prior to servicing any fuse link. Failure to do this will cause personal injury and damage to the vehicle and its electrical and electronic components.**

Maxi-Fuses

Many late-model vehicles use **maxi-fuses** instead of fusible links. Maxi-fuses look and operate like two-prong blade or spade fuses, except they are much larger and are used in higher-current circuits. (Typically, a maxi-fuse is four to five times larger.) Maxi-fuses are located in their own underhood fuse block.

Maxi-fuses are easier to inspect and replace than are fuse links. To check a maxi-fuse, look at the fuse element through the transparent plastic housing. If there is a break in the element, the maxi-fuse has blown. To replace it, pull it from its fuse box or panel. Always replace a blown maxi-fuse with a new one having the same ampere rating. The best way to test any circuit protection device is with a DMM. Looks can be deceiving.

Maxi-fuses allow the vehicle's electrical system to be broken down into smaller circuits that are easy to diagnose and repair. For example, in some vehicles a single fusible link controls one-half or more of all circuitry. If it burns out, many electrical systems are lost. By replacing this single fusible link with several maxi-fuses, the number of systems lost due to a problem in one circuit is drastically reduced. This makes it easy to pinpoint the source of trouble.

Circuit Breakers

Circuit breakers protect circuits against overloads. Circuit breakers are usually labeled as *c.b.* in a fuse chart of a service manual.

Some circuits are protected by **circuit breakers**. They can be fuse panel mounted or inline. Like fuses, they are rated in amperes. Each circuit breaker conducts current through an arm made of two types of metal bonded together (bimetal arm). If the arm starts to carry too much current, it heats up. As one metal expands faster than the other, the arm bends, opening the contacts. Current flow is broken. A circuit breaker will either automatically reset, or it must be manually reset by depressing a button.

Thermistors

Some systems use a positive temperature coefficient (PTC) thermistor as a protection device. When there is high current and heat, the resistance of the thermistor increases and causes a decrease in current flow. These can be checked with an ohmmeter. If an infinite reading is displayed, the thermistor is open. Another way of checking a thermistor is to change its temperature and see if its resistance changes.

TESTING SWITCHES

Classroom Manual Chapter 10, page 252

Switches can be tested with a voltmeter, test light, or ohmmeter. To check the operation of a switch with a voltmeter or test light, connect the meter's positive lead to the battery side of the switch. With the negative lead attached to a good ground, voltage should be measured. Without closing the switch, move the positive lead to the other side of the switch. If the switch is open, no voltage will be present at that point. The amount of voltage present at this side of the switch should equal the amount on the other side when the switch is closed. If the voltage decreases, the switch is causing a voltage drop due to excessive resistance. If no voltage is present on the groundside of the switch with it closed, the switch should be replaced.

If a switch has been removed from the circuit, it can be tested with an ohmmeter or a self-powered test light. By connecting the leads across the switch connections, the action of the switch should open and close the circuit (Figure 10-32).

Clutch Safety Switch

Classroom Manual Chapter 10, page 261

The **clutch safety switch** is connected to the starting circuit. The switch prevents starting of the engine unless the clutch pedal is fully depressed. Note that in vehicles with an automatic transmission, this function is performed by a neutral safety switch. The switch is normally open when the clutch pedal is released. When the clutch pedal is depressed, the switch closes and completes the circuit between the ignition switch and the starter solenoid or PCM.

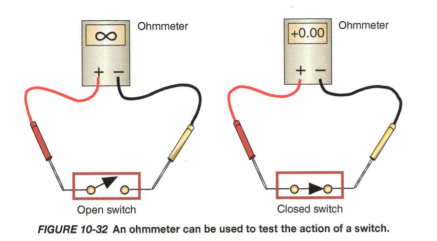

FIGURE 10-32 An ohmmeter can be used to test the action of a switch.

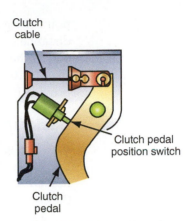

FIGURE 10-33 Location of a clutch switch in a clutch pedal assembly.

This switch can be the cause of no-start problems. Because the clutch pedal switch is wired in series between the starter relay coil and the ignition switch, the engine will not crank if the switch is faulty. Some clutch switches are part of the clutch pedal assembly (Figure 10-33), while others are part of the clutch master cylinder's pushrod (Figure 10-34).

To check a clutch interlock switch, disconnect the connector to the switch at the clutch pedal assembly. With an ohmmeter, check for continuity between the terminals of the switch (Figure 10-35) with the clutch released and with it engaged. With the clutch released, there should be no continuity between the terminals. With the clutch engaged, there should be good continuity between the terminals. If the switch doesn't complete the circuit when the clutch is engaged, check the adjustment of the switch. (Some switches are not adjustable.) If the adjustment is correct, replace the switch. Another way to check the clutch switch is to simply bypass it with a jumper wire. If the engine starts when the switch is jumped, a bad switch is indicated.

After replacing the switch, slowly work the clutch pedal and note any unusual noises or interferences with pedal movement. If the switch is adjustable, make sure the clutch pedal height and pushrod (free) play are correct. Then check the release point of the clutch. Begin by engaging the parking brake and installing wheel chocks. Start the engine and allow it to idle. Without depressing the clutch pedal, slowly shift the shift lever into reverse until the gears contact. Gradually depress the clutch pedal and measure the stroke distance from the

⚠️

CAUTION:
Before replacing the clutch switch or any component connected to the vehicle's control computer, disconnect the negative cable from the battery.

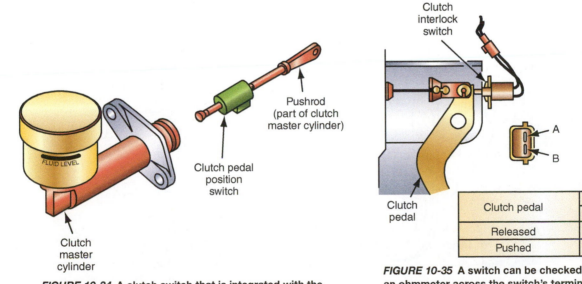

FIGURE 10-34 A clutch switch that is integrated with the clutch's master cylinder pushrod.

Clutch pedal	Terminal	
	A	B
Released		
Pushed	●—●	

FIGURE 10-35 A switch can be checked by placing an ohmmeter across the switch's terminals and moving the switch through its different positions.

point the gear noise stops (this is the release point) up to the full stroke end position. If the distance is not within specifications, check the pedal height, pushrod play, and pedal free-play. If the distance is correct, check the clearance between the switch and the pedal assembly (Figure 10-36) when the clutch is fully depressed. Loosen the switch and adjust its position to provide for the specified clearance.

Back-up Light Switch

Typically the **back-up light fuse** is also the fuse for the turn signals. If the turn signals work, the fuse is probably good.

Classroom Manual

Chapter 10, page 262

Back-up lights illuminate the area behind the vehicle and warn other drivers and pedestrians that the vehicle is moving in reverse. Typically, power for the lamps is supplied through the ignition switch when it is in the run position. When the driver shifts the transmission into reverse, the contacts of the **back-up light switch** (Figure 10-37) are closed, completing the light circuit.

To diagnose nonworking back-up lights, begin by checking the fuse. Then move the gearshift slightly in all directions with the ignition switch on but the engine off. If the back-up lights come on or flicker when the gearshift is moved, the back-up light switch probably needs to be adjusted. Sometimes a loose or worn shift linkage will cause the same problem. Check the linkage for tightness.

Continue your diagnosis with an inspection of the connectors at the lamps and the back-up light switch. Also check the condition of the bulbs. Correct any problem found and recheck the operation of the back-up lights.

If no problems were found, check the switch by measuring the resistance across the switch's terminals. There should be good continuity only when the gearshift lever is in the reverse position. If the switch remains open when the transmission is in reverse, the switch needs to be replaced. If there is continuity across the switch's terminals and the back-up lights don't work, check the circuit for an open.

Other Switches

Other transmission and driveline switches with two terminals can be checked in the same way. At times, there will be more than two terminals attached to the switch. These are tested as other two terminal switches, except the terminals must be identified before testing. The

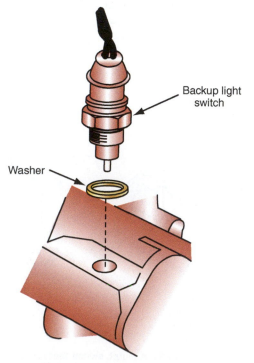

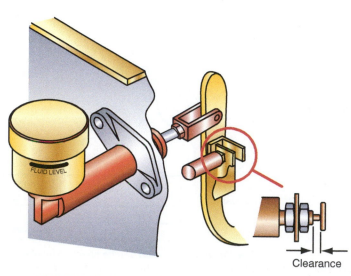

FIGURE 10-36 Clutch switch to pedal assembly clearance.

FIGURE 10-37 A back-up light switch.

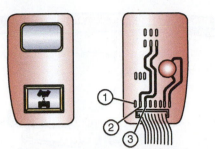

Center diff. lock switch condition	Tester connected terminal number	Specified condition
ON	1 - 2	No continuity
	1 - 3	No continuity
	2 - 3	Continuity
OFF	1 - 2	Continuity
	1 - 3	No continuity
	2 - 3	No continuity

FIGURE 10-38 A switch with multiple terminals and the desired test readings across the terminals.

service manual often has an illustration showing the terminals along with the desired test readings in the various test positions (Figure 10-38).

TESTING SPEED SENSORS

Wheel speed sensors are used in antilock brake systems to measure the speed of the wheels. The tip of the sensor is located near a toothed ring or rotor. The toothed ring is typically part of the outer CV joint or axle assembly or is mounted next to the differential ring gear. Some transmissions are fitted with a vehicle speed sensor (Figure 10-39) and/or output shaft speed sensor.

Classroom Manual
Chapter 10, page 264

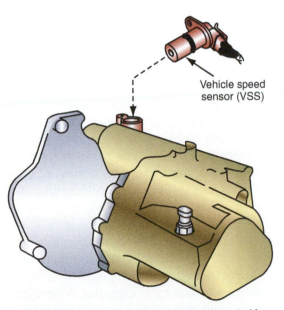

Vehicle speed sensor (VSS)

FIGURE 10-39 A vehicle speed sensor mounted in a transaxle.

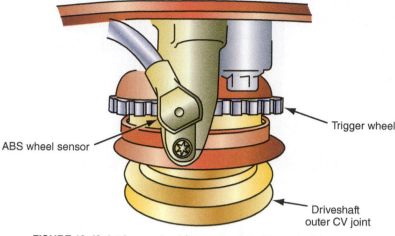

FIGURE 10-40 A trigger wheel (tone wheel) that is part of the outer CV joint assembly.

Extremecare should be taken when servicing CV joints, half-shafts, and differentials so as not to damage the speed sensors or the trigger wheels. For proper ABS operation, the sensors and trigger wheels must be in good condition. The trigger wheels are usually replaced as part of the assembly, with the outer CV joint (Figure 10-40) or the wheel hub and bearing assembly. Others are individually replaceable. These have a *slip-fit* and must be installed by hand or by even pressure on a hydraulic press. Each trigger wheel should be inspected for broken or damaged teeth. If any defects are found, the trigger wheel should be replaced.

Most wheel sensors are mounted on an adjustable mounting or have slotted bolt holes for adjustment. Others are mounted solidly and have no means for adjustment. Adjusting the clearance between the sensor and the trigger wheel is critical for the proper operation of the ABS. If the sensor is adjustable, always follow the procedure recommended by the manufacturer. Some nonadjustable sensors have a polyethylene or paper spacer that provides for the correct gap. When reusing this type of sensor, make sure this spacer is in good condition.

The condition of the wheel sensors can be checked by going through the diagnostic routines for the ABS. If this test indicates that a sensor is faulty, the sensor should be replaced. Wheel sensors can also be checked with an ohmmeter. Most manufacturers list a resistance specification. The resistance of the sensor is measured across the sensor's terminals. The typical range for a good sensor is 800–1,400 ohms of resistance.

SOLENOIDS

Solenoids are commonly found on automatic transmissions. Although not common, manual transmissions equipped with shift blocking use solenoids to control or limit the movement of the gearshift. Two types of shift blocking are currently found. One prevents shifting into reverse when the vehicle is moving forward. The other prevents shifts from first to second and third when the vehicle is not under a load and is moving slowly. A solenoid blocks the shifter from engaging into second or third gears from first gear. This shift-blocking solenoid is controlled by the powertrain control module and is checked in the same way as other computer system components.

Growing numbers of late-model performance vehicles are being equipped with "paddle shifters" and clutchless operation. Actually, these transmissions are fitted to a clutch, in some cases two clutch assemblies, but there is no clutch pedal. Shifting and clutching are automatically performed by solenoids controlled by a computer. A problem with the solenoid or the solenoid's circuit can cause difficult shifting or a no-shift problem. For example, if there is an open in the power wire to the reverse lockout solenoid (Figure 10-41), the driver will not be able to shift into reverse. To diagnose a problem like this, check for voltage at the solenoid.

Classroom Manual

Chapter 10, page 264

A **solenoid** is an electromagnet with a moveable core. The core is used to complete an electrical circuit and/or to cause a mechanical action.

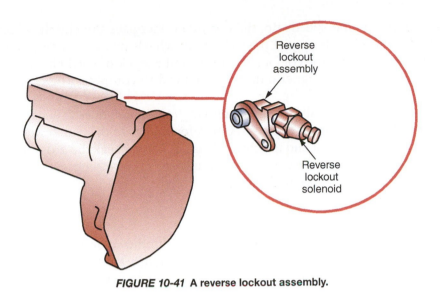

FIGURE 10-41 A reverse lockout assembly.

Then work away from the solenoid until voltage is present. The open will be between the last point in the circuit where voltage is not present and the point at which voltage is measured. To diagnose this problem, you should refer to the wiring diagram for the vehicle first.

If the solenoid is faulty, if there is excessive resistance in the circuit, or if the gear lockout assembly is damaged, the driver may experience very difficult shifting into reverse or second and third gears. If less than battery voltage is available at the solenoid, check the circuit for excessive voltage drops.

If a solenoid is bad, it will either be open, have a short, or have high internal resistance. The exact problem can be identified by checking the solenoid with an ohmmeter. Resistance checks should be made across the terminals of the solenoid and between the terminals and the case. A good solenoid will have low resistance only across the terminals. A resistance reading between the terminals and the case indicates that the windings are shorted to the case. If there is high resistance across the terminals, there is excessive resistance in the solenoid. Also, if there is no continuity across the terminals, the windings are open.

ELECTROMAGNETIC CLUTCHES

The rear axle differential, center differential, and transfer case (Figure 10-42) on some vehicles are equipped with electromagnetic clutches for lockup of the differential. The clutch may be controlled by a switch or by a traction control computer. When one wheel

Classroom Manual
Chapter 10, page 266

FIGURE 10-42 A typical electromagnetic clutch assembly.

or axle loses traction and speeds up, the computer energizes the clutch, which locks up the differential and provides equal power to both wheels or axles. The electromagnetic clutch consists of an actuator coil, an armature, and a stack of steel clutch plates. When the clutch is energized, a magnetic field is produced to compress the clutches, thereby locking the differential.

The clutch can mechanically or electrically fail. If the clutch doesn't engage, the problem is probably an open in its electrical circuit. Check for voltage to the clutch when it should be energized. If voltage is present and the clutch does not work, check the ground. If the ground is good, replace the clutch. If the clutch doesn't fully engage or disengage, suspect an adjustment problem or high resistance in the electrical circuit.

Relays

A relay can be checked with a jumper wire, voltmeter, ohmmeter, or test light. If the terminals are accessible and the relay is *not* controlled by a computer, a jumper wire and test light is the quickest method. The schematic for a relay is typically shown on the outside of the relay. If not, check a wiring diagram to identify its terminals (Figure 10-43). Also check the wiring diagram to determine if the relay is controlled by a power or ground switch.

If the relay is controlled on the ground side, follow this procedure to test the relay:

1. Use a test light to check for voltage at the power feed to the relay. If voltage is not present, the fault is in the battery feed to the relay. If there is voltage, continue testing.
2. Probe for voltage at the coil terminal. If voltage is not present, the coil is faulty. If voltage is present, continue testing.
3. Use a fused jumper wire to connect the control terminal to a good ground. If the relay works, the fault is in the control circuit. If the relay does not work, continue testing.
4. Connect the jumper wire from the battery to the output terminal of the relay. If the device operated by the relay works, the relay is bad. If the device does not work, the circuit between the relay and the device or the device's ground is faulty.

If the relay is controlled by a computer, do not use a test light. Rather, use a high-impedance voltmeter set to the 20V DC scale, then:

1. Connect the negative lead of the voltmeter to a good ground.
2. Connect the positive lead to the output wire. If no voltage is present, continue testing. If there is voltage, disconnect the relay's ground circuit. The voltmeter should now read 0 volts. If voltage is still present, the relay is faulty.
3. Connect the positive lead to the power input terminal. If near battery voltage is not measured there, the relay is faulty. If it is, continue testing.

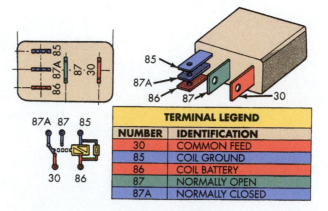

FIGURE 10-43 Before testing a relay, refer to the wiring diagram to identify the terminals and their connections.

4. Connect the positive lead to the control terminal. If near battery voltage is not measured there, check the circuit from the battery to the relay. If it is, continue testing.

5. Connect the positive lead to the relay ground terminal. If more than 1 volt is present, the circuit has a poor ground.

If the relay terminals are not accessible, remove the relay from its mounting. Use an ohmmeter to test for continuity between the relay coil terminals. If the meter indicates an infinite reading, replace the relay. If there is continuity, use a pair of jumper wires to energize the coil. Check for continuity through the relay contacts. If there is an infinite reading, the relay is faulty. If there is continuity, the relay is good and the circuits need to be tested.

Measure the resistance of the coil and compare your readings to the specifications. Low resistance indicates the coil is shorted. If the coil is shorted, the transistors and/or driver circuits in the computer could be damaged due to excessive current flow.

PROTECTING ELECTRONIC SYSTEMS

You should be aware of the ways to protect electrical systems and electronic components during storage and repair. Keep the following in mind at all times:

- Do not connect or disconnect electronic components with the key on.
- Never touch the electrical contacts on any electronic part. Skin oils can cause corrosion and poor contacts.
- Be aware of any part that is marked with a code or symbol to warn that it is sensitive to electrostatic discharge.
- Before touching a computer, always touch a good ground first. This safely discharges any static electricity, which can generate up to 25,000 volts and can easily damage a computer.
- Static-proof work mats allow work inside the vehicle without creating static electricity.
- Wear a grounding wrist strap. A wire connects the wrist strap to a good ground.
- Never allow grease, lubricants, or cleaning solvents to touch the end of the sensor or its electrical connector.
- Be careful not to damage connectors and terminals when removing components.
- Do not connect jumper wires across a sensor unless the service information tells you to.
- Never apply 12 volts directly to an electronic component unless instructed to do so.
- Never use a test light to test electronic components unless instructed to do so.
- Sensor wires should never be rerouted. When replacing wiring, always check the service information and follow the routing instructions.
- Disconnect any module that could be affected by welding, hammering, grinding, sanding, or metal straightening.

ELECTRONIC CIRCUITS

The introduction of powertrain computer controls brought with it the need for tools capable of troubleshooting electronic control systems. There are a variety of computer scan tools available today that do just that. A **scan tool** is a microprocessor designed to communicate with the vehicle's computer. Connected to the computer through diagnostic connectors, a scan tool can access diagnostic trouble codes (DTCs), run tests to check system operations, and monitor the activity of the system. Trouble codes and test results are displayed on an LED screen or printed out on the scanner printer.

A scan tool receives its testing information from one of several sources. Some scan tools have a **programmable read-only memory (PROM)** chip that contains all the information needed to diagnose specific model lines. This chip is normally contained in a cartridge that is plugged into the tool. The type of vehicle being tested determines the appropriate cartridge that should be inserted. These cartridges contain the test information for that particular car.

A cartridge may be needed for each make and model vehicle. As new systems are introduced on new car models, a new cartridge is made available.

LED displays are generally only large enough to display four short lines of information. This feature limits the technician's ability to compare test data. However, most scan tools overcome this inadequacy by storing the test data in **random access memory (RAM)**. The scan tool can then be interfaced with a printer, personal computer, or larger engine analyzer that can retrieve the information stored in the memory.

Trouble codes are only set by the vehicle's computer when a voltage signal is out of its normal range. The codes help technicians identify the cause of the problem when this is the case. If a signal is within its normal range but is still not correct, the vehicle's computer may not display a trouble code. However, a problem still exists. As an aid in identifying this type of problem, most manufacturers recommend that the signals to and from the computer be observed carefully. This is called watching the data stream.

Most diagnostic work on computer-control systems should be based on a description of symptoms. With this description, you can locate any technical service bulletins that refer to the problem. You also can use the description to locate the appropriate troubleshooting sequence in the manufacturer's service manuals.

TESTING ELECTRONIC CIRCUITS AND COMPONENTS

Only high-impedance meters should be used to check electronic circuits. A lab scope is valuable and is primarily used to measure voltages, pulse, and duty cycle.

Whenever using a scope on a circuit, always follow the meter's instruction manual for hookup and proper settings. Most manuals have illustrations of the patterns that are expected from the different electronic components. If the patterns do not match those in the manual, a problem is indicated. The problem may be in the component or in the circuit. Further testing is required to locate the exact cause of the problem.

Checking Diodes

Multimeters can be used to check diodes, including zener diodes and LEDs. Regardless of the diode's bias, it should allow current flow in only one direction. Connect the meter's leads across the diode. Observe the reading on the meter. Then reverse the meter's leads and again observe the reading. The resistance in one direction should be very high or infinite and close to zero in the other direction (Figure 10-44). Any other readings indicate a bad diode. A diode that has low resistance in both directions is shorted. A diode that has high resistance or an infinite reading in both directions is open.

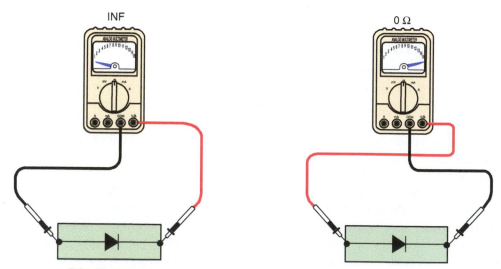

FIGURE 10-44 Using an ohmmeter to check a diode for an open or a short.

You may run into problems when checking a diode with a high-impedance DMM, since many diodes don't allow current flow through them unless the voltage is at least 0.6 volts. This results in readings that indicate the diode is open, when in fact it may not be. Due to this, many DMMs are equipped with a diode-testing feature. This allows for increased voltage at the test leads. Again, continuity should be present in one direction, and not the other. Some meters will make a beeping noise when there is continuity during a diode check.

Diodes can also be tested with a voltmeter. Measure the voltage drop across the diode. The meter should read low voltage in one direction and higher voltage in the other direction. Most automotive diodes will drop 500–650 mVolts.

Shift Lights

Upshift and shift lamps inform the driver when to shift into the next gear in order to maximize fuel economy. These lights are controlled by the PCM, which activates the light according to engine speed, engine load, and vehicle speed. Basically, these light circuits operate like a vacuum gauge. When engine load is low, engine vacuum is high. And when engine load is great, vacuum is low. The shift light will come on whenever there is high vacuum. The shift lamp is lit at those engine speeds and loads in which engine vacuum is high and the transmission is in a forward gear. The shift light stays on until the transmission is shifted or the engine's operating conditions change. This circuit works in all forward gears, except high gear in which a high-gear switch disables the circuit.

Classroom Manual
Chapter 10, page 263

Shift Blocking

Some six-speed transmissions have shift blocking, which prevents the driver from shifting into second or third gear from first gear, when the coolant temperature is above 50°C (122°F), the speed of the vehicle is between 12 and 22 mph (20 and 29 km/h), and the throttle is opened 35 percent or less. **Shift blocking** occurs to ensure good fuel economy and keeps the vehicle in compliance with federal fuel standards. Shift blocking is controlled by the PCM. A solenoid is used to block off the shift pattern from first gear to second or third. The driver moves the gearshift from its up position to a lower position, as if shifting into second, and fourth gear is selected. The solenoid does not impede downshifting.

The shift blocking system consists of two main components: a second and third gear blockout solenoid and a second and third gear blockout relay (Figure 10-45). The PCM uses

The **shift blocking** feature is sometimes referred to as the 1–4 upshift system.

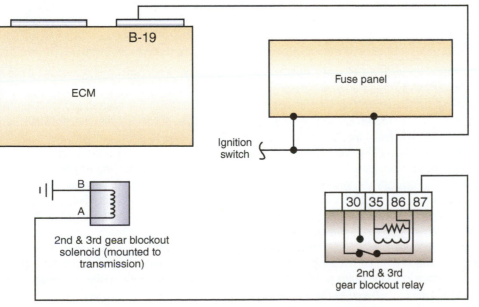

FIGURE 10-45 Simplified diagram for a shift blocking system.

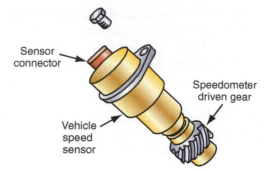

FIGURE 10-46 A voltage-generating VSS.

engine speed and vehicle speed to calculate whether the transmission is in first or fourth gear. The PCM provides a ground for the blockout relay when second and third gears should be bypassed. The blockout solenoid is located in the transmission's shift linkage. When the PCM provides a ground for the blockout relay, the relay sends power to the blockout solenoid. The solenoid mechanically locks out the transmission from being shifted from first gear into any gear other than fourth gear.

The PCM looks at many inputs from a variety of sensors to determine when shift blocking should occur. One of the sensors read by the PCM is the vehicle speed sensor (VSS). This sensor (Figure 10-46) is a voltage generator and is mounted in the transmission. It uses a gear driven by the transmission's output shaft to generate voltage pulses at a rate of 4,000 pulses per mile. When the shift blocking system is activated, a bulb on the dash lights.

Diagnosis of this system is typical of most computer-related components or systems. First, it must be determined if the problem is mechanical or electrical. To do this, attempt to shift the vehicle into first and fourth gears. If you are unable to shift into either gear, the problem is mechanical. If you are able to shift into both gears, proceed with the electrical diagnosis.

Using a scan tool or jumper wires with the engine off and the ignition on, energize the blockout solenoid. Test the shifter. If the blockout did not occur, only detailed testing will identify the exact cause of the problem. The service manual will have a diagnostic chart that will lead you to the cause of the problem.

CASE STUDY

A late-model car with a five-speed transaxle was brought into the shop by the driver with a complaint of an inoperative cruise control unit. The technician went through the normal routine for diagnosing a cruise control system and found that the control unit was not being turned on. Looking over the wiring diagram, she identified the switches and wires that fed power to the unit. She carefully checked the fuse, the wires, and the connectors. No problem was found.

She again referred to the wiring diagram to see what could possibly cause the problem. From this and her previous findings, she suspected the clutch switch. The clutch switch is wired to disengage the cruise control function when the clutch pedal is depressed. This led to a bit of confusion as the engine started when the clutch pedal was depressed. It seemed to be working fine.

The technician referred to the wiring diagram and again suspected the clutch switch. To verify this as a cause of the problem, she rechecked the switch. This time she put the transmission in neutral and attempted to start the engine without depressing the clutch pedal. The engine started. The clutch switch was sending a signal that the clutch pedal was always depressed; it was shorted or the wires to it were shorted together.

The switch was removed and tested with an ohmmeter. It was shorted. A new switch was installed and the car had cruise control once more.

ASE-STYLE REVIEW QUESTIONS

1. When discussing the use of a test light,
   Technician A says that a test light can be used to check a switch while it is mounted on a transmission.
   Technician B says that a self-powered test light can be used to test a switch after it is removed from a transmission.
   Who is correct?
   A. A only
   B. B only
   C. Both A and B
   D. Neither A nor B

2. When discussing measuring resistance,
   Technician A says that an ohmmeter can be used to measure resistance of a component before disconnecting it from the circuit.
   Technician B says that a voltmeter can be used to measure voltage drop. Something that has very little resistance will drop to zero or very little voltage.
   Who is correct?
   A. A only
   B. B only
   C. Both A and B
   D. Neither A nor B

3. When using a voltmeter,
   Technician A connects it across the circuit being tested.
   Technician B connects the red lead of the voltmeter to the most positive side of the circuit.
   Who is correct?
   A. A only
   B. B only
   C. Both A and B
   D. Neither A nor B

4. When using a DMM,
   Technician A uses a DMM to test voltage.
   Technician B uses the same tool to test resistance.
   Who is correct?
   A. A only
   B. B only
   C. Both A and B
   D. Neither A nor B

5. Which of the following is not a typical electrical problem?
   A. high resistance in the circuit
   B. an open circuit
   C. a shorted circuit
   D. a closed circuit

6. When discussing resistance,
   Technician A says that current will increase with a decrease in resistance.
   Technician B says that current will decrease with an increase in resistance.
   Who is correct?
   A. A only
   B. B only
   C. Both A and B
   D. Neither A nor B

7. When discussing shift blocking systems,
   Technician A says that the system is controlled by the powertrain control module.
   Technician B says that a solenoid is used to block or prevent shifting into particular gears when certain conditions are met.
   Who is correct?
   A. A only
   B. B only
   C. Both A and B
   D. Neither A nor B

8. When discussing how to test a switch,
   Technician A says that the action of the switch can be monitored by a voltmeter.
   Technician B says that continuity across the switch can be checked by measuring the resistance across the switch's terminals when the switch is in its different positions.
   Who is correct?
   A. Technician A
   B. Technician B
   C. Both A and B
   D. Neither A nor B

9. While performing electrical tests,
   Technician A uses a test light to detect resistance.
   Technician B uses a jumper wire to test circuit breakers, relays, and lights.
   Who is correct?
   A. Technician A
   B. Technician B
   C. Both A and B
   D. Neither A nor B

10. When discussing electricity,
    Technician A says that an open causes unwanted voltage drops.
    Technician B says that high-resistance problems cause increased current flow.
    Who is correct?
    A. Technician A
    B. Technician B
    C. Both A and B
    D. Neither A nor B

ASE CHALLENGE QUESTIONS

1. *Technician A* says that a 16-gauge wire can be safely replaced with a 14-gauge wire.

   Technician B says that a loose electrical connector will cause the current flowing through a circuit to decrease.

   Who is correct?

   A. Technician A
   B. Technician B
   C. Both A and B
   D. Neither A nor B

2. A voltmeter that is connected across the power and ground terminals of an operating motor indicates 7.5 volts.

   Technician A says that there may be a problem in the ground side of the circuit.

   Technician B says that the power supply of the circuit may be faulty.

   Who is correct?

   A. Technician A
   B. Technician B
   C. Both A and B
   D. Neither A nor B

3. A DVOM that is being used to measure the resistance of a solenoid coil is indicating *OL* in its display.

   Technician A says that this means that the solenoid coil may be open.

   Technician B says that the meter leads may have been connected backward.

   Who is correct?

   A. Technician A
   B. Technician B
   C. Both A and B
   D. Neither A nor B

4. An inoperative back-up light circuit is being tested. After the transmission shifter is placed in reverse, a DVOM connected across the terminals of the circuit fuse indicates 12 volts.

   Technician A says that the fuse is probably blown.

   Technician B says that the circuit ground is probably faulty.

   Who is correct?

   A. Technician A
   B. Technician B
   C. Both A and B
   D. Neither A nor B

5. Vehicle on-board computer troubleshooting is being discussed.

   Technician A says that if no trouble codes are stored by the computer, the system is operating correctly.

   Technician B says that a coolant temperature sensor that is indicating an engine temperature 25 degrees higher than the actual temperature will probably not cause the computer system to store a trouble code.

   Who is correct?

   A. Technician A
   B. Technician B
   C. Both A and B
   D. Neither A nor B

Name _____ Date _____

USING A DIGITAL MULTIMETER

Upon completion of this job sheet, you should be able to use a digital multimeter for diagnosing electrical and electronic circuits.

Tools and Materials

DMM

Procedure

1. List the manufacturer and model number of the DMM you are using for this worksheet. _____

   _____

2. Different DMMs are capable of measuring different things, what can this DMM do?

   _____

   _____

   _____

   _____

3. In order to use this DMM to full capacity, do you need to insert a module or connect certain leads? If so, what are they?

   _____

   _____

4. According to the literature that came with the DMM, what are the special features of this DMM?

   _____

   _____

   _____

   _____

5. What is the impedance of the meter?

   _____

   _____

6. If the meter is auto-ranging, how many decimal points are shown on the meter and what suffixes are used?

_____

_____

_____

_____

7. Does the ohmmeter function of this meter need to be zeroed before use?

_____

_____

8. What are the meter's low and high limits for measuring voltage, amperage, and resistance?

_____

_____

Instructor's Response: _____

_____

_____

Name _____ Date _____

Adjust Clutch Pedal Switch

Upon completion of this job sheet, you should be able to adjust a clutch pedal switch.

ASE NATEF MAST Task Correlation

Clutch Diagnosis and Repair

Task # 2 Inspect clutch pedal linkage, cables, automatic adjuster mechanisms, brackets, bushings, pivots, and springs: perform necessary action.

Tools and Materials

A vehicle with an adjustable clutch switch

A measuring tape

Basic hand tools

Describe the Vehicle Being Worked On

Year _____ Make _____ VIN _____

Model _____

Procedure

Task Completed
_____

1. Find the specifications for the following. (If exact specifications are not available, use the general specifications given in the text.)

   Pedal free play _____

   Clutch pedal height _____

   Pedal travel _____

   Clearance between clutch switch and pedal assembly _____

2. Measure the following and compare to the specifications.

   Pedal free play _____

   Clutch pedal height _____

3. If the measurements from step 2 are not within specifications, correct them before proceeding.

4. Check the release point of the clutch.

   a. Engage the parking brake and install wheel chocks.

   b. Start the engine and allow it to idle.

   c. Without depressing the clutch pedal, slowly move the shift lever into reverse until the gears contact.

   d. Gradually depress the clutch pedal until the noise stops. This is the release point.

5. Measure the stroke distance from the release point up to the full stroke end position.

Your measurement _____

If this distance is not within specifications, check the pedal height, pushrod play, and pedal free play. Adjust the pedal for the correct travel.

6. If the travel is correct, check the clearance between the switch and the pedal assembly when the clutch is fully depressed.

Your measurement _____

☐ 7. If clearance is not within specifications, loosen the switch and adjust its position to provide for the specified clearance.

☐ 8. Tighten switch to specifications.

☐ 9. Work clutch pedal and note any noises. If there are no unusual noises, the task is complete. If there are noises, find the cause and correct.

Instructor's Response: _____

_____

_____

Name _____ Date _____

TESTING SWITCHES, CONNECTORS, RELAYS, AND WIRES

Upon completion of this job sheet, you should be able to inspect and test switches, connectors, relays, and wires of electrical and electronic circuits with a DMM and jumper wires.

ASE NATEF MAST Task Correlation

Four-Wheel Drive/All-Wheel Drive Component Diagnosis and Repair

Task # 6 Diagnose, test, adjust, and replace electrical/electronic components of four-wheel-drive systems.

Tools and Materials

Test light

Jumper wire

DMM

Protective Clothing

Goggles or safety glasses with side shields

Describe the Vehicle Being Worked On

Year _____ Make _____ VIN _____

Model _____

Procedure

Protection Devices

1. Check a fuse with an ohmmeter. If the fuse is good, there will be continuity through it. To test a circuit protection device with a voltmeter, check for available voltage at both terminals of the unit. If the device is good, voltage will be present on both sides. A test light can be used in place of a voltmeter. Summarize the results of this check.

   _____

   _____

2. Measure the voltage drop across a fuse or other circuit protection device. If a fuse, a fuse link, or circuit breaker is in good condition, a voltage drop of zero will be measured. If 12 volts is read, the fuse is open. Any reading between zero and 12 volts indicates some voltage drop. If there is voltage drop across the fuse, it has resistance and should be replaced. Make sure you check the fuse holder for resistance as well. Summarize the results of this check.

   _____

   _____

Switches

1. To check a switch, disconnect the connector at the switch. With an ohmmeter, check for continuity between the terminals of the switch with the switch moved to the ON position and to the OFF position. With the switch in the OFF position, there should be no continuity between the terminals. With the switch on, there should be good continuity between the terminals. Summarize the results of this check.

   _____

   _____

2. If the switch is activated by something mechanical and doesn't complete the circuit when it should, check the adjustment of the switch. If the adjustment is correct, replace the switch.

3. Another way to check a switch is to simply bypass it with a jumper wire. If the component works when the switch is jumped, a bad switch is indicated. Summarize the results of this check.

   _____

   _____

4. Voltage drop across switches also should be checked. Ideally when the switch is closed there should be no voltage drop. Any voltage drop indicates resistance and the switch should be replaced. Summarize the results of this check.

   _____

   _____

Relays

1. Check the wiring diagram for the relay being tested to determine if the control is through an insulated or ground switch. If the relay is controlled on the ground side, continue with this procedure. If the relay is controlled on the positive side, describe the correct way to test it.

   _____

   _____

2. Connect the negative lead of the voltmeter to a good ground.

3. Connect the positive lead to the output wire. If no voltage is present, continue testing. If there is voltage, disconnect the ground circuit of the relay. The voltmeter should now read 0 volts. If it does, the relay is good. If voltage is still present, the relay is faulty and should be replaced.

4. Connect the positive voltmeter lead to the power input terminal. Close to battery voltage should be there. If not, the relay is faulty. If the correct voltage is present, continue testing.

5. Connect the positive meter lead to the control terminal. Close to battery voltage should be there. If not, check the circuit from the battery to the relay. If the correct voltage is there, continue testing.

6. Connect the positive meter lead to the relay ground terminal. If more than 1 volt is present, the circuit has a poor ground. Summarize the results of this check.

_____

_____

Stepped Resistors

1. Remove the resistor from its mounting. ☐

2. Make sure the ohmmeter is set to the correct scale for the anticipated amount of resistance. Connect the ohmmeter leads to the two ends of the resistor. ☐

3. Compare the results against specifications. Summarize the results of this check.

_____

_____

Variable Resistors

1. Identify the input and output terminals and connect the ohmmeter across them. ☐

2. Rotate or move the variable control while observing the meter. The resistance value should remain within the limits specified for the switch. If the resistance values do not match the specified amounts, or if there is a sudden change in resistance as the control is moved, the unit is faulty. Summarize the results of this check.

_____

_____

Instructor's Response: _____

_____

_____

Chapter 11

DIAGNOSING ADVANCED ELECTRONIC SYSTEMS

UPON COMPLETION AND REVIEW OF THIS CHAPTER, YOU SHOULD BE ABLE TO:

- Use the data retrieved from the vehicle's onboard diagnostic system to diagnose drivetrain systems.
- Identify the cause of intermittent problems.
- Reprogram and update control modules.
- Diagnose the cause of communication errors.
- Test the primary sensors in an electronic circuit.

- Diagnose computer voltage supply and ground wires.
- Test computer outputs and actuators.
- Diagnose motor-related problems.
- Test electromagnetic clutches.
- Describe the operational characteristics of an electronically controlled manual transmission/transaxle.

As time goes on, more and more functions of a manual transmission, final drive unit, and four-wheel-drive systems are being controlled and monitored by the vehicle's onboard computers. This is especially true of self-shifting manual transmissions. Their operation depends on precise and immediate engagement and disengagement of clutches, gears, and shafts; this is the role of the electronic control systems.

These systems will also affect the engine's operation, such as the way some self-shifting transmissions will open the throttle during downshifting to match the speed of the engine with the speed of the rotating gears inside the transmission. This is done through electronics.

Diagnosis of these systems begins with checks of the electronic system. This chapter covers those basics, plus a detailed discussion on how to test individual sensors and actuators.

ONBOARD DIAGNOSTICS

The use of **onboard diagnostics II (OBD II)** systems has brought computerization to most systems on a vehicle. Onboard diagnostic capabilities are incorporated into a vehicle's computer to monitor virtually every component that can affect emission performance. Each component is checked by a diagnostic routine to verify that it is functioning properly. In addition to self-diagnostics, OBD provides a means to check the operation of the control systems.

Troubleshooting OBD II Systems

Diagnosing a computer-controlled system is much more than accessing diagnostic trouble codes (DTCs). The importance of logical troubleshooting cannot be overemphasized. Logical diagnosis means following a simple basic procedure. Start with the most likely cause and work to the most unlikely. In other words, check the easiest, most likely problem sources before

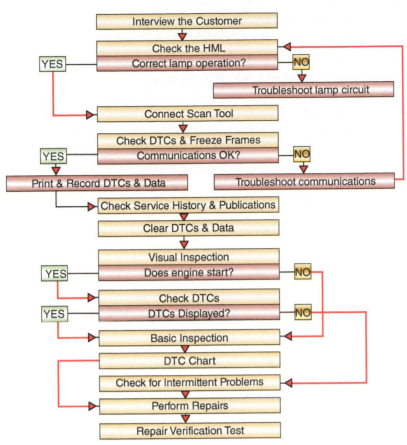

FIGURE 11-1 The steps for troubleshooting OBD II systems.

proceeding to the less likely, and more difficult, sources. Do not guess at the problem or jump to a conclusion while troubleshooting. Always check all traditional nonelectronic systems before attempting to diagnose the electronic control system. For example, low-battery voltage might result in faulty sensor readings.

Always refer to the manufacturer's information before beginning your diagnosis. The steps for diagnosing an OBD II system are listed in Figure 11-1. The following are some things to consider while troubleshooting OBD II systems.

Interview the Customer. Gather as much information as possible from the customer. Get a description of the conditions that are present when the problem occurs, including weather, traffic, and speed.

Check the MIL. The malfunction indicator lamp (MIL) should turn on when the ignition is turned ON and the engine is not running. When the engine is started, the MIL should go off. If this does not occur, troubleshoot the lamp system before continuing.

Check DTC(s) and Freeze-Frame Data. Connect an OBD II-compliant scan tool. Retrieve the DTCs. If an asterisk (*) is next to the DTC, this may indicate there is stored freeze-frame data related to that DTC. If there was a no or poor communication DTC, solve that problem before continuing.

Check Service History and Service Publications. There may be a TSB that may have the necessary repair information. Check them and follow the given procedures. Service history can give clues about cause of the problem, as the problem may be related to a recent repair.

Visual Inspection. Check all wires for frays, looseness, and damage. Also, check the wiring for burned or chafed spots, pinched wires, or contact with sharp edges or hot exhaust parts. Try not to wiggle the wires while doing this; a wiggle may correct an intermittent problem that may be hard to find later. Check the condition of the battery and all sensors and actuators for

The malfunction indicator lamp (MIL) is a warning lamp in a vehicle's instrument panel that lets the driver know when the vehicle's electronic control units detected a problem.

physical damage. Check all connections to sensors, actuators, control modules, and ground points. Also check all vacuum hoses for pinches, cuts, or disconnects. Correct any problems.

Check DTCs. The original DTCs should be cleared from the computer. After the visual inspection, they should be checked again. If there are no DTCs, check the status of the monitors' readiness and the pending codes. Do what is necessary to complete the necessary drive cycles before continuing. For many DTCs, the powertrain control module (PCM) will enter into the fail-safe mode. Refer to the service information to determine if any DTCs indicate the fail-safe mode; if so, follow the appropriate diagnostic steps. Current DCTs indicate a problem that is present. Use the DTC chart to determine what was detected, the probable problem areas, and how to diagnose that DTC.

Basic Inspection. When the DTC is not confirmed in the previous check or if a DTC was not retrieved in the first check, use the problem symptoms chart given in the service information.

Check for Intermittent Problems. If the cause of the problem has not yet been determined, proceed to check for an intermittent problem.

Perform Repairs. Once the cause of the concern has been identified, perform all required services. After repairs have been made, check your work by clearing all codes, checking the MIL, and rechecking for codes.

Diagnostic Trouble Codes

OBD II codes are standardized and most DTCs mean the same thing regardless of the vehicle. However, manufacturers can have additional DTCs and add more data streams, report modes, and diagnostic tests. DTCs are designed to indicate the circuit and system where a fault has been detected. A DTC does not necessarily indicate the faulty component; it only indicates the circuit of the system that is not operating properly.

An OBD II DTC is a five-character code (Figure 11-2). The first character of the code is a letter. It defines the problem system. The codes are "B" for body, "C" for chassis, "P" for powertrain, and "U" for undefined. The U-codes are designated for special or future use.

The second character is a number. This defines the code as being a mandated code or a special manufacturer code. A "0" code means that the fault is mandated by OBD II. A "1" code means the code is manufacturer specific.

The third through fifth characters are numbers and describe the fault. The third character tells you where the fault occurred. The remaining two characters describe the exact condition that set the code. In most cases, the codes are organized so that the various codes related to a particular sensor or system are grouped together.

Not all DTCs will cause the MIL to light; this depends on the problem. DTCs that will not affect emissions will never illuminate the MIL. When some faults are detected for the first time, a DTC is stored as a **pending code**. A pending DTC is a code representing a fault that has occurred, but that has not occurred enough times to illuminate the MIL.

Freeze-Frame Data

A mandated capability of OBD II is the **freeze-frame data** or snapshot feature. Although the regulations mandate just emission-related DTCs, manufacturers can choose to include this feature for other systems. With this feature, the PCM takes a snapshot of the activity of the various inputs and outputs at the time the PCM illuminated the MIL. This feature is valuable to technicians, especially when trying to identify the cause of an intermittent problem. A review of the action of the sensors and actuators when the code was set can be used to identify components that should be tested. The information held in freeze frame are actual values; they have not been altered by the adaptive strategy of the PCM (Figure 11-3).

Once a DTC and related freeze-frame data are stored in memory, they can only be removed with a scan tool. When DTCs are erased, the scan tool will also erase all freeze-frame data.

EXAMPLE: P0137 LOW VOLTAGE BANK 1 SENSOR 2

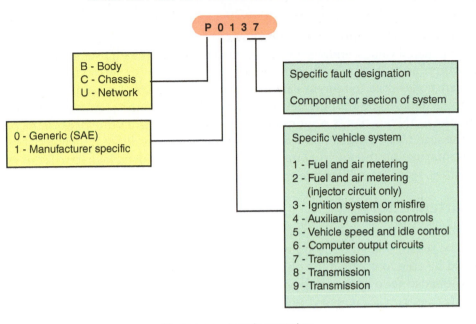

P 0 1 3 7

B - Body
C - Chassis
U - Network

0 - Generic (SAE)
1 - Manufacturer specific

Specific fault designation

Component or section of system

Specific vehicle system

1 - Fuel and air metering
2 - Fuel and air metering
    (injector circuit only)
3 - Ignition system or misfire
4 - Auxiliary emission controls
5 - Vehicle speed and idle control
6 - Computer output circuits
7 - Transmission
8 - Transmission
9 - Transmission

Typical generic code examples:

Mass Airflow	P0102, P0103
Intake Air Temperature	P0112, P0113, P0127
Barometric Pressure	P0106, P0107, P0108, P0109
Engine Coolant Temperature	P0117, P0118
Oxygen Sensor (one of several)	P0131, P0133, P0135, P0136, P0141
Throttle Position	P0121, P0122, P0123
EGR/EVP	P0400, P0401, P0402
Vehicle Speed	P0500, P0501, P0503

Note: Manufacturers will also use specific codes that apply only to their systems.

FIGURE 11-2 Explanation of OBD II DTCs.

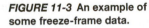

FREEZE FRAME   0

TROUBLE CODE P0304
ENGINE SPD 683RPM
COOLANT TEMP............... 190°F
VEHICLE SPD 0MPH
CALC LOAD 18.0%
FUEL SYS #1 CL
FUEL SYS #2 CL
SHORT FT #1 .... 0.8%
LONG FT #1 .... −5.4%
SHORT FT #2 .... −0.7%
LONG FT #2 .... 12.5%

FIGURE 11-3 An example of
some freeze-frame data.

AUTHOR'S NOTE: Many electronic control systems have adaptive strategies that allow the engine to run when one or more inputs fail. When the computer senses an out-of-limit value from a sensor, it will rely on a predetermined value and ignore the input. Check for an error message when retrieving DTCs.

Serial Data

The scan tool can be set to retrieve Mode 1 data. This mode is also referred to as the **parameter identification (PID)** mode. This mode allows access to current data values of inputs and outputs, calculated values, and system status information. Some PID values are manufacturer specific; others are common to all vehicles. This information is referred to as **serial data**.

PIDs are codes used to request data from a PCM. A scan tool is used for the request and receipt of the data. To do this, the desired PID is entered into the scan tool. The device connected to the CAN bus that is responsible for that PID reports a value back to the bus and is read on the scan tool. Manufacturers list and define the various PIDs in their service information. This information should be compared with the observed data. If an item is not within the normal values, record the difference and diagnose that particular item.

Intermittent Faults

An intermittent fault is a fault that is not always present. It may not activate the MIL or cause a DTC to be set. By studying the system and the relationship of each component to another, you should be able to create a list of possible causes for the intermittent problem. To help identify the cause, follow these steps:

1. Observe the history of DTCs, DTC modes, and freeze-frame data.
2. Combine your knowledge of the system with the available service information. Call technical assistance for possible solutions.
3. Evaluate the symptoms and conditions described by the customer.
4. Use a checklist to identify the circuit or electrical component that may have the problem.
5. Follow the suggestions for intermittent diagnosis found in service material.
6. Visually inspect the suspected circuit or system.
7. Test the circuit's wiring for shorts, opens, and high resistance. This should be done with a DMM, unless instructed differently in the service information.

Most intermittent problems are caused by faulty connections or wiring. Refer to a wiring diagram for each of the suspected circuits or components to identify the connections and components in them. The entire electrical system of those circuits should be carefully inspected. Check for burnt or damaged wire insulation, damaged terminals at the connectors, corrosion at the connectors, loose connectors, loose wire terminals, and loose ground wire or straps.

A voltmeter can be connected to the suspected circuit and the wiring harness wiggled (Figure 11-4). If the voltage changes with the wiggles, the problem is in that circuit. The vehicle can also be taken for a test drive with the voltmeter connected. If the voltmeter readings become abnormal with changing conditions, the circuit being observed probably has the problem.

The vehicle can also be taken for a test drive with the scan tool connected. The scan tool will monitor the activity of a circuit while the vehicle is being driven. This allows a look at a circuit's response to changing conditions. The snapshot or freeze-frame feature stores the conditions and operating parameters at command or when the PCM sets a DTC. If the snapshot is taken when the intermittent problem is occurring, the problem will be easier to diagnose.

With a scan tool, actuators can be activated and their functionality tested. The results of their operation can be monitored. Also the outputs can be monitored as they respond to changes from the inputs. When an actuator is activated, watch the response on the scan tool. Also listen for the clicking of the relay that controls that output. If no clicking is heard, measure the voltage at the relay's control circuit; there should be a change of more than 4 volts when it is activated.

Monitor how the PCM and an output respond to a change in sensor signals by selecting the mode that relates to that circuit. View and record the data for that circuit. Compare them to

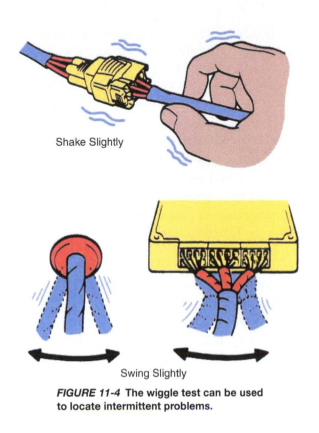

Shake Slightly

Swing Slightly

FIGURE 11-4 The wiggle test can be used to locate intermittent problems.

specifications. Then create a condition that would cause the related inputs to change. Observe the data to see if the change was appropriate.

Service Information

After retrieving the DTCs, find the description of the DTC in the service information. As can be seen in Figure 11-5, there is typically more than one possible cause of the problem. One is the sensor itself and the other two relate to the sensor's circuit.

DTC Code	Reference	Possible Problem Areas
B1214	Short to B+ in Door System Communication Bus Malfunction	1. Theft warning ECU assembly (Theft deterrent ECU) 2. Multiplex network door ECU back (Back door ECU) 3. Wire harness 4. Instrument panel junction block assembly (Body ECU)
B1215	Short to GND in Door System Communication Bus Malfunction	1. Theft warning ECU assembly (Theft deterrent ECU) 2. Multiplex network door ECU back (Back door ECU) 3. Wire harness 4. Instrument panel junction block assembly (Body ECU)
B1269	Theft Deterrent ECU Communication Stop	1. Theft warning ECU assembly (Theft deterrent ECU) 2. Wire harness
B1287	Back Door ECU Communication Stop	1. Multiplex network door ECU back (Back door ECU) 2. Wire harness

FIGURE 11-5 Multiplex communication system DTC chart.

The description of the DTC also leads to pinpoint tests. These are designed to guide you through a step-by-step procedure. To be effective, each step should be performed, in the order given, until the problem has been identified.

Make sure to check all available service information related to the DTCs. There may be a TSB related to the code. This may be a recommendation to reprogram the computer's software.

Reprogramming Control Modules

Reprogramming a computer is typically called **flashing** the computer. When a computer is flashed, the old program is erased and a new one written. Reprogramming is often necessary when the manufacturer discovers a common concern that can be solved through changing the system's software. New programs are downloaded into the scan tool and then downloaded into the computer through a dedicated circuit.

Each type of scan tool has a different procedure for flashing. Always follow the manufacturer's instructions. Some scan tools are connected to a PC and the software is transferred from a CD or a Web site. Photo Sequence 27 shows a typical procedure for flashing a PCM or BCM.

Symptom-Based Diagnosis

At times, no DTCs are set but a problem exists. To discover the cause of these concerns, the description of the problem or its symptoms should be used to determine what systems or components should be checked. Before diagnosing a problem based on its symptoms, make sure:

- The PCM and MIL are operating correctly.
- There are no stored DTCs.
- All data observed on the scan tool is within normal ranges.
- There is communication between the scan tool and the control system.
- There are no TSBs available for the current symptom.
- All of the grounds for the BCM/PCM are sound.
- All vehicle modifications are identified.
- The vehicle's tires are properly inflated and are the correct size.

There is typically a section in the service information dedicated to symptom-based diagnosis. Although a customer may describe a problem in nontechnical terms, you should summarize the concern to match one or more of the various symptoms listed by the manufacturer. Each potential problem area should be checked. It is important to realize that some problems may cause more than one symptom.

COMMUNICATION NETWORKS

Today's vehicles have hundreds of circuits, sensors, and other electrical parts. In order for the control systems to operate correctly, there must be good communication between them. Communication can take place through wires connecting each sensor and its circuit to a control module. If more than one control module is involved, additional pairs of wires must connect the sensor and circuit to the other modules. The result of this communication network is miles of wires and hundreds of connectors. To eliminate this need, manufacturers use multiplexing (Figure 11-6).

A **protocol** is the language control modules speak when they are talking to each other. The differences in protocol are based on the speed and technique used. SAE has classified the different protocols as:

- *Class A* (low-speed communication): This is used for convenience systems, such as entertainment systems, audio, trip computer, seat controls, windows, and lighting. Most Class A functions require inexpensive, low-speed communication and use a generic universal asynchronous receiver/transmitter (UART). These functions are proprietary and have not been standardized by the industry.

Multiplexing is a means of transmitting information between computers. It is a system in which electrical signals are transmitted by a peripheral serial bus instead of common wires, allowing several devices to share signals on a common conductor.

A TYPICAL PROCEDURE FOR FLASHING A PCM OR BCM

P27-1 Use the scan tool to retrieve the BCM's part number or the vehicle's VIN and record them.

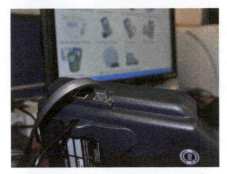

P27-2 Connect the scan tool to a PC that can link the scan tool to the flash software. Some scan tools will connect directly to an Internet site.

P27-3 Enter the BCM part number in the appropriate field and select "Show Updates" on the menu.

P27-4 Select the desired flash line.

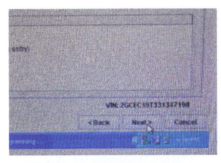

P27-5 Hit the "Next" button to begin downloading of the software.

P27-6 Monitor the progress of the downloading to the scan tool.

P27-7 Once downloading is complete, connect a battery charger to the vehicle's battery. Turn it on and maintain about 13.5 volts at the battery.

P27-8 Connect the scan tool to the DLC and turn the scan tool on.

P27-9 Move through the menus on the scan tool until the desired flash screen is shown. Then follow all instructions given on the tool.

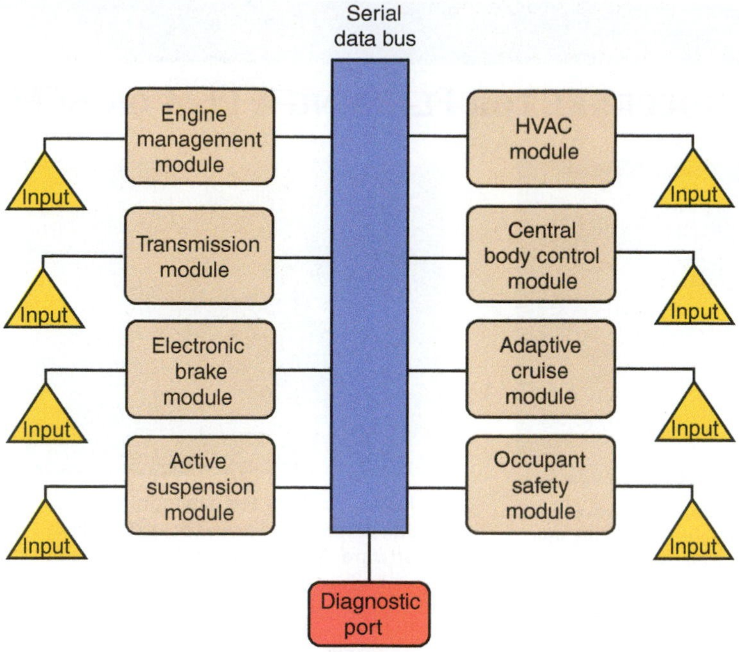

FIGURE 11-6 A multiplexed system uses a serial data bus to allow communications between the various control modules.

- *Class B* (medium-speed communication): Class B is used primarily with the instrument cluster, vehicle speed, and emissions data recording. Included in this classification are different standards, designated by a number. The most common is the SAE J1850 standard. These standards are also divided by their operation. One is a VPW (variable pulse width) type that uses a single-wire bus. Another is a PWM (pulse width modulation) type that uses a two-wire bus.
- *Class C* (high-speed communication): This protocol is for real-time control of the powertrain, vehicle dynamics, and brake-by-wire. This protocol uses a twisted pair of wires, but a shielded coaxial or fiber optics cable may be used for less noise interference. The most common class C protocol is CAN 2.0. (controller area network version 2.0). CAN assigns a unique identifier to every message. The identifier classifies the content of the message and the priority of the message sent. CAN buses are found in nearly all late-model vehicles.

It is common to find a variety of multiplexing classes in a single vehicle. Some systems require high-speed communications, whereas other systems do not.

CAN Buses

The total network in most vehicles comprises two or three CAN buses (Figure 11-7). Each of these networks operates at a different speed. The different buses are identified by a prefix or suffix. Typically, CAN-B is the body bus and CAN-C is the engine and diagnostic bus. A medium-speed bus may be called CAN-B or MS-CAN. Likewise, a high-speed bus can be called CAN-C or HS-CAN. Manufacturers are not consistent with these labels.

These networks are integrated through the use of a gateway. A **gateway** module allows for data exchange between the different buses. It translates a message on one bus and transfers that message to another bus without changing the message. The gateway interacts with each bus according to the protocol of that bus.

Each twisted wire in a CAN bus carries a different voltage (Figure 11-8). The slightest change in voltage will affect the operation of one or more systems; therefore all potential for voltage spikes, electrical noise, or induction must be eliminated. Twisting the wires eliminates the possibility of voltage being induced in one wire as current flows through the other.

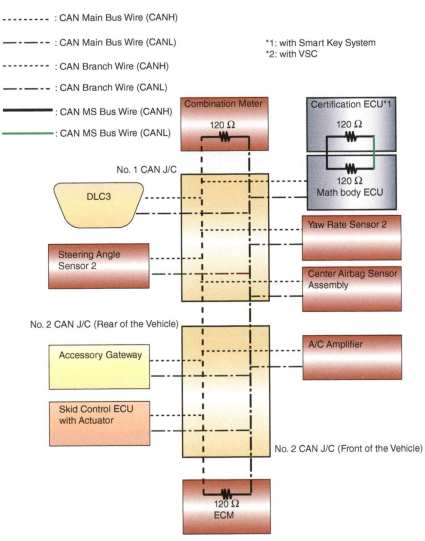

- - - - - - - : CAN Main Bus Wire (CANH)

—·—·— : CAN Main Bus Wire (CANL)

········· : CAN Branch Wire (CANH)

—·—·— : CAN Branch Wire (CANL)

━━━━ : CAN MS Bus Wire (CANH)

━━━━ : CAN MS Bus Wire (CANL)

*1: with Smart Key System
*2: with VSC

Combination Meter
120 Ω

Certification ECU*1
120 Ω

120 Ω
Math body ECU

No. 1 CAN J/C

DLC3

Yaw Rate Sensor 2

Steering Angle
Sensor 2

Center Airbag Sensor
Assembly

No. 2 CAN J/C (Rear of the Vehicle)

Accessory Gateway

A/C Amplifier

Skid Control ECU
with Actuator

No. 2 CAN J/C (Front of the Vehicle)

120 Ω
ECM

FIGURE 11-7 A basic look at a **CAN** communication system.

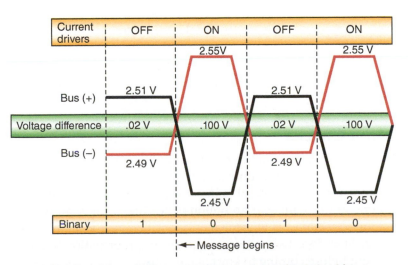

Current drivers	OFF	ON	OFF	ON
Bus (+)	2.51 V	2.55V	2.51 V	2.55 V
Voltage difference	.02 V	.100 V	.02 V	.100 V
Bus (−)	2.49 V	2.45 V	2.49 V	2.45 V
Binary	1	0	1	0

← Message begins

FIGURE 11-8 In order for a message to be transmitted, the drivers
are energized to pull up the bias on the + bus and pull down on
the − bus. There must be a voltage differential between the wires.

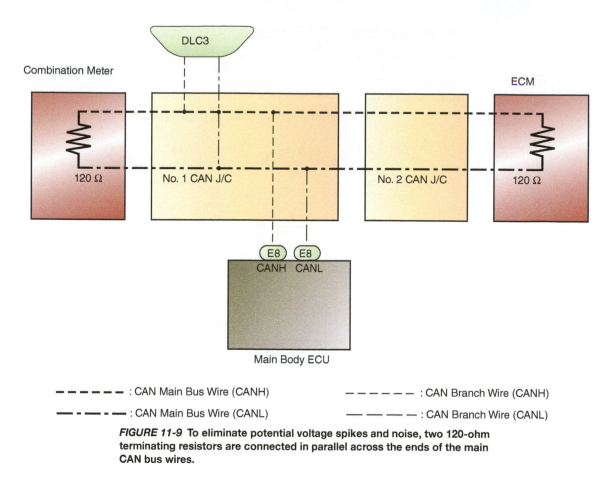

FIGURE 11-9 To eliminate potential voltage spikes and noise, two 120-ohm terminating resistors are connected in parallel across the ends of the main CAN bus wires.

To eliminate other potential spikes and noise, two 120-ohm resistors are connected in parallel across the ends of the CAN bus wires. These are called terminating resistors (Figure 11-9). The location of the resistors varies.

By-Wire Technology. The use of by-wire technology is now a reality because of the high-powered computers used in today's vehicles. Currently this technology has eliminated the mechanical connection from the throttle pedal to the fuel injection system. This capability allows for speed matching in self-shifting transmissions. The driver does not work the throttle, the control system does. Shift-by-wire technology is also used for gear selection.

Drive-by-wire systems use sensors to translate the movement of shifters, pedals, the steering wheel, and other parts into electronic signals. The computer receives these signals and commands electric motors to perform the function ordered by the driver. The systems respond much quicker than mechanical linkages and can send feedback to the computer as they operate.

Communication Checks

Performing diagnostic checks on vehicles should include a communications check. If the different control modules are not communicating with each other, there is no way to properly diagnose the systems. When a scan tool is installed, it will try to communicate with every module that could be in the vehicle. If an option is not there, the scan tool will display "No Comm" for that control module. That same message will appear if the module is present but not communicating. Therefore always refer to the service information to identify what modules should be present before coming to any conclusions.

The system periodically checks itself for communication errors. The different buses send messages to each other immediately after it sends a message. The message between messages

FIGURE 11-10 The bus wires are twisted pairs to protect them from unwanted induced voltages that can change voltage signals.

checks the integrity of the communication network. All of the modules in the network also receive a message within a specific time. If the message is not received, that control module will set a DTC stating it did not receive the message.

There are three types of DTCs set by CAN buses:

- **Loss of communication.** Loss of communication (and bus-off) DTCs are set when there is a problem with the communication between modules. This could be caused by bad connections, wiring, or the module.
- **Signal error.** The control modules can run diagnostics on some input circuits to determine if they are operating normally. If a circuit fails the test, a DTC will set.
- **Internal error.** The modules also run internal checks. If there is a problem, it will set an internal error DTC.

Bus Wire Service. If the bus wire needs repair, it must not be relocated or untwisted. The twisting serves an extremely important purpose (Figure 11-10). After a bus wire has been repaired by soldering, wrap the repair with vinyl tape. Never run the repair wire so it bypasses the twisted sections. CAN bus wires are likely to be influenced by noise if they are not twisted.

A bus is the common connector used as an information source for the vehicle's various control units.

BASIC TESTING

Diagnosis of electronic control systems includes inspecting and testing the individual parts and their circuits. The operation of some components can be monitored with the scan tool; however, additional tests are normally necessary. These tests include:

- *Ohmmeter checks:* Most sensors and output devices can be checked with an ohmmeter.
- *Voltmeter checks:* Many sensors, output devices, and their wiring can be diagnosed by checking the voltage to them and, in some cases, from them. Moreover, voltage drop tests are extremely valuable.
- *Lab scope checks:* The activity of sensors and actuators can be monitored with a lab scope or a graphing multimeter. By watching their activity, abnormal behavior can be observed.

⚠️ **WARNING:** Before disconnecting any electronic component, be sure the ignition is turned off. Disconnecting components may cause high-induced voltages and computer damage.

Testing Sensors

To monitor conditions, the computer uses a variety of sensors. All sensors perform the same basic function. They detect a mechanical condition (movement or position), chemical state, or temperature and change them into electrical signals that can be used by the PCM.

If a DTC directs you to a faulty sensor or sensor circuit, or if you suspect a faulty sensor, it should be tested. Always follow the correct procedures when testing sensors and other electronic components. Also, make sure you have the correct specifications for each part. Sensors are tested with a DMM, scan tool, and/or lab scope or GMM.

Make sure you follow the instructions from the scope's manufacturer. If the scope is set wrong, the scope will not break. It just will not show you what you want to be shown. To help with understanding how to set the controls on a scope, keep the following things in mind. The vertical voltage scale must be adjusted in relation to the expected voltage signal. The horizontal time base or milliseconds per division must be adjusted so the waveform appears properly on the screen.

Minor adjustments of the trigger line may be necessary to position the waveform in the desired vertical position. Trigger slope indicates the direction in which the voltage signal is moving when it crosses the trigger line. A positive trigger slope means the voltage signal is moving upward as it crosses the trigger line, whereas a negative trigger slope indicates the voltage signal is moving downward when it crosses the trigger line.

Software packages, often programmed in a lab scope or GMM, are available to help you properly interpret scope patterns and setup the scope. These also contain an extensive waveform library that you can refer to and find what the normal waveform of a particular device should look like. The library also contains the waveforms caused by common problems. You can also add to the library by transferring waveforms to a PC from the lab scope. After the waveforms have been transferred, notes can be added to the file. The software may also include the theory of operation and diagnostic procedures for common inputs and outputs.

Some sensors are simple on-off switches. Others are variable resistors that change resistance according to temperature changes. Some are voltage or frequency generators, whereas others send varying signals according to the rotational speed of a device. Knowing what they are measuring and how they respond to changes are the keys to accurately testing a sensor.

Some inputs to the PCM come from another control module or are simply a connection from a device. The battery's voltage is available on the data bus, and many control modules need this information. There is no sensor involved, just a connection from the battery to the bus. The heated windshield module tells the computer when the heated windshield system is operating. This helps the PCM to accurately determine engine load and control idle speed.

Testing Switches

A switch can be easily tested with an ohmmeter. Disconnect the connector at the switch. Refer to the wiring diagram to identify the terminals at the switch if there are more than two (Figure 11-11). Connect the ohmmeter across the switch's terminals. Perform whatever action is necessary to open and close the switch.

Switches can also be checked with a voltmeter. The signal to the PCM from supply-side switches should be 0 volts with the switch open and supply voltage when the switch is close. Using a voltmeter is preferred because it tests the circuit as well as the switch. If less than supply voltage is present with the switch closed, there is unwanted resistance in the circuit. Expect the opposite readings on a ground-side switch.

Some switches are adjustable and must be set so they close and open at the correct time. A clutch switch is used to inform the computer when there is no load (clutch pedal depressed) on the engine. The switch is also connected to the starting circuit. The switch prevents starting of the engine unless the clutch pedal is fully depressed. The switch is normally open when the

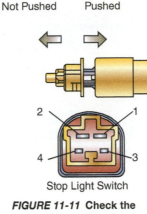

Not Pushed   Pushed

2   1

4   3

Stop Light Switch

FIGURE 11-11 **Check the
service information for the
proper testing points for
a switch.**

clutch pedal is released. When the clutch pedal is depressed, the switch closes and completes
the circuit between the ignition switch and the starter solenoid. It also sends a signal of no-load
to the PCM.

Most grounding switches react to some mechanical action to open or close. There are
some, however, that respond to changes in pressure or temperature. An example of this type
of switch is the power-steering pressure switch. This switch informs the PCM when power-
steering pressures reach a particular point. When the power-steering pressure exceeds that
point, the PCM knows there is an additional load on the engine and will increase idle speed.

To test this type of switch, monitor its activity with a DMM or lab scope. With the engine
running at idle speed, turn the steering wheel to its maximum position on one side. The voltage signal should drop as soon as the pressure in the power-steering unit has reached a high
level. If the voltage does not drop, either the power-steering assembly is incapable of producing high pressures or the switch is bad.

Temperature-responding switches operate in the same way. When a particular temperature is reached, the switch opens. This type of switch is best measured by removing it, connecting it to an ohmmeter, and submerging it in heated water. A good temperature-responding
switch will open (have an infinite reading) when the water temperature reaches the specified
amount. If the switch fails this test, it should be replaced.

Testing Temperature Sensors

The PCM changes the operation of many systems based on temperature. Temperature sensor problems are often caused by wiring faults or loose or corroded connections. Nearly all
temperature sensors are NTC thermistors. The resistance of the sensor changes with a change
in temperature. Typically, the PCM supplies a reference voltage of 5 volts to the sensor. That
voltage is changed by the change of the sensor's resistance and is fed back to the PCM. Based
on the return voltage, the PCM calculates the exact temperature. When the sensor is cold,
its resistance is high, and the return voltage signal is also high. As the sensor warms up, its
resistance drops and so does the voltage signal. Many testers are able to show where to place
the probes of the tester to check sensors, such as an ECT (Figure 11-12).

Temperature sensor circuits should be tested for opens, shorts, and high resistance. Often,
a DTC will be set. Scan tool data should also be looked at. If the observed temperature is the
coldest possible, the circuit is open. If the temperature is the highest possible, the circuit has
a short. Also, if the connector to the sensor is disconnected, the readings should drop to cold.
High-resistance problems will cause the PCM to respond to a lower temperature than the
actual temperature. This can be verified by using a good thermometer (infrared is best) to

A NTC thermistor
is a negative
temperature
coefficient thermistor
and a PTC is a
positive temperature
coefficient thermistor
that reacts in the
opposite way as
an NTC.

FIGURE 11-12 Many modern testers have the capability of noting the terminals of a switch for testing purposes.

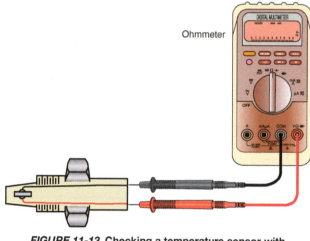

Ohmmeter

FIGURE 11-13 Checking a temperature sensor with an ohmmeter.

measure the temperature and compare it with the scan tool readings. There will be a slight difference between the two if the sensor circuit is working properly. Unwanted resistance in the circuit can cause delayed or poor shifting in 4WD and self-shifting systems.

Temperature sensors can be tested by removing them and placing them into a container of water with an ohmmeter connected across the sensor terminals (Figure 11-13). A thermometer is also placed in the water. When the water is heated, the sensor should have the specified resistance at any temperature. If the sensor does not have the specified resistance, replace the sensor. Manufacturers give a temperature and resistance chart for each of the temperature sensors.

⚠️ **WARNING:** Never apply an open flame to a temperature sensor for test purposes. This action will damage the sensor.

With the wiring connectors disconnected from the sensor and the computer, connect an ohmmeter from each sensor terminal to the computer terminal to which the wire is connected. Both sensor wires should indicate less resistance than specified by the manufacturer. If the wires have high resistance, the wires or wiring connectors must be repaired.

With the sensor installed, its terminals may be backprobed to connect a voltmeter to the sensor terminals. The sensor should provide the specified voltage drop at any temperature (Figure 11-14). Use an infrared temperature probe to measure the temperature.

COLD	HOT
10K-ohm resistor	909-ohm resistor
−20° F 4.7 V	110° F 4.2 V
0° F 4.4 V	130° F 3.7 V
20° F 4.1 V	150° F 3.4 V
40° F 3.6 V	170° F 3.0 V
60° F 3.0 V	180° F 2.8 V
80° F 2.4 V	200° F 2.4 V
100° F 1.8 V	220° F 2.0 V
120° F 1.2 V	240° F 1.6 V

FIGURE 11-14 Typical values of a thermistor-type temperature sensor at a variety of temperatures.

Testing Pressure Sensors

Most pressure sensors are piezometric sensors (Figure 11-15). A silicon chip in the sensor flexes with changes in pressures. One side of the chip is exposed to a reference pressure, which is either a perfect vacuum or a calibrated pressure. The other side is the pressure that will be measured. As the chip flexes in response to pressure, its resistance changes. This changes the voltage signal sent to the PCM. The PCM looks at the change and calculates the pressure change.

Normally, the PCM sends a voltage reference signal to a pressure sensor. With the ignition switch on, backprobe the reference wire and measure the voltage. If the reference wire does not have the specified voltage, check the reference voltage at the PCM. If the voltage is within specifications at the PCM, but low at the sensor, repair the wire. When this voltage is low at the PCM, check the voltage supply wires and ground wires for the PCM. If the wires are good, replace the computer.

The term Piezoresistive represents the characteristic of something that changes resistances in relation to changes in pressure.

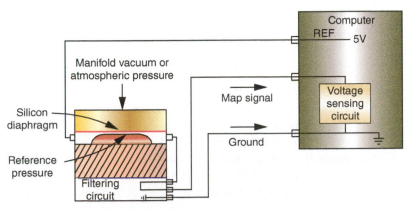

FIGURE 11-15 A piezometric sensor is made up of a silicon diaphragm sealed in a quartz plate.

With the ignition switch on, connect the voltmeter from the sensor's ground wire to the battery ground. If the voltage drop across this circuit exceeds specifications, repair the ground wire from the sensor to the computer. To check the voltage signal of a pressure sensor, turn the ignition switch on and connect a voltmeter to the sensor signal wire. Operate the transmission, engine, or transfer case in a way that should cause the monitored pressure to change. If there was no change in the voltage signal or if it was not what was expected, replace the sensor.

To check a pressure sensor with a lab scope, connect the scope to the sensor's output and a good ground. Operate the transmission, engine, or transfer case in a way that should cause the monitored pressure to change. If there was no change in the voltage signal or if it was not what was expected, replace the sensor. If the signal is erratic, the sensor or sensor wires are defective.

Some pressure sensors use a diaphragm to sense pressure changes. As the diaphragm moves, the resistance of the sensor changes. To test these, remove the sensor and apply varying amounts of air pressure to the port below the diaphragm. Observe the resistance changes with an ohmmeter. Compare your findings to the specifications, if they are available.

Testing Speed Sensors

Speed sensors measure the rotational speed of something. These sensors provide critical input to 4WD/AWD systems and to the controller for electronically shifted manual transmissions and transaxles. Speed sensors are either Hall-effect switches or magnetic pulse generators. Identifying the type of sensor used in a particular application dictates how the sensor should be tested.

Magnetic pulse generator (variable reluctance) sensors generate a waveform at a frequency that is proportional to the speed of an object (Figure 11-16). When moving at a low speed, the sensor produces a low-frequency signal. As speed increases, so does signal frequency. An example of this type of sensor is the vehicle speed sensor (VSS). The PCM uses the VSS signal to control the fuel injection system, ignition system, cruise control, EGR flow, canister purge, transmission shift timing, variable steering, and torque converter clutch lockup timing. The signal is also used to initiate diagnostic routines. On some vehicles, the VSS is used to limit the vehicle's speed. When a predetermined speed is reached, the PCM limits fuel delivery.

If the PCM does not receive a speed signal it will set a DTC; it may also set a code if the signal does not correlate with other inputs. Check the wiring and connectors at the sensor and the control modules. Make sure the connections are tight and not damaged.

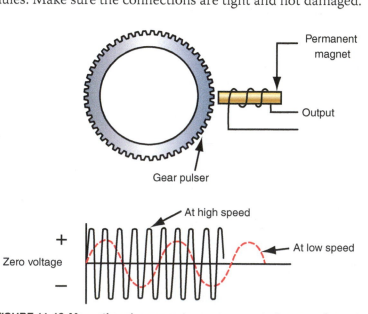

FIGURE 11-16 Magnetic pulse generator sensors generate a waveform at a frequency that is proportional to the speed of the object being measured.

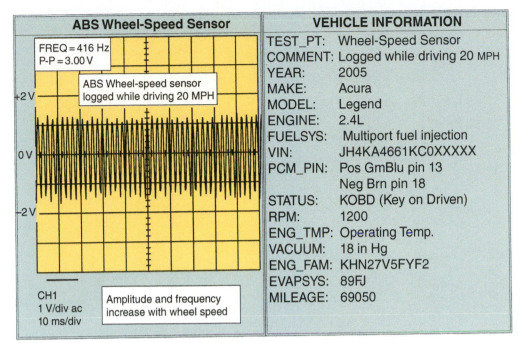

ABS Wheel-Speed Sensor	VEHICLE INFORMATION

FREQ = 416 Hz
P-P = 3.00 V

ABS Wheel-speed sensor
logged while driving 20 MPH

+2 V

0 V

−2 V

CH1
1 V/div ac
10 ms/div

Amplitude and frequency
increase with wheel speed

TEST_PT: Wheel-Speed Sensor
COMMENT: Logged while driving 20 MPH
YEAR: 2005
MAKE: Acura
MODEL: Legend
ENGINE: 2.4L
FUELSYS: Multiport fuel injection
VIN: JH4KA4661KC0XXXXX
PCM_PIN: Pos GmBlu pin 13
                    Neg Brn pin 18
STATUS: KOBD (Key on Driven)
RPM: 1200
ENG_TMP: Operating Temp.
VACUUM: 18 in Hg
ENG_FAM: KHN27V5FYF2
EVAPSYS: 89FJ
MILEAGE: 69050

FIGURE 11-17 The waveform from a wheel speed sensor.

A speed sensor can be tested with a scan tool or lab scope (Figure 11-17). A lab scope is especially helpful in finding "dirty" sensor signals, that is, a signal that drops out so briefly that a scan tool might not catch it. Connect a lab scope, GMM, or scan tool and operate the vehicle at the specified speeds. If the measurements meet the specifications, the sensor is working properly and any speed sensor-related problem is probably caused by the PCM. If the measurements do not match the specifications and the wiring is sound, the sensor is bad.

A speed sensor can also be checked with the vehicle on a hoist. The vehicle should be positioned so the drive wheels are free to rotate. Backprobe the sensor's output wire and connect the voltmeter leads from this wire to ground. Select the 20-volt AC scale on the meter. Then start the engine. Put the transmission in a forward gear and observe the meter. If the sensor's voltage is not 0.5 volts or more, replace the sensor. If the signal is correct, backprobe the sensor's terminal at the PCM and measure the voltage with the wheels rotating. If 0.5 volts is at this terminal, the trouble may be in the PCM.

When 0.5 volts is not available, turn the ignition off and disconnect the wire from the sensor to the PCM. Connect an ohmmeter across the wire. The meter should read 0 ohms. Repeat the test with the leads connected to the sensor's ground and the PCM ground terminal. This wire should also have 0 ohms. If the resistance of these wires is more than specified, repair the wires.

Speed sensors can also be checked with an ohmmeter. Most manufacturers list a resistance specification. The resistance of the sensor is measured across the sensor's terminals. The typical range for a good sensor is 800–1,400 ohms of resistance.

Hall-Effect Sensors. To test a Hall-effect switch, disconnect its wiring harness. Connect a low-voltage source across the positive and negative terminals of the Hall layer. Then connect a voltmeter across the negative and signal voltage terminals. Insert a metal feeler gauge between the Hall layer and the magnet. Make sure the feeler gauge is touching the Hall element. If the sensor is operating properly, the meter will read close to battery voltage. When the feeler gauge blade is removed, the voltage should decrease. On some units, the voltage will drop to near 0. Check the service information to see what voltage you should observe when inserting and removing the feeler gauge.

A Hall-effect switch produces a voltage pulse when there is the presence of a magnetic field. The Hall-effect is the consequence of moving current through a thin conductor that is exposed to a magnetic field; as a result, voltage is produced.

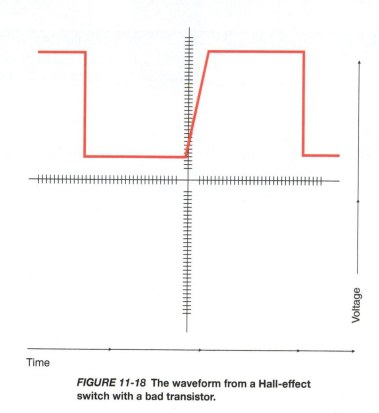

FIGURE 11-18 The waveform from a Hall-effect switch with a bad transistor.

When observing a Hall-effect sensor on a lab scope, pay attention to the downward and upward pulses. These should be straight. If they appear at an angle (Figure 11-18), this indicates the transistor is faulty, causing the voltage to rise slowly. The waveform should be a clean and flat square wave. Any change from a normal trace means the sensor should be replaced.

TESTING THROTTLE POSITION (TP) SENSORS

Throttle position (TP) sensors send a signal to the PCM regarding the rate of throttle opening and relative TP. These sensors provide important input to traction and stability control systems, and to 4WD/AWD and electronically shifted manual transmission systems. In a single potentiometer-style TP sensor, the wiper arm in the sensor is rotated by the throttle shaft (Figure 11-19). As the throttle shaft moves, the wiper arm moves along on the resistor.

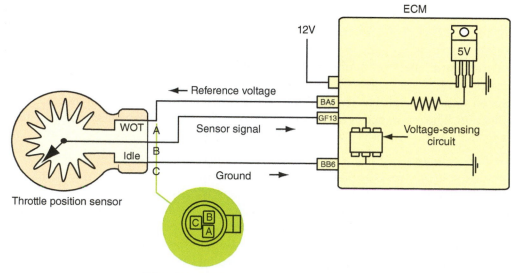

FIGURE 11-19 The basic circuit for a TP sensor.

The TP sensor is mounted on the throttle body. A separate idle contact switch or wide open throttle (WOT) switch may also be used to signal when the throttle is in those positions.

A basic TP sensor has three wires. One wire carries the 5-volt reference signal, another serves as the ground for the resistor, and the third is the signal wire. When the throttle plates are closed, the signal voltage will be around 0.6–0.9 volts. As the throttle opens, there is less resistance and the voltage signal increases. At wide-open throttle the signal will be approximately 3.5–4.7 volts. Often the connector terminals for the sensor are gold plated. The plating makes the connector more durable and corrosion resistant. With the ignition on, connect the voltmeter from the sensor signal wire to ground. Slowly open the throttle and observe the meter. The reading should increase smoothly and gradually. If the TP sensor does not have the specified voltage or if the voltage signal is erratic, replace the sensor.

Connect a voltmeter between the reference wire and ground. Normally, the voltage should be 5 volts. If the reference wire is not supplying the specified voltage, check the voltage on this wire at the computer terminal. If the voltage is within specifications at the computer, but low at the sensor, repair the reference wire. When the voltage is low at the computer, check the voltage supply wires and ground wires on the computer. If these wires are satisfactory, replace the computer.

A TP sensor can also be checked with an ohmmeter. Most often, the total resistance of the sensor is given in the specifications. If the sensor does not meet these, it should be replaced.

TP sensors can be checked with a lab scope. Connect the scope to the sensor's output and a good ground and watch the trace as the throttle is opened and closed. The resulting trace should look smooth and clean, without any sharp breaks or spikes in the signal (Figure 11-20). A bad sensor will typically have a glitch (a downward spike) somewhere in the trace (Figure 11-21) or will not have a smooth transition from high to low. These glitches are an indication of an open or short in the sensor.

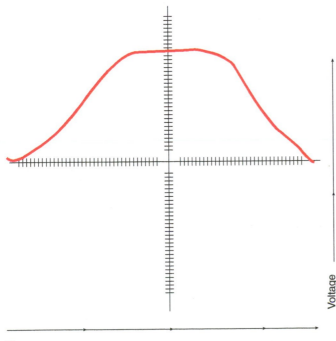

Time

FIGURE 11-20 The waveform from a good TP sensor as the throttle is opened and closed.

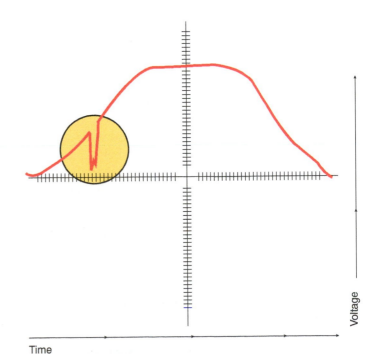

Time

FIGURE 11-21 The waveform of a defective TP sensor. Notice the glitch while the throttle opens.

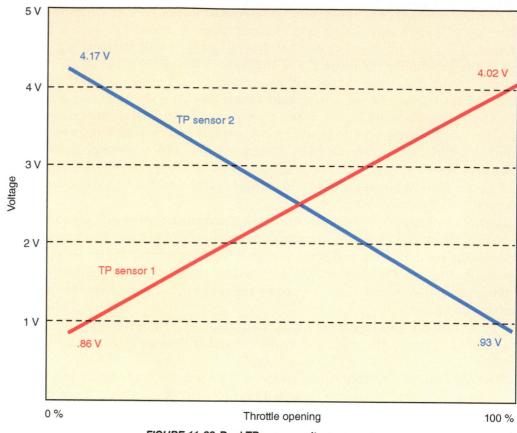

FIGURE 11-22 Dual TP sensor voltage ranges.

The action of a TP sensor can also be monitored on a scan tool. Compare the position, expressed in a percentage, to the voltage specifications for that TP.

Care needs to be taken, as some TP sensors have four wires. The additional wire is connected to an idle switch. Normally, when the switch is closed, there will be 0 volts and battery voltage when the switch is open. Check the wiring diagram before deciding if the switch and circuit is good.

TP Sensors for Electronic Throttle Control. The TP sensor for electronic throttle systems has two wiper arms and two resistors in a single housing (Figure 11-22). Therefore they have two signal wires. This is done to ensure accurate throttle plate position in case one sensor fails. Some of these TP sensors have a different voltage signal on the signal wire. However, they work in the same way. One of the signals starts at a higher voltage and has a different change rate. Other designs of this sensor have a signal that decreases with throttle opening and the other increases with the opening. In either case, the PCM uses both signals to determine throttle opening and depends on one if the other sends an out-of-range signal.

DIAGNOSIS OF COMPUTER VOLTAGE SUPPLY AND GROUND WIRES

> **AUTHOR'S NOTE:** Never replace a computer unless the ground wires and voltage supply wires are proven to be in satisfactory condition.

All PCMs cannot operate properly unless they have good ground connections and the correct voltage at the required terminals. A wiring diagram for the vehicle must be used for the

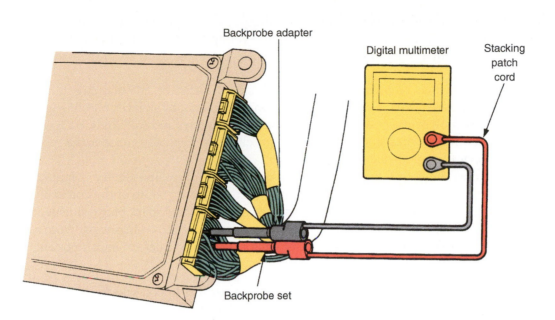

FIGURE 11-23 Using a digital multimeter to check the circuit of the PCM.

following tests. Backprobe the battery terminal at the PCM and connect a digital voltmeter from this terminal to ground (Figure 11-23). Always ground the black meter lead.

The voltage at this terminal should be 12 volts with the ignition switch off. If 12 volts are not available at this terminal, check the computer's fuse and related circuit. Turn the ignition on and connect the red meter lead to the other battery terminals at the PCM, with the black lead still grounded. The voltage measured at these terminals should also be 12 volts. When the specified voltage is not available, test the voltage supply wires to these terminals. These terminals may be connected through fuses, fuse links, or relays.

Ground Circuits

Ground wires usually extend from the computer to a ground connection on the engine or battery. With the ignition on, connect a voltmeter from the battery ground to the computer ground. The voltage drop across the ground wires should be 30 millivolts or less. If the voltage is greater than that, repair the ground wires or connection.

Not only should the computer ground be checked, but so should the ground (and positive) connection at the battery. Checking the condition of the battery and its cables should always be part of the initial visual inspection before diagnosing an engine control system.

A good ground is especially critical for all reference voltage sensors. The problem here is not obvious until it is thought about. A bad ground will cause the reference voltage (normally 5 volts) to be higher than normal. As a result, the computer will be making decisions based on the wrong information. If the output signal is within the normal range for that sensor, the computer will not notice the wrong information and will not set a DTC.

Electrical Noise. Poor grounds can also allow **electromagnetic interference (EMI)** or noise to be present on the reference voltage signal. This noise causes small changes in the voltage going to the sensor. Therefore, the output signal from the sensor will also have these voltage changes. The best way to check for noise is to use a lab scope.

Connect the lab scope between the 5-volt reference signal wire at the sensor and its ground. The trace on the scope should be flat. If noise is present, move the scope's negative

EMI is something that occurs when voltage changes in a nearby wire or component and influences the voltage in the wire or component that it is close to.

probe to a known good ground. If the noise disappears, the sensor's ground circuit is bad or has resistance. If the noise is still present, the voltage feed circuit is bad or there is EMI in the circuit from another source, such as the AC generator. Find and repair the cause of the noise.

Circuit noise may be present at the positive side or negative side of a circuit. It may be evident by a flickering MIL, by a popping noise on the radio, or by an intermittent engine miss. The most common sources of noise are electric motors, relays and solenoids, AC generators, ignition systems, switches, and A/C compressor clutches. Typically, noise is the result of an electrical device being turned on and off.

Sometimes the source of the noise is a defective suppression device. Manufacturers include these devices to minimize or eliminate electrical noise. Some of the commonly used noise suppression devices are resistor-type secondary cables and spark plugs, shielded cables, capacitors, diodes, and resistors. Capacitors or chokes are used to control noise from a motor or generator. If the source of the noise is not a poor ground or a defective component, check the suppression devices.

Clamping Diodes. Clamping diodes are placed in parallel to coil windings to limit high-voltage spikes. These spikes are induced by the collapsing of the magnetic field around a winding in a solenoid, relay, or electromagnetic clutch. The field collapses when current flow to the winding is stopped. The diode prevents the voltage from reaching the computer and other sensitive electronic parts. When the diode fails to suppress the voltage spikes, the transistors inside the computer can be destroyed. If the diode is bad, a negative spike will appear in a voltage trace.

> A clamping diode is typically installed in parallel to a coil, creating a bypass for the electrons during the time the circuit is opened.

Resistors are also used to suppress voltage spikes. They do not eliminate the spikes; rather, they limit the intensity of the spikes. If a voltage trace has a large spike and the circuit is fitted with a resistor to limit noise, the resistor may be bad.

COMPUTER OUTPUTS AND ACTUATORS

Once the PCM determines that a correction or adjustment must be made to the system, an output signal is sent to a control device or actuator. These actuators like solenoids, switches, relays, or motors physically act or carry out the command sent by the PCM.

Actuators are electromechanical devices that convert an electrical current into mechanical action. This action can be used to open and close valves, engage or disengage gears, control vacuum to other components, or open and close switches. When the PCM receives an input signal indicating a change in one or more of the operating conditions, the PCM determines the best strategy for handling the conditions. The PCM then controls a set of actuators to achieve a desired effect or strategy goal. In order for the computer to control an actuator, it must rely on a component called an output driver.

The driver usually completes the ground circuit of the actuator. The ground can be applied steadily if the actuator must be activated for a selected amount of time. Or the ground can be pulsed to activate the actuator in pulses. Output drivers are transistors or groups of transistors that control the actuators. These drivers operate by the digital commands from the PCM. If an actuator can't be controlled digitally, the output signal must pass through an A/D converter before flowing to the actuator.

Most systems allow for testing an actuator through a scan tool. Actuators that are duty cycled are more accurately diagnosed this way. Serial data can be used to diagnose outputs. The data should be compared against specifications to determine the condition of an actuator. Also, when an actuator is suspected as being faulty, make sure the inputs related to the control of that actuator are within normal range. Faulty inputs will cause an actuator to appear faulty.

Many systems have operating modes that can be accessed with a scan tool to control the operation of an output. Common names for this mode are the output state control (OSC) and output test mode (OTM). In this mode, an actuator can be enabled or disabled or the duty cycle or the movement of the actuator can be increased or decreased. While the actuator is being controlled, related PIDs are observed as an indication of how the system reacted to the changes. The actuators that can be controlled by this mode vary. Always refer to the service information to determine what can be checked and how it should be checked.

Testing with a DMM

Some actuators are easily tested with a voltmeter by checking input voltage at the actuator. If there is the correct amount of input voltage, check the condition of the ground. If both of these are good, then the actuator is faulty.

When checking anything with an ohmmeter, logic can dictate good and bad readings. If the meter reads infinite, this means there is an open. Based on what you are measuring across, an open could be good or bad. The same is true for very low-resistance readings. Across some things, this would indicate a short. For example, you do not want an infinite reading across the windings of a solenoid. You want low resistance. However, you want an infinite reading from one winding terminal to the case of the solenoid. If you have low resistance, the winding is shorted to the case.

Testing Actuators with a Lab Scope

Actuators are electromechanical devices, meaning they are electrical devices that cause some mechanical action. Actuators can be electrically faulty or mechanically faulty. By observing the action of an actuator on a lab scope, you will be able to watch its electrical activity. Normally if there is a mechanical fault, this will affect its electrical activity as well. Therefore, you get a good sense of the actuator's condition by watching it on a lab scope.

Most actuators are solenoids. The computer controls the action of the solenoid by controlling the pulse width of the control signal. By watching the control signal, you can see the turning on and off of the solenoid (Figure 11-24). The voltage spikes are caused by the discharge of the coil in the solenoid.

Some actuators are controlled by pulse-width modulated signals (Figure 11-25). These devices are controlled by varying the pulse width, signal frequency, and voltage levels. These waveforms should be checked for amplitude, time, and shape. You should also observe changes to the pulse width as operating conditions change. A bad waveform will have noise, glitches, or rounded corners. You should be able to see evidence that the actuator immediately turns off and on according to the commands of the computer.

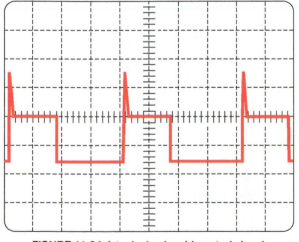

FIGURE 11-24 A typical solenoid control signal.

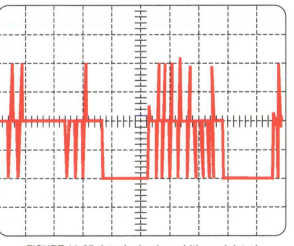

FIGURE 11-25 A typical pulse-width modulated solenoid control signal.

MOTORS

An electric motor converts electric energy into mechanical energy. Through the years, many different designs of motors have been used. All electric motors operate by the same basic principle. That principle is easily observed by taking two bar magnets and placing them end-to-end with the other. If the ends have the same polarity, they will push away from each other. If the ends have the opposite polarity, they will move toward each other and form one magnet.

If we put a pivot through the center of one of the magnets, to allow it to spin, and moved the other magnet toward it, the first magnet will either rotate away from the second or move toward it (Figure 11-26). This is basically how a motor works. Although we do not observe a complete rotation, we do see part of one, perhaps a half turn. If we could change the polarity of the second magnet, we would get another half turn. So in order to keep the first magnet spinning, we need to change the polarity immediately after it moves halfway. If we continued to do this, we would have a motor.

In a real motor, an electromagnet is fitted on a shaft. The shaft is supported by bearings or bushings to allow it to spin and to keep it in the center of the motor. Surrounding, but not touching, this inner magnet is a stationary permanent magnet or an electromagnet. Actually, there are more than one magnets or magnetic fields in both components. The polarity of these magnetic fields is quickly switched and we have a constant opposition and attraction of magnetic fields. Therefore, we have a constantly rotating inner magnetic field, the shaft of which can do work due to the forces causing it to rotate. The torque of a motor varies with rotational speed, motor design, and the amount of current draw the motor has. The rotational speed depends on the motor's current draw, the design of the motor, and the load on the motor's rotating shaft.

The basic components of a motor are the stator or field windings that are the stationary part of the motor and the rotor or armature that is the rotating part. The field windings comprise slotted cores made of thin sections of soft iron wound with insulated copper wire to form one or more pairs of magnetic poles. The armature comprises loops of current-carrying wire. The loops are formed around a metal with low **reluctance** to increase the magnetic field. The magnetic fields around the armature are pushed away by the magnetic field of the field windings, causing the armature to rotate away from the windings' fields.

> **Reluctance** is a term used to indicate a material's resistance to the passage of magnetic lines of flux.

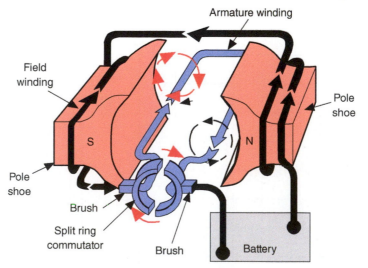

FIGURE 11-26 A simple DC motor.

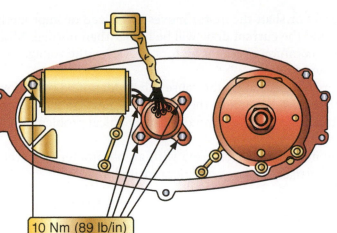

FIGURE 11-27 A small motor is often mounted to a transfer case to engage and disengage gears.

Ultra-thin DC brushless motor

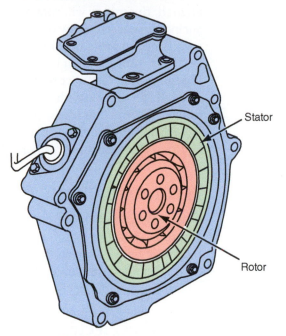

Stator

Rotor

FIGURE 11-28 Hybrid vehicles use sophisticated electric motors.

Brushes ride on the **commutator** of the armature. There are positive and negative brushes. As the armature rotates, the brushes ride on different sets of commutator segments. This action causes the polarity of the armature windings to change. This results in the continuous rotation of the armature.

The field windings or the armature may be made with permanent magnets rather than electromagnets. Both cannot be permanent magnets. An electromagnet allows for a change in the polarity of the magnetic fields, which keeps the armature spinning. By changing the direction of current flow, the magnetic polarities are changed.

The placement and purpose of a motor depends on the vehicle and the purpose of the motor. The actual size of the motor is a reflection of the amount of work it must do. Small motors are used to run accessories and to cause a shaft to move to engage or disengage gears (Figure 11-27). Larger, very powerful motors are used in hybrid and electric vehicles (Figure 11-28). These are powered by more than 12 volts. Different motor designs are used, again, according to the application. Some are powered by DC voltage, and others use AC. DC motors can be powered directly by the batteries, whereas AC motors require converters and inverters to change the DC voltage stored in the batteries into the AC required by the motors.

Testing Motors

The operation of a motor depends on many things, including the amount of voltage and current to it. Begin your diagnosis of a motor with a check of the battery and all wires and cables leading to the motor. A motor that does not operate properly and is receiving the correct amount of voltage is undoubtedly bad. However, what the motor is attempting to move can also result in poor motor performance.

A motor develops the most torque when it is rotating slowly. This is also the time when the motor is drawing the most current. Measuring the current draw of a motor can determine if the motor is working properly. Again, what the motor is moving will also influence the amount

A commutator is at the end of an armature. It is made of heavy copper segments separated from each other and the armature shaft by insulation. Individual segments are connected to the individual windings in the armature. Brushes ride on the surface of the commutator.

523

of current the motor draws. If the gears or shaft the motor moves are seized or improperly aligned, the motor will rotate slowly and the current draw will be higher than normal. Measuring the current draw of a motor is a common check for motors. However, the voltage to it and its ground circuit should be checked first.

If a motor has a higher than normal current draw, the cause can be identified by removing the motor. Connect the appropriate power source to the motor and measure the current draw. If the draw is within specifications, suspect a nonmotor or mechanical problem. If the draw is still outside specs, the motor is faulty.

Although individual parts of the motor can be checked with an ohmmeter and other equipment, if the motor is faulty it is normally replaced and not repaired.

ELECTROMAGNETIC CLUTCHES

> Electromagnetism relies on the passing of electricity through a material to align the electrons and give it magnetic properties.

Gear engagement is often controlled by an **electromagnetic clutch** (Figure 11-29 and Figure 11-30). The clutch is engaged by a magnetic field and disengaged by springs when the magnetic field is broken. When the controls call for clutch engagement, the electrical circuit to the clutch is completed, the clutch is energized, and the clutch engages. When the electrical circuit is opened, the clutch disengages and springs pull it away.

Checking Electromagnetic Clutches

A clutch assembly should be carefully examined for discoloration, peeling, or other damage. If there is damage, replace the assembly. At times, the play and drag of the clutch can be checked by rotating it by hand. Replace the clutch if it is noisy or has excessive play or drag. The clutch's field coil can be checked with an ohmmeter. The exact testing points and acceptable resistance readings are typically given in the service information. If resistance is not within specifications, replace the field coil or clutch.

If the clutch is not operating, check the electrical connections to it. Make sure they are secure and corrosion free. Then, check for power to the clutch. If there is no power, locate and repair the problem. If there is power to the clutch, check the ground circuit with a DMM. If there is power and a good ground, the clutch is defective and must be replaced.

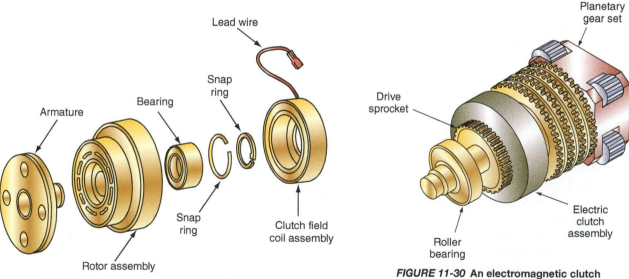

FIGURE 11-29 An electromagnetic clutch assembly.

FIGURE 11-30 An electromagnetic clutch connected to a planetary gear set.

Clutch Clearance

Many clutch assemblies have a specification for the distance between the clutch and its pressure plate. This clearance is measured with a nonmagnetic feeler gauge. If the clearance is too great, the clutch may slip and cause a scraping or squealing noise. If the clearance is too small, the clutch may chatter. As the clutch wears, the clearance increases and should be checked and adjusted whenever symptoms suggest doing so. Always follow the specific procedures for measuring and correcting the gap.

REPAIRING THE SYSTEM

After identifying the source of the problem, repairs should be made. When servicing or repairing OBD II circuits, the following guidelines are important:

- Do not move or alter grounds from their original locations.
- Faulty relays should be thrown away and replaced with an exact replacement.
- Make sure all connector locks are in good condition and are in place.
- After repairing connectors or connector terminals, make sure the terminals are properly retained and the connector is sealed (Figure 11-31).
- When installing a fastener for an electrical ground, tighten it to the specified torque.

After repairs have been made, the system should be rechecked to verify that the repair took care of the problem. This may involve a road test in order to verify that the complaint has been resolved. Record the fail records or freeze-frame data taken before the repair. Use the scan tool to erase any DTCs. Then operate the vehicle within the conditions noted in the fail records or the freeze-frame data. After driving the vehicle through a variety of conditions, recheck the DTCs. If the vehicle worked fine and no new codes were set, the repair has been verified.

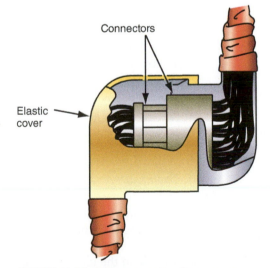

FIGURE 11-31 Some connectors have covers that need to be securely attached to protect the connectors from moisture and dirt.

CASE STUDY

A customer brought his new BMW with an M-DCT transmission back to the dealership. He said he was very unhappy with the harsh shifting. The car was assigned to one of the lead technicians. The technician was very good at diagnostics, especially with electronic controls. This, however, was the first time he would be working on a vehicle with a DCT.

Starting out he knew two things. If the problem was in the transmission, he would need to replace it because the manufacturer did not authorize repairs. He also knew the transmission was totally electronically controlled. He began his diagnosis with a test drive and noticed the transmission did upshift and downshift quite harshly. Not sure what the shifts should feel like, he took another new car with a DCT on a test drive. This transmission was much smoother than the first. Now he knew there was a problem.

He then connected a scan tool to the car and found no DTCs or data that would indicate a problem. He then searched through the literature to see if BMW had any references to this problem. Looking through the TSBs, he found what he was looking for. There had been a software update for the transmission.

Using the scan tool, he checked the software that was currently loaded onto the control module. He compared that to the one listed in the TSB and found that this car needed the updated software. He then flashed the module with the latest software. Then he took the car out on a road test and found the transmission to shift very nicely; he had fixed the problem.

ASE-STYLE REVIEW QUESTIONS

1. While diagnosing a shifting problem in a CVT, *Technician A* interviews the customer to find out as much as possible about the concern.
Technician B checks the MIL and retrieves the DTCs.
Who is correct?
   A. Technician A
   B. Technician B
   C. Both A and B
   D. Neither A nor B

2. While discussing vehicle speed sensor tests, *Technician A* says use an ohmmeter to test the resistance of the coil.
Technician B says the voltage generated by the sensor can be measured by connecting a voltmeter across the sensor's terminals.
Who is correct?
   A. Technician A
   B. Technician B
   C. Both A and B
   D. Neither A nor B

3. After making the necessary repairs, *Technician A* clears all codes, checks the MIL, and then rechecks for codes.
Technician B initially takes the vehicle out on a road test; if everything works fine, he says there is no need for further checks.
Who is correct?
   A. Technician A
   B. Technician B
   C. Both A and B
   D. Neither A nor B

4. While diagnosing an intermittent electrical fault, *Technician A* connects a voltmeter to the suspected circuit and wiggles the wiring harness. If the reading changes, the source of the problem is in that circuit.
Technician B says most intermittent problems are caused by faulty electrical connections or wiring. Therefore, he carefully and thoroughly inspects the wires and connectors in the circuit.
Who is correct?
   A. Technician A
   B. Technician B
   C. Both A and B
   D. Neither A nor B

5. While discussing TP sensor diagnosis, *Technician A* says a four-wire TP sensor contains an idle switch.
Technician B says in some applications, the TP sensor mounting bolts may be loosened and the TP sensor housing rotated to adjust the voltage signal with the throttle in the idle position.
Who is correct?
   A. Technician A
   B. Technician B
   C. Both A and B
   D. Neither A nor B

6. While diagnosing an output in an electronic control system,

Technician A says actuators that are duty cycled are best tested with a scan tool.

Technician B says some scan tools allow a technician to energize and deenergize individual actuators so the technician can observe the activity.

Who is correct?

A. Technician A
B. Technician B
C. Both A and B
D. Neither A nor B

7. While using a lab scope,

Technician A adjusts the horizontal voltage scale in relation to the expected voltage signal.

Technician B adjusts the vertical time base or milliseconds per division to get the clearest view of the waveform as possible.

Who is correct?

A. Technician A
B. Technician B
C. Both A and B
D. Neither A nor B

8. While making repairs to a bus wire,

Technician A replaces the entire harness because these should not be repaired.

Technician B says if a small section of the wire is damaged, a short length of wire can be used to bypass the damaged section.

Who is correct?

A. Technician A
B. Technician B
C. Both A and B
D. Neither A nor B

9. While inspecting the electronic control system for a transmission,

Technician A checks all wires to make sure they are firmly connected and not damaged. To do this he wiggles and gently pulls on all connectors.

Technician B checks the wiring for signs of burned or chafed spots, pinched wires, or contact with sharp edges or hot exhaust parts.

Who is correct?

A. Technician A
B. Technician B
C. Both A and B
D. Neither A nor B

10. While discussing freeze-frame data,

Technician A says control systems only store snapshots of emissions-related concerns.

Technician B says this data shows the activity of a particular sensor or actuator during the time the computer set a DTC.

Who is correct?

A. Technician A
B. Technician B
C. Both A and B
D. Neither A nor B

ASE CHALLENGE QUESTIONS

1. While discussing the process for flashing a control module,

Technician A says the old software must be deleted from the module before connecting it to a scan tool or PC for updating.

Technician B says flashing is typically only necessary when the vehicle's battery has been disconnected and the computer has lost its memory.

Who is correct?

A. Technician A
B. Technician B
C. Both A and B
D. Neither A nor B

2. While discussing the reasons for DTC that denotes a communication problem,

Technician A says the individual control modules are not able to communicate with each other.

Technician B says the scan tool is unable to communicate with each control module.

Who is correct?

A. Technician A
B. Technician B
C. Both A and B
D. Neither A nor B

3. While testing a temperature sensor,

Technician A says if the observed temperature reading from the sensor is the coldest possible value, the circuit is open.

Technician B says these sensors can be tested by removing them and placing them into a container of water with an ohmmeter connected across the sensor terminals. When the water is heated, the sensor should have the specified resistance at the different temperatures.

Who is correct?

A. Technician A
B. Technician B
C. Both A and B
D. Neither A nor B

4. While testing a speed sensor,

Technician A measures the resistance across the sensor and compares the readings to specifications.

Technician B monitors the activity of the sensor while the vehicle is moving. She expects the voltage and frequency of the signal to decrease as speed increases.

Who is correct?

A. Technician A
B. Technician B
C. Both A and B
D. Neither A nor B

5. While discussing the possible causes of a flickering MIL,

Technician A says this is an indication of a pending DTC.

Technician B says this can be caused by electrical noise in the circuit.

Who is correct?

A. Technician A
B. Technician B
C. Both A and B
D. Neither A nor B

Name _____ Date _____

ELECTRONIC CONTROLLED TRANSMISSIONS

Upon completion of this job sheet, you will be able to describe the operation of electronically controlled transmissions and transaxles.

ASE NATEF MAST Task Correlation

Transmission/Transaxle Diagnosis and Repair

Task #2, Describe the operational characteristics of an electronically controlled manual transmission/transaxle.

Tools and Materials

Sales information on the vehicle

Describe the Vehicle Being Worked On

Year _____ Make _____ VIN _____

Model _____

Procedure

1. Explain the differences between a manually shifted automatic transmission and an automatic manual transmission.

   _____
   _____
   _____
   _____

2. How does the manufacturer describe the automatic manual transmission you will be looking at?

   _____
   _____

3. What is the transmission called?

   _____
   _____

4. Briefly describe how the transmission works.

   _____
   _____
   _____
   _____

5. How many different shift modes are available and what are they called?

_____

_____

6. What does the manufacturer claim to be the benefits of this transmission?

_____

_____

_____

Instructor's Response: _____

_____

_____

Name _____ Date _____

Conduct a Diagnostic Check on an Engine Equipped with OBD II

Upon the completion of this job sheet, you should be able to conduct a system inspection and retrieve codes from the PCM of an OBD II-equipped engine.

Tools and Materials

A vehicle equipped with OBD II

Scan tool

Service manual

Describe the Vehicle Being Worked On

Year _____ Make _____ VIN _____

Model _____

Engine size and type _____

Procedure

1. Check all vehicle grounds, including the battery and computer ground, for clean and tight connections. Comments:

   _____

2. Perform a voltage drop test across all related ground circuits. State where you tested and what your findings were:

   _____

   _____

   _____

3. Check all vacuum lines and hoses, as well as the tightness of all attaching and mounting bolts in the induction system. Comments:

   _____

4. Check for damaged air ducts. Comments:

   _____

5. Check the ignition circuit, especially the secondary cables, for signs of deterioration, insulation cracks, corrosion, and looseness. Comments:

   _____

6. Are there any unusual noises or odors?   ☐ Yes   ☐ No

   If there are, describe them and tell what may be causing them.

   _____

   _____

7. Inspect all related wiring and connections at the PCM. Comments:

   _____

   _____

8. Gather all pertinent information about the vehicle and the customer's complaint. This should include detailed information about the symptom from the customer, a review of the vehicle's service history, published TSBs, and the information in the service manual.

9. Are there any vacuum leaks?   ☐ Yes   ☐ No

10. Is the engine's compression normal?   ☐ Yes   ☐ No

11. Is the ignition system operating normally?   ☐ Yes   ☐ No

12. Are there any obvious problems with the air/fuel system?   ☐ Yes   ☐ No

13. Your conclusions from the above:

    _____

    _____

14. Check the operation of the MIL by turning the ignition ON. Describe what happened and what this means.

    _____

    _____

15. Connect the scan tool to the DLC.

16. Enter the vehicle identification information into the scan tool.

17. Retrieve the DTCs with the scan tool.

18. List all codes retrieved by the scan tool.

    _____

    _____

    _____

    _____

19. Conclusion from these tests and checks.

_____

_____

_____

_____

Instructor's Response: _____

_____

_____

Name _____ Date _____

USING A DSO ON SENSORS AND SWITCHES

Upon completion of this job sheet, you should be able to connect a DSO and observe the activity of various sensors and switches.

Tools and Materials

A vehicle with accessible sensors and switches

Service manual for the above vehicle

Component locator manual for the above vehicle

A DSO

A DMM

Describe the Vehicle Being Worked On

Year _____ Make _____ VIN _____

Model _____

Procedure

1. Connect the DSO across the battery. Make sure the scope is properly set. Observe the trace on the scope. Is there evidence of noise? Explain.

   _____
   _____
   _____
   _____

2. Locate the A/C compressor clutch control wires. Start and run the engine. Connect the DMM to read available voltage. Observe the meter and then turn the compressor on. What happened on the meter?

   _____
   _____

   Now connect the DSO to the same point with the compressor turned off. Observe the waveform and then turn the compressor on. What happened to the trace?

   _____
   _____

3. Turn off the engine but keep the ignition on. Locate the TP sensor and identify the purpose of each wire to it. List each wire and describe the purpose of each.

   _____
   _____
   _____

4. Connect the DMM to read reference voltage at the TP sensor. What do you read?

_____

Now move the leads to read the output of the sensor. Starting with the throttle closed, slowly open the throttle until it is wide open. Watch the voltmeter while doing this. Describe your readings.

_____

_____

5. Now connect the DSO to read reference voltage at the TP sensor. What do you see on the trace?

_____

_____

Now move the leads to read the output of the sensor. Starting with the throttle closed, slowly open the throttle until it is wide open. Watch the trace while doing this. Describe your readings.

_____

_____

6. Now raise the vehicle. Locate a wheel speed sensor and identify the purpose of each wire to it. Connect the DMM to read voltage generated by the sensor. (To get a reading, the wheel must be spun.) Watch the meter and describe what happened below.

_____

_____

_____

_____

7. Now connect the DSO to read voltage output from the sensor. It is not important that you spin the wheel at the same speed as you did for the previous test; however, it should be about the same speed. Watch the trace and describe what happened below.

_____

_____

_____

_____

8. Explain what you observed as the difference between testing with a DMM and a DSO.

_____

_____

_____

Instructor's Response: _____

_____

_____

Name _____ Date _____

CHECK SOLENOID OPERATION

Upon completion of this job sheet, you should be able to check the operation of a solenoid.

ASE NATEF MAST Task Correlation

Transmission/Transaxle Diagnosis and Repair

Task #2, Describe the operational characteristics of an electronically controlled manual transmission/transaxle.

Tools and Materials

A vehicle with reverse blockout or a shift blocking system

Jumper wires

A multimeter

Describe the Vehicle Being Worked On

Year _____ Make _____ VIN _____

Model _____

Procedure

Task Completed

1. To determine if there is a shifting problem, the first step would be to decide if the problem is mechanical or electrical. To do this, attempt to shift the vehicle into first and fourth gears. If you are unable to shift into either gear, the problem is mechanical. If you are able to shift into both gears, proceed with the electrical diagnosis. ☐

2. The solenoids can be activated with a scan tool or jumper wires with the engine off and the ignition on. Check the service manual for the procedure to do this.

   List the steps here:

   _____
   _____
   _____
   _____

3. If the blockout occurred with the solenoid energized, the solenoid is fine. ☐

   If the blockout did not occur, proceed with more detailed testing to identify the exact cause of the problem.

4. Check for voltage at the solenoid.

Findings _____

Less than battery voltage indicates either an open or high resistance in the circuit.

5. Refer to the wiring diagram and check for voltage at various spots in the circuit.

Findings _____

An open is indicated by a lack of voltage at the solenoid. The open will be between the last point in the circuit where voltage was not present and the point at which voltage was measured.

To determine the source of high resistance, measure the voltage drop across different parts of the circuit.

6. If battery voltage was present at the solenoid, the solenoid should be checked. Resistance checks should be made on the solenoid. Refer to the service manual and record any specifications given for the solenoids.

Specifications _____

If no specifications are found, use the general guidelines given in this text-book.

7. Measure the resistance across the terminals of the solenoid.

Your measurement _____

8. Measure the resistance between one terminal and the case of the solenoid.

Your measurement _____

9. Measure the resistance between the other terminal and the case of the solenoid.

Your measurement _____

☐ **10.** Compare your measurements to the specifications. A good solenoid will have low resistance only across the terminals. A resistance reading between the terminals and the case indicates that the windings are shorted to the case. If there is high resistance across the terminals, there is excessive resistance in the solenoid. If there is no continuity across the terminals, the windings are open.

Instructor's Response: _____

_____

_____

1. A customer says that he hears the sound of gears clashing when he attempts to shift into reverse gear immediately after disengaging the clutch. He says that he does not hear any noises when he shifts into first gear.
   Technician A says that the customer appears to be shifting into reverse gear too soon after disengaging the clutch.
   Technician B says that there may be a problem with reverse gear.
   Who is correct?
   A. Technician A   C. Both A and B
   B. Technician B   D. Neither A nor B

2. A customer says that the clutch of her car appears to be slipping; after shifting into first gear the vehicle does not begin to move until the clutch is almost completely engaged.
   Technician A says that there may be excessive clutch pedal free play.
   Technician B says the clutch master cylinder primary seal may be leaking.
   Who is correct?
   A. Technician A   C. Both A and B
   B. Technician B   D. Neither A nor B

3. When discussing worn clutch components,
   Technician A says that normal clutch wear will result in a decrease in clutch pedal free play on a vehicle equipped with linkage-type actuation.
   Technician B says that a worn release bearing fork pivot will result in excessive clutch pedal free play on linkage-type clutch systems.
   Who is correct?
   A. Technician A   C. Both A and B
   B. Technician B   D. Neither A nor B

4. When discussing the possible causes of clutch chatter,
   Technician A says that clutch chatter can be caused by a clutch disc that has been saturated with oil.
   Technician B says that a flywheel that has insufficient lateral runout can cause clutch chatter.
   Who is correct?
   A. Technician A   C. Both A and B
   B. Technician B   D. Neither A nor B

5. A vehicle has been towed into the shop because the customer claims the transmission will not shift into any gear when the engine is running. Which of the following could cause this problem?
   A. A clutch disc with worn friction material
   B. A pressure plate with a weak spring diaphragm
   C. A frozen clutch release bearing
   D. A frozen clutch pilot bearing

6. A severe pulsation is felt at the clutch pedal the instant the clutch pedal is touched.
   Technician A says that a flywheel with excessive lateral runout could cause this problem.
   Technician B says that this problem could be caused by a worn transmission input shaft bearing.
   Who is correct?
   A. Technician A   C. Both A and B
   B. Technician B   D. Neither A nor B

7. A clutch master cylinder is being replaced.
   Technician A says that this procedure does not require any adjustments.
   Technician B says that failure to bleed the hydraulic system properly can result in excessive clutch pedal free play.
   Who is correct?
   A. Technician A   C. Both A and B
   B. Technician B   D. Neither A nor B

8. During an engine replacement, an oily fluid is found all around the clutch components.
   Technician A says that the engine rear main oil seal could have been leaking.
   Technician B says that if the parts are not worn out they can be washed in cleaning solvent and then reused.
   Who is correct?
   A. Technician A   C. Both A and B
   B. Technician B   D. Neither A nor B

9. The alignment of the clutch bell housing to the engine block is being discussed.

   Technician A says that a typical misalignment limit is .500 inch.

   Technician B says that bell housing bolts only hold the bell housing to the block, and that dowels actually align the bell housing to the block.

   Who is correct?

   A.   Technician A
   B.   Technician B
   C.   Both A and B
   D.   Neither A nor B

10. A vehicle with a hydraulically actuated clutch is towed into the shop. The customer says the clutch pedal stays on the floor after it has been depressed.

    Technician A says that the clutch master cylinder may be faulty.

    Technician B says that the clutch pedal may be binding.

    Who is correct?

    A.   Technician A
    B.   Technician B
    C.   Both A and B
    D.   Neither A nor B

11. While diagnosing an electronic control system,

    Technician A checks the connections for high resistance.

    Technician B says a bad ground can cause the reference signal to a sensor to be too high.

    Who is correct?

    A.   Technician A
    B.   Technician B
    C.   Both A and B
    D.   Neither A nor B

12. A transmission is hard to shift into fifth gear; all other shifts are fine. All of the following could cause this problem *except*:

    A.   Worn fifth gear blocking ring
    B.   Worn shifter fork
    C.   Excessive clutch pedal free play
    D.   Worn fifth gear synchronizer teeth

13. A bearing-type noise is heard coming from a manual transmission only when the vehicle is in motion; changing the load on the engine does not seem to affect the noise level. The noise goes away completely when the vehicle is at a standstill.

    Technician A says that the transmission input shaft bearing could be worn.

    Technician B says that worn cluster gear bearings could be the source of this noise.

    Who is correct?

    A.   Technician A
    B.   Technician B
    C.   Both A and B
    D.   Neither A nor B

14. When discussing diagnosis of dual-clutch vehicles,

    Technician A says that a growling sound in a dual-clutch vehicle when the engine is running and the vehicle is stopped could be coming from inside the transmission.

    Technician B says a dual-clutch vehicle that slips in all odd gears may have a bad wet clutch.

    Who is correct?

    A.   Technician A
    B.   Technician B
    C.   Both A and B
    D.   Neither A nor B

15. A five-speed transmission that had worn cluster gear bearings would be relatively quiet in which gear?

    A.   First
    B.   Third
    C.   Fourth
    D.   Reverse

16. A vehicle with a five-speed transmission is towed into the shop because it will not move at all when placed in any gear except fourth; when the transmission is in fourth gear, the vehicle attempts to move.

    Technician A says that the cluster gear teeth may be badly worn.

    Technician B says that the clutch may be worn out.

    Who is correct?

    A.   Technician A
    B.   Technician B
    C.   Both A and B
    D.   Neither A nor B

17. The blocking ring on the third-gear side of the 3–4 synchronizer hub is worn beyond specifications.
Technician A says that this will result in a hard 5–4 downshift.
Technician B says that this will result in a hard 3–4 upshift.
Who is correct?

A. Technician A
B. Technician B
C. Both A and B
D. Neither A nor B

18. When discussing manual transmission inspection,
Technician A says that worn teeth on second gear of the cluster gear would result in a hard 1–2 upshift.
Technician B says that worn teeth on third gear of the main shaft would result in a hard 2–3 upshift.
Who is correct?

A. Technician A
B. Technician B
C. Both A and B
D. Neither A nor B

19. The condition of synchronizer assemblies is being discussed,
Technician A says that the internal surface grooves of a blocking ring should be sharp.
Technician B says that the tips of the internal splines of a synchronizer clutch sleeve should be rounded.
Who is correct?

A. Technician A
B. Technician B
C. Both A and B
D. Neither A nor B

20. When discussing manual transmission diagnosis,
Technician A says that a faulty interlock system could result in a transmission being in two gears at the same time.
Technician B says that a worn shift fork could cause the transmission to jump out of gear.
Who is correct?

A. Technician A
B. Technician B
C. Both A and B
D. Neither A nor B

21. When replacing a manual transaxle bearing,
Technician A says that when the service manual specifies that a service shim should be used, the original shim should be installed in addition to the service shim.
Technician B says that the placement of a service shim is between the bearing cup and the transaxle case.
Who is correct?

A. Technician A
B. Technician B
C. Both A and B
D. Neither A nor B

22. A continuous humming sound is heard from the front of a vehicle when it is in motion; the noise seems to go away when the steering wheel is moved from side to side.
Technician A says that a wheel bearing may be worn.
Technician B says that an unevenly worn tire could be the source of the noise.
Who is correct?

A. Technician A
B. Technician B
C. Both A and B
D. Neither A nor B

23. A clunking sound is heard from the front of a vehicle whenever it is in motion and the engine is accelerated.
Technician A says that a worn inner CV joint could be causing this noise.
Technician B says that worn transaxle gears could cause this noise.
Who is correct?

A. Technician A
B. Technician B
C. Both A and B
D. Neither A nor B

24. When servicing a CV joint,
Technician A says that if the boot of a CV joint is not leaking grease then the joint must be in good condition.
Technician B says that one of the first steps in CV drive axle removal is to drain the transaxle after the vehicle has been raised.
Who is correct?

A. Technician A
B. Technician B
C. Both A and B
D. Neither A nor B

25. When discussing CV joint repair,
Technician A says that some discoloration of an outer CV joint housing is acceptable.
Technician B says that it is acceptable to mix different brands of CV joint grease.
Who is correct?
A. Technician A
B. Technician B
C. Both A and B
D. Neither A nor B

26. While discussing the MIL on OBD II systems,
Technician A says the MIL will flash if the PCM detects a fault that would damage the catalytic converter.
Technician B says whenever the PCM has detected a fault, it will turn on the MIL.
Who is correct?
A. Technician A
B. Technician B
C. Both A and B
D. Neither A nor B

27. When discussing drive shaft problem diagnosis,
Technician A says that worn U-joints will cause a vibration that is very speed sensitive: the faster the vehicle is traveling the greater the vibration will be.
Technician B says that an unbalanced drive shaft will cause a vibration that is most noticeable at low speeds.
Who is correct?
A. Technician A
B. Technician B
C. Both A and B
D. Neither A nor B

28. When discussing how other vehicle problems can affect U-joint failure,
Technician A says that raising the suspension height of a vehicle excessively could result in premature U-joint failure.
Technician B says that a very high engine idle speed setting could cause excessive U-joint wear.
Who is correct?
A. Technician A
B. Technician B
C. Both A and B
D. Neither A nor B

29. Failure to index the driving and driven yokes of the rear U-joint before removal could result in which of the following?
A. Oil leak
B. Vibration
C. Noise
D. All of the above

30. When installing a new U-joint, one bearing cup fits into the slip joint easily, while the other needs extreme force to put it in place,
Technician A says that the slip yoke may need to be replaced.
Technician B says that the U-joint may be the wrong one for the application.
Who is correct?
A. Technician A
B. Technician B
C. Both A and B
D. Neither A nor B

31. The front- and rear-drive shaft angles are being discussed,
Technician A says that the angle of the front slip yoke is usually adjusted to match the angle of the rear slip yoke.
Technician B says that ideally both the front- and rear-drive shaft angles should cancel each other out.
Who is correct?
A. Technician A
B. Technician B
C. Both A and B
D. Neither A nor B

32. When diagnosing a differential noise,
Technician A says that a noise heard during deceleration will also be heard on acceleration.
Technician B says that a noise heard during acceleration will not necessarily be heard during a coast condition.
Who is correct?
A. Technician A
B. Technician B
C. Both A and B
D. Neither A nor B

33. The owner of a vehicle equipped with a limited-slip differential says the differential is making a chattering noise when he makes a right- or a left-hand turn.
Technician A says that the differential fluid may need to be replaced.
Technician B says that the fluid used in limited-slip differentials is different from the fluid used in conventional differentials.
Who is correct?

A.	Technician A	C.	Both A and B
B.	Technician B	D.	Neither A nor B

34. When discussing sealers and leak diagnosis,
Technician A says that when using silicone sealer to seal a differential housing cover, a bead that is at least ½ inch wide should be used.
Technician B says that silicone sealer can be used to repair a porous differential housing.
Who is correct?

A.	Technician A	C.	Both A and B
B.	Technician B	D.	Neither A nor B

35. A rear-wheel-drive car exhibits a knocking sound only during turns; the vibration disappears when the vehicle is driven straight.
Technician A says that the differential pinion gears may be damaged.
Technician B says that the carrier bearings may be worn out.
Who is correct?

A.	Technician A	C.	Both A and B
B.	Technician B	D.	Neither A nor B

36. When checking the components of a differential,
Technician A says that the maximum allowable runout of a ring gear is about .003 inch.
Technician B says that excessive ring gear runout may be caused by a worn pinion gear.
Who is correct?

A.	Technician A	C.	Both A and B
B.	Technician B	D.	Neither A nor B

37. When making ring and pinion gear adjustments,
Technician A says that a gauge set (arbor and block gauge) can be used to adjust pinion bearing preload.
Technician B says that the gauge set can be used to determine the thickness of the pinion gear spacer shim.
Who is correct?

A.	Technician A	C.	Both A and B
B.	Technician B	D.	Neither A nor B

38. There is excessive end play on an axle shaft of an integral carrier differential (where the shaft is retained in a C-lock).
Technician A says that the differential side gear thrust washer may be worn.
Technician B says that a worn axle shaft bearing may be the cause of the problem.
Who is correct?

A.	Technician A	C.	Both A and B
B.	Technician B	D.	Neither A nor B

39. While rebuilding a transfer case,
Technician A says pressing a bearing too far into the case can block off oil flow and cause the bearing to overheat.
Technician B coats the thrust washers with petroleum jelly or trans gel before installing them.
Who is correct?

A.	Technician A	C.	Both A and B
B.	Technician B	D.	Neither A nor B

40. When discussing the effect of engine vacuum on transfer case controls,
Technician A says that low manifold vacuum in a poorly running engine can result in the improper operation of some transfer cases.
Technician B says that at least 5 in. Hg are necessary in order to properly operate a vacuum-controlled transfer case.
Who is correct?

A.	Technician A	C.	Both A and B
B.	Technician B	D.	Neither A nor B

41. When discussing viscous clutch service and inspection, *Technician A* says that viscous clutches can be tested on a bench.
Technician B says that a viscous clutch can be serviced in the shop by adding oil to its reservoir.
Who is correct?
A. Technician A
B. Technician B
C. Both A and B
D. Neither A nor B

42. When discussing the unsprung weight of a vehicle, *Technician A* says that installing lightweight aluminum wheels on a vehicle will reduce its sprung weight.
Technician B says that replacing a steel hood with an aluminum hood will reduce a vehicle's unsprung weight.
Who is correct?
A. Technician A
B. Technician B
C. Both A and B
D. Neither A nor B

43. Transfer case problems are being discussed,
Technician A says that incorrect drive shaft angles can result in harsh engagement when shifting into 4WD.
Technician B says that an overheated viscous coupling can cause binding when making a sharp turn on dry pavement.
Who is correct?
A. Technician A
B. Technician B
C. Both A and B
D. Neither A nor B

44. The resistance of an electromagnetic clutch coil is 200 ohms; it should be 4 ohms.
Technician A says that this will result in a blown fuse.
Technician B says that the magnetic field developed by the clutch coil will be excessive.
Who is correct?
A. Technician A
B. Technician B
C. Both A and B
D. Neither A nor B

45. A poorly operating fan motor circuit is being tested. A voltmeter that is placed across the power and ground terminals of connector the motor indicates 0.0 volts when the fan switch is turned on.
Technician A says that this test reading indicates that the connector is probably okay.
Technician B says that this test is referred to as an available voltage test.
Who is correct?
A. Technician A
B. Technician B
C. Both A and B
D. Neither A nor B

46. The amount of current being drawn by a load is about to be measured.
Technician A says that the ammeter can be connected in the positive side of the circuit.
Technician B says that the amperage can be measured by connecting the ammeter to the negative side of the circuit.
Who is correct?
A. Technician A
B. Technician B
C. Both A and B
D. Neither A nor B

47. A digital ammeter being used to measure current drain is indicating 101 mA.
Which of the following represents this reading?
A. .101 amps
B. .00101 amps
C. 1.01 amps
D. 10.1 amps

48. A vehicle towed into the shop because of a no-start condition is found to have an inoperative starter. A voltmeter connected to both of the terminals of the clutch safety switch indicates 8 volts when the clutch pedal is depressed and the ignition switch is placed in the Start position.
Technician A says that the clutch safety switch is faulty.
Technician B says that the resistance of the circuit is higher than normal.
Who is correct?
A. Technician A
B. Technician B
C. Both A and B
D. Neither A nor B

49. A customer complains that her car, which is equipped with an electronically controlled manual transmission, is experiencing difficulty engaging in reverse gear.
Technician A says that there may be excessive voltage drop in the power feed circuit of the reverse lock out solenoid.

Technician B says that a testlight would be the most effective tool to use in order to test for a suspected voltage drop condition.

Who is correct?

A. Technician A
B. Technician B
C. Both A and B
D. Neither A nor B

50. When discussing manual transmission electrical problems,
Technician A says that an inoperative Upshift lamp may be caused by a faulty engine load sensor.

Technician B says that a scan tool may be used to energize the blockout solenoid of a manual transmission equipped with shift-blocking.

Who is correct?

A. Technician A
B. Technician B
C. Both A and B
D. Neither A nor B

to convert these	to these	multiply by
TEMPERATURE		
Centigrade degrees	Fahrenheit degrees	1.8 then +32
Fahrenheit degrees	Centigrade degrees	0.556 then −32
LENGTH		
Millimeters	Inches	0.03937
Inches	Millimeters	25.4
Meters	Feet	3.28084
Feet	Meters	0.3048
Kilometers	Miles	0.62137
Miles	Kilometers	1.60935
AREA		
Square centimeters	Square inches	0.155
Square inches	Square centimeters	6.45159
VOLUME		
Cubic centimeters	Cubic inches	0.06103
Cubic inches	Cubic centimeters	16.38703
Cubic centimeters	Liters	0.001
Liters	Cubic centimeters	1000
Liters	Cubic inches	61.025
Cubic inches	Liters	0.01639
Liters	Quarts	1.05672

to convert these	to these	multiply by
Quarts	Liters	0.94633
Liters	Pints	2.11344
Pints	Liters	0.47317
Liters	Ounces	33.81497
Ounces	Liters	0.02957
WEIGHT		
Grams	Ounces	0.03527
Ounces	Grams	28.34953
Kilograms	Pounds	2.20462
Pounds	Kilograms	0.45359
WORK		
Centimeter Kilograms	Inch-Pounds	0.8676
Inch-Pounds	Centimeter Kilograms	1.15262
Meter-Kilograms	Foot-Pounds	7.23301
Foot-Pounds	Newton-Meters	1.3558
PRESSURE		
Kilograms/Sq. Cm	Pounds/Sq. Inch	14.22334
Pounds/Sq. Inch	Kilograms/Sq. Cm	0.07031
Bar	Pounds/Sq. Inch	14.504
Pounds/Sq. Inch	Bar	0.06895

Association of Automotive aftermarket Distributors

ATEC Trans Tool
San Antonio, TX

Parts Plus
Memphis, TN

Baum Tools Unlimited, Inc.
Longboat Key, FL

Big A Auto Parts, APS Inc.
Houston, TX

Carquest Corp.
Tarrytown, NY

Drivetrain Specialists
Warren, MI

GKN Drivetech Inc.
Walled Lake, MI

Gray Manufacturing
St. Joseph, MO

Hastings Manufacturing Co.
Hastings, MI

KD Tools, Danaher Tool Group
Lancaster, PA

Kent-Moore, Div. SPX Corp.
Warren, MI

Lisle Corp.
Clarinda, IA

Mac Tools
Washington Courthouse, OH

Matco Tools
Stow, OH

NAPA Hand/Service Tools
Lancaster, PA

On Tool.com
Sunnyvale, CA

OTC, Div. SPX Corp.
Owatonna, MN

Snap-on Inc.
Kenosha, WI

TCI Automotive
Ashland, MS

Transtar Industries, Inc.
Cleveland, OH

Trans-Tool
San Antonio, TX

Van Norman Equipment Co.
Winona, MN

Acura Division of American Honda Motor Co.
www.acura.com

American Honda Motor Co.
www.honda.com

Aston Martin
www.astonmartin.com

Audi of America Inc.
www.audi.com

BMW of North America
www.bmwusa.com

Buick Motor Division of GM
www.buick.com

Cadillac Motor Car Division of GM
www.cadillac.com

Chevrolet Motor Division of GM
www.chevrolet.com

Chrysler Division of Fiat Chrysler Automobiles
www.chrysler.com

Dodge Division of Fiat Chrysler Automobiles
www.dodge.com

Ford Division of Ford Motor Company
www.ford.com

General Motors
www.gm.com

Hyundai Motor America
www.hyundaiusa.com

Infiniti Division, Nissan North America Inc.
www.infiniti-usa.com

Jaguar Cars North America
www.jaguar.com/us

Jeep Division of Fiat Chrysler Automobiles
www.jeep.com

Kia Motors America Inc.
www.kia.com/us

Land Rover North America Inc.
www.landrover.com

Lexus Division, Toyota Motor Sales USA
www.lexus.com

Lincoln Mercury Division, Ford Motor Co.
www.lincolnvehicles.com

Mazda North American Operations
www.mazdausa.com

Mercedes-Benz USA Inc.
www.mbusa.com

Mitsubishi Motor Sales of America Inc.
www.mitsubishicars.com

Nissan Division, Nissan North America Inc.
www.nissan-usa.com

GMC Division of GM
www.gmc.com

Porsche Cars North America
www.porsche.com

Saab Cars USA Inc.
www.saabcars.com

Saturn Corp.
www.saturn.com

Subaru of America Inc.
www.subaru.com

Tesla Motors
www.teslamotors.com

Toyota Division, Toyota Motor Sales USA Inc.
www.toyota.com

Volkswagon of America Inc.
www.vw.com

Volvo Cars North America Inc.
www.volvocars.com

GLOSSARY
GLOSARIO

Note: **Terms are highlighted in bold**, followed by Spanish translation in color.

American wire gauge (AWG) The Society of Automotive Engineers (SAE) labeling system for wire size.

Calibre de cables estadounidense (AWG, en inglés) Sistema de etiquetado de tamaños de cable de la Sociedad de Ingenieros Automotrices (SAE, por sus siglas en inglés).

Ammeter A meter used to measure electrical current.

Amperímetro Aparato que se usa para medirla corriente eléctrica.

Ampere The unit of measure for electrical current.

Amperio Unidad de medida de una corriente eléctrica.

Amplitude A measurement of a vibration's intensity.

Amplitud Medida de la intensidad de una vibración.

Automotive Service Excellence (ASE) A national nonprofit organization that promotes automotive education and offers individual and organization certifications in a broad number of specific repair and classification areas.

Automotive Service Excellence (ASE) Organización nacional sin fines de lucro que promueve la educación automotriz y ofrece certificaciones individuales e institucionales en numerosas áreas de reparación y clasificación.

Backlash The amount of clearance or play between two meshed gears.

Contragolpe Cantidad del espacio libre u holgura entre dos engranajes.

Back probing The process of electrical measurement at the wire side (back side) of a connected connector.

Sondeo eléctrico trasero Proceso de medición eléctrica en el lado del cable (parte trasera) de un conector enchufado.

Back-up light fuse A circuit protection device in the electrical circuit for the back-up or reverse lights.

Fusible de las luces de reversa Dispositivo de protección de circuitos en el circuito eléctrico para las luces de reversa.

Back-up light switch A switch completing the reverse or back-up light circuit when the driver places the vehicle into reverse.

Interruptor de luces de reversa Interruptor que completa el circuito de las luces de reversa cuando el conductor coloca el vehículo en reversa.

Ball joint A suspension component that attaches the control arm to the steering knuckle and serves as its lower pivot point. The ball joint gets its name from its ball-and-socket design. It allows both up and down motion as well as rotation. In a MacPherson strut FWD suspension system, the two lower ball joints are nonload carrying.

Junta esférica Componente de la suspensión que fija el brazo de mando al muñón de dirección, y sirve como punto de pivote inferior para el muñón de dirección. Así se le llama a la junta esférica por su diseño de rótula. Permite tanto el movimiento de ascenso y descenso como el de rotación. En un sistema de suspensión de tracción delantera montante MacPherson, las dos juntas esféricas inferiores no portan carga.

Bearing rumble A low-frequency noise caused by damaged bearing races or rollers.

Rechinar del cojinete Ruido de baja frecuencia provocado por pistas o rodillos de cojinetes dañados.

Bearing whine A high-frequency noise caused by a worn or damaged bearing surface.

Silbido del cojinete Ruido de alta frecuencia provocado por la superficie desgastada o dañada del cojinete.

Bleeding The process of removing air from a closed system.

Purga de aire Proceso utilizado para remover el aire de un sistema cerrado.

Blood-borne pathogens Pathogenic microorganisms that are present in human blood and can cause disease in humans. These pathogens include, but are not limited to, hepatitis B virus (HBV) and human immunodeficiency virus (HIV).

Patógenos en la sangre Microorganismos patógenos que están presentes en la sangre humana y pueden causar enfermedades en los humanos. Entre estos patógenos están a modo enunciativo y no limitativo el virus de la hepatitis B (VHB) y el virus de la inmunodeficiencia humana (VIH).

Blowgun A tool connected to a pressurized air source used to clean or dry components.

Pistola de aire Herramienta conectada a una fuente de aire presurizado, que se utiliza para limpiar o secar piezas.

Blueing The discoloration evident on an overheated metal object. A flywheel that has been overheated will have blue areas, also called hot spots or hard spots.

Azulado Decoloración evidente en un objeto metálico sobrecalentado. Un volante sobrecalentado tendrá zonas azules, que también se denominan puntos calientes o puntos duros.

Bolt head The top of a bolt where the socket or wrench is placed to tighten or loosen the bolt.

Cabeza del perno Parte superior de un perno en donde se coloca la llave para ajustar o desajustar el perno.

Bolt shank The section of a bolt on which the threads are cut. Shank size determines the size of the bolt.

Vástago del perno Sección del perno que tiene las marcas de rosca. El tamaño del vástago determina el tamaño del perno.

Caliper A sliding-scale precision measurement tool with variable jaws. Sliding calipers may be plain, dial, or digital.

La mordaza (caliper) Herramienta con escala móvil que se usa para hacer mediciones de precisión y tiene una apertura variable. Las mordazas deslizantes pueden ser simples, a dial o digitales.

Camber The amount that the centerline of the wheel is tilted inward or outward from the vertical plane.

Curvatura Inclinación hacia adentro o hacia afuera de la línea central de la rueda con respecto al plano vertical.

Carbon monoxide (CO) An odorless, colorless, and deadly gas present in the exhaust of engines.

Monóxido de carbono Gas mortífero, inodoro e incoloro presente en los gases de escape de los motores.

Caster A measurement expressed as an angle of the forward or rearward tilt of the top of the wheel spindle.

Ángulo de comba de eje Medida expresada como el ángulo de inclinación hacia adelante o hacia atrás de la parte superior del portamuela.

Caustic The characteristic of a material that has the ability to destroy or eat through something. Caustic materials are considered extremely corrosive.

Cáustico Caracteristica de una materia que tiene el poder de destruir o corroer algo. Se consideran muy corrosivas las materias cáusticas.

Center support plate A bearing support housing between the front and back case sections of a transmission.

Placa de soporte central Carcasa de soporte de cojinetes entre las secciones delantera y trasera de la caja de transmisión.

Chassis ground A ground path in an electrical circuit to the negative side of the battery that is directly connected to the vehicle chassis.

Tierra del chasis Conductor a tierra de un circuito del lado negativo de la batería que se conecta directamente al chasis del vehículo.

Chatter A condition where a vehicle shakes or vibrates as the clutch is being engaged. Also, a condition felt or heard in a limited-slip differential with worn clutches or incorrect lubricant.

Chasquido Estado en el que el vehículo se sacude o vibra al accionar el embrague. Además, se percibe u oye cuando un diferencial autobloqueante tiene los embragues desgastados o lubricante incorrecto.

Chuckle A rattling noise that sounds much like a stick rubbing against the spokes of a bicycle wheel.

Estrépito Ruido muy fuerte o estruendo parecido al sonido de un palo que golpea contra los rayos de la rueda de una bicicleta.

Circuit breakers Electrical circuit protection devices that either automatically or manually reset if they open due to excessive current flow.

Disyuntores Dispositivos de protección de circuitos eléctricos que se restablecen automática o manualmente si se abren por un exceso de corriente.

Class A fires A type of fire in which wood, paper, and other ordinary materials are burning.

Incendio de clase A Incendio en el que se queman la madera, el papel y otros materiales comunes.

Class B fires A type of fire involving flammable liquids, such as gasoline, diesel fuel, paint, grease, oil, and other similar liquids.

Incendio de clase B Incendio en el que se queman líquidos inflamables, como por ejemplo la gasolina, el diesel, la pintura, la grasa, el aceite, y otros líquidos similares.

Class C fires Electrical fires.

Incendio de clase C Incendios eléctricos.

Class D fires A unique type of fire in which the material burning is a metal. An example of this is a burning "mag" wheel; the magnesium used in the construction of the wheel is a flammable metal and will burn brightly when subjected to high heat.

Incendio de clase D Incendio único porque la material que se quema es un metal. Es un ejemplo de esto la quema de una rueda "mag"; el magnesio con el cual se fabrica la rueda es un metal inflamable que resplandece cuando se somete a altas temperaturas.

Clunking A metallic noise most often heard when an automatic transmission is engaged into reverse or drive. It is also often heard when the throttle is applied or released. Clunking is caused by excessive backlash somewhere in the driveline and is felt or heard in the axle.

Sonido sordo Sonido metálico que se escucha con frecuencia cuando se engrana una transmisión automática en marcha atrás o en marcha adelante. Se escucha también cuando se aprieta o se suelta el acelerador. El sondio metálico sordo se debe al contragolpe excesivo en alguna parte de la línea de transmisión, y se siente o escucha en el eje.

Clutch drag A condition where the clutch disc is still contacting the pressure plate and flywheel when the clutch pedal is fully depressed.

Arrastre del embrague Estado en el que el disco del embrague sigue en contacto con la placa de presión y el volante cuando el pedal de embrague se aprieta totalmente.

Clutch reserve The distance between when the clutch pedal is fully depressed and its position when the vehicle starts to move forward while in gear.

Reserva de embrague La distancia entre el momento en que el pedal del embrague está completamente presionado y la posición en que se encuentra cuando el vehículo comienza a avanzar con el cambio acoplado.

Clutch safety switch A switch in series with the starting circuit that prevents completion of the circuit until the clutch pedal is pushed down.

Interruptor de seguridad del embrague Un interruptor en serie con el circuito de arranque que evita que se complete el circuito hasta que se empuje el pedal del embrague.

Clutch slippage Engine speed increases but increased torque is not transferred through to the driving wheels because of clutch slippage.

Deslizamiento del embrague La velocidad del motor aumenta, pero el par de torsión no se transmite a las ruedas motrices a causa del deslizamiento del embrague.

Coast The concave side of a gear tooth.

Costa Lado cóncavo de un diente de engranaje.

Commutator An assembly is made up of heavy-copper segments separated from each other and the armature shaft by insulation. Individual segments are connected to the individual windings in the armature. Brushes ride on the surface of the commutator.

Conmutador Un conjunto que consta de segmentos de cobre pesado separados entre sí y del eje del inducido por aislamiento. Los segmentos individuales están conectados a los bobinados individuales del inducido. En la superficie del conmutador hay escobillas.

Companion flange A mounting flange that attaches a drive shaft to another drivetrain component.

Brida acompañante Una brida de montaje que fija un árbol de mando a otro componente del tren de mando.

Concentric slave cylinder The hydraulic component of a concentric clutch release mechanism.

Cilindro secundario concéntrico Componente hidráulico de un mecanismo de liberación de embrague concéntrico.

Corrosivity The ability of something to dissolve metals and other materials or burn skin.

Corrosividad Capacidad que tiene algo de disolver los metales y otros materiales o quemar la piel.

Current The flow of electricity through a circuit.

Corriente El flujo de la electricidad por un circuito.

Data link connector (DLC) The electrical connector in a vehicle that is accessed by a scan tool to retrieve diagnostic trouble codes and other vehicle information.

Conector para extraer datos (DLC, en inglés) Conector eléctrico de un vehículo al que se accede con una herramienta de exploración para obtener códigos diagnósticos de falla y otra información del vehículo.

Diagnosis A systematic study of a machine or machine parts to determine the cause of improper performance or failure.

Diagnóstico Estudio sistemático de una máquina o de piezas de un máquina para establecer la causa del mal funcionamiento o la falla.

Diagnostic trouble code (DTC) An alpha numerical code used to identify a vehicle problem or component failure. DTC's are stored in the vehicle's control modules and are typically accessed with a scan tool.

Código diagnóstico de falla (DTC, en inglés) Código alfanumérico que se usa para identificar un problema del vehículo o la falla de algún componente. Los códigos diagnósticos de fallas se almacenan en los módulos de control del vehículo y se suele acceder a ellos con una herramienta de exploración.

Dial indicator A measuring instrument with the readings indicated on a dial rather than on a thimble as on a micrometer.

Indicador de cuadrante Instrumento de medida que muestra las lecturas en un cuadrante en vez de un tambor como en el caso de un micrómetro.

Digital multimeter (DMM) A single digital meter that combines the capabilities of an ammeter, ohmmeter, and voltmeter.

Multímetro digital (DMM, en inglés) Medidor digital que combina las funciones de un amperímetro, ohmímetro y voltímetro.

Digital storage oscilloscope (DSO) A fast-reacting meter that measures and displays voltages within a specific time frame. The voltages are displayed as a waveform or trace on the screen of the DSO.

Osciloscopio de almacenamiento digital (DSO, en inglés) Medidor de reacción rápida que mide y visualiza los voltajes dentro de un lapso de tiempo específico. Los voltajes se visualizan en forma de onda o de trazado en la pantalla del osciloscopio de almacenamiento digital.

Dimpling Brinelling or the presence of indentations in a normally smooth surface.

Abolladura Brinelado o formación de hendiduras en una superficie normalmente lisa.

Drive The convex side of a gear tooth.

Tracción Lado convexo de un diente de engranaje.

Drive sprocket A toothed wheel that drives a chain and driven sprocket to turn a component.

Piñón impulsor Rueda dentada que impulsa una cadena y un piñón para mover una pieza.

Dual clutch transmission A manual transmission with two separate input shafts, each with its own clutch or clutch pack.

Transmisión de doble embrague Transmisión manual con dos flechas de entrada separadas, cada una con su propio embrague o paquete de embrague.

Dummy shaft A shaft, shorter than the countershaft, used during disassembly and reassembly in place of the countershaft.

Árbol falso Árbol más corto que el árbol de retorno empleado durante el desmontaje y el remonte en vez del árbol de retorno.

Eccentric washer A normal looking washer with its hole offset from the center.

Arandela excéntrica Arandela de apariencia normal pero cuyo agujero se encuentra fuera del centro.

Electromagnetic clutch A device that relies on electrical current to form a magnetic field. When current is flowing through the clutch, a magnetic field is present and the clutch is engaged.

Embrague electromagnético Un dispositivo que se basa en la corriente eléctrica para formar un campo magnético. Cuando la corriente fluye por el embrague, esta induce un campo magnético que acciona el embrague.

Electromagnetic Interference (EMI) An unwanted disturbance that affects an electrical circuit due to either electromagnetic conduction or electromagnetic radiation emitted from an external source.

Interferencia electromagnética (IEM) Una perturbación no deseada que afecta un circuito eléctrico debid, ya sea a la conducción electromagnética o a la radiación electromagnética que emite una fuente externa.

Electromotive force The pressure that exists between a positive and negative point within an electrical circuit.

Fuerza electromotriz La presión entre un punto positivo y negativo dentro de un circuito eléctrico.

End play The amount of axial or end-to-end movement in a shaft due to clearance in the bearings.

Holgadura Amplitud de movimiento axial o movimiento de extremo a extremo en un árbol debido al espacio libre entre los cojinetes.

EP toxicity A classification of poisonous materials that leach one or more of eight heavy metals in concentrations greater than 100 times primary drinking water standard concentrations.

Toxicidad EP Una clasificación de materias venenosas que filtran uno u más de ocho metales en concentraciones mayores a 100 veces la concentración estándar principal en el agua potable.

Face The front surface of an object.

Frente Superficie frontal de un objeto.

Feeler gauge A metal strip or blade finished accurately with regard to thickness used for measuring the clearance between two parts; such gauges ordinarily come in a set of different blades graduated in thickness by increments of 0.001 inch.

Calibrador de espesores Lámina metálica o cuchilla acabada con precisión de acuerdo al espesor que se utiliza para medir el espacio libre entre dos piezas. Dichos calibradores normalmente están disponibles en juegos de cuchillas con diferente graduación según el espesor, en incrementos de 0,001 pulgadas.

Fillet The smooth curve where the shank flows into a bolt head or a shaft flows into a flange.

Filete Curva suave donde se unen el vástago y la cabeza del perno o el eje se une con una brida.

Fill plug A plug found on most transmissions and transaxles used to allow fluid to be added to a unit and to, when tightened, seal a bore in a fluid reservoir.

Tapón de llenado Un tapón que se encuentra en la mayoría de las transmisiones que se utiliza para permitir que se agregue fluido a la unidad y, al ajustarlo, sellar el diámetro interior de un depósito de fluidos.

Flammability The characteristic of something that makes it burn quickly and easily.

Flameante Característica de algo que le permite quemarse de manera fácil y rápida.

Flange A projecting rim or collar on an object for keeping it in place.

Brida Corona proyectada o collar sobre un objeto que lo mantiene en su lugar.

Flank The area near the bottom of a gear tooth.

Flanco Área cercana a la parte inferior del diente de un engranaje.

Flashing The common name given to the process of reprogramming the memory in a control module.

Flashing El nombre común que se le da al proceso de reprogramar la memoria de un módulo de control.

Free play The clearance between the pressure plate release fingers and the release bearing. This is felt at the top of the clutch pedal travel.

Holgadura El espacio libre entre las uñas de liberación de la placa de presión y el cojinete de desembrague. Esto se siente en la parte superior de la trayectoria de viaje del pedal del embrague.

Freeze-frame data In this feature of OBD II systems, the PCM takes a snapshot of the activity of various inputs and outputs at the time the PCM illuminated the MIL.

Datos de la imagen congelada En esta función de los sistemas OBD II, la PCM toma una instantánea de la actividad de varias entradas y salidas en el momento que la PCM ilumina la MIL.

Frequency The rate at which something occurs.

Frecuencia Velociadad a la cual ocurre algo.

Front-axle disconnect A mechanism on a 4WD front axle that connects or disconnects two sections of a drive axle.

Desconexión del eje delantero Mecanismo del eje delantero de la tracción en las cuatro ruedas (4WD) que conecta o desconecta dos secciones de un eje de mando.

Front probing The process of electrical measurement at the mating (front) side of an electrical connector.

Sondeo eléctrico delantero Proceso de medición eléctrica en el lado (delantero) correspondiente de un conector eléctrico.

Fusible link Circuit protection devices built into the wiring of a particular circuit. Like a fuse, the fusible link is rated at the current at which they are designed to blow.

Eslabón fusible Dispositivos de protección de circuitos incorporados en el cableado de un circuito específico. Al igual que los fusibles, el eslabón fusible está clasificado según la corriente a la cual está diseñado para quemarse.

Galvanic corrosion A type of corrosion that occurs when two dissimilar metals, such as magnesium and steel, are in contact with each other.

Corrosión galvánica Tipo de corrosión producida cuando dos metales distintos, como por ejemplo el magnesio y el acero, entran en contacto el uno con el otro.

Gateway A module that allows for data exchange between different buses. It translates a message on one bus and transfers that message to another bus without changing the message. The gateway interacts with each bus according to the protocol of that bus.

Puerta de enlace Un módulo que permite el intercambio de datos entre diferentes buses. Traduce un mensaje en un bus y lo transfiere a otro bus sin cambiar el mensaje. La puerta de enlace interactúa con cada bus de acuerdo con el protocolo del bus.

Gear clash The noise that results when two gears are traveling at different speeds and are forced together.

Choque de engranajes Ruido producido cuando dos engranajes giran a velocidades diferentes y se unen por fuerza.

Gear knock A sound from a transmission, transfer case, or differential caused by broken or chipped gear teeth.

Golpeteo del engranaje Sonido de la transmisión, la caja reductora o el diferencial provocado por dientes de engranaje rotos o astillados.

Gear noise The howling or whining of the ring gear and pinion due to an improperly set gear pattern, gear damage, or improper bearing preload.

Ruido del engranaje Aullido o silbido de la corona y del piñón debido al montaje incorrecto de los engranajes, a averías en los engranajes o a carga previa incorrecta del cojinete.

Gear rattle A repetitive metallic impact or rapping noise that occurs when the vehicle is lugging in gear. The intensity of the noise increases with operating temperature and engine torque, and decreases with increasing vehicle speed.

Estruendo del engranaje Impacto metálico repetitivo o golpeteo que ocurre cuando el vehículo arrastra los engranajes durante la marcha. La intensidad del ruido aumenta con la temperatura de funcionamiento y el par de torsión del motor, y disminuye al aumentar la velocidad del vehículo.

Glitch A brief malfunction or defect.

Falla imprevista Breve anomalía o defecto.

Grade marks Cast marks on the head of a bolt that signify the tensile strength of the bolt. In the Imperial system the marks are a radial series of lines. In the metric system numbers indicate the tensile strength and yield strength of the bolt.

Marcas de grado Marcas colocadas en la cabeza de un perno e indican la fuerza de tensión del perno. En el sistema imperial, las marcas son una serie radial de líneas. En el sistema métrico, los números indican la fuerza de tensión y resistencia del perno.

Graphing multimeter (GMM) A digital multimeter that displays voltage, resistance, current, and frequency as a waveform.

Multímetro gráfico (GMM) Un multímetro digital que muestra el voltaje, la resistencia, la corriente y la frecuencia en forma de onda.

Hazardous waste Waste is considered hazardous if it has one or more of the following characteristics: ignitability, corrosivity, reactivity, and EP toxicity.

Residuo peligroso Los residuos se consideran peligrosos si tienen una o más de las características siguientes: inflamable, corrosivo, reactivo, toxicidad EP.

Heel The outside, larger half of the gear tooth.

Talón Mitad exterior más grande del diente de engranaje.

Hertz The unit that frequency is most often expressed and is equal to one cycle per second.

Hercio Forma más común de expresar la frecuencia y equivale a un ciclo por segundo.

High pedal A clutch pedal that has an excessive amount of pedal travel.

Pedal alto Pedal del embrague con una trayectoria de recorrido excesivo.

Hot spots The small areas on a friction surface that are a different color, normally blue, or are harder than the rest of the surface.

Zonas de calor Zonas pequeñas sobre una superficie de rozamiento de un color diferente, normalmente azul, o más duras que el resto de la superficie.

Hydrocarbon A substance that is based on petroleum, such as gasoline.

Hidrocarburo Sustanica que se deriva del petróleo, como la gasolina.

Hygroscopic The property of a fluid that has a tendency to absorb moisture from the atmosphere.

Higroscópico Cualidad que posee un fluido que tiende a absorber la humedad de la atmósfera.

Ignitability The property of a liquid with a flash point below 140° or a solid that can spontaneously ignite.

Inflamabilidad La propiedad de un líquido de encenderse a temperaturas menores de 140° o un sólido que se incendia espontáneamente.

Inclinometer A device designed with a spirit level and graduated scale to measure the inclination of a driveline assembly. The inclinometer connects to the drive shaft magnetically.

Inclinómetro Instrumento diseñado con nivel de burbuja de aire y escala graduada para medir la inclinación de un conjunto de la línea de transmisión. El inclinómetro se conecta magnéticamente al árbol de mando.

Jack (safety) stands Welded steel stands designed to safely support the weight of a vehicle.

Plataformas (de seguridad) para gato Plataformas de acero soldado, diseñadas para soportar el peso de un vehículo sin riesgos.

King pin A metal rod or pin on which steering knuckles turn.

Clavija maestra Varilla o chaveta de metal sobre la cual giran los muñones de dirección.

Knock A heavy-metallic sound usually caused by a loose or worn bearing.

Golpeteo Sonido metálico pesado normalmente causado por un cojinete suelto o desgastado.

Lapping The process of fitting one surface to another by rubbing them together with an abrasive material between the two surfaces.

Lapidado Proceso de ajustar una superficie contra otra rozando la una contra la otra con un material abrasivo colocado entre las dos superficies.

Lift kit A group of components used to raise vehicle height for increased ground and tire clearance.

Kit de elevación Grupo de componentes utilizados para elevar al vehículo con el fin de aumentar el espacio libre entre las ruedas y el suelo.

Load A term normally used to describe an electrical device that is operating in a circuit. Load can also be used to describe the relative amount of work a driveline must do.

Carga Un término que normalmente describe un dispositivo eléctrico operando en un circuito. La carga también puede describir la cantidad relativa de trabajo que debe efectuar una flecha motriz.

Locking A condition of a bearing caused by large particles of dirt that become trapped between a bearing and its race.

Bloqueo Condición de un cojinete causada por partículas de polvo que quedan atrapadas entre un cojinete y su anillo.

Lugging A term used to describe an operating condition in which the engine is operating at too low an engine speed for the selected gear.

Arrastre Término utilizado para describir una condición en la que el motor funciona a una velocidad demasiado baja para la marcha elegida.

Machinist's rule A steel ruler that is precisely graded to allow accurate measurements.

Regla técnica Regla de acero graduada con exactitud para realizar mediciones precisas.

Material Safety Data Sheet (MSDS) Information sheets containing chemical composition and precautionary information for all products that can present a health or safety hazard.

Hoja de información de seguridad (MSDS) Hojas de información sobre la composición química e información de advertencia de todos los productos que pueden ser peligrosos para la salud o la seguridad.

Maxi-fuses Large two-prong blade or spade fuses used to protect a high-current circuit. Maxi-fuses are typically found in their own underhood fuse block.

Maxifusibles Fusibles de pala o paleta grande de dos terminales que se usan para proteger un circuito de alta corriente. Los maxifusibles se suelen encontrar en su propio bloque de fusibles debajo del capó.

Micrometer A precision measuring device used to measure small bores, diameters, and thicknesses. Also called a mike.

Micrómetro Instrumento de precisión utilizado para medir calibres, espesores y diámetros pequeños. Llamado también mic.

Modular clutch A complete pre-assembled clutch assembly including the flywheel, pressure plate, and clutch disc.

Embrague modular Conjunto completo de embrague premontado, que incluye volante, placa de presión y disco de embrague.

Multiviscosity A lubricant rating that indicates the oil changes viscosity depending on temperature.

Viscosidad múltiple Medidor de lubricante que indica los cambios en la viscosidad del aceite según la temperatura.

Neutral rollover rattle A sound coming from a transmission running in neutral caused by the backlash between the input gear and the front gear of the countershaft.

Traqueteo de volteo en velocidad neutral Sonido que proviene de la transmisión en velocidad neutral, causado por el juego libre entre el engranaje de entrada y el engranaje delantero del eje intermedio auxiliar.

Noise An audible sound.

Ruido Sonido audible.

Ohm A unit of measurement for electrical resistance.

Ohmio Unidad de medida de la resistencia eléctrica.

Ohm's law A statement that describes the characteristics of electricity as it flows in a circuit.

Ley de Ohm Declaración que describe las características de la electricidad cuando esta recorre un circuito.

Onboard diagnostics II (OBD II) The second generation of the mandated onboard diagnostic systems designed to minimize the emissions of vehicles.

Diagnóstico incorporado II (OBD II, en inglés) La segunda generación del sistema de diagnóstico incorporado obligatorio diseñado para minimizar las emisiones de los vehículos.

Open A break in an electrical circuit that prevents the circuit from being completed.

Abierto Corte de un circuito eléctrico que evita que se complete un circuito.

Occupational Safety and Health Administration (OSHA) A government agency charged with ensuring safe work environments for all workers.

Division de Seguridad y Salud en el Trabajo (OSHA) Agencia gubernamental que tiene la responsabilidad de proteger la seguridad del ambiente donde trabajan todos los empleados.

Parameter identification (PID) A scan tool display for OBD-II systems that shows current emission-related data values of inputs and outputs, calculated values, and system status information.

Identificación de parámetros (PID) Un despliegue de herramienta de escaneo para sistemas OBD II que muestra los valores de la información relacionada a la emisión actual de entradas, salidas, valores calculados e información del estado del sistema.

Pawl A lever that pivots on a shaft. When lifted, it swings freely and when lowered, it locates in a detent or notch to hold a mechanism stationary.

Trinquete Una palanca que gira sobre un eje. Cuando se lo levanta, se balancea libremente y cuando se lo baja, se ubica en un retén o en una muesca para mantener fijo un mecanismo.

Peening Stretching or clinching metal over by pounding with the rounded end of a hammer. Also a description of gear tooth damage caused by impact load.

Granallar Estirar o remachar golpeando con el extremo redondo de un martillo. También se usa esta palabra para describir el daño a un diente de engranaje causado por la carga de impacto.

Pending code The term used to define a DTC that has resulted from insufficient occurrences to cause the MIL to be lit.

Código pendiente El término se utiliza para definir un DTC que se produjo como consecuencia de acontecimientos insuficientes para hacer que se encienda la MIL.

Pinch bolt A bolt used to clamp a tapered stud. These bolts are specially designed for the particular application.

Perno de fijación Perno utilizado para sujetar un espárrago ahusado. Estos pernos están diseñados especialmente para una aplicación específica.

Pinion flange A splined flange that connects the drive shaft to the differential drive pinion. Also called a companion flange or yoke.

Brida del piñón Brida estriada que conecta el eje de transmisión con el piñón de mando del diferencial. También se denomina horquilla o pestaña plana del tren de potencia.

Pitman arm A short arm in the steering linkage that connects the steering gear to other steering components.

Brazo pitman Brazo corto en el cuadrilátero de la dirección que articula el mecanismo de dirección a los otros componentes de dirección.

Pitting A condition normally associated with a thin oil film, possibly due to high-oil temperatures. A very small amount of pitting gives the surface of the gear teeth a gray appearance.

Picadura de metales Ocurre generalmente cuando la capa de aceite es demasiado delgada, posiblemente a causa de temperaturas muy altas del aceite. Una cantidad pequeña de picadura da una apariencia gris a la superficie de los dientes del engranaje.

Preload A load applied to a part during assembly so as to maintain critical tolerances when the operating load is applied later.

Carga previa Carga aplicada a una pieza durante su montaje para mantener tolerancias críticas cuando más tarde se aplique la carga de funcionamiento.

Press-fit Forcing a part into an opening that is slightly smaller than the part itself to make a solid fit.

Ajuste en prensa Forzar una pieza dentro de una apertura un poco más pequeña que la pieza misma para lograr un ajuste sólido.

Pressure plate That part of the clutch that exerts force against the friction disc; it is mounted on and rotates with the flywheel. A heavy-steel ring pressed against the clutch disc by spring pressure.

Placa de presión Pieza del embrague que ejerce fuerza contra el disco de fricción; se monta encima y gira con el volante. Un anillo pesado de acero, comprimido contra el disco de embrague mediante presión elástica.

Programmable read-only memory (PROM) A chip found in some scan tools that contains all of the necessary information needed to diagnose specific model lines.

Memoria de solo lectura programable (PROM, en inglés) Chip que se encuentra en algunas herramientas de exploración que contiene toda la información necesaria para diagnosticar modelos específicos.

Protocol The name for the language different control modules speak when they are talking to each other. The differences in protocol are based on the speed and the technique used.

Protocolo El nombre del lenguaje que usan diferentes módulos de control cuando se comunican entre sí. Las diferencias de protocolo se basan en la velocidad y la técnica que se utilizan.

Pulsation To move or beat with rhythmic impulses.

Pulsación Mover o golpear con impulsos rítmicos.

Quadrant A section of a gear. A term sometimes used to identify the shift lever selector mounted on the steering column.

Cuadrante Sección de un engranaje. Término utilizado en algunas ocasiones para identificar el selector de la palanca de cambio de velocidades montado sobre la columna de dirección.

Radius arm A longitudinal suspension arm that locates an axle.

Brazo radial Brazo de suspensión longitudinal que ubica un eje.

Random access memory (RAM) Memory circuits in a processor or scan tool that store information and test data.

Memoria de acceso aleatorio (RAM, en inglés) Circuitos de memoria de un procesador o herramienta de exploración que almacenan información y evalúan datos.

Reactivity A statement of how easily a substance can cause or be part of a chemical reaction.

Reactividad Declaración sobre cuán fácilmente una sustancia puede causar o formar parte de una reacción química.

Reluctance A term used to indicate a material's resistance to the passage of magnetic lines of flux.

Reluctancia Un término que se utiliza para indicar la resistencia de un material al pasaje de líneas magnéticas de flujo.

Resistance An obstruction or impedance to flow of any kind. Resistance in an electrical circuit is measured in ohms.

Resistencia Obstrucción o impedancia en cualquier tipo de corriente. La resistencia en un circuito eléctrico se mide en ohmios.

Rolling A result of overload and sliding, which leaves a burr on the tooth edge. Insufficient bearing support results in rolling of the metal due to the sliding pressure.

Rodamiento Causado por la sobrecarga o el deslizamiento que deja una rebaba en el borde del diente. La falta de soporte por parte de los cojinetes causa el rodamiento del metal por la presión de deslizamiento.

Runout Deviation of the specified normal travel of an object. The amount of deviation or wobble a shaft or wheel has as it rotates. Runout is measured with a dial indicator.

Desviación Desalineación del movimiento normal indicado de un objeto. Cantidad de desalineación o bamboleo que tiene un árbol o una rueda mientras gira. La desviación se mide con un indicador de cuadrante.

Scan tool A microprocesser used to communicate with a vehicle's computer.

Herramienta de escaneo Microprocesador empleado para comunicarse con la computadora de un vehículo.

Schematics Wiring diagrams that are used to show how circuits are constructed and how the components are connected.

Esquemas Diagramas de cableado que indican de qué modo están diseñados los circuitos y cómo están conectados los componentes.

Score A scratch, ridge, or groove marring a finished surface.

Muesca Rayado, rotura o ranura que estropea una superficie acabada.

Self-powered test light An electrical diagnostic tool used to check for continuity in a circuit. Also called a continuity tester.

Luz de prueba autosuficiente Herramienta eléctrica de diagnóstico utilizada para verificar la continuidad de un circuito. También se denomina probador de continuidad.

Serial data The term for the communications to and from the computer.

Datos seriales El término que se usa para las comunicaciones hacia y desde la computadora.

Shank A reinforcement inserted in the bottom of a shoe to prevent sharp objects from penetrating into the sole of the shoe or boot.

Fuste Refuerzo insertado al fondo de la zapata de freno para impedir la penetración de objetos agudos en la suela de la zapata o del fuelle.

Shift blocking A mechanism that prevents the driver from shifting into second or third gear, forcing a direct shift from first to fourth gear under certain circumstances. Shift blocking is designed to help improve fuel economy.

Bloqueo de cambios Mecanismo que evita que el conductor pase al segundo o tercer cambio y lo obliga a cambiar directamente de primera a cuarta, en ciertas circunstancias. El bloqueo de cambios está diseñado para mejorar el rendimiento del combustible.

Shift mode sleeve A sliding sleeve that engages/disengages four-wheel drive in a transfer case. The sleeve is moved into position by a shift mode fork.

Manguito de modo de cambios Manguito deslizante que activa/desactiva la tracción en las cuatro ruedas en una caja reductora. El manguito es colocado en posición mediante una horquilla de modo de cambios.

Shock absorber A tubular hydraulic suspension dampening device.

Amortiguador Dispositivo tubular lubricado de suspensión hidráulica.

Short An unwanted path for current flow in an electrical circuit.

Corto Recorrido no deseado de la corriente en un circuito eléctrico.

Shudder A shake or shiver movement.

Estremecimiento Sacudida o temblor.

Solenoid An electromagnet with a moveable core. The core is used to complete an electrical circuit or to cause a mechanical action.

Solenoide Un electroimán de núcleo móvil. El núcleo sirve para completar un circuito eléctrico o para causar una acción mecánica.

Spalling A condition of a bearing caused by overloading the bearing that is evident by pits on the bearings or their races.

Esquirla Condición de un cojinete provocada al ser sobrecargado, que se manifiesta a través de hendiduras en los cojinetes o sus anillos.

Spontaneous combustion Combustion that occurs without introducing heat or flame. It occurs when the materials heat themselves enough to cause combustion due to decomposing or decay.

Combustión espontánea Combustión que se produce sin calor ni llamas. Se produce cuando los materiales se calientan lo suficiente para causar combustión debido a la descomposición o desintegración.

Spring pad A mounting point for a leaf spring.

Almohadilla de muelle Punto de montaje para una hoja de muelle.

Sprocket carrier A bearing or bushing-supported component that is splined to a sprocket and is connected to a housing.

Portador de piñón Componente sostenido por un cojinete o buje, que está fijado con lengüeta en un piñón y conectado a una carcasa.

Stagger gauge A sliding measuring device used to determine the diameter of a tire.

Calibrador escalonado Dispositivo deslizante de medición utilizado para determinar el diámetro de un neumático.

Steering damper A hydraulic component used to stabilize the movement of the front tires during turning.

Amortiguador de dirección Componente hidráulico empleado para estabilizar el movimiento de los neumáticos delanteros al girar.

Sway bar Also called a stabilizer bar, it prevents the vehicle's body from diving into turns.

Barra de oscilación lateral Llamada también barra estabilizadora. Impide que la carrocería del vehículo se desestabilice durante los virajes.

Tail shaft A commonly used term for a transmission's extension housing.

Extremo del árbol Término comúnmente utilizado para la carcasa de la extensión de la transmisión.

Tapered shim An angled metal component used under the leaf spring to adjust pinion nose angle.

Lámina de ajuste ahusada Componente metálico angular utilizado debajo de la hoja de muelle para ajustar el ángulo de la punta del piñón.

Technical Service Bulletin (TSB) A type of automotive service information that describes a particular problem or change in procedure.

Boletín de servicio técnico (TSB) Un tipo de información de servicio automotriz que describe un problema o cambio de procedimiento específico.

Test light A circuit tester used to check for voltage at various points in a circuit.

Lámpara de prueba Probador de circuito que se usa para verificar el voltaje en varios puntos de un circuito.

Thread pitch The distance from the top of one bolt thread to the top of the next. In the USCS system, thread pitch is determined by the number of threads per 1 inch of bolt length.

Paso de la rosca Distancia desde la parte superior de una rosca de un perno hasta la parte superior de la siguiente. En el sistema USCS, el paso de la rosca se determina contando la cantidad de roscas en 1 pulgada de largo del perno.

Throttle position (TP) sensor A sensor that tracks the position of the throttle opening. The output by this sensor is used by that of the vehicle's control modules and is available on the CAN bus.

Sensor de posición del acelerador (TP, en inglés) Un sensor que hace un seguimiento de la posición de apertura del acelerador. La información que provee este sensor es utilizada por los módulos de control del vehículo y se encuentra disponible en el bus del CAN.

Thrust bearing race A polished steel race on one or both sides of a thrust bearing.

Pista del cojinete de empuje Pista de acero pulido en un lado o ambos lados de un cojinete de empuje.

Toe A suspension dimension that reflects the difference in the distance between the extreme front and extreme rear of the tires. Also a term used to indicate the inside portion of a tooth on a ring gear.

Tope Dimensión de la suspensión que refleja la diferencia de la distancia entre los extremos delantero y trasero de las ruedas. También es un término que se utiliza para indicar la porción interior de un diente o engranaje de anillo.

Toe-out on turns An alignment setting that allows for the front wheels to be at different angles on turns.

Divergencia en giros Configuración de alineación que permite que las ruedas delanteras tengan ángulos diferentes en los giros.

Torque wrench A wrench having a dial or other indicator indicationg the amount of torque being applied.

Llave de torque Llave que tiene un dial u otro indicador que señala el nivel de par de torsión que se aplica.

Total pedal travel The total amount the pedal moves from no free play to complete clutch disengagement.

Avance total del pedal Distancia total a la que el pedal se mueve desde cero hasta completar el desengrane total del embrague.

Total runout The sum of the maximum readings below and above the zero line on the indicator.

Desviación total Suma de las lecturas máximas bajo y sobre la línea de cero en el indicador de cuadrante.

Unsprung weight The weight of the tires, wheels, axles, control arms, and springs.

Peso no suspendido Peso de las ruedas, llantas, ejes, brazos de mando y muelles.

Vehicle identification number (VIN) The number assigned to each vehicle by its manufacturer, primarily for registration and identification purposes.

Número de identificación del vehículo Número asignado por el fabricante a cada vehículo principalmente para su registro e identificación.

Vibration A quivering, trembling motion felt in the vehicle at different speed ranges.

Vibración Estremecimiento y temblor que se advierte en el vehículo a diferentes rangos de de velocidad.

Volatility The tendency for a fluid to evaporate rapidly or pass off in the form of vapor. For example, gasoline is more volatile than kerosene because it evaporates at a lower temperature.

Volatilidad Tendencia de un fluído a evaporarse rápidamente o transformarse en vapor. Por ejemplo la gasolina es más volátil que el kerosén porque se evapora a una temperatura más baja.

Voltage Electrical pressure that causes current to flow.

Voltaje Presión eléctrica que causa que fluya la corriente.

Voltmeter The instrument used to measure electrical pressure or potential.

Voltímetro Instrumento para medir la presión eléctrica o la energía potencial.

Volts The units of measurement for electrical pressure or EMF.

Voltio Unidad de medida de la presión eléctrica o EMF (la fuerza electromagnética).

Wheel shimmy The wobble of a tire.

Bailoteo de la rueda Movimiento lateral de una rueda.

Zerk fitting A common name for grease fittings. A very small check valve that allows grease to be injected into a part but keeps the grease from squirting out again.

Conexión Zerk Término común que se utiliza para las conexiones de engrase. Válvula de retención sumamente pequeña que permite inyectar la grasa en un componente, y que a la vez impide que esa grasa se derrame nuevamente.

INDEX

Note: Page numbers followed by "f" indicate material in a figure.